Herpes Zoster: Postherpetic Neuralgia and Other Complications

C. Peter N. Watson • Anne A. Gershon
Michael N. Oxman
Editors

Herpes Zoster: Postherpetic Neuralgia and Other Complications

Focus on Treatment and Prevention

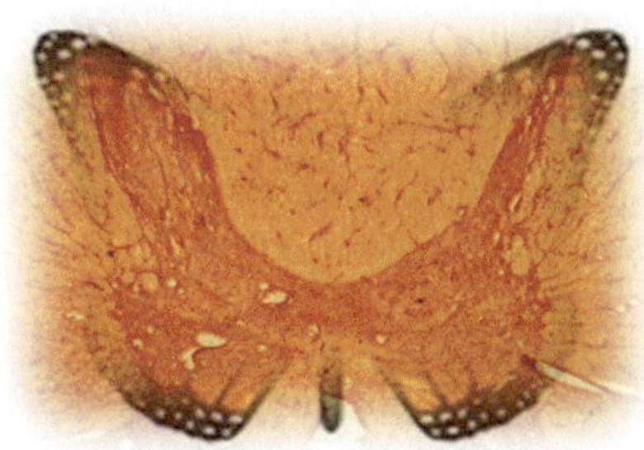

Adis

Editors
C. Peter N. Watson
Department of Medicine
University of Toronto
Toronto, Ontario
Canada

Anne A. Gershon
College of Physicians and Surgeons
Columbia University College
of Physicians and Surgeons
New York, NY
USA

Michael N. Oxman
Department of Medicine
Division of Infectious Diseases
University of California
San Diego, La Jolla, California
USA

Medicine Service
Infectious Diseases Section
VA San Diego Healthcare System
San Diego, California
USA

ISBN 978-3-319-83042-1 ISBN 978-3-319-44348-5 (eBook)
DOI 10.1007/978-3-319-44348-5

Illustration page iii: Composition of coronal section of the spinal cord at T7 showing no difference in dorsal horn substance P staining and dorsal horn atrophy on the right side of the image using immunocytohistochemistry superimposed on a monarch butterfly.

Printed on acid-free paper

This Adis imprint is published by Springer Nature
The registered company is Springer International Publishing AG
The registered company address is: Gewerbestrasse 11, 6330 Cham, Switzerland

"If I have seen further it is by standing on the shoulders of giants."
(Sir Isaac Newton 1643–1727)

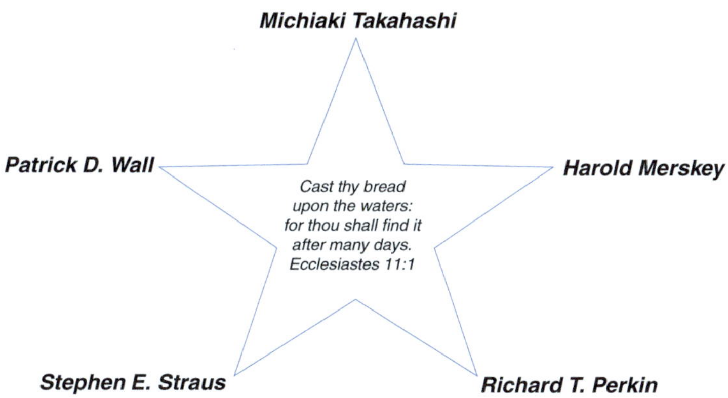

This book is dedicated to individuals who have played important roles in modern times in preventing, treating, and understanding the varicella zoster virus (VZV) and its complications:

Michiaki Takahashi (1928–2013):
Dr. Takahashi's vision and courage inspired him to attenuate VZV over 40 years ago, in order to prepare a live vaccine against this virus. Guided by his experience with live attenuated measles and polio vaccines, he isolated VZV from a child with varicella and attenuated it by serial passage in human and guinea pig cells at reduced temperature. When first developed, the live attenuated VZV vaccine was highly criticized by many scientists. However, Dr. Takahashi and his colleagues persevered, demonstrating the vaccine's safety and efficacy in preventing varicella in healthy children exposed to VZV at home and in school, in children hospitalized with a variety of illnesses, and, finally, in children with leukemia in whom varicella is associated with significant mortality. Dr. Takahashi and his colleagues also demonstrated the vaccine's safety and immunogenicity in older adults already latently infected with VZV, an important preamble to its use to prevent herpes zoster.

Reassured by Dr. Takahashi's pioneering studies in Japan and aided by Dr. Takahashi's extraordinary generosity, physicians in the United States and throughout the world began to study the live attenuated Oka VZV vaccine, which has proven to be one of the safest and most effective vaccines in use today. It has saved millions of individuals worldwide from morbidity and mortality due to VZV. Its success in preventing herpes zoster and postherpetic neuralgia has served as a model for the development of newer VZV vaccines and encouraged investigators attempting to develop therapeutic vaccines for latent and persisting infections by other human herpesviruses. At this writing, VZV remains the only human herpesvirus for which there are licensed vaccines. Dr. Takahashi was an inspiring colleague and a warm friend.

Dr. Stephen E. Straus (1946–2007) *was a consummate physician-scientist whose career was devoted to innovative research on the molecular biology, pathogenesis, treatment, and prevention of human viral infections and immunological diseases. His research extended from the bedside to the laboratory bench and back to the bedside. He was a leading contributor to our understanding of VZV and the diseases it causes. Steve Straus and his colleagues cloned and mapped the complete VZV genome and constructed a detailed map of VZV's transcripts. He was the first to prove by molecular techniques that the viruses causing varicella and a subsequent episode of herpes zoster in the same individual were identical. His studies of VZV latency in human sensory ganglia characterized the cells that were latently infected, determined the VZV copy number in individual neurons, mapped the VZV transcripts expressed during latency, and demonstrated fundamental differences between HSV and VZV latency. Steve played a leading role in the planning and execution of the Shingles*

Prevention Study, which demonstrated the safety and efficacy of live attenuated Oka VZV vaccine in reducing the incidence and severity of herpes zoster and postherpetic neuralgia, and led to licensure of zoster vaccine in 2006. He served on the study's planning and executive committees, the writing committee, and the clinical evaluation committee, which adjudicated every suspected case of herpes zoster, and he was the initial principal investigator of the National Institutes of Health study site. In 1999, Steve was appointed the first director of the National Center for Complementary and Alternative Medicine, where he established the primacy of rigorous scientific research to ensure that therapies in this challenging area were evidence based. Steve was diagnosed with a brain tumor in November 2004; yet, despite his illness, he continued to make major contributions to both basic and clinical research until his death in 2007. Steve was a wise mentor, a sterling role model, and a dear friend to scores of fellows, students, and colleagues, many of whom pursued successful careers in medical research inspired by his example.

Harold Merskey (1929–) *Harold is an inspiration and source of wisdom for his many friends, students, and colleagues because of the breadth of his scholastic ability, leadership, generosity, courage, feisty defense of the vulnerable, ongoing sage advice, and friendships over many decades. He collaborated on the original identification by randomized controlled trial of the independent analgesic effect of the antidepressant amitriptyline in postherpetic neuralgia. He has published as well 10 books and over 400 articles on various aspects of pain. The following quotation from Virgil is reminiscent of Harold: "It was his part to learn the powers of medicine and the practice of healing, and careless of fame, to exercise the quiet art."*

Richard T. Perkin (1931–2003): *Dick Perkin was well known to many of the authors of this volume. The eldest son of an American entrepreneur who co-founded the company Perkin-Elmer in 1937, Dick like his father had a deep interest in science. A crater of the moon was named after his father. A graduate of Harvard College in 1954, Dick was also passionate about music, and he was strongly motivated to heal the world and make it a better place. After retiring from various executive positions, he devoted his life to philanthropic causes, particularly through Rockefeller University, the Wildlife Conservation Society, and the Juilliard School of Music. In 1991, at the age of 60, when his elderly mother developed vision-threatening severe ophthalmic zoster with PHN, he decided to establish a research group dedicated to developing means to prevent zoster, through research and education. He assembled a group of internationally recognized academic physician-investigators with a similar interest and called it the Varicella Zoster*

Virus (VZV) Research Foundation, which he successfully directed for the next 12 years, until his death. This group conducted international research meetings, raised and discussed questions about VZV, and supported the research of highly qualified young investigators with an interest in the virus. The Foundation, which met in such venues as Paris, Osaka, New York, La Jolla, and Washington, was instrumental in supporting universal vaccination against varicella and played a critical role in the development and testing of the first vaccine to prevent zoster, which was Dick Perkin's dream. In addition to his intellectual side, Dick had a quick wit and an abiding interest in people, and he deeply loved the members of his family. He was beloved by his office staff, the VZV Board of Directors, and the many and diverse scientists involved in VZV research. Reflecting his modesty, the VZV Research Foundation that Dick established did not bear his name, but it will forever evoke his memory and remind us all of his good works and dedication.

Patrick Wall (1925–2001) *An early and memorable grand rounds by Professor Wall was on postherpetic neuralgia when he was a visiting professor in Toronto. It must have been surprising for clinicians that a basic scientist would discuss this condition with such knowledge of the clinical manifestations of the disease. Dr. Wall felt strongly that his postdoctoral basic science students should attend the pain clinic and see patients with intractable neuropathic pain such as postherpetic neuralgia. It was from the nerve fiber spectrum from four nerve biopsies of postherpetic neuralgia by the neurosurgeon William Noordenbos that Patrick Wall and Ronald Melzack suggested the gate control theory of pain. They thought from the preponderance of small slowly conducting fibers (pain excitatory) and reduction in large fibers (inhibition) that "fast blocks slow." Pat Wall continued to be a world leader and an inspiration for his many*

students and colleagues over many years via fellowships and his editorship of the journal PAIN.

C. Peter N. Watson,
Anne A. Gershon,
Michael N. Oxman

The Patients Speak in Poetry, Art, and Prose

This book begins and ends with patients' perceptions first with Elizabeth MacCallum with poetry and last with Susan Telling's drawings. In between, Chap. 3 is a narrative by Anne Tuzi regarding the impact of postherpetic neuralgia on her life.

My Constant Companion

By Elizabeth MacCallum

Before the first rays of light
Warning of the day to come
My friend is here.
The darkness in the light
Snickering
My constant companion.
The devil incarnate
Burrowing, gnawing, snarling
My sine qua non.
The claws in the caress
Needling the would-be bliss
My most faithful servant.
But at the last I will escape.
Promising the world to come
Grace is here
Unsullied kindly light.

Preface

A quarter of a century has passed since the publication of Peter Watson's first volume *Herpes Zoster and Postherpetic Neuralgia* [9]. During this interval, there have been numerous advances in pain science and medicine, including in our *understanding* of the causes of chronic pain, in *treatment modalities* and in *attitudes* about what pain is, and how we should relate to it. The most dramatic change has been in the basic sciences, where we know a great deal more than we did in 1993 about the biological mechanisms that underlie chronic pain. Then, almost every journal article and book chapter on neuropathic pain (and there were plenty in the early 1990s) opened with the statement that pain following nerve injury is fundamentally a mystery. Today this is no longer the case, albeit much still needs to be learned. On the clinical front, advances have been modest in comparison, but some new treatment modalities have come online (e.g., the lidocaine patch for postherpetic neuralgia) while others have faded away. Substantial advances have also occurred in the realm of changed attitudes. The public campaign launched by EFIC at the European Parliament in Brussels in October 2001 declared that chronic pain needs to be viewed not as a symptom, but as a disease, a major biomedical problem, and a healthcare priority in its own right [7]. A Google search today for "Chronic pain is a disease" turned up over 300,000 web pages referring to this phrase. Perhaps there is a little hyperbole here, but the slogan and the concept have resonated broadly. This, together with related campaigns promoting pain monitoring as the "fifth vital sign" and the need to overcome the fear of using opiates for palliative care and beyond, has significantly changed attitudes among pain professionals and the public [2]. Increasingly it is held that when pain relief can be achieved, it constitutes a human right.

But despite progress, painful conditions continue to rank high among the contributors to the global burden of disease. Pain is also big business. Pain relieving drugs are among the most prescribed (and lucrative) in the pharmacopeia [5, 8], http://www.medscape.com/viewarticle/844317. Yet several large members of Big-Pharma have closed down their research programs on analgesic drugs in the past few years. This is after failures of vastly expensive development efforts such as tachykinin receptor antagonists and COX-2 inhibitors. In the specific niche of drugs approved for neuropathic pain, today's most widely used compounds provide only fractional pain relief for a small minority of the patients who take them (e.g., for gabapentin NNT = 7 [4]). Their effect is mostly placebo. Indeed, whole sectors of widely used pain remedies are based entirely on the placebo effect, homeopathy being a prime example. The placebo effect and more broadly, context-related modulation of pain, is a powerful analgesic modality that some investigators feel should be introduced knowingly into pain practice [6]. Our general failure to crack the problem of chronic pain, including pain in shingles and postherpetic neuralgia, has led some opinion leaders to conclude that reducing the intensity of pain is not even the endpoint we should be shooting for. Rather, we should focus more on the psychosocial factors that exacerbate pain [1].

Personally, I remain optimistic. Historically speaking, it is not so long ago that pain was taken for granted as a normal part of life including in the most enlightened parts of the world. Temporal and religious leaders saw pain, and even torture, as legitimate means of seeking truth (e.g., trial by ordeal), educating children, entertaining the public, and redeeming souls. Today, gladiatorial sport, public burning of nonbelievers, and the like have been eradicated from most of the planet. Indeed, in many countries large fortunes are spent, by public demand, to minimize suffering even in farm animals and laboratory rodents. Standing back, the distance we have travelled in the realm of attitudinal change regarding pain is stupendous [3].

Taking the long view, there are also successes in the realm of treatment modalities. We have four broad-spectrum, tried-and-true drugs that are so good that we almost take them for granted. First are opiates which, with all their side effects and abuse potential, are a gift from the Gods of antiquity. Second is aspirin and the related over-the-counter NSAID and non-anti-inflammatory pain relievers. They are cheap, safe enough for short-term use without a doctor's prescription, and they work well for pain of modest intensity. Third is lidocaine and the other local anesthetics. Safe, pretty easy to use, and for short-term nerve and regional blocks, they work dramatically almost every time (no NNT = 7 for lidocaine!). Fourth are general anesthetics. Until their introduction, limbs were sawed off with the patient only a shot of whiskey away from being wide awake. The advent of ether and chloroform heralded today's essentially pain-free surgery. Anesthesia was hailed at the time, and quite correctly, as "the conquest of pain." But we now take it so much for granted that, despite the fact that we have virtually no idea how propofol and the other anesthetics eliminate pain (and consciousness), it is rare to see even a single lecture addressing this subject at major international pain conferences. Over the years, these four modalities have steadily improved as new agents, e.g., anticonvulsants and antidepressants, have come on line. Perhaps further down the road, when Peter Watson once again gathers his troops to review herpes zoster and postherpetic neuralgia, some of the remarkable discoveries currently coming out of basic science laboratories will already have been translated into new families of drugs and procedures that really work in the clinic.

Marshall Devor
Israel

References

1. Ballantyne JC, Sullivan MD (2015) Intensity of chronic pain–The wrong metric? N Engl J Med 373:2098–2099
2. Campbell JN (2016) The fifth vital sign revisited. Pain 157:3–4
3. Devor M, Rappaport I, Rappaport ZH (2015) Does the Golem feel pain? Moral instincts and ethical dilemmas concerning suffering and the brain. Pain Pract 15:497–508
4. Finnerup NB, Attal N, Haroutounian S, McNicol E, Baron R, Dworkin RH, Gilron I, Haanpaa M, Hansson P, Jensen TS, Kamerman PR, Lund K, Moore A, Raja SN, Rice AS, Rowbotham M, Sena E, Siddall P, Smith BH, Wallace M (2015) Pharmacotherapy for neuropathic pain in adults: a systematic review and meta-analysis. Lancet Neurol 14:162–173
5. Institute of Medicine (2011) Relieving pain in America: a blueprint for transforming prevention, care, education, and research, 2012/05/04 ed. National Academies Press (US), Washington, DC
6. Klinger R, Colloca L, Bingel U, Flor H (2014) Placebo analgesia: clinical applications. Pain 155:1055–1058
7. Niv D, Devor M (2004) Chronic pain as a disease in its own right. Pain Pract 4:179–181
8. Vos BP et al (2015) Global, regional, and national incidence, prevalence, and years lived with disability for 301 acute and chronic diseases and injuries in 188 countries, 1990–2013: a systematic analysis for the Global Burden of Disease Study 2013. Lancet 386:743–800
9. Watson CPN (1993) Herpes zoster and postherpetic neuralgia. Elsevier, New York

Preface and Acknowledgments

> "Life is short and the Art is long; the occasion fleeting; experience fallacious; and judgment difficult. The physician must not only do what is right himself, but also to make ….the externals cooperate" – attributed to Hippocrates, the first aphorism

All of the above is true of herpes zoster (and life). Although the evidence base for this disease is increasing, treatment remains very much an Art (sic). Herpes zoster has been known since antiquity [1]. The *occasion* suggested for effective treatment of zoster with antivirals is indeed *fleeting* (72 h). Favorable *experience* with the treatment of herpes zoster and postherpetic neuralgia based on the many early uncontrolled interventions is often *fallacious* because of the natural history of resolution that occurs in many patients. *Judgment*, for example, regarding the use of opioids for intractable postherpetic neuralgia is indeed *difficult*. The last sentence of the aphorism refers to a variety of "externals" but includes prevention [2] of herpes zoster by vaccination.

Herpes zoster, overall, is the most common neurological disease. Postherpetic neuralgia is its most feared and common complication with its potential accompaniment of visual loss and facial disfigurement, which can completely change functioning and quality of life. Progress regarding this virus has been slow in the prevention of postherpetic neuralgia at the acute stage of the painful rash and also in the treatment of established postherpetic neuralgia. An important advance is the promise of prevention by the current live attenuated zoster vaccine and the possibility of other new, more effective, and more broadly applicable vaccines. New information is available regarding epidemiology, pathophysiology, neuropathology, antiviral drugs for treatment of acute zoster, and pharmacotherapy of postherpetic neuralgia. The evidence for viral persistence supports the important role of this virus in vascular diseases, such as stroke, myocardial infarction, and temporal arteritis (granulomatous angiitis).

This book has been written for a wide readership. There is a broad interest in herpes zoster and its complications, not only from infectious disease and pain specialists but also among general internists, general practitioners, neurologists, oncologists, public health departments, the pharmaceutical industry, clinical trial experts, and the general public.

The book has been organized into five parts. After the introduction, a chapter on varicella is followed by a section on the acute illness of herpes zoster with its complications and management. Part III focuses on the assessment, pathology, and pathophysiology of postherpetic neuralgia. This is important because postherpetic neuralgia has been one of the main neuropathic pain conditions targeted by clinical research in neuropathic pain (together with painful diabetic neuropathy) and ideas about its pathophysiology have broader ramifications for neuropathic pain in general. Part IV on the treatment of postherpetic neuralgia is comprehensive and also cautionary because once the disease is established, treatment is difficult. The chapter on improving clinical trial design is critical in looking for new and more effective treatments for this terrible neuropathic pain. Part V focuses on methods for preventing the disease in the first place which is our best hope of dealing with this virus and its complications. Prevention will hopefully occur chiefly by means of the currently licensed live attenuated zoster vaccine and the promise of new, more effective, and more widely applicable recombinant varicella-zoster virus glycoprotein vaccines.

I have been very fortunate in eliciting the collaboration of two senior world authorities on infectious diseases, particularly diseases caused by varicella-zoster virus. Dr. Anne Gershon was invaluable as coeditor of my previous volume on this topic. Dr. Michael Oxman was the first author of a seminal article on the first vaccine to prevent herpes zoster. Together we have been fortunate to assemble an exceptional group of chapter authors in various fields who are at the cutting edge of the many different facets of the varicella-zoster virus.

As well as these coeditors, I want to acknowledge the invaluable assistance and generosity of the many and eminent chapter authors who have contributed to this book. Further, the support of my family, Dr. Judy Watt Watson, and our children, Simon and Emily, has been vital.

Toronto, ON, Canada C. Peter N. Watson

References

1. Abraham N, Murray J (1993) The belt of roses from hell: historic aspects of herpes zoster and postherpetic neuralgia. In: Watson CPN (ed) Herpes Zoster and postherpetic neuralgia. Pain research and clinical management, 1st edn, vol 8, Elsevier, Amsterdam, New York
2. Nuland Sherwin B (2008) The uncertain art: thoughts on a life in medicine. Random House, New York, pp 140–147

Patrick D. Wall (1925–2001): An Appreciation

One of the dedications of this book is to the memory of Professor Patrick D. Wall FRS (1925–2001). It is timely since 2015 the time of this writing, marks not only the 90th anniversary of his birth but also the 50th anniversary of the publication of his seminal collaborative paper with Ronald Melzack on the "gate control theory of pain" [1]. It is a testament to Pat Wall's (and Ron Melzack's) capability as insightful thinkers that this theory continues to resonate and inspire despite 50 years of intense scientific scrutiny. Many scientific competitors attacked specific aspects of the hypothesis in the first few years after its publication, and Wall relished engaging in the resulting debates. With the perspective of time, however, the overall influence of the gate control theory on the landscape of pain science and medicine is undeniable.

It is also pertinent to recall Pat Wall's life and scientific contributions at this time since there is now a whole generation engaged in pain research and clinical management who never witnessed this great man in action and never directly benefited from his wisdom, critical advice, and encouragement. Hopefully, this dedication will draw attention to Wall's fundamental contributions to current pain research.

Pat Wall was born in Nottingham, United Kingdom. His father Thomas was a schoolteacher who went on to become a school inspector. Thomas Wall was a Cambridge University graduate who had served as a decorated artillery officer in the First World War [2]. Pat was also physically active as a young man, a fact which may come as a surprise to many who knew him in later years. His father Thomas was an athlete who participated in international competition in field and track and in soccer. Wall described his childhood thus: "I was brought up in a family full of adventure. My father's extrovert character effectively submerged my mother's [Ruth] covert Puritanism. My older brother's obsession with cars and airplanes so successfully distracted my parents that I grew up in a wonderful calm" [3]. Pat was educated at St. Paul's School, London, where one particularly influential teacher, Tony Barnett, was fundamental in sowing the seeds of Pat's iconoclastic character, including the often fierce and usually gladiatorial deployment of reasoned argument to seek out and challenge ill-founded authoritarian pronouncements, scientific "facts," and seeming paradoxes. These early influences were also reflected in his left-wing politics. As one of his obituaries noted: "Although he possessed immense charm, he could be intolerant and critical of scientists who made statements which he considered unjustified, and often ruffled feathers by challenging 'received wisdom'."

Wall studied medicine at Oxford University and the Middlesex Hospital, qualifying in 1948. During this time, he spent a summer working in the laboratory of Sir Alexander Fleming at St. Mary's Hospital. Like many London medical students of that generation, he volunteered to go to Europe and assist in caring for surviving Holocaust victims. While he rarely spoke of this experience, it undoubtedly shaped his humanitarian and his political views later in life. Although he never practiced clinical medicine for any length of time, the ethos of guiding his laboratory research by lessons learned from careful clinical observation of patients was one that he carried through his scientific career. His first two scientific papers, one on brain connectivity and one on a novel experimental method, were published in *Brain and Nature* when he was only 21. After graduating, he moved to the United States and passed through a number of major universities including Yale, Chicago, and Harvard before ending up at the Massachusetts Institute of Technology. It was when he was at MIT that he met his collaborator, the Canadian psychologist Ronald Melzack. Together they set about developing the gate control theory published in *Science* in 1965 [1]. This seminal paper is a masterpiece of scientific reasoning which leaned heavily on clinical observations in chronic pain patients. It elaborated the concept of sensory-sensory modulation of pain at the level of the spinal dorsal horn and laid the foundations for the study of spinal gating by pathways descending from the brain. The clinical community embraced the idea of gate control quickly, encouraged by a productive collaboration between Pat and the well-known neurosurgeon Bill Sweet which yielded the methods of TENS (transcutaneous electrical nerve stimulation) and spinal cord stimulation. The embrace by clinicians still endures half a century later. In time the research community also came on board. Today, spinal modulation and descending control are central themes in the neuroscience of pain.

In 1967 Wall returned to the United Kingdom to a Chair in the Anatomy Department at University College London, where he remained until his retirement. In his early days at UCL, he was heavily influenced by the then head of department, the great neuroanatomist JZ Young.

From his base in London, Pat continued to contribute to the understanding of pain. His oft-quoted advice to those embarking on a scientific career was "One has to choose an important subject that no one else is working on, write a book about it and start a journal for it." His own career followed such a path. His life's work revolved around elucidating the mechanisms of pain, a largely ignored topic at the time. He co-edited what remains the major textbook in the field *Wall & Melzack's Textbook of Pain* (Elsevier), now in its sixth edition. And he founded and edited *PAIN*, still the premier journal in the field. One of the likely reasons for the success of the journal and of the textbook is that they both provided a welcoming forum for the views of laboratory scientists, clinical researchers, and practitioners from a wide range of disciplines. In 1973, Pat also played a key role in the establishment of the

International Association for the Study of Pain (IASP), in the "guise" of its scientific study officer.

One of Wall's major contributions while at UCL was to change the one-dimensional way that pain was then viewed. He argued strongly that "pathological" pain, which had no physiological or survival benefit to the organism, had to be viewed differently from acute pain. This approach drew attention to the fundamental importance of plasticity in the nervous system. He was particularly angered by pain from the prostate cancer which, in the end, took his life. This was pain which, in the pre-medical world at least, served no purpose. He pioneered research on aberrant responses of peripheral nerves and of the central nervous system (CNS) to nerve injury. Such responses, he argued, underlie neuropathic pain. Characteristically, his views that there are dramatic alterations in the CNS in response to peripheral nerve injury or inflammation were fiercely attacked at the time. Again, he relished the debate. Equally characteristically, in the fullness of time, his forward thinking and challenging ideas on neuroplasticity in the pain system came to form a central focus of current day pain research. Were he here with us today, Wall would probably still be raising challenges and fighting dragons.

As noted, throughout his life, Pat's ideas were fomented by observations in patients. For much of his early time at UCL, he had an honorary appointment at the Hebrew University of Jerusalem, Israel, where he made frequent scientific trips and developed a number of collaborations. The development of his ideas concerning neuropathic pain was strongly influenced by clinical observations of post-amputation pain in Yom Kippur war veterans. This led to his work on experimental neuromas and the publication of the first animal model of neuropathic pain [4]. While full-thickness sciatic nerve axotomy has since been largely succeeded by more refined partial nerve injury models, this paper remains important, perhaps mostly for what was in it, but has been forgotten and is only now being rediscovered. We refer to the recognition that neuropathy is often characterized by spontaneous pain in the presence of sensory loss, "anesthesia dolorosa." It is now becoming apparent from the increasing number of sensory profiling studies of neuropathic pain patients that sensory loss (hypesthesia) accompanied by ongoing pain is a far more frequent clinical presentation in neuropathic pain patients than sensory gain phenomena (e.g., allodynia and hyperalgesia). The partial exceptions are postherpetic neuralgia, CRPS, and traumatic nerve injury, where sensory gain is frequent. Sensory gain is also the endpoint that has been most commonly used in animal models for the past decades as we struggle to find ways of assessing ongoing pain.

The sciatic nerve neuroma model in rodents also drew attention to the importance of measuring ethologically appropriate complex behavioral responses in animals, as opposed to altered thresholds for reflexive responses to simple sensory stimuli, the endpoint of convenience that dominates the field today. In his 1979 paper, Wall highlighted a complex self-mutilation behavior that developed following nerve injury that he termed "autotomy." He believed that autotomy behavior reflects ongoing pain perceived by the animal, akin to phantom limb pain or anesthesia dolorosa. The precise relevance of autotomy behavior to clinical pain is still debated; indeed the authors of this appreciation chapter do not fully agree on its interpretation. The debate about autotomy was relished and encouraged by Wall, and it is still going on today. Wall stressed the importance of understanding the normal behavior of an animal as a framework for interpreting pathological responses. He sometimes had a laboratory rat freely roaming in his office at UCL. This was both to learn firsthand about his experimental subjects and to make the point among his scientific colleagues. Pat's insights concerning spontaneous *versus* evoked pain are currently being rediscovered, decades after he first drew attention to them.

Wall remained active in pain research in his retirement, moving his laboratory from UCL to St. Thomas' Hospital London, adjacent to the laboratory of his protégé Steve McMahon. He continued to conduct his own electrophysiological recordings and produced new and important work right up to the time of his death.

A testament to his remarkable abilities as a teacher and mentor is the legacy of his students and associates, many of whom have gone on to become world-leading researchers in the field of pain. Examples include Howard Fields, Allan Basbaum, Jonathan Dostrovsky, Steve McMahon, Marshall Devor, Maria Fitzgerald, and Clifford Woolf. He also profoundly influenced others who worked in different aspects of neuroscience, such as the 2014 Nobel laureate John O'Keefe. Many, including the authors of this volume, continue to directly and indirectly benefit from the downward dissemination of his influence through his many academic offspring.

Andrew SC Rice
Marshall Devor

Patrick Wall characteristically engaging an audience

Obituaries:

- Woolf CJ. Patrick D. Wall (1925–2001). Nature. 2001;413(6854):378.
- Devor M. Obituary. Patrick David Wall, 1925–2001. Pain. 2001;94(2):125–9.
- The Times (15 Aug 2001)
- The Guardian (16 Aug 2001)
- The Independent (18 Aug 2001)
- Daily Telegraph (23 Aug 2001)

References

1. Melzack R, Wall PD (1965) Pain mechanisms: a new theory. Science 150(3699):971–979
2. The National Archives. Captain Thomas Wall, Royal Garrison Artillery. Armed Forces Service Records 1914–1922. WO 339/59688. Kew
3. Wall PD, Patrick D Wall (2001) In: Squire LR (ed) The history of neuroscience in autobiography, 3rd edn. Academic Press, San Diego, pp 6–34
4. Wall PD, Devor M, Inbal R, Scadding JW, Schonfeld D, Seltzer Z, Tomkiewicz MM (1979) Autotomy following peripheral nerve lesions: experimental anaesthesia dolorosa. Pain 7(2):103–111

Contents

Part III Postherpetic Neuralgia: Assessment, Pathology, Pathophysiology

Part IV Postherpetic Neuralgia: Treatment

Part V The Prevention of Herpes Zoster and Postherpetic Neuralgia

About the Editors

Dr. C. Peter N. Watson is a neurologist in the Department of Medicine, Division of Neurology, University of Toronto, Ontario, Canada. Dr. Watson is a long-standing member of the Canadian Pain Society and International Association for the Study of Pain. He has worked in the field of pain since 1973 as a neurologist and clinician-scientist. From his base of busy clinical work, he has made outstanding scientific contributions to the pharmacological treatment of pain and to the understanding and treatment of postherpetic neuralgia. Despite his busy clinical practice, he is well published in journals such as *Pain*, *Neurology*, *The New England Journal of Medicine*, *The Journal of Pain and Symptom Management*, and *The Clinical Journal of Pain*, to name a few.

Dr. Watson published the first well-conducted trial in postherpetic neuralgia, demonstrating the efficacy of amitriptyline as an analgesic independent of its effects on depression. That design and those findings have been well replicated in other leading centers, and the results were of great importance to our understanding of the pathophysiological and psychological aspects of pain. His paper on postmortem findings in postherpetic neuralgia was regarded as a classic by Noordenbos. Dr. Watson has achieved an international reputation for these studies and his many other contributions in the field. These include the two widely used editions of *Herpes Zoster and Postherpetic Neuralgia*, which he both edited and wrote a considerable part of (Watson 1993; Watson and Gershon 2001). He wrote the chapter on this topic for Bonica's textbook (3rd edition) and contributed chapters to other leading volumes such as *The Textbook of Pain*. His seminal study on amitriptyline was followed by a series of pioneering and definitive trials of other oral medications, including maprotiline (1992), nortriptyline (1998), and oxycodone (1998, 2003). In addition, following from a work on the topical application of capsaicin (Watson et al. 1988, 1992) and the combination of lidocaine and prilocaine, he established his preeminent expertise in the use of topical local analgesics, an important field currently undergoing active study. He has participated extensively in the teaching of medical students and other healthcare practitioners. Dr. Watson has had a significant impact on the treatment of pain in Canada and throughout the world and has been recognized by receipt of the Canadian Pain Society's highest honor, the Distinguished

Career Award, in 2003. Dr. Watson has also received the medal of the Varicella Zoster Research Foundation on May 8, 2007, for pioneering research in postherpetic neuralgia.

Dr. Anne Gershon is a professor of pediatrics at Columbia University College of Physicians and Surgeons. She is a graduate of Cornell Medical School. Her research over the past 40 years has included epidemiology, diagnosis, immunology, latency, prevention, and treatment of varicella and zoster. Her studies with varicella vaccine were critical for its licensure in the USA. She is continuing to study the safety and efficacy of varicella vaccine in the "vaccine era." She has also focused on HIV infection in children, particularly opportunistic infections. She has received research funding from NIH for the past 40 years. Dr. Gershon has served on numerous national and international medical committees. She was president of the Infectious Diseases Society of America (IDSA) in 2009. She has received many professional awards including the gold medal of the Sabin Vaccine Institute and the Fleming Award of the IDSA. She is the author of over 300 publications and has edited 11 books as an infectious disease specialist. She has a major interest in varicella-zoster virus infections and vaccines.

Dr. Michael N. Oxman received his medical degree from Harvard Medical School in 1963 and trained in internal medicine, infectious diseases, and virology with Drs. Maxwell Finland, Wallace P. Rowe, and John F. Enders. He was a faculty member in the Childrens-Beth Israel Hospital Infectious Diseases Training Program, Associate Professor of Microbiology and Molecular Genetics at Harvard Medical School, director of the Clinical Virology Laboratory at Boston Children's Hospital, and recipient of an American Cancer Society Faculty Research Award. He moved to the University of California, San Diego, in 1976 as Professor of Medicine and Pathology and Chief of the Infectious Diseases Section at the Department of Veterans Affairs San Diego Healthcare System.

Dr. Oxman's research during the past 50 years has involved the mechanism of action of interferon; the control of SV40 gene expression and the integration of SV40 DNA into cellular and adenovirus DNA, respectively, in SV40-transformed cells and adenovirus-SV40 hybrid viruses; and the pathogenesis, diagnosis, treatment, and prevention of viral diseases, especially diseases caused by herpes simplex virus (HSV) and varicella-zoster virus (VZV). He has published more than 160 books, book chapters, and manuscripts in prestigious peer-reviewed medical journals, including the *Proceedings of the National Academy of Sciences*, *The New England Journal of Medicine*, *Virology*, the *Journal of Virology*, the *Journal of Infectious Diseases*, *Clinical Infectious Diseases*, *Infection and Immunity*, the *Journal of Immunology*, the *Journal of Clinical Microbiology*, *The Lancet*, the *Journal of General Virology*, *Annals of Internal Medicine*, and the *American Journal of Medicine*. He has played a major role in training and mentoring physician-scientists at Harvard and the University of California and pioneered the use of large double-blind placebo-controlled multicenter clinical trials to evaluate treatment and prevention of HSV and VZV infections. Dr. Oxman is national chairman of the landmark Shingles Prevention Study (VA Cooperative Study #403), which demonstrated the efficacy of live attenuated Oka zoster vaccine, leading to its licensure by the US FDA in 2005 and its use in the USA and many other countries to prevent herpes zoster and its complications, principally postherpetic neuralgia. Dr. Oxman has been a visiting professor at the University of British Columbia, University of Cincinnati, University of Geneva, University of Minnesota, Mayo Clinic College of Medicine, and Memorial Sloan Kettering Cancer Center and has served as a member of the NIAID Task Force on Virology, the NIH Experimental Virology Study Section, the Armed Forces Epidemiological Board, the NIAID Collaborative Antiviral Study Group Advisory Committee, the Science Board of the Biomedical Research Centre at the University of British Columbia, the Microbiology Test Committee of the National Board of Medical Examiners, the Steering Committee of MASTER study (Canadian prospective herpes zoster natural history study), the Planning Committee for an American College of Rheumatology Trial of Zoster Vaccine in Patients with Rheumatoid Arthritis, and the Defense Health Board. He is a fellow of the Infectious Diseases Society of America and a member of the American Society for Virology, the American Society for Microbiology, the Society for General Microbiology, and the American Society for Clinical Investigation. Dr. Oxman is the recipient of the 2005 VZV Research Foundation Lifetime Achievement Award, the 2005 Harold L. Stewart Lectureship at the Uniformed Services of the Health Sciences, the 2006 Plenary Lectureship honoring Professor Michiaki Takahashi at the 10th Annual Meeting of the Japanese Society of Vaccinology, the 2009 Stephen E. Straus Memorial Lectureship in Infectious Diseases at the National Institutes of Health, the 2014 Veterans Research Alliance Award For Excellence, the 2014 Abraham I. Braude Visiting Professorship at UCSD, and the 2014 Department of Veterans Affairs John Blair Barnwell Award for outstanding scientific achievements in clinical research.

Contributors

Jane Ballantyne Anesthesiology and Pain Medicine, University of Washington, Seattle, WA, USA

Ralf Baron Division of Neurological Pain Research and Therapy, Department of Neurology, University Hospital Schleswig-Holstein, Campus Kiel, Kiel, Germany

Judith Breuer Division of Infection and Immunity, University College London, Cruciform Building, WC1E 6BT, UK

Shane Brogan Pain Service, Huntsman Cancer Institute, University of Utah, Salt Lake City, UT, USA

James Chodosh Massachusetts Eye & Ear, Harvard Medical School, Boston, MA, USA

Emma Davies Massachusetts Eye & Ear, Harvard Medical School, Boston, MA, USA

Marshall Devor, PhD Department of Cell & Developmental Biology, Institute of Life Sciences, and Center for Research on Pain, The Hebrew University of Jerusalem, Jerusalem, Israel

Mélanie Drolet Centre de recherche du CHU de Québec – Université Laval, Laval, QC, Canada

Robert H. Dworkin, PhD Departments of Anesthesiology, Neurology, and Psychiatry, University of Rochester School of Medicine and Dentistry, Rochester, NY, USA

James Elliott Toronto Regional Wound Healing Clinic, Mississauga, ON, Canada

Perry Fine Anesthesiology, Pain Research and Management Centers, School of Medicine, University of Utah, Salt Lake City, UT, USA

Anne A. Gershon Professor of Pediatrics, Columbia University College of Physicians and Surgeons, New York, NY, USA

Columbia University, New York, NY, USA

Don Gilden Department of Neurology, University of Colorado School of Medicine, Aurora, CO, USA

Ian Gilron Department of Anesthesiology and Perioperative Medicine, Queens University, Kingston, ON, Canada

Maija Haanpaa Department of Neurosurgery, Helsinki University Hospital, Helsinki, Finland

Mutual Insurance Company Etera, Helsinki, Finland

Ruth Harbecke Department of Medicine, Division of Infectious Diseases, University of California, San Diego, CA, USA

Research Service, VA San Diego Healthcare System, San Diego, CA, USA

Philipp Hüllemann Division of Neurological Pain Research and Therapy, Department of Neurology, University Hospital Schleswig-Holstein, Campus Kiel, Kiel, Germany

Deborah Pavan Langston Massachusetts Eye & Ear, Harvard Medical School, Boston, MA, USA

Elizabeth MacCallum Massey College, University of Toronto, Toronto, Canada

Paul Möller Division of Neurological Pain Research and Therapy, Department of Neurology, University Hospital Schleswig-Holstein, Campus Kiel, Kiel, Germany

Maria A. Nagel Department of Neurology, University of Colorado School of Medicine, Aurora, CO, USA

Anne Louise Oaklander Harvard Medical School and Massachusetts General Hospital, Boston, MA, USA

Michael N. Oxman Department of Medicine, Division of Infectious Diseases, University of California, San Diego, CA, USA

Medicine Service, Infectious Diseases Section, VA San Diego Healthcare System, San Diego, CA, USA

Andrew S. C. Rice Pain Research, Department of Surgery and Cancer, Faculty of Medicine, Imperial College, London, UK

Michael C. Rowbotham, MD California Pacific Medical Center Research Institute, San Francisco, CA, USA

Kenneth Schmader Department of Medicine, Division of Geriatrics, Duke University Medical Center, Durham, NC, USA

Geriatric Research Education and Clinical Center (GRECC), Durham VA Medical Center, Durham, NC, USA

Sigrun Alba Johannesdottir Schmidt Department of Clinical Epidemiology, Aarhus University Hospital, Aarhus, Denmark

Barry Sessle University of Toronto, Toronto, ON, Canada

Gary Sibbald Faculty of Medicine, University of Toronto, Toronto, ON, Canada

Ranjani Somayaji Division of Infectious Disease, Department of Medicine, University of Calgary, Calgary, AB, Canada

Susan Telling Patient and Artist

Anne Tuzi Patient

Charlotte Warren-Gash Institute of Health Informatics, University College London, London, UK

C. Peter N. Watson Department of Medicine, University of Toronto, Toronto, ON, Canada

Chapter 1
Introduction

The Varicella Zoster Virus: Fascination, Frustration and Excitement

C. Peter N. Watson

At length, his senses were overpowered; his eyes swam in his head, his head gradually declined, and he fell into a deep sleep. On waking, it was a bright sunny morning. He looked around for his gun, but in place of the clean well-oiled fowling piece, he found an old firelock lying by him, the barrel encrusted with rust, the lock falling off, and the stock worm-eaten.

Rip Van Winkle by Washington Irving

For me, after more than four decades, the varicella zoster virus continues to fascinate, frustrate, and excite.

1.1 Fascination

The fascination has always been how, by natural selection, this virus has survived over the years in isolated communities by causing two diseases: varicella (chickenpox) and herpes zoster (shingles). First the virus spreads like wildfire through children as varicella, seldom killing or causing serious complications but resulting in viral persistence and host immunity. The virus then disappears into the nervous system (dorsal and trigeminal ganglia) lying dormant for a half century contained by a vigilant immune system. Like Rip Van Winkle, the virus awakens in the face of declining immunity, and erupting in the aged and vulnerable body of a 60-year-old

C.P.N. Watson
Department of Medicine, University of Toronto,
Toronto, ON, Canada
e-mail: peter.watson@utoronto.ca

C.P.N. Watson et al. (eds.), *Herpes Zoster: Postherpetic Neuralgia and Other Complications*, DOI 10.1007/978-3-319-44348-5_1

grandmother in an often preferred site, one side of the forehead or thorax, causing (with no intent save its own survival), a severe varicelliform rash, ongoing terrible burning and shock-like pain, sometimes blindness, facial scarring and other complications. Her grandchildren visit, and now, never being exposed to the virus, they and their friends complete the cycle and develop varicella.

1.2 Frustrations

My personal frustrations have come from the ongoing difficulty of preventing and treating the complication of post-herpetic neuralgia and of preventing it by even aggressive treatment of the acute rash and pain of herpes zoster by antivirals and analgesics and by the difficulties of clinical research seeking better pharmacotherapy for the persistent pain.

1.3 Excitement

The excitement comes from the increasingly important and protean manifestations of the virus, now suggesting that it is of even greater impact regarding vascular disease (Chaps. 7 and 8), the vaccines for varicella (Chap. 2) and the increasing efficacy and applicability of vaccines to prevent herpes zoster (Chap. 24).

1.4 Epidemiology, Complications, Pathology

In Canada, there are 150,000 cases of herpes zoster and 17,000 cases of post-herpetic neuralgia per year [1]. There are well over a million new cases of herpes zoster each year in the United States with an increasing prevalence of patients with post-herpetic neuralgia [4]. It is likely that the incidence of herpes zoster and post-herpetic neuralgia will increase because the population is ageing and the incidence is age related. Because post-herpetic neuralgia may persist for years, the prevalence of this condition will also increase. The increased incidence of zoster may also be associated with immunosuppression, such as occurs with HIV, the use of immunosuppressant agents to treat autoimmune diseases and immunosuppression caused by chemotherapy and radiation treatment of malignancies.

Apart from post-herpetic neuralgia, blindness or loss of the eye and facial scarring may occur when zoster affects the ophthalmic division of the trigeminal nerve. Facial paralysis may occur with the involvement of the seventh cranial nerve (the Ramsay Hunt syndrome). Meningitis, encephalitis and spinal cord involvement may be complications (Chap. 6). Quality of life is often greatly impacted ([2], Chap. 11).

Pathologically, herpes zoster has been shown to damage the nervous system, including the peripheral nerve [6], the dorsal and trigeminal ganglia and the dorsal horn of the spinal cord [3, 8]. In these locations there can be alterations in pain pathophysiology [5], there may be ectopic discharges resulting in pain, and these sites are also where analgesic drugs have their action, thus potentially rendering the relief of pain in this way less effective.

1.5 Vascular Disease

Increasing evidence suggests an association with the risk of heart disease and stroke even in younger ages, both before and after the eruption of zoster (Chap. 8). There is also evidence of persistent and smouldering activity which may be the cause of other common chronic diseases such as granulomatous angiitis (temporal arteritis) (Chap. 7). Corticosteroids are not always successful for preventing blindness in temporal arteritis, and adding a course of an antiviral agent (oral valacyclovir or famciclovir or even intravenous aciclovir) is reasonable and safe pending further research.

1.6 Three Therapeutic Approaches

It is increasingly likely that the most effective approach and first line of defence is vaccination, both for varicella and for preventing herpes zoster later in life at the age of highest risk, particularly after the age of 50.

A hopeful but probably limited second preventive measure, not conclusively effective, is an aggressive treatment of acute zoster at the first sign of the disease with antiviral drugs and various analgesics, aiming to attenuate the condition and prevent severe post-herpetic neuralgia and other complications (Chap. 23).

A third approach is to treat established post-herpetic neuralgia, which continues to be difficult. Since the first randomized trial of amitriptyline over 30 years ago [7], we have a variety of therapeutic options of moderate efficacy, which follow guidelines for neuropathic pain in general (Chap. 19). In terms of vascular disease, one of the most exciting future prospects is the potential for the prevention of heart disease, stroke and other diseases (granulomatous angiitis) by preventing varicella and herpes zoster in the first place (Chaps. 7 and 8).

All of these aspects of the virus will be discussed in this book.

1.7 Book Dedications and Caveats

This book begins with a patient's poem and ends with another patient's drawings and includes a chapter with an account of another sufferer's experience (Chap. 3). It is dedicated to five remarkable individuals Michiaki Takahashi, Stephen E. Straus, Harold Merskey, Richard T. Perkin and Patrick D. Wall (see Dedications).

In this book the reader will find: (1) redundancies of text and images (each chapter is meant to be stand alone), (2) the best science from experts on their subjects, (3) ideas (to stimulate further much needed research), (4) opinion (since randomized controlled trials often exclude complicated cases or older subjects resulting in limited external validity or generalizability to practice, thus clinical experience and observational data can be important with difficult complex patients and over the long term) and (5) that there will be differences in opinion and ideas (but that is a healthy thing and this editor makes no apology for this).

This book, of course, is not the final answer but, hopefully, another stepping stone towards obliteration of the problems caused by this virus.

You are not required to complete the work, but neither are you free to desist from it
Rabbi Tarphon, Talmud, Avot, 2;21.

References

1. Brisson M, Pellissier JM, Camden S et al (2008) The potential cost-effectiveness of vaccination against herpes zoster and post-herpetic neuralgia. Hum Vaccin 4(3):238–245
2. Drolet M, Brisson M, Schmader KE et al (2010) The impact of herpes zoster and postherpetic neuralgia on health-related quality of life: a prospective study. CMAJ 182(16):1731–1736
3. Head H, Campbell AW (1900) The pathology of herpes zoster and its bearing on sensory localization. Brain 23:353–523
4. Insinga DRP, Itzler RF, Pellissier JM et al (2005) The incidence of herpes zoster in a United States administrative base. J Gen Intern Med 20(98):748–753
5. Melzack R, Wall PD (1965) Pain mechanisms: A new theory. Science 150:971–979
6. Noordenbos W (1959) Pain: problems pertaining to the transmission of nerve impulses which give rise to pain. Amsterdam, Elsevier: Ch 1, pp 4–1; Ch 10, pp 68–80
7. Watson CPN, Merskey H et al (1982) Amitriptyline versus placebo in postherpetic neuralgia. Neurology 32:671–673
8. Watson CPN, Morshead C, Van der Kooy D, Deck JH, Evans RJ (1988) Postherpetic neuralgia: post-mortem analysis of a case. Pain 34:129–138

Part I
Varicella (Chickenpox)

Chapter 2
Varicella

Anne A. Gershon

2.1 Introduction

For many years, no one seems to have paid much attention to varicella-zoster virus (VZV); this pathogen seemed not to be taken very seriously. VZV was long known to cause varicella, also known as chickenpox, an anodyne name. How could one take chickenpox seriously in comparison to the already recognized and very serious diseases, the great pox (syphilis) and smallpox? The source of the name "chicken" is lost in history but may refer to the farmyard bird or to the chickpea which varicella's typical vesicular lesions are said to resemble. Varicella was thought to be an inconsequential disease; it was considered a rite of childhood that everyone experienced, from which they recovered after a few days.

Slowly in the twentieth century, however, two medical problems began to emerge: varicella was related to another disease the laity called shingles (derived from the Latin for belt); subsequently it was realized that patients with shingles (medically known as zoster) could transmit varicella to children and adults who had not previously had chickenpox. That observation suggested that VZV might somehow persist after varicella and then become rejuvenated to cause zoster. Until the late 1960s, however, the public were unprepared to conceive of a virus that could silently persist for years in the body without symptoms. It was generally accepted that people became infected with a virus and then either died or recovered and that the virus then vanished. At that time no one had heard of human immunodeficiency virus (HIV), Ebola, or Zika viruses which can be highly persistent. There was, however, an emerging inchoate concept of herpes simplex virus (HSV) infections becoming latent or persistent in nervous tissue. This was recognized in the early twentieth century and was largely investigated by Goodpasture, Cushing, and their colleagues [39]. In the 1940s, Garland correctly proposed that VZV was a virus capable of a latent infection that developed after varicella and that it could persist silently in the human body for decades [17].

A.A. Gershon
Department of Pediatrics, Columbia University College of Physicians and Surgeons,
630 West 168th St., New York, NY 10128 , USA
e-mail: aag1@columbia.edu

C.P.N. Watson et al. (eds.), *Herpes Zoster: Postherpetic Neuralgia and Other Complications,* DOI 10.1007/978-3-319-44348-5_2

Everything changed radically for VZV, though, when chemotherapy for cancer was being developed in the early 1950s. Suddenly it was noted that children being treated for leukemia could manifest extensive, unusually severe, and even fatal varicella [10]. The Nobel Laureate Thomas Weller was then at Harvard University Medical School, where Sidney Farber was strenuously investigating cancer chemotherapy. Weller, who was a pediatrician, isolated VZV in cell culture for the very first time [53, 54] and, while he was at it, developed immunofluorescence (known then as "the fluorescent antibody technique") to demonstrate the presence of the virus in cell cultures [55]. After Weller and his colleagues described children with fatal varicella, it was suddenly clear that this virus was not chicken, and it had to be managed scientifically in the modern world. It also began to dawn on medical researchers that zoster in the growing ranks of the elderly was an emerging scourge as well.

2.2 Clinical Features of Varicella

Varicella disease in unvaccinated populations presents with a generalized, pruritic, vesicular rash distributed mainly on the head and trunk with some sparing of the extremities. It is usually seen in children; it is unusual in adults in whom may be severe and even fatal. The rash classically consists of lesions in various stages of development from maculopapules to vesicles followed by crusts. They are typically from 250 to 500 skin lesions, which last for about a week [18].

Varicella is one of the most contagious diseases, just a little less so than smallpox and measles. It is mainly spread from skin lesions of patients with varicella or zoster, infecting the respiratory tract of an individual who has not previously been infected with VZV [11]. It has an incubation period of 2–3 weeks, during which time both innate and adaptive immunity develop. Innate immunity is mediated by the production of interferon gamma in epidermal cells which delays viremic viral spread by CD4 and CD8 lymphocytes somewhat, as specific antibodies and cellular immunity develop. Eventually the interferon response in the skin is overcome and the rash of varicella develops as VZV begins to multiply there [1]. Cellular immune responses are of the greatest importance in combating VZV and recovery from the illness [18]. Unfortunately, before these responses are fully developed, lymphocytes have already transferred VZV to neurons where the virus establishes latent infection.

Complications of varicella are mainly those involving the nervous system and bacterial superinfections [18]. These include cerebellar ataxia, encephalitis, strokes/vasculopathy, various palsies (paralysis), and pneumonia, cellulitis, and sepsis. Other rare complications include gastrointestinal infections, arthritis, hepatitis, glomerulonephritis, and primary viral pneumonia. In addition to adults, infants and immunocompromised patients are at risk to develop severe varicella. Prior to universal vaccination in the United States, which began in 1995, about 100–125 individuals perished annually from varicella, despite the availability of specific antiviral therapy [18]. Severe varicella typically consists of an extensive rash that may be hemorrhagic, often accompanied by pneumonia and hepatitis. Patients being treated for cancer and human immunodeficiency virus (HIV) or receiving high doses of corticosteroids are at high risk. Pregnant women who contract varicella may develop severe infections and may infect their fetus, resulting in the congenital varicella syndrome, with severe damage to the nervous system. [18].

2.3 Treatment

Varicella in otherwise healthy children is usually treated symptomatically with therapies designed to decrease itching such as topical calamine lotion rather than with antivirals. Regular bathing is recommended in order to minimize cutaneous bacterial superinfections.

The first successful specific treatment for VZV infections, acyclovir, was licensed for use by the US Food and Drug Administration in 1982 [58]. Over 30 years of experience with oral or intravenous acyclovir have indicated that this drug is highly effective against VZV (both varicella and zoster) and that it is extremely safe and well tolerated [57]. The development of additional and improved oral drugs valacyclovir and famciclovir followed acyclovir. These medications have been especially important for treatment of children at high risk to develop severe varicella and also for most patients with zoster.

In countries where varicella vaccine is routinely administered, there is less and less need for antiviral treatment. Additional details regarding specific therapy of VZV infections is presented in Chap. 8.

2.4 Prevention: Live Attenuated Varicella Vaccine

While passive immunization is useful to modify varicella in high-risk susceptibles, it rarely if ever prevents varicella. Similarly prevention of varicella by isolation of cases is also not very successful. Antiviral therapy is useful to treat varicella, but there is little experience with its success in preventing the disease. The only known effective way to prevent varicella from occurring is by the use of an effective vaccine [21].

Live attenuated varicella vaccine was developed by Professor Michiaki Takahashi in Osaka, Japan, beginning about 45 years ago [46]. Takahashi became interested in how viruses are attenuated while investigating poliovirus. He realized the importance of viral propagation at different temperatures and in cell cultures from different animals, which had been used by others to attenuate viruses. He found that VZV grew to higher titers at 37°C than at 39°C and that infectivity was higher in guinea pig embryo cells than in human embryonic cells. These observations led him to passage serially VZV in human embryonic lung cells at 34°C (11 times), in guinea pig embryo cells at 34°C (12 times), and human diploid cells (WI-38; three times) at 37°C. Three additional passages in human diploid cells (MRC-5) followed [45]. He was also cognizant of the following: in the first quarter of the twentieth century, there were five published studies in which healthy children were inoculated with vesicular fluid from patients with either zoster or varicella, in early attempts to make a vaccine against varicella [9, 23, 24, 27, 30]. None of these children were reported to experience adverse events, although none seemed to become immune to varicella. Takahashi felt that one success of these experiments was that it made it very unlikely that children would be harmed by his candidate live varicella vaccine, which had been administered safely to a number of different animals. There was no animal model of varicella on which to test the safety or efficacy of the vaccine. Takahashi and his clinical colleagues administered the vaccine to healthy children and then to children with various other illnesses for which they were hospitalized [45, 46]. They demonstrated the immunogenicity and efficacy of the vaccine by immunizing children in families where a child had recently developed varicella. The vaccinees developed specific antibodies and did not become ill with varicella. They also immunized children who were exposed to varicella in their schools; vaccinees appeared to be protected from varicella. They measured cellular immunity in vaccinees using a skin test for varicella. Since children with leukemia were at high risk to develop fatal varicella, they administered varicella vaccine to children with leukemia who were in remission for at least 6 months but still receiving maintenance chemotherapy. Of 11 such children, only two developed rashes, which were not severe [45]. These studies in Japan led to investigations of live attenuated varicella vaccine in the United States.

In 1975, about 80 % of American children with acute leukemia were being cured of this disease. These were mainly young children who were susceptible to varicella, and it was not uncommon for them to develop this infection while being treated for leukemia. In one report, 7 % of these children died from varicella and many had terrible morbidity associated with the illness [14]. Interestingly, the record did not improve significantly even after the introduction of passive immunization and antiviral therapy [15]. In the late 1970s, the Japanese were publishing their research investigations on live attenuated varicella vaccine in American pediatric journals, and it was extremely difficult for academic physicians to ignore these successes [29]. The varicella vaccine however became highly controversial in the United States [6–8, 40]. There was fear of the vaccine causing severe latent infections, adverse events directly from the vaccine itself, and failure of the vaccine to stimulate long-term immunity. Finally, a meeting was organized by the National Institute of Allergy and Infectious Disease in 1977, to decide what to do. The meeting was attended by diverse groups such as older highly respected virologists and younger hematologic and infectious disease academicians who cared for the leukemic children who might die from varicella. Included was Professor Takahashi himself, who was the first speaker. By the end of the meeting, it was agreed that it was time to investigate live attenuated varicella vaccine and that the best risk/benefit ratio was in children with leukemia in remission. A large, collaborative study of vaccination of 191 varicella-susceptible children with leukemia in remission for at least 1 year eventually was organized as a result of this meeting. Maintenance chemotherapy was withheld for 2 weeks after the vaccination was carried out; the vaccine was well tolerated except for some patients with rashes for which oral acyclovir was sometimes given. Twenty-two vaccinees had household exposures to varicella over a period of several years, and because they were known to have detectable antibodies to VZV, no intervention was taken. The attack rate of varicella in these children was 18 %, not the historical rate of 85 % in susceptibles. All case of varicella were very mild. The vaccine was 80 % effective in preventing varicella in these children and 100 % effective in preventing severe varicella [22]. Following the success in children with underlying leukemia, numerous studies in healthy children took place, all of which indicated that the vaccine was safe and highly effective for healthy children [31, 32, 41, 43, 49–51, 56]. Live attenuated varicella vaccine became licensed for routine use in healthy children in 1995 [21].

The face of varicella changed dramatically after universal vaccination for all healthy varicella susceptibles was introduced into the United States. At first only one dose of vaccine was given, but introduction of a second dose was administered months to years after the first improved immune responses [42]. At present the rate of receipt of two doses of varicella vaccine in children approaches 90% [35], and the incidence of hospitalizations and deaths from varicella has fallen by over 90 % [4, 33, 37]. There is evidence of decreased circulation of wild-type VZV [44], and it is now unusual to see children or adults with this infection. Most cases are traceable to patients with zoster. Vaccinated children may develop a mild case of varicella termed "breakthrough varicella," but this is unusual in those who have received two doses of vaccine [18, 42]. The incidence of zoster after vaccination is lower than that following the natural infection [26, 52]. Interestingly, zoster in as many as 50 % of vaccinees is caused by wild-type VZV [52].

The live attenuated varicella vaccine is one of the safest vaccines available today. It is used worldwide. The countries in which it is licensed include: Australia, Canada, Costa Rica, Germany, Greece, Israel, Japan, Qatar, Latvia, Luxembourg, Taiwan, Saudi Arabia, South Korea, United Arab Emirates, Uruguay, and some areas of Italy and Spain.

There is little or no evidence of waning immunity after vaccination [2, 3, 48]. It may be that periodic boosting of immunity due to asymptomatic or very mild cases of reactivation/zoster serve to boost immune responses to VZV, as was originally proposed by Hope-Simpson years ago. Hope-Simpson was the first to propose that zoster develops when immunity to VZV decreases due to aging [28]. He also hypothesized that immunity to VZV following varicella was restimulated when there was an external exposure to the virus, such as from close contact with a child with active varicella. He furthermore proposed that immunity was stimulated when VZV underwent low-level internal reactivation without symptoms. Over the years studies have demonstrated that exposures to patients with VZV disease increases asymptomatic specific immunity [19, 47]. In addition, subclinical VZV reactivation has been demonstrated along with increases in immunity [20, 34, 36] [38, 59] [12].

In 2002, it was hypothesized that where there was extensive varicella vaccination, there would be decreased circulation of VZV, resulting in an increase in zoster due to lack of immune stimulation [5]. Based on mathematical computer modeling, a zoster epidemic was predicted in young adults, with numerous deaths. This prediction led a number of countries not to use varicella vaccine in children routinely. Fifteen years later, however, there has been no such zoster epidemic in the United States. The mathematical model accounted for external exposures to VZV but did not include the possibility of internal subclinical reactivation of VZV with specific immune stimulation. Internal subclinical reactivation of VZV probably explains the results of a retrospective study of the incidence of zoster in French sequestered religious groups who were not exposed to children. The incidence of zoster in sequestered nuns and monks was compared with that of the nearby general public, by evaluating medical records over a 30-year period. The incidence of zoster (about 15 %) was the same in all groups. This study demonstrated that exposure to individuals who spread varicella is not required to maintain long-term immunity to zoster [13, 16].

It is clear, however, that the incidence of zoster is on the rise in the United States. The cause of this increase is probably multifactorial. It includes an increase in ascertainment of the disease, the aging of our population, increasing numbers of cancer and transplant survivors, and widespread stresses in our complex world. There are now ample publications in the literature indicating that there is no direct correlation of the development of varicella and the increase in zoster, which began in the United States long before varicella vaccine was developed [21, 25].

In summary, the live attenuated varicella vaccine is in use all over the world due to its high degree of safety and effectiveness. One could think of the live attenuated varicella vaccine as the parent of the zoster vaccine. Its success led not only to control of varicella but also to a similar but more powerful vaccine that would prove to be not only safe but highly effective in preventing zoster.

Grant Support R01DK093094; Biken Institute of Osaka University

References

1. Arvin AM, Moffat JF, Sommer M, Oliver S, Che X, Vleck S, Zerboni L, Ku CC (2010) Varicella-zoster virus T cell tropism and the pathogenesis of skin infection. Curr Top Microbiol Immunol 342:189–209
2. Baxter R, Ray P, Tran TN, Black S, Shinefield HR, Coplan PM, Lewis E, Fireman B, Saddier P (2013) Long-term effectiveness of varicella vaccine: a 14-year, prospective cohort study. Pediatrics 131:e1389–e1396
3. Baxter R, Tran TN, Ray P, Lewis E, Fireman B, Black S, Shinefield HR, Coplan PM, Saddier P (2014) Impact of vaccination on the epidemiology of varicella: 1995–2009. Pediatrics 134:24–30
4. Bialek SR, Perella D, Zhang J, Mascola L, Viner K, Jackson C, Lopez AS, Watson B, Civen R (2013) Impact of a routine two-dose varicella vaccination program on varicella epidemiology. Pediatrics 132:e1134
5. Brisson M, Gay N, do WJ, Andrews NJ (2002) Exposure to varicella boosts immunity to herpes-zoster: implications for mass vaccination against chickenpox. Vaccine 20:2500–2507
6. Brunell PA (1977) Commentary: protection against varicella. Pediatrics 59:1–2
7. Brunell PA (1975) Live varicella vaccine. Lancet i:98

8. Brunell PA (1978) Varicella vaccine: the crossroads is where we are not! Pediatrics 62:858–859
9. Bruusgaard E (1932) The mutual relation between zoster and varicella. Brit J Derm Syph 44:1–24
10. Cheatham WJ, Weller TH, Dolan TF, Dower JC (1956) Varicella: report of two fatal cases with necropsy, virus isolation, and serologic studies. Am J Pathol 32:1015–1035
11. Chen JJ, Zhu Z, Gershon AA, Gershon MD (2004) Mannose 6-phosphate receptor dependence of varicella zoster virus infection in vitro and in the epidermis during varicella and zoster. Cell 119:915–926
12. Cohrs RJ, Mehta SK, Schmid DS, Gilden DH, Pierson DL (2008) Asymptomatic reactivation and shed of infectious varicella zoster virus in astronauts. J Med Virol 80:1116–1122
13. Crumpacker C (2011) Absence of exposure to varicella does not increase the risk of zoster. Clin Infect Dis Off Publ Infect Dis Soc Am 53:412–413
14. Feldman S, Hughes W, Daniel C (1975) Varicella in children with cancer: 77 cases. Pediatrics 80:388–397
15. Feldman S, Lott L (1987) Varicella in children with cancer: impact of antiviral therapy and prophylaxis. Pediatrics 80:465–472
16. Gaillat J, Gajdos V, Launay O, Malvy D, Demoures B, Lewden L, Pinchinat S, Derrough T, Sana C, Caulin E, Soubeyrand B (2011) Does monastic life predispose to the risk of saint anthony's fire (Herpes Zoster)? Clin Infect Dis Off Publ Infect Dis Soc Am 53:405
17. Garland J (1943) Varicella following exposure to herpes zoster. N Engl J Med 228:336–337
18. Gershon A, Breuer J, JI C, Cohrs R, Gershon MD, Gilden D, Grose C, Hambleton S, Kennedy P, Oxman M, Seward J, Yamanishi. K (2015) Varicella zoster virus infection. Nat Rev Primers 1:15016
19. Gershon A, LaRussa P, Steinberg S, Lo SH, Mervish N, Meier P (1996) The protective effect of immunologic boosting against zoster: an analysis in leukemic children who were vaccinated against chickenpox. J Infect Dis 173:450–453
20. Gershon A, Steinberg S (1981) Antibody responses to varicella-zoster virus and the role of antibody in host defense. Am J Med Sci 282:12–17
21. Gershon A, Takahashi M, Seward JF (2013) Live attenuated varicella vaccine. In: Plotkin S, Orenstein W, Offit P (eds) Vaccines, 6 edn. WB Saunders, Philadelphia, pp. 837–869
22. Gershon AA, Steinberg S, Gelb L, NIAID-Collaborative-Varicella-Vaccine-Study-Group (1984) Live attenuated varicella vaccine: efficacy for children with leukemia in remission. JAMA 252:355–362
23. Greenthal RM (1926) The prophylaxis of varicella with vesicle fluid. Am. J Dis Child 31:851–855
24. Gulacsy Z (1933) Aktive Schutzimpfung gegen varizellen. Arch Kinderheik 100:75–78
25. Hales CM, Harpaz R, Joesoef MR, Bialek SR (2013) Examination of links between herpes zoster incidence and childhood varicella vaccination. Ann Intern Med 159:739–745
26. Hardy, I., A. A. Gershon, S. P. Steinberg, P. LaRussa, et. al. 1991. The incidence of zoster after immunization with live attenuated varicella vaccine. A study in children with leukemia. Varicella Vaccine Collaborative Study Group. N Engl J Med 325:1545-1550.
27. Hess AF, Ungar LJ (1918) A protective therapy for varicella, and a consideration of the pathogenesis. Am J Dis Child 16:34–38
28. Hope-Simpson RE (1965) The nature of herpes zoster: a long term study and a new hypothesis. Proc R Soc Med 58:9–20
29. Kempe CH, Gershon AA (1977) Varicella vaccine at the crossroads. Pediatrics 60:930–931
30. King CA (1915) Technik der Schutzimpfung gegan Varicella. Berl Klin Wschr 52:13–15
31. Kuter B, Matthews H, Shinefield H, Black S, Dennehy P, Watson B, Reisinger K, Kim LL, Lupinacci L, Hartzel J, Chan I (2004) Ten year follow-up of healthy children who received one or two injections of varicella vaccine. Pediatr Infect Dis J 23:132–137
32. Kuter BJ, Weibel RE, Guess HA, Matthews H, Morton DH, Neff BJ, Provost PJ, Watson BA, Starr S, Plotkin S (1991) Oka/Merck varicella vaccine in healthy children: final report of a 2-year efficacy study and 7-year follow-up studies. Vaccine 9:643–647

33. Leung J, Bialek SR, Marin M (2015) Trends in varicella mortality in the United States: Data from vital statistics and the national surveillance system. Hum Vaccin Immunother 11:662–668
34. Ljungman P, Wilczek H, Gahrton G, Gustavsson A, Lundgren G, Lonnqvist B, Ringden O, Wahren B (1986) Long-term acyclovir prophylaxis in bone marrow transplant recipients and lymphocyte proliferation responses to herpes virus antigens in vitro. Bone Marrow Transplant 1:185–192
35. Lopez AS, Cardemil C, Pabst LJ, Cullen KA, Leung J, Bialek SR, D. Division of Viral, C. Centers for Disease, and Prevention (2014) Two-dose varicella vaccination coverage among children aged 7 years – six sentinel sites, United States, 2006-2012. MMWR Morb Mortal Wkly Rep 63:174–177
36. Luby J, Ramirez-Ronda C, Rinner S, Hull A, Vergne-Marini P (1977) A longitudinal study of varicella zoster virus infections in renal transplant recipients. J Infect Dis 135:659–663
37. Marin M, Zhang JX, Seward JF (2011) Near elimination of varicella deaths in the US after implementation of the vaccination program. Pediatrics 128:214–220
38. Mehta SK, Cohrs RJ, Forghani B, Zerbe G, Gilden DH, Pierson DL (2004) Stress-induced subclinical reactivation of varicella zoster virus in astronauts. J Med Virol 72:174–179
39. Paine TF Jr (1964) Latent herpes simplex infection in man. Bacteriol Rev 28:472–479
40. Plotkin SA (1977) Varicella vendetta: plotkin's plug. Pediatrics 59:953–954
41. Reuman PD, Sawyer MH, Kuter BJ, Matthews H, G. MMRV Study (1997) Safety and immunogenicity of concurrent administration of measles-mumps-rubella-varicella vaccine and PedvaxHIB^{R} vaccines in healthychildren twelve to eighteen months old. Pediatr Infect Dis J 16:662–667
42. Shapiro ED, Vazquez M, Esposito D, Holabird N, Steinberg SP, Dziura J, Larussa PS, Gershon AA (2011) Effectiveness of 2 doses of varicella vaccine in children. J Infect Dis 203:312–315
43. Shinefield H, Williams WR, Marchant C, Reisinger K, Stewart T, Meissner HC, Guerrero J, Klopfer SO, Xu J, Schodel F, Kuter BJ (2005) Dose-response study of a quadrivalent measles, mumps, rubella and varicella vaccine in healthy children. Pediatr Infect Dis J 24:670–675
44. Son M, Shapiro ED, LaRussa P, Neu N, Michalik DE, Meglin M, Jurgrau A, Bitar W, Vasquez M, Flynn P, Gershon AA (2010) Effectiveness of varicella vaccine in children infected with HIV. J Infect Dis 201:1806–1810
45. Takahashi M, Asano Y, Kamiya H, Baba K, Yamanishi K (1981) Active immunization for varicella-zoster virus. In: Nahmias A, Dowdle W, Schinazi R (eds) The human herpesviruses. Elsevier, New York, pp. 414–431
46. Takahashi M, Otsuka T, Okuno Y, Asano Y, Yazaki T, Isomura S (1974) Live vaccine used to prevent the spread of varicella in children in hospital. Lancet 2:1288–1290
47. Thomas S, Wheeler J, Hall AJ (2002) Contacts with varicella or with children and protection against herpes zoster in adults: a case-control study. Lancet 360:678–682
48. Vazquez M, LaRussa PS, Gershon AA, Niccolai LM, Muehlenbein CE, Steinberg SP, Shapiro ED (2004) Effectiveness over time of varicella vaccine. JAMA 291:851–855
49. Watson B, Laufer D, Kuter B, Staehle B, White CJ (1996) Safety and immunogenicity of a combined live attenuated measles, mumps, rubella, and varicella vaccine ($RRR_{II}V$) in healthy children. J Infect Dis 173:731–734
50. Weibel R, Kuter B, Neff B, Rothenberger C, Fitzgerald A, Connor K, Morton D, McLean A, Scolnick E (1985) Live Oka/Merck varicella vaccine in healthy children: further clinical and laboratory assessment. JAMA 245:2435–2439
51. Weibel R, Neff BJ, Kuter BJ, Guess HA, Rothenberger CA, Fitzgerald AJ, Connor KA, McLean AA, Hilleman MR, Buynak EB, Scolnick EM (1984) Live attenuated varicella virus vaccine: efficacy trial in healthy children. N Engl J Med 310:1409–1415
52. Weinmann S, Chun C, Schmid DS, Roberts M, Vandermeer M, Riedlinger K, Bialek SR, Marin M (2013) Incidence and clinical characteristics of herpes zoster among children in the varicella vaccine era, 2005–2009. J Infect Dis 208:1859–1868
53. Weller T, Stoddard MB (1952) Intranuclear inclusion bodies in cultures of human tissue inoculated with varicella vesicle fluid. J Immunol 68:311–319

54. Weller TH (1953) Serial propagation in vitro of agents producing inclusion bodies derived from varicella and herpes zoster. Proc Soc Exp Biol Med 83:340–346
55. Weller TH, Coons AH (1954) Fluorescent antibody studies with agents of varicella and herpes zoster propagated in vitro. Proc Soc Exp Biol Med 86:789–794
56. White CJ, Kuter BJ, Hildebrand CS, Isganitis KL, Matthews H, Miller WJ, Provost PJ, Ellis RW, Gerety RJ, Calandra GB (1991) Varicella vaccine (VARIVAX) in healthy children and adolescents: results from clinical trials, 1987 to 1989. Pediatrics 87:604–610
57. Whitley RJ (2009) A 70-year-old woman with shingles: review of herpes zoster. JAMA 302:73–80
58. Whitley RJ, Gnann JW (1992) Acyclovir: a decade later. N Engl J Med 327:782–789
59. Wilson A, Sharp M, Koropchak CM, Ting SF, Arvin AM (1992) Subclinical varicella-zoster virus viremia, herpes zoster, and T lymphocyte immunity to varicella-zoster viral antigens after bone marrow transplantation. J Infect Dis 165:119–126

Part II
Herpes Zoster (Shingles)

Chapter 3
Herpes Zoster: A Patient's Perspective

Anne Tuzi and C. Peter N. Watson

3.1 About My Pain

My name is Ann Tuzi; I am 57 years old and I would like to tell my story. First of all I want to thank God for giving me a second chance at life although it left me with a condition, which I will tell you about. My journey started 13 years ago in 2002 when I was diagnosed with Burkitt's lymphoma/leukemia (a rare cancer of the blood). I fought a big fight with that type of cancer, and it managed to go into remission, but in October 2003 I got shingles on the right side of my abdomen during one of my chemotherapy treatments. At first I didn't know what it was but the pain was out of this world. We had to stop my chemo treatment and I was bedridden for 6 months. By the time I was referred to a pain clinic, it was too late; all the nerves of my waist had been damaged. So I was put on gabapentin 600 mg/day which helped a little but didn't get rid of the pain. If that wasn't enough, my shingles then spread to my right eye. So I was put in isolation for a week where they were able to treat that with famciclovir.

Right now, I am still suffering with postherpetic neuralgia nowhere near as painful as it was, although there are times during the day when it gets so bad that I become frozen, so I can't talk or move. On a scale from one to ten, I am usually a seven out of ten. The only way I can describe it is that it's like hundreds of knives being stabbed in my waist. I wake up several times a night with major pain, and the only way I make it through the night is with an ice pack on my side. It hasn't been easy, and my life has been turned around 360°. However, I have learned to work around my pain instead of allowing my pain to control me.

A. Tuzi • C.P.N. Watson (✉)
Department of Medicine, University of Toronto,
Toronto, ON, Canada
e-mail: peter.watson@utoronto.ca

C.P.N. Watson et al. (eds.), *Herpes Zoster: Postherpetic Neuralgia and Other Complications*, DOI 10.1007/978-3-319-44348-5_3

3.2 Getting Treatment

I'd like to tell you about some of the treatments that my pain specialist has tried. Some have been temporary fixes. I have had over 50 different drugs, acupuncture/laser, naturopathic treatments, physiotherapy, and TENS treatments, all of which didn't work. I have taken pain management programs, a mindfulness meditation program, and laughter program. I have seen over ten doctors, and the waiting time to see a specialist was anywhere between 1 and 4 months. My costs have been high with very limited income and limited assistance.

3.3 Impact on Life, Family, Work, and Finances

I would like explain to you how shingles has affected my life. My life has been turned around tremendously. I haven't been able to return to work since 2002 (14 years). I've gone from a fairly active, energetic, and involved member of society to having to give up my stained glass hobby, to sell my standard convertible car, and to give up exercise. I have gained weight (50 lbs) and limited my traveling. In order to get things done, I have to think outside of the box, like using a cane, a walker, and a scooter. Thank God for computers where I am able to do my grocery shopping and banking and make other purchases. My social life has been drastically affected. I went from being a social butterfly, who couldn't be stopped for anything or anybody, to someone who is limited to a few close friends and family socializing primarily through telephone and the Internet. The rest of my circle of friends simply didn't know how to deal with me and my pain, eventually drifting away. My finances have taken a big beating. My costs are high and my income is 50 % of what I was making. My cost for drugs is high. Lucky for me that insurance absorbs 80 % of the cost. Taxis are costing $2000/year. Due to my illness, I am no longer able to take care of my daily tasks; therefore, I have had to hire a cleaning lady and a gardener and contract snow shoveling. I am getting only the cost of living increase on my government disability pension but no increase for 13 years with private insurance. I am not eligible for other government programs. On top of all of this, I am caring for a 93-year-old mother, who is high maintenance. This is a full-time job in itself. I am taking care of myself and I am a caregiver.

In conclusion it's a shame when a very active member of society, who has done everything right, has been very productive, and has worked so hard all her life, is reduced to a social outcast and is barely able to make ends meet, simply because she is afflicted by a disability that few people know and understand.

Commentary

C. Peter N. Watson

Ms. Tuzi has been my patient for 11 years. She has had medically intractable severe postherpetic neuralgia for 13 years, located in the right upper abdomen in the ninth thoracic dermatome region, which has left her with a severe, steady, burning pain, electric shock-like jabbing pain, and exquisite sensitivity of the skin. This is associated with Burkitt's lymphoma and chemotherapy. Recurrence of zoster as occurred on the right forehead is uncommon and happens in about 5 % [1] and may attack the same dermatome [2]. In this instance this recurrence close to the first episode may reflect her state of immunosuppression.

On physical examination (Fig. 3.1), she has pale postherpetic scarring in the right ninth thoracic dermatome. There is very widespread loss of sensation to pin, cold, and touch but marked sensitivity of the skin to touch (dynamic mechanical allodynia) over only a rather small area, mostly over the worst scar. Over the decade that she has been my patient, she has had drug trials with gabapentinoids (pregabalin and gabapentin), tricyclic antidepressants (amitriptyline, nortriptyline, desipramine, and doxepin), serotonin

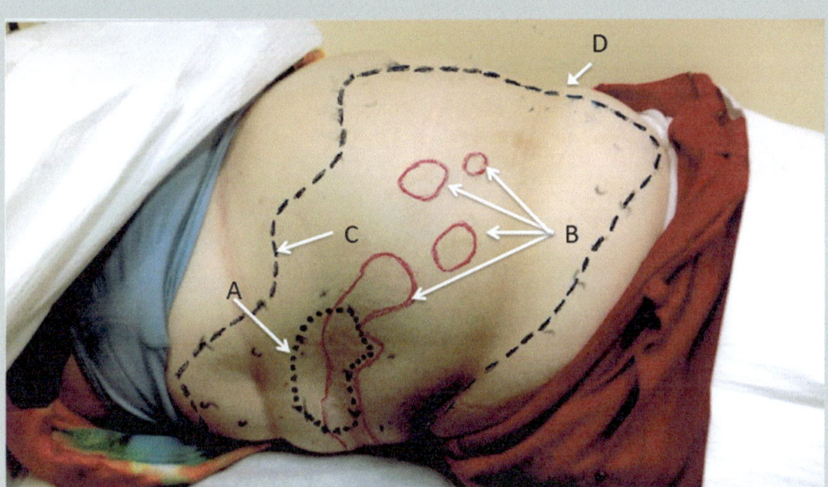

Fig. 3.1 Anterolateral view of the abdomen: *A* (*dotted line*) = area of dynamic mechanical allodynia (pain from skin stroking with cotton), *B* (*solid line*) = scarred areas, *C* (*interrupted line*) = area of sensory loss to pin, cold, and touch, *D* = umbilicus

norepinephrine re uptake inhibitors (duloxetine and venlafaxine), and a variety of opioids (morphine, oxycodone, hydromorphone, transdermal fentanyl, tramadol, and methadone) in keeping with guidelines for treating neuropathic pain. She has also had three oral cannabinoids, other anticonvulsants (phenytoin, lamotrigine, carbamazepine, and oxcarbazepine), and the antispasticity agent, baclofen. She has also used topical agents such as capsaicin and lidocaine. She has additionally had trials of acupuncture, laser therapy, and transcutaneous electrical nerve stimulation.

The only thing that has helped her and that takes the pain from 10/10 to about 7/10 is a high dose of gabapentin in the form of 1200 mg every 6 h for a total of 4800 mg a day. She has been on that drug and dose unchanged for 5 years. The intractability of her postherpetic neuralgia may relate to the singular physical findings of an unusually extensive area of sensory loss (Fig. 3.1) which since only one ganglion is usually affected (Fig. 3.2) could relate to a

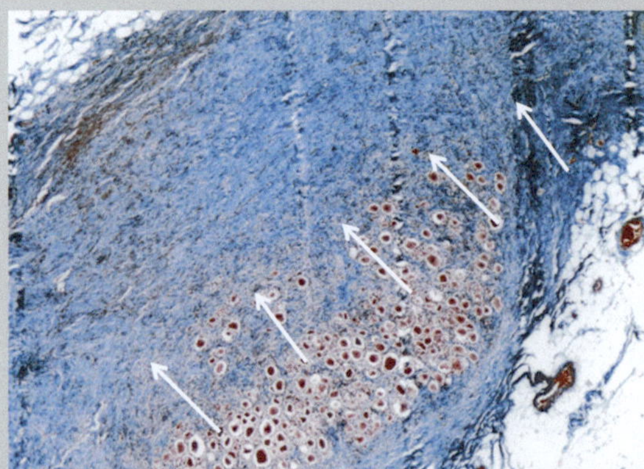

Fig. 3.2 Postherpetic neuralgia with fibrosis and nerve cell loss in the dorsal root ganglion, surviving ganglion cells in red, scarring in upper part of ganglion (*arrows*) (Masson trichrome × 10) [3]

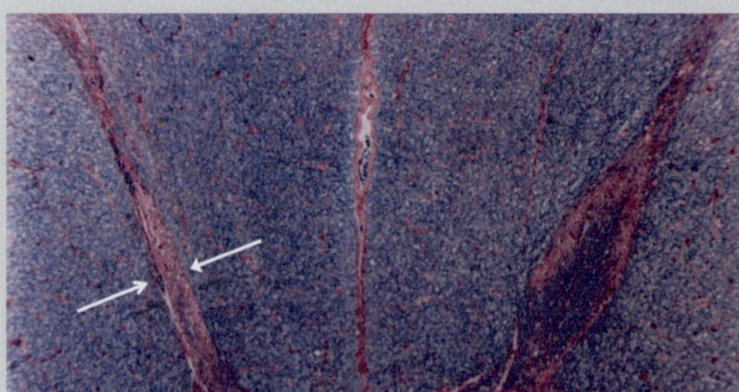

Fig. 3.3 Postherpetic neuralgia showing atrophy of the dorsal horn of the spinal cord on the left side of the image (*arrows*) (H & E LFB × 2.5) [3]

long extent of rostrocaudal involvement of affected and atrophic dorsal horn of the spinal cord (Fig. 3.3) where many analgesics have an action. Immunosuppression associated with both Burkitt's lymphoma and Epstein-Barr virus and also chemotherapy may play a role in this intractability as well as for the second episode of trigeminal zoster. Ms. Tuzi's case illustrates the need for vaccination and particularly the use of a vaccine that can be used safely in patients who are immunosuppressed and which is currently undergoing clinical trials [4].

References

1. Ragozzino MW, Melton LJ, Kurland LT (1982) Population-based study of herpes zoster and its sequelae. Medicine 61:310–316
2. Hope Simpson RE (1965) The nature of herpes zoster: a long term study and a new hypothesis. Proc R Soc Med 58:9–20
3. Watson CPN, Morshead C, Van Der Kooy D, Deck JH, Evans RJ (1988) Postherpetic neuralgia: post-mortem analysis of a case. Pain 34:129–138
4. Lal H, Cunningham AL, Godeaux O et al (2015) Efficacy of an adjuvanted herpes zoster subunit vaccine in older adults. New Engl. J. Med. 372:2087–2096

Chapter 4
The Epidemiology and Natural History of Herpes Zoster and Postherpetic Neuralgia

Kenneth E. Schmader and Robert H. Dworkin

4.1 Introduction

The objective of this chapter is to review the epidemiology and natural history of herpes zoster and postherpetic neuralgia. Epidemiology seeks to explain and comprehend diseases by studying the characteristics of diseases in populations. For this chapter, these characteristics include morbidity, mortality, incidence and prevalence rates, and risk factors for herpes zoster and postherpetic neuralgia. The sources of epidemiological data in this chapter include cohort studies, case-control studies, clinical trials, large case series, and case reports.

4.2 Herpes Zoster

The varicella-zoster virus (VZV) establishes latency in sensory ganglia following primary varicella infection. Herpes zoster (shingles) is the reactivation of the virus and its spread from a single ganglion of the dorsal root, cranial nerve, or autonomic nervous system to the corresponding dermatome and neural tissue of the same segment [18, 40]. The presentation of herpes zoster is variable. In many patients, a prodrome of pain or abnormal sensations in the affected dermatome precedes the

K.E. Schmader (✉)
Division of Geriatrics, Department of Medicine and the Center for the Study of Aging and Human Development, Duke University Medical Center, Durham, NC 27710, USA

Geriatric Research, Education and Clinical Center, Durham VA Medical Center, Durham, NC 27710, USA
e-mail: kenneth.schmader@duke.edu

R.H. Dworkin
Departments of Anesthesiology, Neurology and Psychiatry, University of Rochester School of Medicine and Dentistry, Rochester, NY 14642, USA

Center for Human Therapeutics, University of Rochester School of Medicine and Dentistry, Rochester, NY 14642, USA
e-mail: robert_dworkin@urmc.rochester.edu

C.P.N. Watson et al. (eds.), *Herpes Zoster: Postherpetic Neuralgia and Other Complications,* DOI 10.1007/978-3-319-44348-5_4

appearance of a characteristic unilateral rash. The prodromal symptoms are usually aching, burning, or lancinating pain or itching or tingling. Prodromal symptoms confuse patients and providers alike by imitating other painful conditions in older persons (i.e., migraine headaches, myocardial infarction, cholecystitis, biliary or renal colic, muscle ache, etc.). Although this prodrome begins several days before rash onset in almost all cases, a series of patients with prodromal pain preceding the appearance of the rash by seven to more than 100 days has been reported [36].

Thoracic dermatomes are the most commonly affected sites in herpes zoster and account for 50–70 % of all cases [55, 93, 121]. Cranial (especially the ophthalmic division of the trigeminal nerve), cervical, and lumbar dermatomes each account for 10–20 % of cases, and sacral dermatomes are affected in 2–8 % of cases [93, 121]. The rash is unilateral, dermatomal, red, and maculopapular and usually develops vesicles. It is not uncommon for patients to develop lesions in adjacent dermatomes. The rash generally starts crusting over in a week to 10 days and heals within 2–4 weeks. Atypical rashes may occur. The rash may be limited to a small patch located within a dermatome or may remain maculopapular without ever developing vesicles. Conversely, vesicles may form for several days and involve several dermatomes.

Pain in the affected dermatome accompanies the rash in most patients. Those who did not have a painful prodrome typically begin to experience pain at rash onset or shortly afterwards, although some patients experience the delayed onset of pain days or weeks after rash onset and a few patients never develop pain. This acute herpes zoster pain gradually resolves during or shortly after rash healing in many patients. Dermatomal pain without a rash, referred to as zoster sine herpete, has also been described [37, 40, 41, 101]. The finding of VZV DNA in the cerebrospinal fluid of two patients with prolonged radicular pain and no rash provides evidence for this syndrome [38].

The acute neuritis of herpes zoster produces dermatomal pain that is described as burning, deep aching, or stabbing. A subset of patients may develop severe disabling pain, while some patients describe uncomfortable sensations that they do not label pain such as tingling or numbness. Distracting and sometimes disabling itching is one particularly troubling dysesthesia that commonly accompanies pain or occurs instead of pain and seems to occur most often with herpes zoster involvement of the head, face, and neck [80]. Acute herpetic neuralgia has a profound negative impact on functional status and quality of life and usually results in substantial health service utilization [61, 103].

4.2.1 Incidence of Herpes Zoster in General Population-Based Studies

Investigators have published cohort studies of the incidence of herpes zoster in the general population of at least 22 different countries. The incidence of herpes zoster in these mostly immunocompetent community dwelling populations ranges from 2 to 5 per 1000 person-years [34, 62, 91, 103, 121, 123]. These studies differed in their study populations and methods of case ascertainment which may explain, in part, the differences in the incidence of herpes zoster between studies. Most studies relied on administrative databases or health records which will miss herpes zoster cases that do not come to medical attention. The incidence of herpes zoster in populations limited to immunosuppressed patients is substantially higher which is discussed below. The likelihood of recurrent herpes zoster was 4.1, 5.2, and 6.2 % in the studies of Hope-Simpson [55], Ragozzino et al. [93], and Ywn et al. [122], respectively.

Several investigators have examined longitudinal trends in herpes zoster incidence over the past 20 years and reported that incidence rates of herpes zoster are increasing across all age groups over time. The increase over time in herpes zoster incidence rate occurred in countries without varicella vaccination programs [10, 88, 96, 107]. The increase in herpes zoster incidence rate over time has also been observed before the onset of varicella vaccination programs, and the increase continued after the introduction of varicella vaccination programs [48, 59, 66]. These data indicate that varicella vaccination is not an explanation for the increases in herpes zoster incidence.

4.2.2 *Morbidity and Mortality of Herpes Zoster*

The morbidity of herpes zoster includes acute pain and chronic pain (discussed below), other neurological disorders, and ophthalmological, cutaneous, and visceral complications. In cohort studies [32, 121], the types and estimated frequencies of non-pain neurological complications included motor neuropathy (including Bell's-like palsy), cranial polyneuritis, Ramsay Hunt (3 %), transverse myelitis (less than 1 %), meningoencephalitis (less than 1 %), and cerebral angiitis and stroke after ophthalmic zoster (less than 1 %). Herpes zoster is a significant risk factor for stroke especially in persons with ophthalmic herpes zoster where the risk of stroke is increased threefold to fourfold compared to individuals without herpes zoster [65, 76]. Ophthalmological complications have been described in 2–6 % of herpes zoster cases, including keratitis, uveitis, iridocyclitis, panophthalmitis, and glaucoma [32, 121]. Bacterial infection of the rash and herpes gangrenosum occurred in zero to 2 % in these series. Visceral involvement was not noted in these studies. Elderly and immunosuppressed patients with herpes zoster are at greater risk for these complications. Death due to herpes zoster is uncommon. A systematic review that included ten studies with data on mortality due to herpes zoster found rates ranging from 0.017 to 0.465/100,000 person-years [62]. After adjusting for errors in death certificates, Mahamud et al. [68] estimate the true number of deaths due to herpes zoster in the USA to be 78 (31–118) per year. Nearly all deaths occur in older and/or immunosuppressed persons.

4.2.3 *Risk Factors for Herpes Zoster*

4.2.3.1 Age

A fundamental epidemiological feature of herpes zoster is a marked increase in incidence with aging [102]. This fact was demonstrated elegantly by Hope-Simpson [55] over 50 years ago in a UK general practice. He documented an incidence of 0.74 per 1000 person-years in children under 10 years old, 2.5 per 1000 person-years in adults aged 20–50 years, and 7.8 per 1000 person-years in those older than 60 years old. Many subsequent studies have confirmed the increasing incidence of herpes zoster with aging in different population groups and countries. The incidence of herpes zoster in older adults ranges from 8 to 12 per 1000 person-years in population and health record-based studies in the USA, UK, Australia, Netherlands, Japan, Canada, Italy, Spain, France, Israel, Belgium, Germany and Taiwan [62]. The age at which the sharpest increase in herpes zoster occurs is 50–60 years of age. Furthermore, the slope continues a marked upward course in the decades above 60 years. From these studies it is estimated that the lifetime incidence of herpes zoster is 20–30 % in the general population and as high as 50 % of a cohort surviving to age 85 years old [10, 55].

4.2.3.2 Sex

The results of several large cohort studies are conflicting as to whether there are differences in the incidence of herpes zoster by sex. Some studies found no difference by sex. For example, Hope-Simpson [55], Ragozzino et al. [93], and Donahue et al. [20] reported that the incidence of herpes zoster among females versus males was 3.2 vs 3.6, 1.26 vs. 1.34 vs., and 2.11 vs. 2.19 per 1000 person-years, respectively, not statistically significant differences. Among elderly persons in the Duke Established Populations for the Epidemiological Study of the Elderly (EPESE) [100, 101], being female did not significantly affect lifetime herpes zoster risk or incidence. However, several more recent studies demonstrate that the age-adjusted incidence of herpes zoster is significantly higher in women compared to men. For example, Gauthier et al. [34], Yawn et al. [121], and Opstelten et al. [83] reported the incidence of herpes zoster among females versus males to be 6.0 (95 % CI 5.95–6.14) vs. 4.3, (95 % CI 4.22–4.38), 3.9 (95 % CI 3.7–4.1) vs 3.2 (95 % CI 2.9–3.4), and 3.9 (95 % CI 3.6–4.2), vs. 2.5 (95 % CI, 2.3–2.8) per 1000 person-years. In the study by Opstelten et al. [83], after adjustment for potential confounders, the adjusted OR was 1.38 (95 % CI, 1.22–1.56) favoring women. The differences favoring females was consistent across age groups in these studies.

4.2.3.3 Race and Ethnicity

White populations in the USA and Europe have been the focus of most studies of herpes zoster epidemiology. The Duke EPESE achieved an equal mix of black and white participants by using a stratified, random sampling technique in community dwelling elderly. After controlling for age, cancer, and demographic factors, blacks remained four times less likely than whites to have experienced herpes zoster in this population (adjusted odds ratio, 0.25, 95 % CI, 0.18–0.35) [100]. In a study of the prospective occurrence of herpes zoster in the Duke EPESE [101], 4.3 % of blacks and 10.9 % of whites developed herpes zoster over 6 years. After controlling for the above variables, blacks remained significantly less likely to develop herpes zoster (adjusted risk ratio, 0.35, 95 % CI, 0.24–0.51). Herpes zoster case ascertainment in these studies had a false-positive rate of 3 % and false-negative rate of 0 %, which was not enough to explain these striking racial differences [99]. These findings have been confirmed in more recent studies. For example, in a retrospective cohort study of 2,848,765 Medicare beneficiaries older than 65 years from 1992 to 2010, the annual herpes zoster incidence was lower in black persons (RR, 0.51 [CI, 0.48–0.53]) and Hispanic persons (RR, 0.76 [CI, 0.72–0.81]) than white persons [48]. Among over one million cancer patients from 1991 to 2007 in the Surveillance, Epidemiology, and End Results (SEER) cancer registry-Medicare linked database, black patients were less likely to develop herpes zoster than white patients (IRR = 0.64, $p < 0.001$) [124]. Hypothesized reasons for the differences include racial differences in VZV immunity, age at onset of varicella, and exposure to varicella over the life course [23].

4.2.3.4 Cellular Immune Deficiency

The frequency and complications of herpes zoster are significantly increased in patients with deficient cell-mediated immunity compared to immunocompetent individuals, including persons living with human immunodeficiency virus infection (HIV) and/or acquired immunodeficiency syndrome (AIDS), hematologic and solid cancers, organ transplants, immune-mediated diseases, and immunosuppressive treatments.

Human Immunodeficiency Virus (HIV) Infection and Acquired Immunodeficiency Syndrome (AIDS)

The incidence of herpes zoster in HIV-infected individuals not treated with highly active antiretroviral therapy (HAART) varied from 29 to 51 per 1000 person-years in cohort studies of herpes zoster in HIV-infected individuals [12, 29, 39, 75]. HIV infection increases herpes zoster risk even in untreated HIV-infected children, 27 % of whom developed zoster an average of 1.9 years after varicella in one study [35]. The incidence of herpes zoster appears to be lower in HIV-infected individuals on HAART compared to untreated individuals. In a cohort study of HIV-infected individuals from the HIV Atlanta VA Cohort Study between 1982 and 2011, the incidence of herpes zoster per 1000 person-years decreased from 60.3 in 1987 pre-HAART to 10 in 2011 on HAART [74]. A cohort study of incident herpes zoster cases in an urban HIV-infected population on HAART between 2002 and 2009 discovered an incidence rate of 9.3 per 1000 person-years [6]. Similar decreases have been documented in HIV-infected children [67].

Zoster may be the first clue to underlying HIV infection when it occurs in groups at high risk for HIV infection [17, 73, 108]. Zoster can occur at any stage of HIV/AIDS [12, 29].

Cancer

Malignancy is another important risk factor for zoster [20, 47, 93, 97, 100, 124]. Among malignancies, zoster is more frequent in patients with hematological cancers compared to those individuals with solid tumors [47, 97, 124]. In a cohort study of 82,832 hematologic and 944,777 solid cancer patients ≥65 years in 1991–2007 from the Surveillance, Epidemiology and End Results (SEER) cancer registry-Medicare linked database, 9.2 % of those with hematologic and 6.3 % of solid cancer patients were diagnosed with herpes zoster [124]. The adjusted incidence rate versus a random group of non-cancer elderly patients was 2.4 per 1000 patient years in persons with hematologic cancer and 1.2 per 1000 patient years in persons with solid cancer. The incidence rate of herpes zoster was significantly higher in the hematologic group compared to the solid cancer group (31.0 vs. 14.9 per 1000 patient years, $p < 0.01$). Increasing age, female sex, white race, and immunosuppressive therapy and advanced stage of cancer were significant HZ risk factors within this cancer population.

Organ Transplants

Zoster commonly occurs in bone marrow, stem cell, kidney, liver, and heart transplant recipients. Data on the incidence and risk factors for zoster in these patients are limited by studies that come from single site transplant centers with small numbers of patients and variable underlying diseases and treatment regimens as well as variable length of follow-up. In a commercial insurance, Medicare and Medicaid claims database analysis involving 90.2 million person-years at risk, the incidence of zoster was highest among patients with bone marrow or stem cell transplant (43 per 1000 person-years) followed by solid organ transplant (17 per 1000 person-years) [14]. In a VA study of solid organ transplants, the incidence of zoster was 22 per 1000 patient years [90]. In this study, the incidence of zoster in heart transplant recipients was 40 per 1000 patient years; in kidney transplant recipients, 24 per 1000 patient years; and in liver transplant recipients, 18 per 1000 patient years. In both of these studies, older transplant recipients were at greater risk for developing herpes zoster.

Immune-Mediated Diseases

Studies of herpes zoster and immune-mediated diseases are limited by medical record reviews, referral bias, variable treatment regimens and length of follow-up, and small samples. Despite these limitations, many studies have reliably documented a significant association between herpes zoster and systemic lupus erythematous (SLE) and rheumatoid arthritis. Manzi et al. [69] reported that 15 % of 321 SLE patients developed zoster a median of 6.2 years (range 1 month to 29 years) after the diagnosis of SLE for an estimated incidence of 22 cases per 1000 person-years. A prospective cohort study of 1485 SLE patients between 2001 and 2010 in one center found an age-adjusted herpes zoster incidence of 12.0 per 1000 person-years [13]. Increasing age and reduced functional status were predictors of zoster in SLE patients. The appearance of herpes zoster was not related to SLE activity because cases occurred when SLE disease activity was severe, mild, or inactive.

The risk of herpes zoster was increased in rheumatoid arthritis patients (adjusted OR 1.46) compared to age-, sex-, and practice-matched controls in primary care data (2000–2011) from the UK Clinical Practice Research Datalink [30]. The incidence of herpes zoster per 1000 patient years ranged from 9.9 to 12.1 to 19.7 in cohort studies, respectively, in the VA health-care system (1998–2005) [71], Olmsted County, Minnesota (1980–2007) [110], and the US Medicare population (2006–2011) [125]. The incidence of herpes zoster in individuals with rheumatoid arthritis was significantly higher than herpes zoster incidence rates in individuals without rheumatoid arthritis. Increasing age and corticosteroid drug use were risk factors for herpes zoster among rheumatoid arthritis patients in these studies.

The associations between zoster and systemic lupus erythematosus and rheumatoid arthritis were attenuated after adjustment for immunosuppressive treatments in the UK Clinical Practice Research Datalink study but remained significant [30].

Immunosuppressive Treatments

Immunosuppressive treatments may increase the risk of zoster although it is often difficult to disentangle the effects of the underlying disease and its treatment on the risk of zoster. Cancer chemotherapeutic agents, corticosteroids, posttransplant immunosuppressants, and agents for immune-mediated diseases have been linked to herpes zoster. The most comprehensive analysis of the association of multiple immunosuppressant agents and herpes zoster involved bone marrow, stem cell, and solid organ transplant, HIV infection, SLE, RA, any cancer, inflammatory bowel disease, multiple sclerosis, and psoriasis patients in a commercial insurance, Medicare, and Medicaid claims database analysis involving 90.2 million person-years at risk noted above [14]. The incidence rates of herpes zoster were significantly higher among users of immunosuppressants or chemotherapy than among nonusers. Patients on immunosuppressant or chemotherapy across all these conditions had an approximately 50 % higher risk of herpes zoster compared to individuals without these interventions. The increased risk of herpes zoster associated with immunosuppressants or chemotherapy ranged from 1.7 for bone marrow transplant recipients to 1.46 for SLE patients [14].

Whether anti-TNF therapy increases the risk of herpes zoster is unclear. A prospective cohort study of rheumatoid arthritis patients using the British Society for Rheumatology Biologics Register analyzed 11,881 anti-TNF and 3673 nonbiological disease modifiers and the development of herpes zoster (2001–2009). The adjusted hazard ratio for risk of zoster in anti-TNF compared to nonbiological disease modifier group was 1.8 (95 % CI 1.2–2.8) [33]. A cohort study of patients with rheumatoid arthritis, inflammatory bowel disease, and psoriasis, psoriatic arthritis, or ankylosing spondylitis using a large US multi-institutional collaboration combining data from Kaiser Permanente Northern California, Pharmaceutical Assistance Contract for the Elderly, Tennessee Medicaid, and national Medicaid/Medicare programs analyzed 33,324 new users of anti-TNF therapy and the development of herpes zoster (1998–2007). In this analysis, anti-TNF therapies were not associated with a higher risk of herpes zoster compared with nonbiological treatment regimens (adjusted hazard ratio 1.00, 95 % CI 0.77–1.29) [117].

4.2.3.5 Other Risk Factors

Medical Comorbidities

Chronic medical conditions may increase the risk for herpes zoster. For example, a case-control study using the UK Clinical Practice Research database found that, after adjusting for immunosuppressive conditions, multiple comorbidities, and immunosuppressant medications, the risk of herpes zoster was increased in chronic obstructive pulmonary disease (adjusted odds ratio (aOR) 1.22, 95 % CI 1.17–1.28), asthma (aOR 1.11, 95 % CI 1.06–1.16), chronic kidney disease (aOR 1.12, 95 % CI 1.08–1.17), and type I diabetes (aor 1.26, 95%CI 1.06–1.49) [30]. A retrospective matched cohort study using the Integrated Health Care Information Services database that includes more than 74 million individuals from 46 health plans in the USA from 1997 to 2006 found similar increases in zoster risk in persons with COPD, diabetes mellitus type II, and cardiac disease [43].

Physical Trauma

Clinicians and patients have observed herpes zoster occurring in temporal association with physical trauma often involving the traumatized dermatome. Two case-control studies appear to confirm those observations. Thomas et al. [105] performed an interview study of herpes zoster cases and matched non-zoster controls and compared occurrence of trauma in 6 and 1 month before zoster or an index date. The adjusted odds ratio for having physical trauma in herpes zoster cases in the month before rash onset at the same site of the rash was 12.07 (95 % CI 1.49–97.63). A much larger age-matched case-control study from a 5 % random sample of Medicare beneficiaries (2006–2007) discovered an adjusted odds of having had trauma in the week preceding the herpes zoster rash onset, or index date was 3.4 times (95 % CI, 2.8–4.2) as high among herpes zoster cases as compared to controls [126]. Patients with cranial herpes zoster were 27.5 times more likely than controls to have had cranial trauma in the week before herpes zoster rash onset.

The adjusted odds ratios declined over time from trauma onset in these studies.

Psychosocial Factors

Psychological stress is often thought to play a role in the development of herpes zoster. Stress and herpes zoster was studied in 101 community dwelling elders over age 50 with herpes zoster (cases) and 101 randomly sampled controls without zoster [98]. Stressful life events were identified in these participants with the Geriatric Scale of Recent Life Events. Cases experienced negative life events significantly more often than controls at 2 months (26 vs. 10, odds ratio 2.64, 95 % CI 1.13, 6.27, $p = 0.012$), 3 months (29 vs. 11, odds ratio 2.64, 95 % CI 1.20, 6.04, $p = 0.007$), and 6 months (35 vs. 16, odds ratio 2.00, 95 % CI 1.04, 3.93, $p = 0.012$) prior to herpes zoster onset. The effect of acute (negative life events) and chronic (lack of social support) psychological stress on the risk of herpes zoster in the elderly was studied prospectively in the Duke EPESE. After controlling for multiple demographic, health, and social factors, negative life events increased the risk of herpes zoster, but the result was borderline for statistical significance (adjusted RR = 1.38, 95 % CI 0.96–1.97, $P = 0.078$). No measures of social support were significantly associated with zoster [101]. A prospective case-control study in a French Sentinel Network of ambulatory care practices found that recent negative life events (OR 3.40, 95 % CI 1.67–6.93) were significantly associated with herpes zoster [64].

Family History
Genetic susceptibility to herpes zoster may vary among individuals which raises the question as to whether family history is related to the development of herpes zoster. Three case-control studies have addressed this issue and all discovered a positive relationship. The adjusted odds ratios for first-degree relatives with past herpes zoster in cases versus controls were 4.35 (95 % CI 3.11–6.09) [51], 4.44 (95 % CI 3.11–6.35) [50], and 3.69 (95 % CI 1.81–7.51) [64] in these studies.

4.2.4 Transmission of Varicella-Zoster Virus (VZV) and Herpes Zoster

Herpes zoster epidemiology is ultimately determined by the transmission and spread of VZV in populations. The most important condition in the spread of VZV is primary varicella (chickenpox) infection but the incidence of varicella has declined markedly in countries with varicella vaccination programs. Latent and reactivated VZV infections play important roles in maintaining VZV infection in populations. Latently infected elderly adults and immunosuppressed patients are important reservoirs of virus because VZV is more likely to reactivate in these groups. When herpes zoster does occur, VZV can be transmitted during the vesicular phase of the rash and cause primary varicella infection when the zoster exposure consists of person-to-person contact with a VZV seronegative individual. VZV exposure in a seropositive, latently infected individual may result in a subclinical reinfection and boost of humoral and cellular VZV immunity, but it is unlikely to cause varicella or herpes zoster [1]. Investigators have reported clusters of zoster cases over a short period of time in the workplace and have reported zoster after prior exposure to varicella [3, 86]. It is not clear whether these episodes are coincidence, a clinical manifestation of exogenous reinfection, or stimulation of endogenous VZV reactivation. It is clear that the great majority of exposures of latently infected individuals with herpes zoster or varicella do not result in herpes zoster or varicella. However, herpes zoster or varicella exposure in latently infected individuals may possibly prevent herpes zoster as a consequence of the boosting of VZV immunity.

4.3 Postherpetic Neuralgia

In a considerable percentage of patients with herpes zoster, pain persists following healing of the rash. Persisting herpes zoster pain is termed postherpetic neuralgia (PHN), a chronic pain syndrome which can last for years and cause substantial suffering and reduction in quality of life. Three different types of pain have been distinguished in research on PHN—a steady ongoing throbbing, aching, or burning pain; an intermittent shooting or shock-like pain; and allodynia [5, 8, 58]. As is true of other chronic pain syndromes, PHN patients can develop depression and other types of psychological distress as well as physical, occupational, and social disability as a consequence of their unremitting pain [9, 21, 58]. For example, in a prospective study of health-related quality of life in herpes zoster patients, a high proportion of patients who went on to develop PHN reported anxiety, depression, and problems with mobility and self-care on the EuroQol EQ-5D [21]. PHN interfered mostly with enjoyment of life, mood and sleep as well as activities of daily living because of pain as measured by the Zoster Brief Pain Inventory. Studies using generic health-related quality of life measures such as the SF-36, SF-12, or EQ-5D find that physical and mental health component scores are consistently lower (worse) in PHN patients than general population norms [9, 109]. Not surprising, the magnitude of impact of PHN on the quality of life is directly proportional to increasing pain severity.

Pain in PHN can be discontinuous, with pain-free intervals of varying durations occurring [57, 70, 112]. In a study of 156 patients with PHN, Watson et al. [112] noted that "25 % of patients with a poor outcome said that they could recall a time after the rash when they had little or no pain for a period of weeks to as much as 12 months." Consistent with these findings are the results of a long-term follow-up study of patients originally enrolled in an antiviral trial conducted in the UK; 16 of 132 patients who had reported no pain upon completion of participation in the trial 9 years earlier reported pain within the preceding year [70]. Moreover, PHN can develop even in herpes zoster patients who have not had acute pain [57].

4.3.1 Definitions of PHN

A variety of definitions of PHN have been used by clinicians and investigators, ranging from any pain persisting after rash healing to pain that has persisted at least 60, 90, 120 days or 6 months after rash onset to clinically meaningful pain (e.g., three or more on ten-point pain scale) 3 months or more after rash onset [24, 25, 27, 58, 85, 106]. Some authorities view any pain at least 90 days after rash onset to represent a conventional definition of PHN [58]. An analysis of pain trajectories associated with herpes zoster suggests three phases—an acute herpetic neuralgia that accompanies the rash and lasts for approximately 30 days after rash onset, a subacute herpetic neuralgia that lasts from 30 to 120 days after rash onset, and PHN, defined as pain that persists for at least 120 days after rash onset [19].

Pain in herpes zoster can also be considered as a zoster-associated pain continuum where no distinction is made between acute pain and PHN [114, 118]. Zoster-associated pain has been used as the primary endpoint in antiviral trials where treatment efficacy is the time from enrollment in the trial to complete cessation of all herpes zoster-associated pain. Making a diagnosis of PHN and examining pain as a continuum are not mutually exclusive approaches to studying the persistence of zoster pain; pain data collected on multiple occasions beginning during the acute infection and continuing for several months thereafter can be examined by using a continuum of pain duration as well as by analyzing the incidence and duration of PHN. Accordingly, in the following discussion of prolonged pain in patients with herpes zoster, findings based on a diagnosis of PHN, however defined, will be considered together with findings based on analyses in which zoster pain is examined as a continuum of overall pain duration.

4.3.2 Incidence and Prevalence of PHN

The number of herpes zoster patients with pain declines with time, and estimates of the proportion of patients who develop PHN therefore vary depending on the definition of PHN used as noted above. The prevalence of PHN varied from 5 to 30 % in a systematic review of 49 studies from 1945 to 2012 that used a wide variety of PHN definitions and ages [62]. Among these studies, the results of recent large cohort studies are worth noting. In an analysis of 1669 herpes zoster patients in a population-based study of adult residents of Olmsted County, Minnesota (1996–2001), 10 % of patients developed PHN, defined as zoster-associated pain persisting at least 90 days [121]. In an analysis cohort of 27,225 herpes zoster patients ≥50 years old in the UK General Practice Research Database (2000–2006), 13.7 % of patients developed PHN defined as any pain at least 3 months after herpes zoster diagnosis [34]. Of those patients, 58.5 % had moderate to severe pain. PHN occurred in 33 % of individuals aged 79 years and older. In a multicenter primary care-based study in France, (2007–2008) of 1358 herpes zoster patients ≥50 years old, the prevalence of PHN was 11.6 % defined as any pain 3 months from rash onset [9]. An analysis of six prospective cohort studies revealed that approximately 30–50 % of patients with PHN experienced pain lasting for more than 1 year [62].

4.3.3 Risk Factors for PHN

There are a considerable number of studies in which risk factors for PHN have been investigated. The impetus for these studies has come from several sources, including attempts to identify risk factors as a means of increasing understanding of the natural history and pathogenesis of PHN and to design interventions intended to prevent the development of PHN. Studies of risk factors for PHN make it possible to identify those zoster patients who have the greatest need for preventive efforts such as vaccines because of their increased risk of chronic pain. Although well over 40 risk factors have been investigated in various studies, major risk factors that have been investigated in at least several cohort studies are presented below [31].

4.3.3.1 Greater Age

Increasing age is among the stronger and most consistent risk factors for PHN. This fact has been demonstrated in cohort studies and placebo groups of clinical trials in studies in Europe, Asia, and North America [31]. In children with herpes zoster, the risk of PHN approaches zero [42, 56, 89]. The increasing risk of PHN with increasing age was perhaps first best demonstrated by Hope-Simpson [56] in Cirencester, England; the prevalence of any pain one or more months after rash onset was 3–4 % in age groups 30–49 years but 21 %, 29 %, and 34 % in the age groups 60–69, 70–79, and greater than 80 years, respectively [56]. Even among older adults, the risk of PHN increases with increasing age [104]. In the UK general practice database study, PHN defined as any pain 3 months from rash onset was 11 %, 13 %, 15 %, 18 %, and 21 % in age groups 65–69, 70–74, 75–79, and 80–84 years, respectively. A meta-analysis of 18 cohort studies found that the 10-year increase in age on PHN risk was 1.22–3.11 per 10 years [31].

4.3.3.2 Greater Acute Pain Severity

The possibility that there may be a correlation between the duration of pain and the severity of pain on presentation in herpes zoster was proposed as a hypothesis several years ago [119]. There are now a considerable number of independent studies that indicate that patients with more severe acute pain are at greater risk for both prolonged zoster pain assessed as a continuum and for PHN [31]. In addition, acute pain severe enough to interfere with activities of daily living and significantly impact quality of life has also been found to be a risk factor for PHN [9, 15, 61]. In a meta-analysis of eight cohort studies, severe acute pain increased the risk of PHN by more than twofold (rate ratio 2.23, 95 % CI 1.71–2.92) [31]. The relationship between acute pain severity and prolonged pain has been found in immunocompromised patients [49] as well as immunocompetent patients.

The fact that greater acute pain severity increases risk for PHN has implications for the prevention of PHN. The focus of future research should therefore become the identification of mechanisms accounting for this relationship. In pursuing this task, a priority should be to conduct a careful examination of the specific aspects of acute zoster pain that predict the development of chronic pain. For example, is the predominant quality of pain in zoster—whether burning, throbbing, stabbing, or allodynia—associated with the risk of developing PHN? Stretch and brush induced allodynia as measured on a careful physical examination by pain specialists was associated with a fourfold to fivefold increase in risk of PHN in a study of 93 patients with herpes zoster [46]. Allodynia as assessed on the Neuropathic Pain Symptom Inventory questionnaire was not independently associated with PHN [9]. Burning pain was much less likely to be reported by PHN patients whose zoster had been treated with acyclovir than by PHN patients who had not been treated with an antiviral drug [7, 8]. Additional prospective studies are needed to clarify the relationship between pain quality in zoster and the development of PHN.

In several, but not all studies, the presence of prodromal pain was found to be associated with prolonged pain and the development of PHN [4, 15, 31, 60, 61, 72, 115].

4.3.3.3 Greater Rash Severity

Several studies have reported that greater severity of the zoster cutaneous eruption is associated with prolonged pain and the development of PHN [15, 28, 49, 52–54, 60, 72, 84, 114, 116]. The severity of the zoster rash has been assessed using a variety of methods in these studies, including counts of the number of vesicles and ratings of the proportion of the dermatome affected. In addition, the duration of time until the occurrence of various aspects of rash healing has been examined, including assessments of time to cessation of new vesicle formation and time to complete crusting. Few studies, however, have reported assessments of rash severity on multiple occasions, which would allow rash progression from onset to healing to be examined. Even fewer studies have evaluated the inter-rater reliability of their ratings of rash severity, which involve judgments with a subjective component.

4.3.3.4 Sensory Dysfunction

Herpes zoster patients with greater sensory abnormalities in the affected dermatome, compared to the contralateral unaffected dermatome, were found to be at greater risk for PHN in several studies [11, 44–46, 77, 79, 87, 92, 111]. Evaluations of sensory dysfunction in the affected dermatome included clinical assessments of hypoesthesia as well as quantitative sensory testing (e.g., elevated thermal and vibration thresholds). In a prospective study of 113 herpes zoster patients, pinprick hypoesthesia as well as dynamic and static mechanical allodynia were associated with the later development of PHN [44–46]. Thermal and tactile thresholds, however, did not predict PHN in this study. In one prospective study of 94 herpes zoster patients, eventual PHN subjects had significantly more impairment in detecting warmth and cold, a larger area of altered sensation, a larger area of allodynia, and more severe allodynia [92, 94, 106]. The authors concluded that these results support the idea that severity of initial injury predicts PHN. Elevated vibration thresholds outside the affected dermatome (i.e., in the hands and feet) in zoster patients predicted PHN [2]. These investigators concluded that a generalized subclinical large fiber polyneuropathy is a risk factor for PHN.

It has also been reported that there is greater sensory dysfunction in the affected dermatome in patients with PHN than in zoster patients whose pain did not persist [78]. The results of these studies indicate that sensory dysfunction can persist well beyond the acute phase of zoster and that it is a frequent concomitant of prolonged pain. This sensory dysfunction seems to be associated with ipsilateral [95] as well as bilateral [81] loss of cutaneous innervation.

4.3.3.5 Immune Response

Greater magnitude and duration of the immune response in zoster patients has been reported to predict prolonged pain [52, 54]. Significant relationships were found between the immune response measures and rash severity, suggesting that a more pronounced VZV-specific immune response during zoster may predict prolonged pain because it reflects a more severe acute infection. The largest prospective study to date examined immune response and pain in 981 elderly persons who developed herpes zoster during a zoster vaccine efficacy trial and 1362 without herpes zoster. In this study, greater VZV cell-mediated immunity (CMI) responses in the first week after herpes zoster rash onset correlated with decreased herpes zoster pain severity and with lower occurrence of PHN, whereas higher VZV antibody levels were associated with higher herpes zoster pain severity and PHN [113]. These findings suggest that higher levels of VZV CMI soon after VZV reactivation result in reduced viral replication and a lower incidence of complications, such as pain and PHN. Conversely, a weak VZV CMI response allows the reactivated virus to replicate unchecked, resulting in greater antigenic stimulation of VZV immune responses as noted by increased VZV antibody levels and more pain. These observations are consistent with the notion that more severe acute infection is associated with PHN development.

4.3.3.6 Sex

Several investigators have examined whether there is a relationship between the patient's sex and the risk of prolonged pain, and the majority of these studies have found that there were no significant differences in the risk for PHN among men and women [4, 15, 22, 28, 31, 56, 61, 82, 115, 120]. Among two well-conducted cohort studies that did find a significant sex difference in PHN risk, the findings were opposite with one study reporting that males were at higher risk [9], whereas another study found that females were at higher risk [60].

4.3.3.7 Dermatome

The relationship between the specific dermatome affected in zoster and the risk of prolonged pain has been examined in a number of studies [15, 16, 22, 46, 53, 56, 60, 61, 82, 93, 111]. Although the results of some of these studies suggested that the likelihood of PHN is greater in patients with ophthalmic or trigeminal zoster, this relationship was not found as an independent risk factor in several studies [31].

4.3.3.8 Psychosocial Risk Factors

The risk factors for PHN discussed to this point have consisted of demographic and biomedical characteristics of zoster patients. It has also been suggested that psychosocial factors might play a role in determining which patients will have pain that persists. Prospective studies are necessary to determine whether variables that may plausibly be either antecedents or consequences of chronic pain are risk factors. In a prospective study of a small sample of zoster patients, those who developed PHN had lower life satisfaction and greater depression, anxiety, and disease conviction during their acute infection than patients who did not develop PHN [26]. A larger prospective study of 110 herpes zoster patients found that measures of role functioning, personality disorder symptoms, disease conviction hypochondriasis, somatosensory amplification, and somatic symptoms during herpes zoster each made independent contributions to predicting either the presence or intensity of PHN [61]. Depression and anxiety were not independently associated with PHN in this study and two other large prospective studies [9, 61, 84]. However, affective distress was a risk factor for PHN in a natural history study of 94 patients following herpes zoster [106].

4.4 Conclusions

As discussed previously, there is evidence that the incidence of zoster has increased in recent decades. Despite the availability of an effective live attenuated zoster vaccine, vaccine uptake by the target populations has been low. Therefore, it is possible that the number of patients suffering from herpes zoster and PHN will continue to be a significant problem. PHN could become more prevalent not only because of this increased incidence of zoster but also because PHN is more likely to develop in the older individuals whose numbers are increasing. Because PHN patients suffer from physical and social disability and psychological distress and have greatly increased health-care utilization as a result of their chronic pain, PHN could continue to have a major impact on public health unless we better utilize preventive measures. The newly described highly effective (over 90 %) subunit glycoprotein E vaccine, which is adjuvanted, not infectious, and effective even in the elderly, may, however, change this paradigm if it is well received and utilized widely [63].

References

1. Arvin AM, Koropchak CM, Wittek AE (1983) Immunological evidence of reinfection with varicella-zoster virus. J Infect Dis 148:200–205
2. Baron R, Haendler G, Schulte H (1997) Afferent large fiber polyneuropathy predicts the development of postherpetic neuralgia. Pain 73:231–238
3. Berlin BS, Campbell T (1970) Hospital-acquired herpes zoster following exposure to chickenpox. JAMA 211:1831–1832
4. Beutner KR, Friedman DJ, Forszpaniak C et al (1995) Valaciclovir compared with acyclovir for improved therapy for herpes zoster in immunocompetent adults. Antimicrob Agents Chemother 39:1546–1553
5. Bhala BB, Ramamoorthy C, Bowsher D et al (1988) Shingles and postherpetic neuralgia. Clin J Pain 4:169–174
6. Blank LJ, Polydefkis MJ, Moore RD et al (2012) Herpes zoster among persons living with HIV in the current antiretroviral therapy era. J Acquir Immune Defic Syndr 61:203–207

7. Bowsher D (1992) Acute herpes zoster and postherpetic neuralgia: effects of acyclovir and outcome of treatment with amitriptyline. Br J Gen Pract 42:244–246
8. Bowsher D (1993) Sensory change in postherpetic neuralgia. In: Watson CPN (ed) Herpes Zoster and Postherpetic Neuralgia. Elsevier, Amsterdam, pp. 97–107
9. Bouhassira D, Chassany O, Gaillat J et al (2012) Patient perspective on herpes zoster and its complications: an observational prospective study in patients aged over 50 years in general practice. Pain 153:342–349
10. Brisson M, Edmunds WJ, Law B et al (2001) Epidemiology of varicella zoster virus infection in Canada and the United Kingdom. Epidemiol Infect 127:305–314
11. Bruxelle J (1995) Prospective epidemiologic study of painful and neurologic sequelae induced by herpes zoster in patients treated early with oral acyclovir. Neurology 45:S78–S79
12. Buchbinder SP, Katz MH, Hessol NA et al (1992) Herpes zoster and human immunodeficiency virus infection. J Infect Dis 166:1153–1156
13. Chakravarty EF, Michaud K, Katz R et al (2013) Increased incidence of herpes zoster among patients with systemic lupus erythematous. Lupus 22:238–244
14. Chen SY, Suaya JA, Li Q et al (2014) Incidence of herpes zoster in patients with altered immune function. Infection 42:325–334
15. Choo PW, Galil K, Donahue JG et al (1997) Risk factors for postherpetic neuralgia. Arch Intern Med 157:1217–1224
16. Coen PG, Scott F, Leedham-Green M et al (2006) Predicting and preventing post Herpetic neuralgia: are current risk factors useful in clinical practice? Eur J Pain 10:695–700
17. Colebunders R, Mann JM, Francis H et al (1988) Herpes zoster in African patients: a clinical predictor of human immunodeficiency virus infections. J Infect Dis 157:314–318
18. Cohen JI (2013) Herpes Zoster. N Engl J Med 369:255–263
19. Desmond RA, Weiss HL, Arani RB et al (2002) Clinical applications for change-point analysis of herpes zoster pain. J Pain Symptom Manag 23:510–516
20. Donahue JG, Choo PW, Manson JE et al (1995) The incidence of herpes zoster. Arch Intern Med 155:1605–1609
21. Drolet M, Brisson M, Schmader KE et al (2010a) The impact of herpes zoster and postherpetic neuralgia on quality of life. CMAJ 182:1731–1736
22. Drolet M, Brisson M, Schmader K et al (2010b) Predictors of postherpetic neuralgia among patients with herpes zoster: a prospective study. J Pain 11:1211–1221
23. Dworkin RH (1996) Racial differences in herpes zoster and age at onset of varicella. J Infect Dis 174:239–241
24. Dworkin RH, Gnann JW, Oaklander AL et al (2008) Diagnosis and assessment of pain associated with herpes zoster and postherpetic neuralgia. J Pain 9:S37–S44
25. Dworkin RH, Portenoy RK (1996) Pain and its persistence in herpes zoster. Pain 67:241–251
26. Dworkin RH, Hartstein G, Rosner HL et al (1992) A high-risk method for studying psychosocial antecedents of chronic pain: the prospective investigation of herpes zoster. J Abnorm Psychol 101:200–205
27. Dworkin RH, Carrington D, Cunningham A et al (1997) Assessment of pain in herpes zoster: lessons learned from antiviral trials. Antivir Res 33:73–85
28. Dworkin RH, Boon RJ, Griffin DRG et al (1998) Postherpetic neuralgia: impact of famciclovir, age, rash severity, and acute pain in herpes zoster patients. J Infect Dis 178: S76–S80
29. Engels EA, Rosenberg PS, Biggar RJ (1999) Zoster incidence in human immunodeficiency virus-infected hemophiliacs and homosexual men, 1984–1997: district of Columbia Gay Cohort Study, Multicenter Hemophilia Cohort Study. J Infect Dis 180:1784–1789
30. Forbes HJ, Bhaskaran K, Thomas SL et al (2014) Quantification of risk factors for herpes zoster: population based case-control study. BMJ 348:g2911. doi:10.1136/bmj.g2911
31. Forbes HJ, Thomas SL, Clayton T et al (2016) A systematic review and meta-analysis of risk factors for postherpetic neuralgia. Pain 157:30–54
32. Galil K, Choo PW, Donahue JG et al (1997) The sequelae of herpes zoster. Arch Intern Med 157:1209–1213

33. Galloway JB, Mercer LK, Moseley A et al (2013) Risk of skin and soft tissue infections (including shingles) in patients exposed to anti-tumour necrosis factor therapy: results from the British Society for Rheumatology Biologics Register. Ann Rheum Dis 72:229–234
34. Gauthier A, Breuer J, Carrington D et al (2009) Epidemiology and cost of herpes zoster and post-herpetic neuralgia in the United Kingdom. Epidemiol Infect 137:38–47
35. Gershon AA, Mervish N, LaRussa P et al (1997) Varicella-zoster virus infection in children with underlying human immunodeficiency virus infection. J Infect Dis 176:1496–1500
36. Gilden DH, Dueland AN, Cohrs R et al (1991) Preherpetic neuralgia. Neurology 41:1215–1218
37. Gilden DH, Dueland AN, Devlin ME et al (1992) Varicella zoster virus reactivation without rash. J Infect Dis 166:S30–S34
38. Gilden DH, Wright RR, Schneck SA et al (1994) Zoster sine herpete, a clinical variant. Ann Neurol 35:530–533
39. Glesby M, Moore RD, Chaisson RE (1995) Clinical spectrum of herpes zoster in adults with HIV. Clin Infect Dis 21:370–375
40. Gershon AA, Breuer J, Cohen JI et al (2015a) Varicella zoster virus infection. Nat Rev Primers 1:15016
41. Gershon AA, Chen J, Gershon MD (2015b) Use of saliva to identify varicella-zoster virus (VZV) infection of the gut. Clin Infect Dis 61:536–544
42. Guess HA, Broughton DD, Melton LJ III et al (1985) Epidemiology of Herpes zoster in children and adolescents: a population-based study. Pediatrics 76:512–517
43. Guignard AP, Greenberg M, Lu C et al (2014) Risk of herpes zoster among diabetics: a matched cohort study in a US in a US insurance claim database before introduction of vaccination, 1997–2006. Infection 42:729–735
44. Haanpää ML, Laippala PA, Nurmikko TJ (1999a) Thermal and tactile perception thresholds in acute herpes zoster. Eur J Pain 3:375–386
45. Haanpää M, Laippala P, Nurmikko T (1999b) Pain and somatosensory dysfunction in acute herpes zoster. Clin J Pain 15:78–84
46. Haanpää M, Laippala P, Nurmikko T (2000) Allodynia and pinprick hypesthesia in acute herpes zoster and the development of postherpetic neuralgia. J Pain Symptom Manag 20:50–58
47. Habel LA, Ray GT, Silverberg MJ et al (2013) The epidemiology of herpes zoster in patients with newly diagnosed cancer. Cancer Epidemiol Biomark Prev 22:82–90
48. Hales CM, Harpaz R, Joesoef MR et al (2013) Examination of links between herpes zoster incidence and childhood varicella vaccination. Ann Intern Med 159:739–745
49. Harrison RA, Soong S, Weiss HL et al (1999) A mixed model for factors predictive of pain in AIDS patients with herpes zoster. J Pain Symptom Manag 17:410–417
50. Hernandez PO, Javed S, Mendoza N et al (2011) Family history and herpes zoster risk in the era of shingles vaccination. J Clin Virol 52:344–348
51. Hicks LD, Cook-Norris RH, Mendoza N et al (2008) Family history as a risk factor for herpes zoster: a case–control study. Arch Dermatol 144:603–608
52. Higa K, Dan K, Manabe H et al (1988) Factors influencing the duration of treatment of acute herpetic pain with sympathetic nerve block: importance of severity of herpes zoster assessed by the maximum antibody titers to varicella zoster virus in otherwise healthy patients. Pain 32:147–157
53. Higa K, Mori M, Hirata K et al (1997) Severity of skin lesions of herpes zoster at the worst phase rather than age and involved region most influences the duration of acute herpetic pain. Pain 69:245–253
54. Higa K, Noda B, Manabe H et al (1992) T-lymphocyte subsets in otherwise healthy patients with herpes zoster and relationships to the duration of acute herpetic pain. Pain 51:111–118
55. Hope-Simpson RE (1965) The nature of herpes zoster: a long-term study and a new hypothesis. Proc R Soc Med 58:9–20
56. Hope-Simpson RE (1975) Postherpetic neuralgia. J R Coll Gen Pract 25:571–575
57. Huff JC, Drucker JL, Clemmer A et al (1993) Effect of oral acyclovir on pain resolution in herpes zoster: a reanalysis. J Med Virol 41:93–96

58. Johnson RW, Rice AW (2014) Postherpetic Neuralgia. N Engl J Med 371:1526–1533
59. Jumaan AO, Yu O, Jackson LA et al (2005) Incidence of herpes zoster, before and after varicella-vaccination associated decreases in the incidence of varicella, 1992–2002. J Infect Dis 191:2002–2007
60. Jung BF, Johnson RW, Griffin DR et al (2004) Risk factors for postherpetic neuralgia in patients with herpes zoster. Neurology 62:1545–1551
61. Katz J, McDermott MP, Cooper EM et al (2005) Psychosocial risk factors for postherpetic neuralgia: a prospective study of patients with herpes zoster. J Pain 6:782–790
62. Kawai K, Gebremeskel BG, Acosta C (2014) Systematic review of incidence and complications of herpes zoster: towards a global perspective. BMJ Open 4:e004833
63. Lal H, Cunningham AL, Godeaux O et al (2015) Efficacy of an adjuvanted herpes zoster subunit vaccine in older adults. N Engl J Med 372:2087–2096
64. Lasserre A, Blaizeau F, Gorwood P et al (2012) Herpes zoster: family history and psychological stress-case-control study. J Clin Virol 55:153–157
65. Langan SM, Minassian C, Smeeth L et al (2014) Risk of stroke following herpes zoster: a self-controlled case-series study. Clin Infect Dis 58:1497–1503
66. Leung J, Harpaz R, Molinari NA et al (2011) Herpes zoster incidence among insured persons in the United States, 1993–2006: evaluation of impact of varicella vaccination. Clin Infect Dis 52:332–340
67. Levin MJ, Anderson JP, Seage GR 3rd et al (2009) Short-term and long-term effects of highly active antiretroviral therapy on the incidence of herpes zoster in HIV infected children. J Acquir Immune Defic Syndr 50:182–191
68. Mahamud A, Marin M, Nickell SP et al (2012) Herpes zoster-related deaths in the United States: validity of death certificates and mortality rates, 1979–2007. Clin Infect Dis 55:960–966
69. Manzi S, Kuller LH, Kutzer J et al (1995) Herpes zoster in systemic lupus erythematosus. J Rheumatol 22:1254–1258
70. McKendrick MW, Wood MJ (1995) Acyclovir and post-herpetic neuralgia. BMJ 310:1005
71. McDonald JR, Zeringue AL, Caplan L et al (2009) Herpes zoster risk factors in a national cohort of veterans with rheumatoid arthritis. Clin Infect Dis 48:1364–1371
72. Meister W, Neiß A, Gross G et al (1998) A prognostic score for postherpetic neuralgia In ambulatory patients. Infection 26:359–363
73. Melby M, Grossman RJ, Goedert JJ et al (1987) Risk of AIDS after herpes zoster. Lancet 1:728–371
74. Moanna A, Rimland D (2013) Decreasing incidence of herpes zoster in the highly active antiretroviral therapy era. Clin Infect Dis 57:122–125
75. Moore RD, Chaisson RE (1996) Natural history of opportunistic disease in an HIV-infected urban clinical cohort. Ann Intern Med 124:633–642
76. Nagel MA, Gilden D (2015) The relationship between herpes zoster and stroke. Curr Neurol Neurosci Rep 15:16
77. Noda B, Dan K, Manabe H et al (1987) Prognostic clinical signs in herpes zoster pain. Pain Suppl 4:S382
78. Nurmikko TJ, Bowsher D (1990) Somatosensory findings in postherpetic neuralgia. J Neurol Neurosurg Psychiatry 53:135–141
79. Nurmikko TJ, Rasanen A, Hakkinen V (1990) Clinical and neurophysiological observations on acute herpes zoster. Clin J Pain 6:284–290
80. Oaklander AL, Bowsher D, Galer BS et al (2003) Herpes zoster itch: preliminary epidemiologic data. J Pain 4:338–343
81. Oaklander AL, Romans K, Horasek S et al (1998) Unilateral postherpetic neuralgia is associated with bilateral sensory neuron damage. Ann Neurol 44:789–795
82. Opstelten W, Mauritz JW, de Wit NJ et al (2002) Herpes zoster and postherpetic neuralgia: incidence and risk indicators using a general practice research database. Fam Pract 19:471–475
83. Opstelten W, Van Essen GA, Schellevis F et al (2006) Gender as an independent risk factor for herpes zoster: a population-based prospective. Ann Epidemiol 16:692–695

84. Opstelten W, Zuithoff NPA, van Essen GA et al (2007) Predicting postherpetic neuralgia in elderly primary care patients with herpes zoster: prospective prognostic study. Pain 132:S52–S59
85. Oxman MN, Levin MJ, Johnson GR et al (2005) A vaccine to prevent herpes zoster and postherpetic neuralgia in older adults. N Engl J Med 352:2271–2284
86. Palmer SR, Caul EO, Donald DE et al (1985) An outbreak of shingles? Lancet 2:1108–1111
87. Park J, Jang WS, Park KY et al (2012) Thermography as a predictor of postherpetic neurgalia in acute herpes zoster patients: a preliminary study. Skin Res Technol 18:88–93
88. Pérez-Farinós N, Ordobás M, García-Fernández C et al (2007) Varicella and Herpes Zoster in Madrid, based on the Sentinel General Practitioner Network: 1997–2004. BMC Infect Dis 7:59
89. Petursson G, Helgason S, Gudmundsson S et al (1998) Herpes zoster in children and adolescents. Pediatr Infect Dis J 17:905–908
90. Pergam SA, Forsberg CW, Boeckh MJ et al (2011) Herpes zoster incidence in a multicenter cohort of solid organ transplant recipients. Transpl Infect Dis 13:15–12
91. Pinchinat S, Cebrian-Cuenca AM, Bricout H et al (2013) Similar herpes zoster incidence across Europe: results from systematic literature review. BMC Infect Dis 13:170
92. Petersen KL, Rowbotham MC (2010) Natural history of sensory function after herpes zoster. Pain 150:83–92
93. Ragozzino MW, Melton LF III, Kurland LT (1982) Population-based study of herpes zoster and its sequelae. Medicine 61:310–316
94. Reda H, Greene K, Rice FL et al (2013) Natural history of herpes zoster: Late follow up of 3.9 years (n = 43) and 7.7 years (n = 10). Pain 154:2227–2233
95. Rowbotham MC, Yosipovitch G, Connolly MK et al (1996) Cutaneous innervation density in the allodynic form of posthereptic neuralgia. Neurobiol Dis 3:205–214
96. Russell ML, Schopflocher DP, Svenson L et al (2007) Secular trends in the epidemiology of shingles in Alberta. Epidemiol Infect 135:908–913
97. Rusthoven JJ, Ahlgren P, Elhakim T et al (1988) Varicella-zoster infection in adult cancer patients: a population study. Arch Intern Med 148:1561–1566
98. Schmader K, Studenski S, MacMillan J et al (1990) Are stressful life events risk factors for herpes zoster? J Am Geriatr Soc 38:1188–1195
99. Schmader K, George LK, Newton B et al (1994) The accuracy of self reports of herpes zoster. J Clin Epidemiol 47:1271–1276
100. Schmader K, George LK, Hamilton JD (1995) Racial differences in the occurrence of herpes zoster. J Infect Dis 171:701–704
101. Schmader K, George LK, Burchett BM et al (1998) Race and stress in the incidence of herpes zoster in the elderly. J Am Geriatr Soc 46:973–977
102. Schmader KE, Gnann JW, Watson CPN (2008) The epidemiological, clinical and pathological rationale for the zoster vaccine. J Infect Dis 197:S207–S215
103. Schmader KE, Sloane R, Pieper C et al (2007) The impact of acute herpes zoster pain and discomfort on functional status and quality of life in older adults. Clin J Pain 23:490–497
104. Scott FT, Leedham-Green ME, Barrett-Muir WY et al (2003) A study of zoster and the development of postherpetic neuralgia in East London. J Med Virol Suppl 1:S24–S30
105. Thomas SL, Wheeler JG, Hall AJ (2004) Case control study of the effect of mechanical trauma on the risk of herpes zoster. BMJ 328:7437–7439
106. Thyregod HG, Rowbotham MC, Peters M et al (2007) Natural history of pain following herpes zoster. Pain 128:148–156
107. Toyama N, Shiraki K, Society of the Miyazaki Prefecture Dermatologists (2009) Epidemiology of herpes zoster and its relationship to varicella in Japan: a 10-year survey of 48,388 herpes zoster cases in Miyazaki prefecture. J Med Virol 81:2053
108. Tyndall MW, Nasio J, Agoki E et al (1995) Herpes zoster as the initial presentation of human immunodeficiency virus type 1 infection in Kenya. Clin Infect Dis 21:1035–1037
109. Van Seventer R, Sadosky L, Lucero M et al (2006) A cross-sectional survey of health state impairment and treatment patterns in patients with postherpetic neuralgia. Age Ageing 35:135–137

110. Veetil BM, Myasoedova E, Matteson EL et al (2013) Incidence and time trends of herpes zoster in rheumatoid arthritis: a population-based cohort study. Arthritis Care Res 65:854–861
111. Volpi A, Gatti A, Pica F et al (2008) Clinical and psychosocial correlates of post-herpetic neuralgia. J Med Virol 80:1646–1652
112. Watson CPN, Deck JH, Morshead C et al (1991) Post herpetic neuralgia: further post mortem studies of cases with and without pain. Pain 44:105–117
113. Weinberg A, Zhang JH, Oxman MN et al (2009) VZV-specific immune responses to Herpes Zoster in elderly participants in a trial of a clinically effective Herpes Zoster Vaccine. J Infect Dis 200:1068–1077
114. Whitley RJ, Weiss H, Gnann JW Jr et al (1996) Acyclovir with and without prednisone for the treatment of herpes zoster: a randomized, placebo-controlled trial. Ann Intern Med 125:376–383
115. Whitley RJ, Shukla S, Crooks RJ (1998) The identification of risk factors associated with persistent pain following herpes zoster. J Infect Dis 178:S71–S75
116. Wilson JB (1986) Thirty one years of herpes zoster in a rural practice. BMJ 293:1349–1351
117. Winthrop KL, Baddley JW, Chen L et al (2013) Association between the initiation of anti-tumor necrosis factor therapy and the risk of herpes zoster. JAMA 309:887–895
118. Wood MJ (1995) For debate: how should zoster trials be conducted? J Antimicrob Chemother 36:1089–1101
119. Wood MJ, Johnson RW, McKendrick MW et al (1994) A randomized trial of acyclovir for 7 days or 21 days with and without prednisolone for treatment of acute herpes zoster. N Engl J Med 330:896–900
120. Wood MJ, Kay R, Dworkin RH et al (1996) Oral acyclovir therapy accelerates pain resolution in patients with herpes zoster: a meta-analysis of placebo-controlled trials. Clin Infect Dis 22:341–347
121. Yawn BP, Saddier P, Wollan P et al (2007) A population-based study of the incidence and complication rates of herpes zoster before zoster vaccine introduction. Mayo Clin Proc 82:1341–1349
122. Yawn BP, Wollan PC, Kurland MJ et al (2011) Herpes zoster recurrences more frequent than previously reported. Mayo Clin Proc 86:88–93
123. Yawn BP, Gilden D (2013) The global epidemiology of herpes zoster. Neurology 81:928–930
124. Yenikomshian MA, Guignard AP, Haguinet F et al (2015) The epidemiology of herpes zoster and its complications in medicare cancer patients. BMC Infect Dis 15:106
125. Yun H, Xie F, Delzell E, Le C et al (2015) Risks of herpes zoster in patients with rheumatoid arthritis according to biologic disease-modifying therapy. Arthritis Care Res 67:731–736
126. Zhang JX, Joesoef RM, Bialek S et al (2013) Association of physical trauma with risk of herpes zoster among medicare beneficiaries in the United States. J Infect Dis 207: 1007–1011

Chapter 5
Herpes Zoster Ophthalmicus

Emma Davies, James Chodosh, and Deborah Pavan-Langston

5.1 Epidemiology

Herpes zoster ophthalmicus (HZO) is the reactivation of latent varicella zoster virus (VZV) in the first and/or second division of the trigeminal nerve. The dermatome served by the trigeminal nerve is a common site for reactivation, second only to the thoracic dermatomes, with approximately 250,000 cases of HZO annually in the United States [1, 2]. Several studies have demonstrated a steadily increased incidence of HZO over the past two decades. In one study of Medicare enrollees, the incidence of HZO was found to increase from 1.7 per 1000 in 1993 to 4.4 per 1000 in 2006 [3]. There are several possible explanations for this rise in incidence including the aging population, increased number of immunocompromised patients, and introduction of the varicella vaccination. It has been hypothesized that VZV reactivates in the setting of waning T cell immunity, as occurs with aging (greatest decline in immune system response occurs in the fifth to seventh decade of life) and immunocompromised (from HIV, blood dyscrasias, and the use of immunosuppressant medications) [4]. While the introduction of a mandatory two-dose varicella vaccination program in the United States in 2006 has been suggested to reduce repeat exposure to VZV and therefore limit natural immune boosting against VZV in adults [5–8], this is by no means an accepted theory. Additional details on this subject are presented in Chap. 2. Herpes zoster reactivation has never reliably been shown to have a racial or gender predilection.

In the face of increased incidence of herpes zoster infection, the US FDA approved the use of zoster vaccination in persons 60 years of age or older in 2006 and extended approval to persons aged 50–59 in 2011 [9]. The zoster vaccination has been demonstrated to reduce the incidence of herpes zoster by 51.3 % as well as

---

Adapted from: Pavan-Langston D (2001) Ophthalmic zoster. In: Watson C, Gershon A (eds). Herpes zoster and Postherpetic Neuralgia, 2nd Revised and Enlarged Edition, Pain Research and Clinical Management, Vol. 11, Chapter 11, Elsevier, Amsterdam, pp 119–129

E. Davies • J. Chodosh (✉) • D. Pavan-Langston
Massachusetts Eye and Ear Infirmary, Cornea and External Diseases, Harvard Medical School, Department of Ophthalmology, 243 Charles Street, Boston, MA 02114, USA
e-mail: James_Chodosh@MEEI.HARVARD.EDU

C.P.N. Watson et al. (eds.), *Herpes Zoster: Postherpetic Neuralgia and Other Complications*, DOI 10.1007/978-3-319-44348-5_5

reduce the severity (including eye complications) of the cases that occur [10]. However, the rates of varicella vaccination in the United States have remained disappointingly low at around 14.4 % of the eligible population by 2010 [1, 11].

5.2 Acute HZO

Initial symptoms of HZO are usually fever, malaise, chills, and severe pain in the distribution of the ophthalmic division of the trigeminal nerve (V-1). Approximately 93 % of patients with HZO suffer from acute pain with 33 % having persistence of pain for 6 months or longer [12]. A dermatomal rash develops up to 7 days after the onset of pain in 90 % of patients [13]. The rash appears as multiple crops of vesicles on an erythematous base ("water drops on a rose petal") distributed along the V-1 dermatome and sharply respecting the vertical midline (Fig. 5.1). Over a period of 2–4 weeks, these lesions rupture and then crust over to form eschars. Scarring with hypopigmentation or hyperpigmentation may occur in deep lesions and persist. Rarely, in <1 % of cases, HZO occurs without a dermatomal rash in a variant of zoster termed zoster sine herpete [14].

There are three major branches of V-1, namely, the frontal, lacrimal, and nasociliary nerves. Since the nasociliary nerve innervates the cornea, the most serious ocular involvement develops if this branch is affected by VZV. Classically, involvement of the tip of the nose (Hutchinson's sign) was thought to be a clinical predictor of ocular involvement. However, while patients with a positive Hutchinson's sign have twice the incidence of ocular involvement, one-third of patients without the sign also develop ocular manifestations [15]. A summary of ocular findings in patients with HZO is presented in Table 5.1.

One of the most common acute ocular manifestations of HZO is conjunctivitis, typically a diffuse follicular conjunctivitis. Other manifestations in the conjunctiva

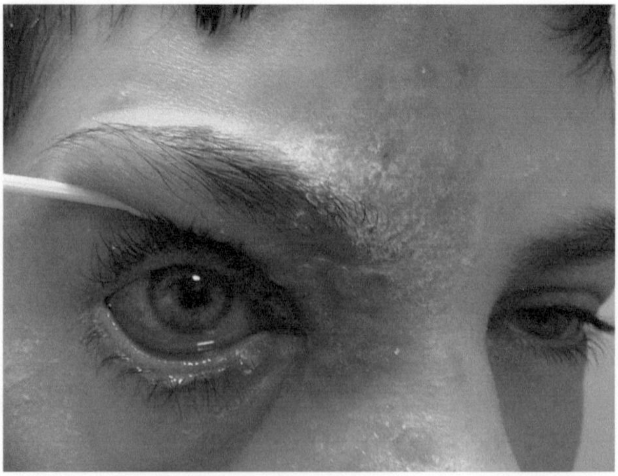

Fig. 5.1 Image of a patient with vesicular lesions in the right V1 distribution that respect the vertical meridian, consistent with a diagnosis of acute herpes zoster ophthalmicus

Table 5.1 Ocular involvement in herpes zoster ophthalmicus

Structure involved	Signs	Onset
Eyelid/conjunctiva		
Blepharoconjunctivitis	Cutaneous macular rash of eyelids respecting midline	Day 1
	Papillary conjunctivitis	Day 2–3
	Vesicular lesions	Day 4–6
Secondary *Staph. aureus* infection	Honey-yellow crusting along lesions	Week 1–2
Cicatricial entropion/ectropion	Eyelid malposition	Month >3
Episclera/sclera		
Episcleritis/scleritis	Diffuse or localized redness and pain	Week 1
Cornea		
Punctate epithelial keratitis	Swollen epithelial cells	Day 1–2
Pseudodendritiform keratitis	Raised infected epithelium with tapered endings	Day 4–6
Stromal keratitis	Nummular subepithelial infiltrates with well-delineated borders	Months 1–3
	Disk-shaped stromal edema with underlying keratic precipitates	Months 3–6
	Corneal edema with deep corneal neovascularization and lipid	Months 3–6
Mucous plaque keratopathy	Gray linear/branching plaque adherent to the surface	Months 3-year 2
Peripheral ulcerative keratitis	Peripheral ulcerative keratitis adjacent to the zone of limbal vasculitis	Months-years
Neurotrophic keratitis	Punctate epithelial defects	Months-years
	Punched-out oval nonhealing epithelial defect	Months-years
	Corneal ulcers	Months-years
Anterior chamber		
Iridocyclitis	Raised intraocular pressure, keratic precipitates, sectoral iris atrophy	Week 1–3
Retina/vitreous		
Acute retinal necrosis	Peripheral necrosis, vitritis, rapid spread, immunocompetent patient	Varies
Progressive outer retinal necrosis	Necrosis typically in posterior pole, immunocompromised patient	Varies
Occlusive vasculitis	Vascular leakage on fluorescein angiography	Varies
Cranial nerves		
Optic neuritis	Swollen, edematous optic nerve head	Varies
Oculomotor palsies	Extraocular motility abnormalities	Varies
Facial palsy	Lagophthalmos	Week 2-months

have also been documented in HZO, ranging from simple papillary conjunctivitis to rare membrane formation with cicatrizing conjunctivitis with permanent lid scarring [16]. Episcleritis and scleritis may ensue rapidly after the onset of acute HZO and represent an inflammatory response to VZV. The scleritis is usually anterior and may be diffuse or nodular (Fig. 5.2). Severe, recurrent scleritis can occur with HZO and result in sclera thinning and atrophy [16].

Corneal involvement occurs in about 65 % of HZO cases and assumes three general forms: epithelial keratitis, neurotrophic keratopathy, and stromal keratitis. Since corneal complications can result in significant visual loss, it is important to treat corneal involvement in HZO.

Epithelial keratitis in HZO may occur acutely or be chronic with frequent recurrences. The earliest manifestation of corneal involvement is punctate epithelial keratitis, characterized by swollen infected epithelial cells. These lesions may either resolve or progress into pseudodendrites, presenting typically 4–6 days after the initial HZO rash [17]. The pseudodendrites appear as elevated plaques of infected epithelium with tapered ends (Fig. 5.3). These can be differentiated from herpes simplex virus dendrites, which are frank epithelial defects with terminal bulbs. A less common form of chronic epithelial keratitis, occurring in approximately 13 % of HZO cases, is characterized by mucous plaque keratopathy and typically appears between 3 months and 2 years after reactivation of VZV [18]. The epithelial plaques appear as white-gray raised regions with well-delineated borders arranged in a linear or branched pattern that migrate across the cornea. Mucous plaque keratopathy is typically accompanied by a limbitis, stromal keratitis, diminished corneal sensation, or iritis [18]. HZO may also present as peripheral ulcerative keratitis, previously termed "serpiginous keratitis," with infiltration and thinning adjacent to a

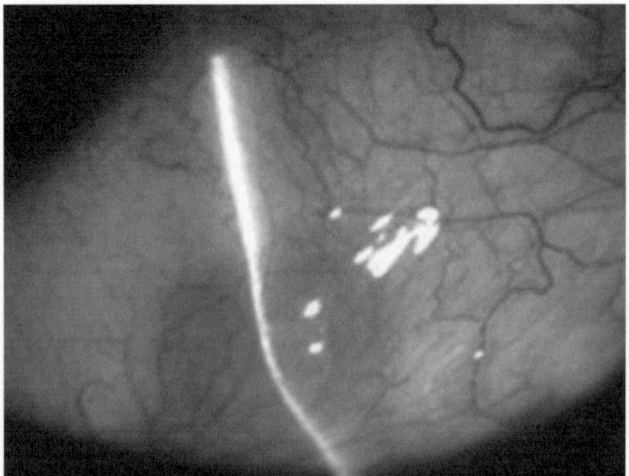

Fig. 5.2 Slit lamp photograph of nodular anterior scleritis associated with herpes zoster ophthalmicus, visualized by a bend in the narrow slit beam overlying the inflammatory nodule

zone of limbal vasculitis [19]. This may progress to corneal neovascularization or perforation and needs to be treated aggressively.

Neurotrophic ulcers occur in more than 25 % of patients with HZO and are characterized in the advanced state by an inferior, horizontally oval epithelial defect with rolled-under edges (Fig. 5.4) and eventually by underlying stromal thinning and scar [12]. However, neurotrophic keratopathy in the earliest stage presents as

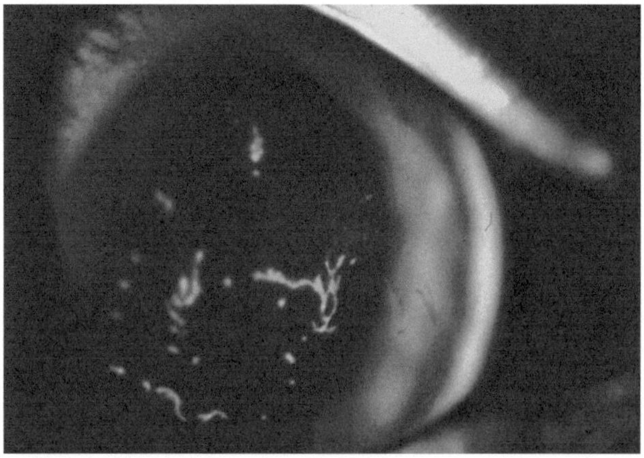

Fig. 5.3 Pseudodendrites on the cornea, as seen by slit lamp biomicroscopy. These appear as elevated plaques of infected epithelium with tapered ends

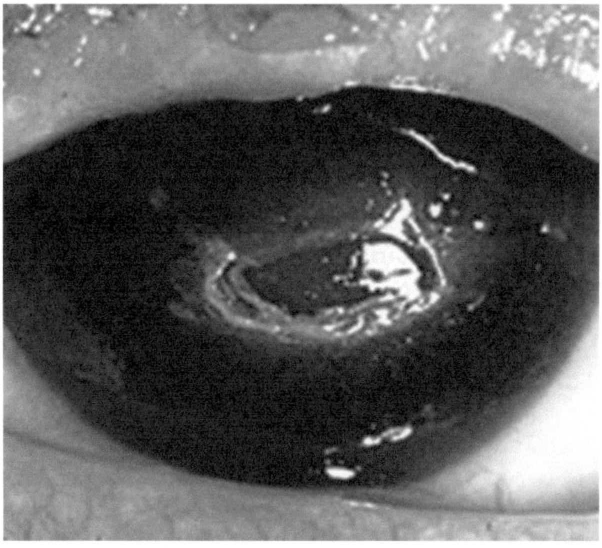

Fig. 5.4 Slit lamp photograph of a neurotrophic corneal ulcer in a patient with a history of herpes zoster ophthalmicus. A neurotrophic ulcer is characterized by a horizontally oval epithelial defect with rolled epithelial edges, with or without underlying stromal ulceration

punctate epithelial keratopathy in the same location. Neurotrophic ulcers are the direct result of VZV ganglionitis and manifest from a combination of factors, including loss of corneal epithelial subbasal nerve plexus with associated loss of trophic epithelial growth factors, corneal hypoesthesia and the inability to self-detect corneal trauma, aqueous tear deficiency due to reduced reflex tearing, and an abnormal blink reflex associated with reduced sensation [12]. Neurotrophic ulcers typically develop several months to years after acute HZO and if not treated aggressively can lead to keratolysis, descemetocele formation, and ultimately corneal perforation.

The pathogenesis of stromal keratitis in HZO is unknown. Dogma holds that HZO stromal keratitis is an immune reaction to viral antigen deposited either during acute VZV reactivation or from late subclinical migration of VZV from the sensory ganglion. The earliest form of stromal keratitis is nummular keratitis, which typically presents about 4 weeks after disease onset in about 25–30 % of patients with HZO [20]. Nummular keratitis is characterized by multiple, fine, coin-shaped, granular infiltrates in the subepithelial stroma [20]. Although nummular keratitis typically occurs acutely, it may have a prolonged and/or recurrent course. Peripheral ulcerative keratitis in HZO ("serpiginous keratitis") presents adjacent to a zone of limbal vasculitis [21]. Without intensive treatment, this may progress to neovascularization or perforation. Endothelial ("disciform") keratitis appears as central, disk-shaped stromal edema with underlying keratic precipitates (collections of lymphocytes along the endothelium). If untreated, endothelial keratitis produces scarring, neovascularization, and lipid deposition [16]. Endothelial keratitis due to HZO cannot be distinguished from endothelial keratitis due to herpes simplex virus or other causes.

Iritis or iridocyclitis affects approximately 43 % of HZO patients acutely and may recur years later [22]. HZO can cause either a non-granulomatous or granulomatous iridocyclitis with extensive keratic precipitates and posterior synechiae. On histopathology, HZO iridocyclitis is characterized by an ischemic occlusive vasculitis that typically results in sectoral iris atrophy [23]. This sectoral iris atrophy differs from the diffuse iris atrophy seen in HSV-related iridocyclitis. Approximately 56 % of patient with HZO iridocyclitis develop glaucoma [22]. Glaucoma may result from a variety of factors including blockage of the trabecular meshwork with cellular debris and iris pigment, pupillary block secondary to posterior synechiae formation, and intrinsic damage to the trabecular meshwork [24].

Retinitis in HZO is rare and can occur during acute reactivation of VZV or up to several decades afterward [20]. There are two general forms of retinitis, acute retinal necrosis (ARN) typically in immunocompetent patients and progressive outer retinal necrosis (PORN) typically in immune-suppressed patients. ARN is characterized by vitritis and a rapidly progressive peripheral retinal necrosis with hemorrhages and occlusive arteriolitis. ARN has been estimated to have an incidence of one case per 2 million people per year, with approximately 50 % of these cases attributed to VZV infection [25, 26]. PORN is characterized by the absence of vitritis and scattered foci of deep retinal lesions classically in the posterior pole. The majority, approximately 70 %, of PORN cases are attributed to VZV infection [26]. Both ARN and PORN can have devastating consequences on vision due to occlusive vasculitis threatening the macula and retinal detachment, particularly if not treated early and aggressively.

HZO can affect several cranial nerves. Optic neuritis associated with HZO is rare, documented in about one in 400 HZO cases, and presents as either an isolated finding or in association with other neurological signs or macular edema from neuroretinitis [27]. Optic neuritis does not necessarily indicate central nervous system involvement because VZV can spread directly from the trigeminal nerve to the optic nerve in the orbit. However, all patients should undergo a lumbar puncture to evaluate for pleocytosis indicative of meningeal involvement in order to initiate appropriate treatment. Extraocular muscle palsies are a fairly common complication of HZO, occurring in approximately 31 % of HZO cases [28]. The oculomotor nerve is the most common extraocular nerve affected by HZO and typically presents in isolation (Fig. 5.5), but the trochlear nerve and abducens nerve can also be involved. Most cases of oculomotor nerve palsy in HZO resolve over several weeks to months leaving a minimal amount of residual ptosis.

Neuralgia, typified by constant boring pain, transient stabbing pain, or pain elicited by superficial stimuli in the distribution of the affected dermatome, is more common in HZO than other forms of zoster. Acute neuralgia, lasting usually 2–3 weeks after rash onset, is seen in less than 15 % of patients under 20 years, about 40 % in patients between 30 and 50 years, and only 20 % of patients over 50 years [29]. In contrast, chronic postherpetic neuralgia (defined as pain lasting more than 1 year after rash onset) is seen in less than 4 % of patients under 20 years, 10 % in those up to age 50 years, and 50 % or higher in those older than 50 years [29]. Post-herpetic neuralgia deeply impacts the quality of life of patients affected and may lead to suicide in elderly patients. The pathogenesis of postherpetic neural-

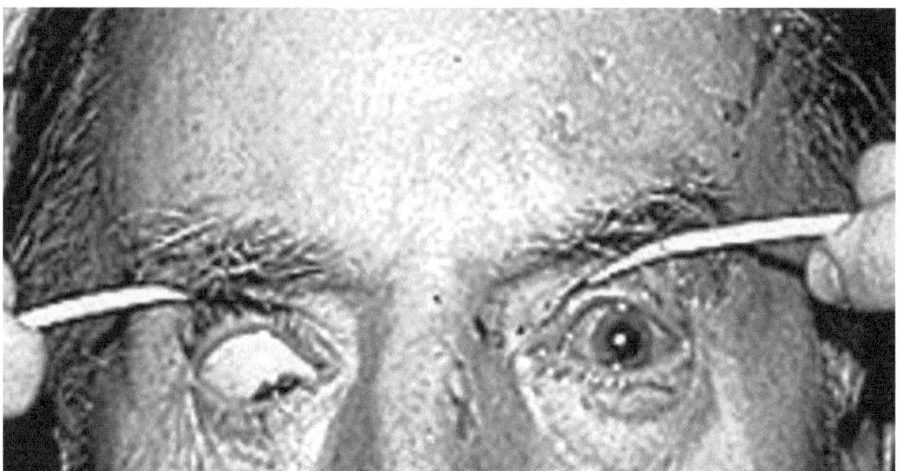

Fig. 5.5 Patient with an acute left oculomotor nerve (CN III) palsy, as demonstrated in this image by the inability for the left eye to infraduct (*look down*), in the setting of herpes zoster ophthalmicus

gia is not completely elucidated, but it may result from acute damage to the sensory nervous system or chronic inflammation persisting in the trigeminal pathways after acute VZV infection has resolved.

5.3 HZO in HIV Patients

HZO can be an early clinical marker for human immunodeficiency virus (HIV) infection, particularly in patients younger than 40 years of age. Screening for HIV in every HZO patient younger than 40 years old is no longer recommended, but rather should be considered if high-risk factors are present. The clinical course of HZO in HIV-infected patients is often prolonged and more severe compared to the general population. The dermatitis in HIV-infected patients tends to involve multiple dermatomes. Several studies have documented that HIV-infected patients with HZO have an increased rate of keratitis, uveitis, ARN, PORN, and postherpetic neuralgia (after adjustment for age) [30, 31]. Overall, ocular complications and subsequent vision loss are estimated to occur in 50–78 % of HIV-infected patients compared to 14–28 % in immunocompetent patients [32]. Recurrence of HZO is also more frequent in HIV-infected patients compared to immunocompetent patients, in whom recurrence is very uncommon [31]. Such chronic, recurrent disease often requires intravenous antiviral therapy to induce resolution. Treatment of HZO in HIV-infected patients can be further complicated by the presence of acyclovir-resistant VZV infection, requiring the use of alternative drugs such as foscarnet [33].

5.4 Neurologic Complication

Stroke is a rare but serious complication of HZO that has been reported in both immunocompetent and immunocompromised patients. The pathogenesis is thought to be direct VZV invasion of the large cerebral arteries by extension from the trigeminal-innervated meninges [34]. The onset of neurologic symptoms, typically headache and contralateral hemiplegia, occurs typically about 7 weeks after acute onset of HZO [34]. Imaging studies show changes consistent with brain infarction, and arteriography demonstrates segmental inflammation of the proximal branches of the middle cerebral artery and anterior cerebral artery. The mortality rate among adults is 20 % to 25 % with a high probability of permanent neurologic sequelae [35].

5.5 Management of HZO

Treatment of acute HZO is aimed at cessation of viral replication with systemic antiviral medications and reduction of pain with tricyclic antidepressants (TCAs) and anticonvulsants (Table 5.2).

Table 5.2 Guidelines to therapy of HZO

Manifestation	Treatment recommendations
Dermatitis	*Treat for 7 days, preferably starting within 72 h of rash onset*
	Antiviral medication
	Famciclovir 500 mg PO TID
	Valacyclovir 1 g PO TID
	Acyclovir 800 mg PO 5× daily
Acute and chronic pain	*Start with antiviral medication in high-risk patient, slowly up-titrate*
	Tricyclic antidepressants
	Nortriptyline 10 mg PO QHS, increase by 25 mg/week up to 75 mg daily
	Desipramine 10 mg PO QHS, increase by 25 mg/week up to 100 mg daily
	Anticonvulsants
	Gabapentin 300 mg PO TID, increase 1800–3600 mg/day
	Use for mild to moderate pain
	Topical agents
	Lidocaine 2 % or 10 % gel, 5 % patch
	Capsaicin 0.05 % ointment, 0.075 % cream, or 8 % patch
Episcleritis/scleritis	Ibuprofen 400 mg PO TID with omeprazole 20 mg PO daily
Cicatrizing entropion/ectropion	Lid repair surgery
Epithelial keratitis	If persistent despite systemic antiviral, consider ganciclovir 0.15 % gel 5× daily
Trophic ulcer	Topical antibiotic ophthalmic ointment and lubricating ointment
	Lateral tarsorrhaphy to promote healing
	Therapeutic soft contact lens to promote healing
	Cyanoacrylate glue with contact lens if progressive thinning
	Boston keratoprosthesis if significant scarring
Stromal keratitis/iritis	*Dose depending on the severity of inflammation*
	Topical steroids
	1 % prednisolone maximum every 2 h, taper slowly by 50 % increments
	0.1 % dexamethasone maximum every 2 h, taper slowly by 50 % increments
Glaucoma	Topical aqueous suppressants
	Beta-blockers (timolol, betaxolol, levobunolol)
	Alpha-adrenergic agonists (brimonidine)
	Carbonic anhydrase inhibitors (brinzolamide, dorzolamide)
ARN	*Standard therapy*
	Acyclovir 10 mg/kg IV every 8 h × 5–10 days then
	Acyclovir 400–800 mg PO 5× daily × 6–12 weeks
	Nonconventional therapy
	Valacyclovir 2 g PO TID
	Famciclovir 500 mg PO TID
PORN	*Antivirals (as above) plus*
	Intravitreal ganciclovir injection (2 mg in 0.8 mL)
	Initiation of HAART

Antiviral therapy is significantly more effective if begun within 72 h of rash onset. Acyclovir (800 mg five times daily for 7–10 days), famciclovir (500 mg three times daily for 7 days), and valacyclovir (1000 mg three times daily for 7 days) have been approved by the US Food and Drug Administration for the treatment of herpes zoster. These antiviral agents are phosphorylated by viral thymidine kinase to a triphosphate form that inhibits viral replication. The most common side effects of antiviral medications are nausea and headache, which occur in no more than 10–20 % of patients [34]. These medications are renally excreted, and so dosage adjustment is required for patients with renal insufficiency. Some preparations contain lactose, but lactose-free versions are available and can be prescribed for lactose-intolerant patients.

Acyclovir was the first antiviral agent developed to treat zoster reactivation. In trials, the use of acyclovir compared to placebo in HZO patients decreases the severity and time with dermatitis, pain, and viral shedding [36, 37]. The use of acyclovir also significantly reduces the incidence and severity of keratopathy, scleritis, and iritis [37]. Furthermore, acyclovir limits the incidence of stromal immune keratitis and late-onset ocular inflammatory disease [37]. The effect of acyclovir on postherpetic neuralgia is variable with some reports showing no efficacy in pain treatment, but a meta-analysis suggests significant decrease in the incidence and severity of chronic pain with the early use of acyclovir [34].

Famciclovir has been compared to acyclovir in the treatment of HZO. Famciclovir demonstrates similar reduction in the severity and duration of the dermatitis and ocular complications compared to acyclovir [38]. Furthermore, multiple studies have shown that famciclovir significantly reduces the incidence and severity of postherpetic neuralgia after HZO [34]. Famciclovir seems to be the most effective antiviral medication in reducing postherpetic neuralgia symptoms [39].

Valacyclovir appears to be as effective as acyclovir in limiting the dermatomal rash and reducing the severity and incidence of ocular complications keratitis, uveitis, and episcleritis [39]. Furthermore, valacyclovir has been demonstrated to be superior to acyclovir in decreasing the severity and incidence of postherpetic neuralgia [39].

Treatment of acute and chronic pain in herpes zoster is discussed elsewhere in this book in greater detail, and the principles do not differ greatly in HZO. Tricyclic antidepressants (TCAs), preferably nortriptyline starting at 10 mg orally daily slowly up-titrated by 25 mg each week up to 75 mg orally daily, should be started along with antiviral therapy to treat acute pain symptoms and reduce the development of postherpetic neuralgia in high-risk patients [40]. The most well-established risk factors for postherpetic neuralgia are older age, greater severity of acute pain during zoster, severe dermatitis, and a prodrome of dermatomal pain before rash onset [41]. The major side effects of TCAs are anticholinergic, including dry mouth, constipation, urinary retention, and drowsiness [40]. A less common but more significant side effect is arrhythmia, and patients with a significant cardiac history should be started on a TCA cautiously [40]. Systemic therapy with anticonvulsants, such as gabapentin or pregabalin, is effective at controlling postherpetic neuralgia. Gabapentin can be started at a dose of 300 mg orally three times daily and slowly up-titrated to a dose between 1800 and 3600 mg/day for full effect [41]. Gabapentin requires dose adjustment in renal insufficiency. The most common side effects are somnolence, dizziness, and peripheral edema. Gabapentin may be continued for months to years in order to control postherpetic neuralgia symptoms with periodic attempts at tapering.

If the combination of TCAs and anticonvulsants is not effective at treating postherpetic neuralgia symptoms, opioids such as tramadol, morphine, oxycodone, and methadone may be considered. If an opioid appears necessary, referral to a pain specialist for management should be considered. Topical agents, such as lidocaine (available as 2 % or 10 % gel and a 5 % patch) and capsaicin (available as an 8 % patch or a 0.075 % cream), are useful in the treatment of mild to moderate pain symptoms [41]. Both lidocaine cream and capsaicin cream should not be applied close to the eyes given the risk of ocular irritation.

5.6 Management of Ocular Complications

Episcleritis and scleritis respond slowly to nonsteroidal anti-inflammatory agents, such as ibuprofen 400 mg orally three times daily, with a concurrent proton pump inhibitor, such as omeprazole 20 mg orally daily, for prophylaxis against gastritis.

Lid reconstruction to repair cicatrizing ectropion or entropion can be performed to alleviate exposure keratopathy and trichiasis, respectively. The goal of these surgical procedures is to limit corneal scarring and associated vision loss.

Epithelial keratitis typically resolves with the use of systemic antiviral medication alone. In a recent study of four immunocompromised patients with persistent pseudodendrites despite systemic antiviral medication, topical antiviral medication (ganciclovir 0.15 % gel administered five times daily) was successful in promoting healing of the corneal lesions in 7 days in all cases [42].

Trophic sterile ulcerations in the anesthetic zoster cornea are initially treated with frequent lubricating ointment and punctal occlusion. If conservative measures are unsuccessful, lateral tarsorrhaphy to narrow the lid opening (Fig. 5.6) or constant-wear, therapeutic contact lens can be utilized to promote healing. If the cornea begins to thin in an open ulcer, cyanoacrylate glue may be used to seal the ulcer (Fig. 5.7), and a therapeutic contact lens is placed over the eye to prevent the adhesive from dislodging with eyelid movement. Corneal transplantation (penetrating keratoplasty) has a limited role in HZO, because wound healing after surgery is compromised by the anesthetic state. Transplanted eyes heal poorly and have an increased chance of wound dehiscence. The use of the Boston keratoprosthesis in the eyes with severely scarred corneas due to HZO has been reported as an alternative to keratoplasty [43].

Stromal keratitis is treated with topical corticosteroids. Depending on the severity, starting doses may range from 1 % prednisolone or 0.1 % dexamethasone every 2 h while awake to once daily. Tapering the frequency begins as the immune disease begins to lessen, with reduction in dose by 50 % increments over several weeks to

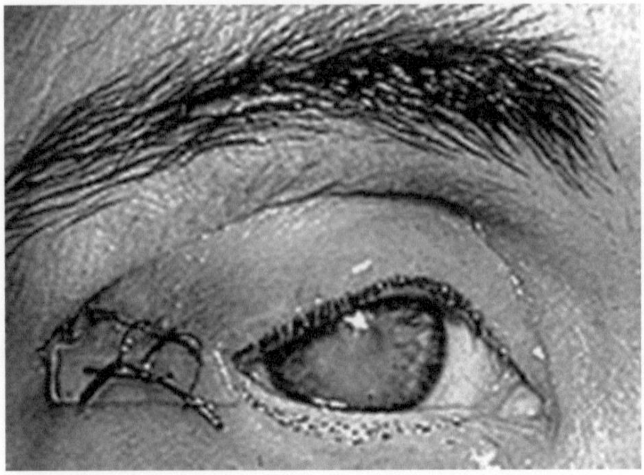

Fig. 5.6 Image of a patient with a temporary lateral tarsorrhaphy (shown by the sutures and bolster) along the lateral one-third of the eyelids in a patient with a neurotrophic corneal ulcer in the setting of herpes zoster ophthalmicus

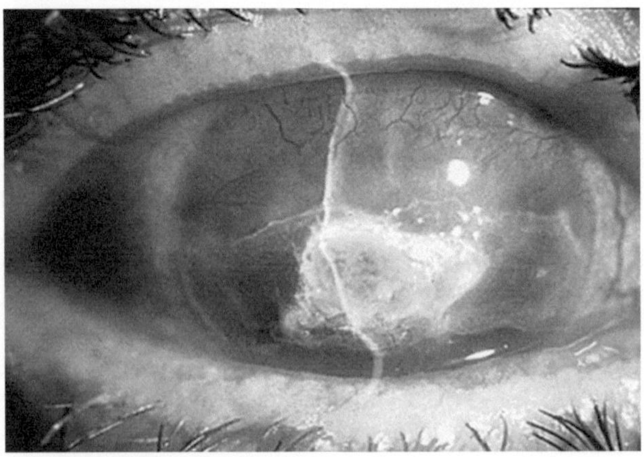

Fig. 5.7 Slit lamp photograph of a neurotrophic corneal ulcer after cyanoacrylate glue patch (as highlighted by the outward bend in the narrow slit beam) for progressive thinning

months. Typically, the lower the dose, the longer the taper, in order to prevent rebound inflammation. Lower-potency corticosteroids can be used to further taper without complete cessation of anti-inflammatory therapy. Intraocular pressure should be checked within 4–6 weeks of starting corticosteroid therapy and at minimum every 4 months while on chronic treatment, because of the risk of glaucoma both from HZO and from the corticosteroids. If a corticosteroid-induced elevation of intraocular pressure occurs, alternative corticosteroids with less effect on intraocular pressure, such as rimexolone 1 %, may be substituted.

Iritis is treated with corticosteroids in a similar fashion to stromal keratitis. Dosage of the topical corticosteroid should be tailored to the severity of the anterior chamber inflammation. Elevated intraocular pressure can be treated with aqueous suppressant medications such as beta-blockers (timolol, betaxolol, levobunolol), alpha-adrenergic agonists (brimonidine), or carbonic anhydrase inhibitors (brinzolamide, dorzolamide). The use of prostaglandin analogues (latanoprost, travoprost, bimatoprost) in the setting of uveitis should be avoided given concerns regarding prostaglandin-induced uveitis. Elevated intraocular pressure in HZO can also occur due to trabeculitis and requires reinstitution of oral antiviral and treatment with topical corticosteroids. Thus, elevated intraocular pressure in patients with HZO on topical corticosteroids can present a diagnostic dilemma.

ARN and PORN are difficult to treat given the aggressive nature of zoster retinitis. The standard of care for the treatment of ARN consists of intravenous acyclovir 10 mg/kg every 8 h for 5–10 days to treat active retinitis, followed by oral acyclovir 400–800 mg five times daily for an additional 6–12 weeks to reduce the risk of involvement of the fellow eye [44]. Recent studies have demonstrated that the use of oral antiviral agents with superior bioavailability to acyclovir (valacyclovir 2 g three times daily or famciclovir 500 mg three times daily) in the treatment of ARN results in resolution of retinitis after 21 days, similar to cases treated with intravenous acyclovir [44, 45]. The treatment of PORN with systemic antiviral therapy alone, ranging from acyclovir to foscarnet, does little to diminish the retinitis. More recently, a combination of systemic antiviral medications with intravitreal ganciclovir injections (2 mg in 0.8 mL) was shown to result in improved visual outcomes [46]. The initiation of highly active antiretroviral therapy (HAART), in PORN cases occurring in HIV-infected patients, in combination with antiviral medications, can also improve visual outcomes [47]. One of the most frequent complications of both ARN and PORN, which occurs in approximately 50 % of cases, is retinal detachment [45, 46]. Retinal detachment is treated with a variety of methods such as sclera buckling, vitrectomy, air-fluid exchange, endolaser photocoagulation, cryotherapy, and retinal tamponade with C_3F_8 gas or SF_6 gas, but in general carries a poor prognosis [48].

5.7 Summary

The first division of the fifth cranial nerve is a common site for VZV reactivation and manifests as HZO. The myriad complications of HZO present risk to vision and quality of life. As the incidence of HZO continues to rise, clinicians need to be aware of the different manifestations and treatment. Indeed, early initiation of antiviral medication (such as famciclovir or valacyclovir) and tricyclic antidepressants (such as nortriptyline) may significantly alter the disease course. Increased utilization of the herpes zoster vaccine in patients over the age of 50 should reduce the frequency and severity of the disease when the new subunit gE vaccine to prevent zoster is licensed [49].

References

1. Edell A, Cohen E (2013) Herpes simplex and herpes zoster eye disease: presentation and management at a city hospital for the underserved in the United States. Eye Contact Lens 39:311–314
2. Jung J, Elkin Z, Xiaochun L et al (2013) Increasing use of the vaccine against zoster through recommendation and administration by ophthalmologists at a city hospital. Am J Ophthalmol 155:787–795
3. Leung J, Harpaz R, Molinari N et al (2011) Herpes zoster incidence among insured persons in the United States, 1993–2006: evaluation of impact of varicella vaccination. Clin Infect Dis 52:332–340
4. Cohen E (2013) Prevention of herpes zoster. JAMA Ophthalmol 131(3):396–398
5. Brisson M, Gay N, Edmunds W et al (2002) Exposure to varicella boosts immunity to herpes-zoster: implications for mass vaccination against chickenpox. Vaccine 20:2500–2507
6. Thomas S, Hall A (2004) What does epidemiology tell us about risk factors for herpes zoster? Lancet Infect Dis 4:26–33
7. Reynolds M, Chaves S, Harpaz R et al (2008) The impact of the varicella vaccination program on herpes zoster epidemiology in the United States: a review. J Infect Dis 197(Suppl 2):S224–S227
8. Bloch K, Johnson J (2012) Varicella zoster virus transmission in the vaccine era: unmasking the role of herpes zoster. J Infect Dis 205:1331–1333
9. Chaudhry R, Schietel S, North F et al (2013) Improving rates of herpes zoster vaccination with a clinical decision support system in a primary care practice. J Eval Clin Pract 19(2):263–266
10. Oxman M, Levin M, Johnson G et al (2005) Shingles prevention study group: a vaccine to prevent herpes zoster and postherpetic neuralgia in older adults. N Engl J Med 352(22):2271–2284
11. Morrison V, Johnson G, Schmader K et al (2014) Long-term persistence of zoster vaccine efficacy. Clin Infect Dis 60(6):900–909
12. Pavan-Langston D (2000) Viral disease of the cornea and external eye. In: Albert D, Jakobiec F (eds) Principles and practice of ophthalmology, 2nd edn. W.B. Saunders, Philadelphia, PA, pp. 846–893
13. Goh C, Khoo L (1997) A retrospective study of the clinical presentation and outcome of herpes zoster in a tertiary dermatology outpatient referral clinic. Int J Dermatol 36:667–672
14. Yamamoto S, Shimomura Y, Pavan-Langston D et al (1995) Detecting varicella-zoster virus DNA in iridocyclitis using polymerase chain reaction: a case of zoster sine herpete. Arch Ophthalmol 113:1358–1359
15. Harding S, Lipton J, Wells J (1987) Natural history of herpes zoster ophthalmicus: predictors of postherpetic neuralgia and ocular involvement. Br J Ophthalmol 71:353–358
16. Kaufman S (2008) Anterior segment complications of herpes zoster ophthalmicus. Ophthalmology 115(Suppl 2):S24–S32
17. Jones D (1974) Herpes zoster ophthalmicus. In: Golden B (ed) Symposium on ocular inflammatory disease. Thomas, Springfield
18. Marsh R, Cooper M (1987) Ophthalmic zoster: mucous plaque keratitis. Br J Ophthalmol 71:725–728
19. Rousseau A, Bourcier T, Colin J et al (2013) Herpes zoster ophthalmicus-diagnosis and management. US Ophthalmic Review 6(2):1–6
20. Womack L, Liesegang T (1983) Complications of herpes zoster ophthalmicus. Arch Ophthalmol 101:42–45
21. Cobo L (1988) Corneal complications of herpes zoster ophthalmicus: prevention and treatment. Cornea 7:50–56
22. Thean J, Hall A, Stawell R (2001) Uveitis in herpes zoster ophthalmicus. Clin Experiment Ophthalmol 29:406–410
23. Van der Lelij A, Ooijman F, Kijlstra A et al (2000) Anterior uveitis with sectoral iris atrophy in the absence of keratitis: a distinct clinical entity among herpetic eye diseases. Ophthalmology 107(6):1164–1170

24. Raber I, Laibson P (1984) Herpes zoster ophthalmicus. In: Liebowitz H (ed) Corneal disorders: clinical diagnosis and management. WB Saunders, Philadelphia, p. 404
25. Cochrane T, Silvestri G, McDowell C et al (2012) Acute retinal necrosis in the United Kingdom: results of a prospective surveillance study. Eye (Lond) 26:370–378
26. Wong R, Jumper J, McDonald H et al (2013) Emerging concepts in the management of acute retinal necrosis. Br J Ophthalmol 97:545–552
27. Lee M, Cooney E, Stoessel K et al (1998) Varicella zoster virus retrobulbar optic neuritis preceding retinitis in patients with acquired immune deficiency syndrome. Ophthalmology 105:467–471
28. Marsh R, Dulley B, Kelly V (1977) External ocular motor palsies in ophthalmic zoster: a review. Br J Ophthalmol 61:677–682
29. Donahue J, Choo P, Manson J et al (1995) The incidence of herpes zoster. Arch Intern Med 155:1605–1609
30. Glesby M, Moore R, Chaisson R (1995) Clinical spectrum of herpes zoster in adults infected with human immunodeficiency virus. Clin Infect Dis 21:370–375
31. Cole E, Miesler D, Calabrese L et al (1984) Herpes zoster ophthalmicus and acquired immune deficiency syndrome. Arch Ophthalmol 102:1027–1029
32. Nithyanandam S, Joseph M, Stephen J (2013) Ocular complications and loss of vision due to herpes zoster ophthalmicus in patients with HIV infection and a caparison with HIV-negative patients. Int J STD AIDS 24:106–109
33. Gnann J, Whitley R (2002) Herpes zoster. N Engl J Med 347:340–346
34. Dworkin R, Johnson R, Breuer J et al (2007) Recommendations for the management of herpes zoster. Clin Infect Dis 44:S1–S26
35. Moriuchi H, Rodriguez W (2000) Role of varicella-zoster virus in stroke syndromes. Pediatr Infect Dis J 19:648–653
36. Hoang-Xuan T, Buchi E, Herort C et al (1992) Oral acyclovir for herpes zoster ophthalmicus. Ophthalmology 99:1062–1071
37. Pavan-Langston D (2008) Herpes zoster antivirals and pain management. Ophthalmology 115(2Suppl):S13–S20
38. Tyring S, Engst R, Corriveau C et al (2001) Famciclovir for ophthalmic zoster: a randomized acyclovir controlled study. Br J Ophthalmol 85:576
39. McDonald E, Kock J, Ram F (2012) Antivirals for management of herpes zoster including ophthalmicus: a systematic review of high-quality randomized controlled trials. Antivir Ther 17(2):255–264
40. Johnson R, Dworkin R (2003) Treatment of herpes zoster and postherpetic neuralgia. BMJ 326(7392):748–750
41. Writers A (2013) Treat herpes zoster with systemic antivirals and post-herpetic neuralgia with various agents depending on its severity. Drugs Ther Perspet 29:348–352
42. Agarwal S, Cavalcanti B, Pavan-Langston D (2014) Treatment of pseudodendrites in herpes zoster ophthalmicus with topical ganciclovir 0.15 % gel. Cornea 33(20):109–113
43. Pavan-Langston D, Dohlman CH (2008) Boston Keratoprosthesis treatment of herpes zoster neurotrophic keratopathy. Ophthalmology 115(2):S21–S23
44. Taylor S, Hamilton R, Hooper C et al (2012) Valacyclovir in the treatment of acute retinal necrosis. BMC Ophthalmol 12:48–52
45. Tibbetts M, Shah C, Young L et al (2010) Treatment of acute retinal necrosis. Ophthalmology 117(4):818–824
46. Gore D, Gore S, Visser L (2012) Progressive outer retinal necrosis. Arch Ophthalmol 130(6):700–706
47. Kim S, Equi R, Belair M et al (2007) Longer-term preservative of vision in progressive outer retinal necrosis treated with combination antiviral drugs and highly active antiretroviral therapy. Ocul Immunol Inflamm 15(6):425–427
48. McDonald H, Lewis H, Kreiger A et al (1991) Surgical management of retinal detachment associated with the acute retinal necrosis syndrome. Br J Ophthalmol 75(8):455–458
49. Lal H et al (2015) Efficacy of an adjuvanted herpes zoster subunit vaccine in older adults. N Engl J Med 372(22):2087–2096 , 2137-2121

Chapter 6
Neurological Complications of Herpes Zoster

Maija Haanpää

6.1 Introduction

Varicella zoster virus (VZV) is an exclusively human neurotropic, double-stranded DNA alphaherpesvirus that resides dormant in sensory ganglia along the whole neuraxis after the primary infection, varicella. The most typical clinical manifestation of the reactivation is herpes zoster (HZ), usually painful dermatomal rash, which represents ganglionitis, peripheral neuritis, and dermatitis caused by VZV. Somatosensory abnormalities, caused primarily by neural inflammation, are common in acute HZ [36]. About one-third of the patients have mechanical allodynia and one-fourth of patients have pinprick hypoesthesia, and the presence of these abnormalities elevates the risk of postherpetic neuralgia (PHN) [38]. In quantitative somatosensory testing, warm and cold thresholds are elevated in one-fifth and tactile thresholds are elevated in one-fourth of patients with HZ. All somatosensory abnormalities tend to normalize with time [37], and most patients recover from zoster neuritis without any complications in some weeks.

Although HZ is a benign and self-limited disease in most patients, about 12 % develop at least one complication (Table 6.1). Of these, PHN is by far the most common, whereas neurological complications other than PHN are rare. A recent population-based retrospective database study from Israel reported that 2.5 % of patients with HZ had ophthalmic complications and 2.7 % had nerve palsies, whereas encephalitis, myelitis, and delayed contralateral hemiparesis were encountered by 0.03 % each [94]. The risk of neurological complications is increased in patients who are immunocompromised or have a disseminated form of the disease. Central nervous system (CNS) complications include meningitis, encephalitis, myelitis, and cerebral vasculitis. The most common mechanism of viral spread seems to be the direct extension of the infection into cerebrospinal fluid (CSF) spaces from the infected ganglia via the meningeal ramus, leading to local meningitis [85, 88]. Because viremia is not uncommon in HZ [56], hematogenous spread to

M. Haanpää, MD, PhD
Department of Neurosurgery, Helsinki University Hospital, Helsinki, Finland

Mutual Insurance Company Etera, Helsinki, Finland
e-mail: maija.haanpaa@etera.fi

C.P.N. Watson et al. (eds.), *Herpes Zoster: Postherpetic Neuralgia and Other Complications*, DOI 10.1007/978-3-319-44348-5_6

Table 6.1 Complications of herpes zoster in two population-based studies from Rochester, Minnesota

Complication	Galil et al. [24]	Ragozzino et al. [79]
Any complication	11.6 %	12.0 %
Postherpetic neuralgia[a]	7.9 %	9.3 %
Ocular complications	2.2 %	1.9 %
Motor deficit	0.9 %	1.0 %
Meningitis, encephalitis, or CNS vasculitis	0.5 %	0.2 %
Herpes zoster oticus	0.2 %	0.2 %

[a]Defined as pain continuing after healing of rash

CNS can take place especially in the immunocompromised patients [88]. In trigeminal HZ, transaxonal spread from the infected ganglia into the brainstem and the cranial arteries can cause brainstem encephalitis and arteritis, respectively. In immunocompromised patients, a focal CNS infection can disseminate widely into the CNS parenchyma, and hematogenous infection of the subependymal microvasculature can lead to periventriculitis and ventriculitis with subsequent spread of the virus to distant parts of CNS [51, 88].

In addition to the CNS complications and peripheral sensory neuritis, lower motor neuron-type (i.e., peripheral) paresis can occur in HZ. Motor weakness can also be a part of the symptoms of encephalitis, myelitis, or cerebrovascular event caused by vasculitis. In these cases the paresis is of upper motor neuron type (i.e., central). The most common neurological complication of HZ, namely, PHN, is discussed in detail on other chapters. This chapter summarizes neurological manifestations of reactivation of VZV with or without rash focusing on immunocompetent patients.

6.2 Subclinical Spread of Inflammation into the Central Nervous System in Herpes Zoster

HZ has been regarded as a simple model of peripheral unilateral neuritis caused by VZV, but evidence from neuropathological reports suggests that viral invasion and inflammation may extend to the CNS as well. In addition to leptomeningitis, inflammatory changes in the anterior and posterior horn of the spinal cord and in the brainstem have been described [11, 20, 81, 88, 93].

A study describes a series of 56 immunocompetent HZ patients without clinical symptoms of the CNS infection [35]. CSF were obtained from 46 patients on day 1 to 18 from the eruption of rash, and 16 consecutive patients with cranial or cervical HZ underwent magnetic resonance imaging (MRI) 1–5 weeks following rash. In 14/46 (35 %) patients, there was evidence of VZV in the CSF either in the form of a positive PCR or anti-VZV-IgG. Leukocytosis (range 5–1440/μL) was found in 21/46 (46 %)

patients. Zoster-related MRI changes were found in the brainstem in 9/16 (56 %) patients and in the cervical cord in 2. This suggests an inflammatory process in the CNS parenchyma. Three patients had enhancement of the trigeminal nerve in addition to the brainstem lesions. According to these results, the entrance of the VZV into the CSF space is common in HZ, and MRI changes can often be detected in patients with cranial or cervical HZ without any symptoms of a CNS infection.

An earlier report compares the EEG findings in 42 patients with HZ and six patients with HZ-associated encephalitis [77]. Thirty-one percent of patients with HZ had EEG changes with reduced rhythm frequency ranging from 7 to 2 Hz activity. The location of HZ did not influence the frequency of EEG abnormalities. When compared to EEG in HZ-associated encephalitis, the findings were qualitatively the same but tended to be more severe in the encephalitis cases. The abnormal EEG findings suggest subclinical encephalitis in HZ patients, and the general symptoms such as prolonged fatigue after HZ may be explained by this.

6.3 Neurological Complications due to Reactivation of Varicella Zoster Virus Without Rash

It has been shown that VZV can reactivate subclinically [13, 27, 56] or without rash. Pain in a dermatomal distribution without rash due to reactivation of VZV is named "zoster sine herpete" [28]. Three virologically confirmed cases with radicular pain in thoracic segments have been reported [2, 28]. Cases of trigeminal "zoster sine herpete" have also been published [5, 15]. However, reactivation of VZV without rash encompasses a much broader syndrome complex than localized pain. In fact, VZV can affect all levels of the nervous system in the absence of skin lesion. It has been reported to cause multiple cranial neuropathies [31, 45, 58, 74], recurrent cranial polyneuritis [21], persistent hiccups and vomiting with multiple cranial nerve palsy [97], abdominal wall paresis [7], polyneuritis [58], radiculoneuritis [21], meningoencephalitis [78], encephalitis [41, 58] myelitis [58], and cranial vasculitis [68, 71] without rash in immunocompetent individuals. In a Spanish series, VZV was the most common cause of acute aseptic meningitis in adults [17]. The diagnosis of nervous system infection by VZV may be confirmed by the presence of antibody to VZV in CSF even without detectable VZV D [21, 50]. A prospective study of 56 patients with pain in a dermatomal distribution without rash in a general practice setting, together with 81 matched control blood donors, showed that the presence of positive serologic test results and positive PCR results from circulating leukocytes was very similar to that seen in the control group, indicating that the presence of unexplained dermatomal pain did not identify clinical or subclinical reactivations. Thus, the routine use of antiviral agents for this purpose is not supported, and there is the added concern that such therapy might abort appropriate investigation of the etiology of the pain [60].

6.4 Peripheral Motor Paresis

Although HZ affects primarily the sensory nervous system, segmental zoster paresis has been recognized as a part of the HZ syndrome. Motor involvement following HZ was first described in 1866 [6]. Segmental zoster paresis is characterized by focal motor weakness that appears in the same segment where the skin eruptions occur. Motor deficit may involve the muscles of the head, trunk, or extremities or may present as a visceral motor complication, such as colonic pseudo-obstruction or bladder. Cervical zoster may be associated with diaphragmatic paralysis.

According to population-based studies, the incidence of lower motor neuron-type paresis in HZ is about 1 % [24, 79], whereas in a study of 1210 zoster patients at the Mayo Clinic, incidence of segmental paresis was 5 % [90]. The low recorded incidence of motor deficit in thoracic HZ is probably related to the difficulty in diagnosing weakness of intercostal and abdominal muscles.

The pathogenesis of motor involvement is still controversial, but it is believed to be due to a direct viral spreading of the infection from the dorsal root ganglion to the anterior horn cells, adjacent motor nerve roots, or peripheral nerves. Immune-mediated mechanism with aseptic inflammation has also been suggested [7]. Motor damage may affect the root, plexus, or peripheral nerve.

Mononeuropathies of the median, ulnar, long thoracic, recurrent laryngeal and phrenic nerves have been described [61].

A review on segmental zoster paresis in the limbs including 138 patients reported that in the upper limbs weakness was most frequent in C5–C7 segments, and in lower limbs in L1–L4 segments [49]. In a retrospective series of zoster-associated mononeuropathies, most patients were immunocompetent. In MRI, enlargement of the affected nerves, T2 signal hyperintensity, and gadolinium enhancement of the affected nerve were found, and the changes were longitudinally extensive along the length of the affected nerve. Compared to more common radiculopathy as a cause of segmental motor paresis, zoster-associated mononeuropathy was associated with prolonged symptoms and a high rate of PHN [80].

In a patient with a rash in the groin and a weakness of the iliopsoas and quadriceps muscles, contrast enhancement of the spinal nerve roots over several segments was observed on MRI, but on subsequent scanning, the enhancement was found to be confined to the rash segment [40]. These findings suggest a more widespread virus-induced inflammation than would have clinically been expected.

According to a recent literature review on segmental zoster abdominal wall paresis, one-fifth of the patients have also visceral neuropathy that affected the gastrointestinal system such as constipation and colonic pseudo-obstruction [7]. The diagnosis of segmental abdominal zoster paresis is clinical by recognition of an abdominal or flank bulge after HZ. To confirm the diagnosis of segmental abdominal paresis or to exclude other etiologies related to abdominal wall protrusion, different diagnostic tests such as electrodiagnostic studies, MRI, computed tomography, and ultrasonography have been used.

In a small series, EMG revealed fibrillation in the paraspinal muscles in up to 70 % of patients with HZ [33], which suggests that subclinical motor involvement is common in HZ. In our own series of 40 immunocompetent patients with HZ, EMG abnormalities were detected in 21/40 (53 %) patients, subclinical in 13 and associated with motor paresis in 8 cases [34]. EMG changes were confined to the myotomes suggested by rash in nine patients and were more widespread in 12 patients. In five patients, the EMG changes became progressively more widespread in repeated examinations in spite of good clinical recovery. Although the rash was unilateral in all patients, six patients had bilateral EMG changes in their paraspinal muscles. These findings suggest that the virus spreads centrally from ganglia into the spinal cord and further to anterior roots to cause peripheral motor paresis. EMG findings reported in HZ patients include fibrillation and positive sharp wave potentials, compatible with Wallerian degeneration; fasciculation potentials, compatible with axonal or anterior horn cell dysfunction; and polyphasic potentials, compatible with reinnervation [8, 25, 34]. The onset of motor weakness is usually rapid, reaching peak levels within hours or days. The interval between the eruption of rash and the onset of weakness ranges from 1 day to several weeks [90]. The prognosis of clinical paresis is good, with a complete or nearly complete recovery of function in 75–100 % of patients [34, 61, 90]. The prognosis is independent of the initial degree of weakness.

Some uncommon motor manifestations associated with HZ have been described in literature. Hemidiaphragmatic paresis associated with cervical zoster, even causing respiratory failure [12], and detrusor paralysis associated with sacral zoster and leading to urinary retention [25, 46] have been reported. Cases of the Guillain-Barré syndrome following shingles have also been described [9, 63, 73].

6.5 Involvement of Cranial Nerves

Cranial or cervical HZ may affect the cranial nerves, leading to the development of a characteristic catalogue of symptoms [4, 76, 84]. Zoster may be accompanied by ophthalmoplegia (most commonly affecting the third cranial nerve), optic neuritis, or both. Palsies of the lower cranial nerves occur less frequently. Cranial neuropathy often occurs weeks after acute zoster has developed. Because all cranial nerves are supplied by blood from the circulation of the carotid artery through small branches that supply groups of two or three cranial nerves, the occurrence of concurrent, contiguous cranial neuropathies suggests infection mediated by small vessels. Varicella zoster may spread transaxonally along trigeminal and other ganglionic afferent fibers from the carotid arteries to the vasa vasorum of small nerves [29].

6.5.1 Ophthalmoplegia

In an extensive review of 2250 cases of ophthalmic HZ, ocular muscle palsies (most frequently third nerve palsy) occurred in 13 % of cases [18]. A recent review of 110 immunocompetent patients presenting with HZ ophthalmicus reported that four of them (4 %) experienced isolated cranial motor nerve palsies [39].

Ocular motor cranial nerve paresis is reported in 5–31 % of patients with HZ ophthalmicus, but the occurrence of complete unilateral ophthalmoplegia (i.e., concurrent unilateral impairment of ocular ductions in all directions) is rarer [96]. Orbital apex syndrome involves dysfunction of ophthalmic division of the trigeminal nerve, the oculomotor nerve (III), the trochlear nerve (IV), and the abducens nerve (VI), as well as dysfunction of the optic (II) nerve. HZ is one possible reason for orbital apex syndrome. Other causes, such as inflammatory, infectious, neoplastic, traumatic, and vascular conditions, need to be excluded with MRI when HZ-caused orbital apex syndrome is suspected. Possible mechanisms of HZ-caused orbital apex syndrome include extensive inflammation around the posterior ciliary nerves and vessels with ocular ischemia; orbital soft tissue edema with direct compression of cranial nerves III, IV, and VI; and direct spread of VZV from cranial nerve V to cranial nerves III, IV, and VI [54].

6.5.2 Herpes Zoster Oticus

In HZ oticus, also called the Ramsay Hunt syndrome, reactivation of VZV occurs in the geniculate ganglion, causing otalgia, auricular vesicles, and peripheral facial paralysis. However, the classic vesicular eruption of the pinna is not always present; some 2–23 % of unilateral facial palsies without vesicles are actually zoster sine herpete [95]. If vesicles appear, they can appear before, during, or after the facial paralysis [82]. HZ oticus accounts for about 12 % of cases of facial palsy. In severe cases of HZ oticus, involvement of vestibulocochlear nerve leads to sensorineural hearing loss in 10 % and vestibular symptoms in 40 % [32]. HZ oticus is diagnosed clinically and is based on unilateral facial weakness plus vesicular lesions in ipsilateral ear, hard palate, or anterior 2/3 of the tongue [95]. The traditional diagnostic triad (otalgia, auricular vesicles, and facial paralysis) cannot be applied to the atypical presentations of HZ oticus such as multiple cranial nerve involvement or zoster sine herpete. The gold standard for diagnosing VZV reactivation is PCR of skin, saliva, or middle ear fluid samples, but this is rarely done clinically [64].

In HZ oticus, multiple cranial nerves are often involved, particularly the eight, producing hearing impairment and vertigo, but cranial nerves V, IX, X, XI, and XII may also be affected [1, 4, 32, 84, 95]. On MRI, enhancement of the VII and VIII cranial nerves and the labyrinth is typical of HZ oticus [52, 75]. Cases of HZ oticus with a pontine lesion, suggesting concomitant brainstem encephalitis, have been reported [42, 62, 87]. Using PCR, VZV has been detected in the facial nerve sheath, mucosa of the middle ear, posterior auricular muscles, temporal bone, and CSF [65, 92]. The prognosis of the facial paralysis in HZ oticus is poorer than in Bell's palsy (i.e., idiopathic facial paralysis), but a meta-analysis of 12 Ramsay Hunt syndrome articles concluded that antiviral therapy plus steroids compared to steroids alone significantly improved facial nerve function recovery [10].

6.6 The Central Nervous System Complications of Herpes Zoster

After zoster develops, VZV usually remains within ganglia. Sometimes, after reactivation in either immunocompetent or, especially, immunocompromised patients, the virus spreads to the spinal cord or brain. Severely immunocompromised patients have the most severe complications of reactivation, the greatest depth of tissue penetration, and the greatest amounts of recoverable virus [29].

6.6.1 *Meningitis*

VZV is a common cause of aseptic meningitis, which can present itself with or without rash. The rash may precede the meningeal symptoms or follow it. The course of the disease is benign; complete recovery may be expected in one to 2 weeks [17].

6.6.2 *Encephalitis*

Encephalitis occurs in up to 5 % of the patients hospitalized because of HZ [59]. In addition to symptoms and signs of meningeal irritation, patients show signs of cerebral involvement. They may be disoriented, confused, somnolent, or agitated, but their mental state improves during the following weeks in most cases [3, 41, 47]. Neuropsychological findings soon after HZ encephalitis include slowing of cognitive processes, memory impairment, and emotional and behavioral changes, which is typical of subcortical-type cognitive impairment [41]. In the acute phase, EEG shows a generalized disturbance of background activity, and SPECT reveals bilateral, mostly frontal, perfusion defects in many patients [41, 47, 77]. In neuropathological examination, mononuclear leptomeningitis and localized or generalized alterations of the CNS parenchyma are seen [85]. The tendency for lesions of encephalitis to occur at gray-white matter junctions is compatible with the neuroradiological and neuropsychological findings [41]. The presence of these lesions warrants the diagnosis of multifocal leukoencephalitis and is regarded to indicate viral invasion to small intraparenchymal arterioles causing small ischemic foci and subsequent development necrosis and demyelination [51]. The proportion of necrosis and demyelination depends on the degree of additional oligodendrocyte infection by the virus [12]. Although the prognosis of HZ encephalitis is good in most cases, cognitive impairment can remain permanent in some patients, and the disease may be even fatal in the immunocompromised.

6.6.3 Myelitis

Symptoms of myelitis appear usually from days to weeks after the appearance of cutaneous zoster ipsilaterally to the rash, with motor dysfunction predominating and evolving into paraplegia, followed by spinothalamic and to a lesser extent posterior column sensory deficit and sphincter disturbances [12, 17]. The outcome ranges from complete recovery to death, and the course of the disease may be acute, remitting-exacerbating, or chronic [28]. Pathological and virological analyses of the spinal cord from fatal cases have revealed frank invasion of VZV in the parenchyma [50] and, in some instances, spread of virus to adjacent nerve roots. The dorsal root entry zone and the posterior horn of the spinal cord segment corresponding to the affected dermatome are most severely involved. Spread of VZV into the spinal cord has been demonstrated in all patients in one series [12].

Forms of VZV myelopathy include a postinfectious process, direct infection of the spinal cord, or VZV vasculopathy [12, 17, 67]. Postinfectious VZV presents as a self-limiting, monophasic spastic paraparesis, with or without sensory features and sphincter problems, and usually occurs in immunocompetent patients days to weeks after acute varicella or zoster. VZV myelopathy can also present as an insidious, progressive, and sometimes, fatal myelitis, mostly in immunocompromised individuals [67]. VZV can also produce spinal cord infarction, identified by diffusion-weighted MRI and confirmed virologically [72]. Thus, VZV vasculopathy can cause stroke in the spinal cord as well as in the brain.

In a study of 31 patients with VZV myelitis (17 immunocompromised and 14 immunocompetent), focal weakness, sensory impairment, and urinary dysfunction were the most common symptoms, being observed in 97 %, 81 %, and 58 %, respectively [43]. Atypical presentation (zoster sine herpete, skin lesions developing after myelopathy, or irrelevant anatomical distribution of myelopathy and zoster) was observed in 55 %, and immunocompromised patients were prone to atypical presentations. Outcome was good (ability to walk independently or with an aid) in all immunocompetent patients but rare in immunocompromised patient.

Diagnosis is confirmed by the presence of VZV DNA or anti-VZV IgG or both in CSF. MRI of the spinal cord may show local swelling and hyperintense lesions in T2-weighted images [22, 44, 91]. Gadolinium enhancement on T1-weighted images may be seen, suggesting severe inflammation leading to subsequent scarring [19]. Early diagnosis and aggressive treatment with intravenous acyclovir have been helpful, even in immunocompromised patients.

6.6.4 Cerebral Vasculopathy

VZV vasculopathy is caused by productive virus infection of cerebral arteries, resulting in pathologic vascular remodeling and transient ischemic attacks, ischemic or hemorrhagic stroke [67–69]. VZV is the only virus proven to cause transient ischemic attacks and stroke in humans [89]. In addition to ischemic infarct, VZV vasculopathy has been reported to cause aneurysm, subarachnoid and cerebral hemorrhage, carotid dissection, and peripheral arterial disease [30].

While the exact incidence is unknown, it is most likely more common than previously believed given recent studies which showed that patients with HZ have a 30 % increased risk of stroke within the following year [48] and a 4.5-fold increased disk with ophthalmic-distribution zoster [55]. The neurological syndrome of thrombotic cerebral vasculopathy appears from days to months after the onset of rash, usually after ophthalmic HZ [57]. Cerebrovascular events after cervical [23, 83] or lingual [26] HZ have also been reported. Patients have clinical stroke-like symptoms, which usually suggest involvement of the territory of the ipsilateral carotid artery. The presentation of the neurological symptoms is usually monophasic, but recurrent ischemic episodes, either transient ischemic attacks or cerebral infarcts, can be seen [86]. Imaging studies reveal an infarction in the distribution of the involved vessels, and angiography or magnetic resonance angiography show the constriction or occlusion of the arteries, usually the anterior or middle cerebral artery [14, 53, 91]. In a recent series of 30 patients with VZV vasculopathy, large and small arteries were involved in 50 %, small arteries in 37 %, and large arteries in only 13 % [68]. Neuropathological evaluation reveals localized arteritis with thrombosis, infarcts, and hemorrhagic lesions in the brain, and VZV antigens or virus-like particles have been found in the wall of the affected vessels [14, 57]. Direct spread of the virus from the ganglionic reactivation sites to the arterial wall by neural pathways is regarded as the main cause of the vascular changes [53, 57, 86]. VZV must be considered as a possible cause of neurological disease in any patient with idiopathic vasculopathy. Intravenous acyclovir is recommended as treatment. One case report is published of rapidly progressive multi-infarct dementia after 2 years' duration which was successfully treated with intravenous acyclovir [89].

Recently, a new variant of VZV vasculopathy, that is, multifocal VZV vasculopathy with temporal artery infection, was reported in three case reports similar to those of giant cell arteritis [67]. All three patients presented with ischemic optic neuropathy, and VZV infection of the ipsilateral temporal artery was confirmed in all three patients. These cases raise the possibility that patients with suspected giant cell arteritis but whose arteries are pathologically negative for giant cell arteritis may have multifocal VZV vasculopathy with temporal artery infection, which was found to be true in one-fifth of 24 patients in a recent series [70].

6.6.5 Diagnosis and Treatment of the Central Nervous System Complications of Herpes Zoster

In cases with recent rash associated with neurological symptoms, there is a clinical suspicion of a zoster-associated neurological complication, but in patients without rash, the diagnosis is more difficult. If the possibility of VZV as a causative agent comes into the clinician's mind, both VZV PCR and antibodies against VZV should be determined in the CSF. MRI and, if necessary, magnetic resonance angiography are the most appropriate imaging methods. Either positive PCR finding or evidence of the intrathecal antibody production against VZV suggests that VZV is the causative agent of the syndrome. In these cases, antiviral treatment with intravenous acyclovir (10–15 mg per kilogram of body weight three times daily for 7–10 days), and in cases of cerebrovascular vasculitis or Ramsay Hunt syndrome, a short-course prednisone treatment (1 mg/kg/day) is recommended [29].

6.7 Neurological Complications of the Fetus Following Maternal Herpes Zoster

Cases of congenital varicella syndrome following maternal HZ have been reported [66]. It includes cutaneous scars, eye abnormalities, limb hypoplasia, brain abnormalities (i.e., cortical atrophy, mental retardation, and seizures) and poor sphincter control. In a series of 201 neurologically symptomatic neonates, four (2 %) of them had intrathecal production of antibodies to VZV [66]. Chickenpox or HZ had not been observed in any of the mothers during pregnancy. The neonates with intrathecal VZV antibody production had seizures as their only neurological symptom. This finding suggests that intrauterine VZV infection with neurological complications of the fetus can be acquired without cutaneous symptoms in the mother. It also suggests that the clinical spectrum of congenital VZV infection seems to be broader than expected. The diagnosis is based on antibody measurements from serum and CSF. Antiviral treatment may prevent recurrent symptoms and progression of neurological injury.

6.8 Summary

According to studies on HZ patients without symptoms of neurological complications, entrance of VZV into the CSF space is common, and MRI changes in the brainstem can also be detected quite often in patients with cranial or cervical zoster. Subclinical lower motor neuron involvement, detected by EMG, is not uncommon in patients with HZ. Aseptic meningitis caused by VZV is common, but the other forms of the CNS complications, i.e., encephalitis, myelitis, and cerebral vasculopathy, are rare. Involvement of cranial nerves can cause cranial polyneuritis, of which Ramsay Hunt syndrome (vesicular rash on the auricle and facial paralysis) is the most well known. All these complications are caused mainly by the direct spread of the virus from the infected ganglion into the spinal cord or brainstem. As the spread of VZV and development of neurological symptoms can occur also in the absence of skin lesion, one must bear in mind VZV as a possible causative agent of various neurological symptoms. An early search for VZV viral DNA by PCR or antibody in the CSF is essential for diagnosis. If either of these is found to be positive, antiviral treatment should be commenced without delay. The prognosis of neurological complications is usually good in immunocompetent patients, but neurological sequelae such as cognitive impairment after encephalitis, weakness of the lower limbs or sphincter disturbances after myelitis, or spastic hemiparesis or aphasia after cerebral infarction can remain in some patients. In the immunocompromised patients, the neurological complications of HZ can be even fatal, but even in this group of patients, early diagnosis and aggressive treatment may produce a favorable response.

References

1. Adour KK (1994) Otological complications of herpes zoster. Ann Neurol 35:S62–S64
2. Amlie-Lefond C, Mackin GA, Ferguson M, Wright RR, Mahalingham R, Gilden DH (1996) Another case of virologically confirmed zoster sine herpete, with electrophysiologic correlation. J Neurovirol 2:136–138
3. Appelbaum E, Kreps SI, Sunshine A (1962) Herpes zoster encephalitis. Am J Med 32:25–31
4. Aviel A, Marshak G (1982) Ramsay Hunt syndrome: a cranial polyneuropathy. Am J Otolaryngol 3:61–63
5. Barrett AP, Katelaris CH, Morris JGL, Schifter M (1993) Zoster sine herpete of the trigeminal nerve. Oral Surg Oral Med Oral Pathol 75:173–175
6. Broadbent WH (1866) Case of herpetic eruption in the course of branches of the brachial plexus, followed by partial paralysis in corresponding motor nerves. BMJ 2:460
7. Chernev I, Dado D (2013) Segmental zoster abdominal paresis (zoster pseudohernia): a review of the literature. PM R 5:786–790
8. Cioni R, Giannini F, Passero C, Paradiso C, Rossi S, Fimiani M, Battistini N (1994) An electromyographic evaluation of motor complications in thoracic herpes zoster. Electromyogr Clin Neurophysiol 34:125–128
9. Dayan AD, Ogul E, Graveson GS (1972) Polyneuritis and herpes zoster. J Neurol Neurosurg Psychiatry 35:170–175
10. de Ru JA, van Benthem PP (2011) Combination therapy is preferable for patients with Ramsay Hunt syndrome. Otol Neurotol 32:852–855
11. Denny-Brown D, Adams RD, Fitzgerald PJ (1944) Pathologic features of herpes zoster. A note on geniculate herpes. Arch Neurol Psychiatry 51:216–231
12. Derveaux L, Lacquet LM (1982) Hemidiaphragmatic paresis after herpes zoster. Thorax 37:870–871
13. Devinsky O, Cho ES, Petito CK, Price RW (1991) Herpes zoster myelitis. Brain 114:1181–1196
14. Devlin ME, Gliden DH, Mahalingham R, Dueland AN, Cohrs R (1992) Peripheral blood mononuclear cells of the elderly contain varicella-zoster virus DNA. J Infect Dis 165:619–622
15. Doyle PW, Gibson G, Dolman CL (1983) Herpes zoster ophthalmicus with contralateral hemiplegia: identification of cause. Ann Neurol 14:84–85
16. Easton HG (1970) Zoster sine herpete causing acute trigeminal neuralgia. Lancet 2:1065–1066
17. Echevarria JM, Casas I, Martinez-Martin P (1997) Infections of the nervous system caused by varicella-zoster virus: a review. Intervirology 40:72–84
18. Edgerton AE (1945) Herpes zoster ophthalmicus (Part II). Arch Ophthalmol 34:114–153
19. Esposito MB, Arrington JA, Murtaugh FR, Coleman JM, Sergay SM (1993) MRI of the spinal cord in a patient with herpes zoster. AJNR Am J Neuroradiol 14:203–204
20. Fabian VA, Wood B, Crowley B, Kakulas BA (1997) Herpes zoster brachial neuritis. Clin Neuropathol 16:61–64
21. Fox RJ, Galetta SL, Mahalingam R, Wellish M, Forghani B, Gilden DH (2001) Acute, chronic, and recurrent varicella zoster virus neuropathy without zoster rash. Neurology 57:351–354
22. Friedman DP (1992) Herpes zoster myelitis: MR appearance. AJNR Am J Neuroradiol 13:1404–1406

23. Fukumoto S, Kinjo M, Hokamura K, Tanaka K (1986) Subarachnoidal hemorrhage and granulomatous angiitis of the basilar artery: demonstration of the varicella-zoster-virus in the basilar artery lesions. Stroke 17:1024–1028
24. Galil K, Choo PW, Donahue DVM, Platt R (1997) The sequelae of herpes zoster. Arch Intern Med 157:1209–1213
25. Gardner-Thorpe C, Foster JB, Barwick DD (1976) Unusual manifestations of herpes zoster. J Neurol Sci 28:427–447
26. Geny C, Yulis J, Azoulay A, Brugiers P, Saint-Val C, Degos JD (1991) Thalamic infarctation following lingual herpes zoster. Neurology 41:1846
27. Gershorn A, Steinberg S, Borkowsky E, Lennette D, Lennette E (1982) IgM to varicella-zoster virus: demonstration in patients with and without clinical zoster. Pediatr Infect Dis 1:164–167
28. Gilden DH, Wright RR, Schneck SA, Gwaltney JM, Mahalingham R (1994) Zoster sine herpete, a clinical variant. Ann Neurol 35:530–533
29. Gilden DH, Kleinschmidt-De-Masters BK, LaGuardia JJ, Mahalingham R, Cohrs RJ (2000) Neurologic complications of the reactivation of varicella-zoster virus. N Engl J Med 342:635–645
30. Gilden D, Cohrs RJ, Mahalingam R, Nagel MA (2009) Varicella zoster virus vasculopathies: diverse clinical manifestations, laboratory features, pathogenesis, and treatment. Lancet Neurol 8:731–740
31. Golden LI, Deeb ZE, deFries H (1990) Atypical findings in cephalic herpes zoster polyneuritis: case reports and radiographic findings. Laryngoscope 100:494–497
32. Gondivkar S, Parikh V, Parikh R (2010) Herpes zoster oticus: a rare clinical entity. Contemp Clin Dent 1:127–129
33. Greenberg MK, McVey AL, Hayes T (1992) Segmental motor involvement in herpes zoster: an EMG Study. Neurology 42:1122–1123
34. Haanpää M, Häkkinen V, Nurmikko T (1997) Motor involvement in acute herpes zoster. Muscle Nerve 20:1433–1438
35. Haanpää M, Dastidar P, Weinberg A, Levin M, Miettinen A, Lapinlampi A, Laippala P, Nurmikko T (1998) Cerebrospinal fluid and magnetic resonance imaging findings in patients with acute herpes zoster. Neurology 51:1405–1411
36. Haanpää M, Laippala P, Nurmikko T (1999a) Pain and somatosensory dysfunction in acute herpes zoster. Clin J Pain 15:78–84
37. Haanpää M, Laippala P, Nurmikko T (1999b) Thermal and tactile perception thresholds in patients with herpes zoster. Eur J Pain 3:375–386
38. Haanpää M, Laippala P, Nurmikko T (2000) Allodynia and pinprick hypoaesthesia in acute herpes zoster, and the development of postherpetic neuralgia. J Pain Symptom Manage 20:50–58
39. Haargaard B, Lund-Andersen H, Milea D (2008) Central nervous system involvement after herpes zoster ophthalmicus. Acta Ophthalmol 86:806–809
40. Hanakawa T, Hashimoto S, Kawamura J, Nakamura M, Suenaga T, Matsuo M (1997) Magnetic resonance imaging in a patient with segmental zoster paresis. Neurology 49:631–632
41. Hokkanen L, Launes J, Poutiainen E, Valanne L, Salonen O, Sirén J, Iivanainen M (1997) Subcortical type cognitive impairment in herpes zoster encephalitis. J Neurol 244:239–245
42. Hu S, Walker M, Czartoski T, Cheng A, Forghani B, Gilden DH, Garden GA (2004) Acyclovir responsive brain stem disease after the Ramsay Hunt syndrome. J Neurol Sci 217:111–113
43. Hung CH, Chang KH, Kuo HC, Huang CC, Liao MF, Tsai YT, Ro LS (2012) Features of varicella zoster virus myelitis and dependence on immune status. J Neurol Sci 318:19–24
44. Hwang YM, Lee BI, Chung JW, Ahn JH, Kim KW, Kim DI (1991) A case of herpes zoster myelitis: positive magnetic resonance imaging finding. Eur Neurol 31:164–167
45. Izzat M, Sharma PD (1992) Isolated bilateral paralysis of the soft palate in an adult. J Laryngol Otol 106:839–840
46. Jellenik EH, Tulloch WS (1976) Herpes zoster with dysfunction of bladder and anus. Lancet 2:1219–1222

47. Jemsek J, Greenberg SB, Taber L, Harvey D, Gershorn A, Couch RB (1983) Herpes zoster-associated encephalitis: clinicopathologic report of 12 cases and review of the literature. Medicine 62:81–97
48. Kang JH, Ho JD, Chen YH, Lin HC (2009) Increased risk of stroke after a herpes zoster attack: a population-based follow-up study. Stroke 40:3443–3448
49. Kawajiri S, Tani M, Noda K, Fujishima K, Hattori N, Okuma Y (2007) Segmental zoster paresis of limbs: report of three cases and review of literature. Neurologist 13:313–317
50. Kleinschmidt-DeMasters BK, Gilden DH (2001) The expanding spectrum of herpesvirus infections of the nervous system. Brain Pathol 11:440–451
51. Kleinschmidt-DeMasters BK, Amlie-Lefond C, Gilden DH (1996) The patterns of varicella zoster virus encephalitis. Hum Pathol 27:927–938
52. Kuo MJ, Drago PC, Prooprs DW, Chavda SV (1995) Early diagnosis and treatment of Ramsay Hunt syndrome: the role of magnetic resonance imaging. J Laryngol Otol 109:777–780
53. Kuroiwa Y, Furukawa T (1981) Hemispheric infarction after herpes zoster ophthalmicus: computed tomography and angiography. Neurology 31:1030–1032
54. Lee CY, Tsai HC, Lee SS, Chen YS (2015) Orbital apex syndrome: an unusual complication of herpes zoster ophthalmicus. BMC Infect Dis 15:33
55. Lin HC, Chien CW, Ho JD (2010) Herpes zoster ophthalmicus and the risk of stroke: a population-based follow-up study. Neurology 74:792–797
56. Mainka C, Fuss B, Harmut G, Höfelmayr H, Wolff MH (1998) Characterization of viremia at different stages of varicella-zoster virus infection. J Med Virol 56:91–98
57. Martin JR, Mitchell WJ, Henken DB (1990) Neurotropic herpesviruses, neural mechanism and arteritis. Brain Pathol 1:6–10
58. Mayo DR, Booss J (1989) Varicella zoster-associated neurologic disease without skin lesions. Arch Neurol 46:313–315
59. Mazur DH, Dolin R (1978) Herpes zoster at the NIH: a 20 year experience. Am J Med 65:738–744
60. McKendrick MW, Care CC, Kudesia G, Bates CJ, Oxley MK, Eley A (1999) Is VZV reactivation a common cause of unexplained unilateral pain? Results of a prospective study of 57 patients. J Infect 39:209–212
61. Merchut MP, Gruener G (1996) Segmental zoster paresis of limbs. Electromyogr Clin Neurophysiol 36:369–375
62. Mizock BA, Bartt R, Agbemazdo B (2000) Herpes zoster oticus with pontine lesion: segmental brain-stem encephalitis. Clin Infect Dis 30:229–231
63. Mondelli M, Scarpini C, Malandrini A, Romano C (1997) Painful polyneuropathy after diffuse herpes zoster. Muscle Nerve 20:229–231
64. Murakami S, Honda N, Mizobuchi M, Nakashiro Y, Hato N, Gyo K (1998a) Rapid diagnosis of varicella zoster virus infection in acute facial palsy. Neurology 51:1202–1205
65. Murakami S, Nakashiro Y, Mizobuchi M, Hato N, Honda N, Gyo K (1998b) Varicella-zoster virus distribution in Ramsay Hunt syndrome revealed by polymerase chain reaction. Acta Otolaryngol (Stockh) 118:145–149
66. Mustonen K, Mustakangas P, Smeds M, Mannonen L, Uotila L, Vaheri A, Koskiniemi M (1998) Antibodies to varicella zoster virus in the cerebrospinal fluid of neonates with seizures. Arch Dis Child Fetal Neonatal Ed 78:F57–F61
67. Nagel MA, Gilden D (2014) Neurological complications of varicella zoster virus reactivation. Curr Opin Neurol 27:356–360
68. Nagel MA, Cohrs RJ, Mahalingam R, Wellish MC, Forghani B, Schiller A, Safdieh JE, Kamenkovich E, Ostrow LW, Levy M, Greenberg B, Russman AN, Katzan I, Gardner CJ, Häusler M, Nau R, Saraya T, Wada H, Goto H, de Martino M, Ueno M, Brown WD, Terborg C, Gilden DH (2008) The varicella zoster virus vasculopathies: clinical, CSF, imaging, and virologic features. Neurology 70:853–860
69. Nagel MA, Traktinskiy I, Azarkh Y, Kleinschmidt-DeMasters B, Hedley-Whyte T, Russman A, Van Egmond EM, Stenmark K, Frid M, Mahalingam R, Wellish M, Choe A, Cordery-Cotter R, Cohrs RJ, Gilden D (2011) Varicella zoster virus vasculopathy: analysis of virus-infected arteries. Neurology 77:364–370

70. Nagel MA, Bennett JL, Khmeleva N, Choe A, Rempel A, Boyer PJ, Gilden D (2013) Multifocal VZV vasculopathy with temporal artery infection mimics giant cell arteritis. Neurology 80:2017–2021
71. Nau R, Lantsch M, Stiefel M, Polak T, Reiber H (1998) Varicella zoster virus-associated focal vasculitis without herpes zoster: recovery after treatment with acyclovir. Neurology 51:914–915
72. Orme HT, Smith AG, Nagel MA, Bert RJ, Mickelson TS, Gilden DH (2007) VZV spinal cord infarction identified by diffusion-weighted MRI (DWI). Neurology 69:398–400
73. Ormerod IEC, Cockerell OC (1993) Guillain-Barré syndrome after herpes zoster infection: a report of 2 cases. Eur Neurol 33:156–158
74. Osaki Y, Matsubayashi K, Okumiya K, Wada T, Doi Y (1995) Polyneuritis cranialis due to varicella-zoster virus in the absence of rash. Neurology 45:2293
75. Osumi A, Tien RD (1990) MR findings in a patient with Ramsay Hunt syndrome. J Comput Assist Tomogr 14:901–903
76. Payten RJ, Dawes DK (1972) Herpes zoster of the head and neck. J Laryngol Otol 86:1031–1055
77. Peterslund NA, Hansen JA (1989) Electroencephalographic changes in patients with herpes zoster. Acta Neurol Scand 79:407–411
78. Powell KF, Wilson HG, Croxson MC, Marshall MR, Wong EH, Anderson NE, Thomas MG (1995) Herpes zoster meningoencephalitis without rash: varicella zoster virus DNA in CSF. J Neurol Neurosurg Psychiatry 59:198–199
79. Ragozzino MW, Melton LJ, Kurland LT, Chu CP, Perry HO (1982) Population-based study on herpes zoster and its sequelae. Medicine 61:310–316
80. Reda H, Watson JC, Jones LK Jr (2012) Zoster-associated mononeuropathies (ZAMs): a retrospective series. Muscle Nerve 45:734–739
81. Reske-Nielsen E, Oster S, Pedersen B (1986) Herpes zoster ophthalmicus and the mesencephalic nucleus. Acta Pathol Microbiol Immunol Scand 94:263–269
82. Robillard RB, Hilsinger AL Jr, Adour KK (1986) Ramsay Hunt facial paralysis: clinical analysis of 185 patients. Otolaryngol Head Neck Surg 95:292–297
83. Ross MH, Abend WK, Schwartz RB, Samuels MA (1991) A case of C2 herpes zoster with delayed bilateral pontine infarction. Neurology 41:1685–1686
84. Rothschild MA, Drake W III, Scherl M (1994) Cephalic zoster with laryngeal paralysis. ENT J 73:850–852
85. Ruppenthal M (1980) Changes of the central nervous system in herpes zoster. Acta Neuropathol (Berl) 52:59–68
86. Sarazin L, Duong H, Bourgouin PM, Melanson M, Chalk C, Richardson J, Vézina JL (1995) Herpes zoster vasculitis: demonstration by MR angiography. J Comput Assist Tomogr 19:624–627
87. Sartoretti-Schefer S, Kollias S, Valavanis A (1999) Ramsay Hunt syndrome associated with brain stem enhancement. Am J Neuroradiol 20:278–280
88. Schmidbauer M, Budka H, Pilz P, Kurata T, Hondo R (1992) Presence, distribution and spread of productive varicella zoster virus infection in nervous tissue. Brain 115:383–398
89. Silver B, Nagel MA, Mahalingam R, Cohrs R, Schmid DS, Gilden D (2012) Varicella zoster virus vasculopathy: a treatable form of rapidly progressive multi-infarct dementia after 2 years' duration. J Neurol Sci 323:245–247
90. Thomas JE, Howard FM (1972) Segmental zoster paresis – a disease profile. Neurology 22:459–466
91. Tien RD, Felsberg GJ, Osumi AK (1993) Herpesvirus infections of the CNS: MR findings. AJR Am J Roentgenol 161:167–176
92. Wackym PA (1997) Molecular temporal bone pathology: II. Ramsay Hunt syndrome (herpes zoster oticus). Laryngoscope 107:1165–1175
93. Watson CPN, Deck HJ, Morshead C, Van der Kooy D, Evans RJ (1991) Post-herpetic neuralgia: further post-mortem studies of cases with and without pain. Pain 44:105–117

94. Weitzman D, Shavit O, Stein M, Cohen R, Chodick G, Shalev V (2013) A population based study of the epidemiology of herpes zoster and its complications. J Infect 67:463–469
95. Worme M, Chada R, Lavallee L (2013) An unexpected case of Ramsay Hunt syndrome: case report and literature review. BMC Res Notes 6:337
96. Xiao Z, Lu Z, Pan S, Liang J, Liu Z (2015) Orbital apex syndrome and meningoencephalitis: a rare complication of herpes zoster. Int J Clin Exp Med 8:14260–14263
97. Yoshida T, Fujisaki N, Nakachi R, Sueyoshi T, Suwazono S, Suehara M (2014) Persistent hiccups and vomiting with multiple cranial nerve palsy in a case of zoster sine herpete. Intern Med 53:2373–2376

Chapter 7
The Role of Varicella Zoster Virus in Giant Cell Arteritis

Maria A. Nagel and Don Gilden

Abbreviations

APD	Afferent pupillary defect
CSF	Cerebrospinal fluid
GCA	Giant cell arteritis
HSV-1	Herpes simplex virus type 1
ION	Ischemic optic neuropathy
OD	Right eye
OS	Left eye
TA	Temporal artery
VZV	Varicella zoster virus

7.1 Introduction

Varicella zoster virus (VZV) is an exclusively human neurotropic alphaherpesvirus. Primary VZV infection causes varicella (chickenpox), after which virus becomes latent in ganglionic neurons along the entire neuraxis. As VZV-specific cell-mediated immunity wanes in elderly and immunocompromised individuals, virus reactivates, resulting in herpes zoster (shingles). Zoster may be complicated by persistent dermatomal distribution pain for months or years (postherpetic neuralgia), with increasing age as a significant risk factor. Zoster may be further complicated

M.A. Nagel, MD (✉)
Department of Neurology, University of Colorado School of Medicine,
12700 E. 19th Avenue, Mail Stop B182, Aurora, CO 80045, USA
e-mail: maria.nagel@ucdenver.edu

D. Gilden, MD
Departments of Neurology and Immunology and Microbiology, University of Colorado School of Medicine, 12700 E. 19th Avenue, Mail Stop B182, Aurora, CO 80045, USA
e-mail: don.gilden@ucdenver.edu

C.P.N. Watson et al. (eds.), *Herpes Zoster: Postherpetic Neuralgia and Other Complications,* DOI 10.1007/978-3-319-44348-5_7

by other serious neurological diseases such as meningoencephalitis, cerebellitis, isolated or multiple cranial nerve palsies (polyneuritis cranialis), myelitis, and vasculopathy, as well as multiple ocular disorders. Importantly, these neurological and ocular disorders can also develop in the absence of zoster rash. The incidence and severity of zoster is best viewed as a continuum in immunodeficient individuals, ranging from a natural decline in VZV-specific immunity with advancing age to more serious host immune deficits encountered in organ transplant recipients, as well as patients with cancer or AIDS or on immunosuppressive therapies.

VZV is the only human virus that has been shown to replicate in arteries and cause stroke (VZV vasculopathy). VZV reactivation from cranial nerve ganglia is followed by transaxonal spread of virus to cerebral arteries, leading to productive virus infection, inflammation, pathological vascular remodeling, and stroke [1, 8]. Examination of virus-infected arteries from patients with virologically confirmed VZV vasculopathy reveals transmural inflammation, medial damage, multinucleated and epithelioid cells, disrupted internal elastic lamina, and a thickened intima composed of myofibroblasts [9]. Inflammatory cells in VZV-infected arteries consist predominantly of CD4 and CD8 T cells and macrophages, with neutrophils more prominent during early disease [10].

Giant cell arteritis (GCA) is the most common systemic cause of vasculitis in the elderly and is characterized by head/scalp pain, temporal artery (TA) tenderness, jaw claudication, vision loss, and elevated inflammatory markers (erythrocyte sedimentation rate (ESR) and C-reactive protein). The gold standard for diagnosis is TA biopsy which reveals transmural inflammation, medial necrosis, and macrophage/epithelioid cells in noncontiguous "skip" regions; importantly, many patients with clinical features of GCA are "biopsy negative" which has been attributed to missed pathology due to skip regions. Standard treatment for GCA is long-term corticosteroids, although up to 50 % of patients relapse during taper or progress to vision loss or stroke. The cause of GCA was unknown. Importantly, the pathologies of intracerebral VZV vasculopathy and GCA are identical, indicating that GCA is primarily extracerebral VZV vasculopathy. Recent correlative clinicopathological-virological case reports demonstrated an association of VZV infection with GCA, thus expanding the spectrum of intracerebral VZV vasculopathy to extracerebral VZV vasculopathy.

7.2 Case Reports of GCA and VZV Vasculopathy

One of the first cases that emerged was of an 80-year-old man who developed left ophthalmic-distribution zoster and ipsilateral ischemic optic neuropathy (ION) [16]. A TA biopsy revealed inflammation, but not the more extensive histopathology characteristic of GCA, as well as abundant VZV antigen in the arterial adventitia and less in the media. No improvement was seen on corticosteroid treatment; when VZV antigen was reported in the TA, the patient was treated with intravenous acyclovir and vision recovered. The final diagnosis was VZV-induced ION and subclinical TA infection after zoster rash. The presence of VZV predominantly in the adventitia supported the notion that virus entered arteries through nerve fibers within the adventitia and spread transmurally.

Another patient was a 75-year-old woman who developed periorbital pain and blurred vision OS without zoster rash [11]. Visual acuity was 20/40 OD, 20/400 OS, with a mild left relative afferent pupillary defect (APD); the left optic nerve was swollen and hyperemic with peripapillary flame hemorrhages. Erythrocyte sedimentation rate (ESR) was elevated (124 mm/h). She was treated with intravenous methylprednisolone, 250 mg q 6 h, with improvement of headaches and vision by day 3. Lab results revealed an ESR of 98 mm/hr and C-reactive protein of 1.40 mg%; rheumatoid factor, ANA, and ANCA titers were negative. On day 4, left TA biopsy was GCA negative and steroids were changed to oral prednisone, 60 mg daily. On day 7, brain MRI with gadolinium was negative. On day 9, pain and vision worsened. On day 11, orbital CT and head CT angiography were negative. On day 15, visual acuity was 20/400 OS with relative left APD. On day 17, the OS became blind and nonreactive to light; fundus was obscured by vitreous hemorrhage. Cerebrospinal fluid (CSF) contained 8 WBCs/mm^3, protein 72 mg%, and glucose 54 mg%. CSF cultures for bacteria, fungi, AFB, and cytology were negative. VZV ION was considered and intravenous acyclovir was started, 10 mg/kg q 8 h for 7 days. On day 31, CSF contained anti-VZV IgG but not anti-herpes simplex virus type 1 (HSV-1) IgG antibody, and serum-to-CSF ratio of anti-VZV IgG was reduced compared to ratios for total IgG and albumin, indicating intrathecal synthesis of antibody to VZV. Immunohistochemistry and pathology revealed VZV antigen and neutrophils in the original left TA specimen. On day 31, she was treated with oral valacyclovir, 1 g TID for 6 weeks; prednisone was reduced to 20 mg daily and tapered to 5 mg/week. Six weeks later, pain resolved and visual acuity improved to finger-counting. The left optic nerve was pale, with clear margins and resolution of hemorrhage. Overall, another case of VZV ION with subclinical TA infection was found similar to the previous case, yet without a history of zoster rash.

Another remarkable case of VZV vasculopathy with clinical features of GCA was a 54-year-old diabetic woman who developed an ION followed by acute retinal necrosis and multiple areas of focal venous beading [6]. The vitreous fluid contained amplifiable VZV DNA, but not HSV-1, CMV, or toxoplasma DNA. The patient also complained of jaw claudication and intermittent scalp pain, prompting a TA biopsy that was pathologically negative for GCA but notable for the presence of VZV antigen. The case added to the clinical spectrum of multifocal VZV vasculopathy in patients with TA biopsies that were histopathologically negative for GCA.

A more comprehensive study to address the incidence of VZV infection in archived GCA biopsy-negative TAs from subjects with clinically suspected GCA revealed VZV, but not HSV-1 antigen, in 5/24 (21 %) TAs of these patients [12]. Thirteen normal TAs did not contain VZV or HSV-1 antigen. Interestingly, all five subjects whose TAs contained VZV antigen presented with clinical and laboratory features of GCA, including early visual disturbances. Thus, it was evident that multifocal VZV vasculopathy can present with the full spectrum of clinical features and laboratory abnormalities characteristic of GCA and that in GCA-negative/VZV-positive TAs, viral antigen predominated in the arterial adventitia.

These studies demonstrating a link between biopsy-negative GCA and VZV vasculopathy were followed by a final important case linking biopsy-positive GCA and VZV vasculopathy. During the continuing search for VZV antigen in GCA-negative TAs, the TA of one subject revealed abundant VZV antigen and VZV DNA in multiple skip areas spanning 350 μm, as well as in skeletal muscle adjacent to the infected TA; pathological analysis of sections adjacent to those containing viral antigen revealed inflammation involving the arterial media and abundant multinucleated giant cells characteristic of GCA [13]. The detection of VZV followed by additional pathological studies led to a change in diagnosis from biopsy-negative GCA to biopsy-positive GCA. In three other such cases, the detection of VZV in a GCA-negative TA led to more extensive pathological studies and also a change in diagnosis to GCA.

7.3 Retrospective Analysis of Archived GCA-Positive and GCA-Negative Temporal Arteries

The case studies demonstrating a link between biopsy-negative and biopsy-positive GCA with VZV vasculopathy led to the hypothesis that VZV infection of arteries triggers the inflammatory cascade of GCA. To test this hypothesis, formalin-fixed, paraffin-embedded GCA-positive TA biopsies (50 sections/TA), including adjacent skeletal muscle, and normal TA biopsies from subjects >50 years of age were examined by immunohistochemistry for the presence and distribution of VZV antigen, as well as by electron microscopy for virions; sections adjacent to those containing VZV antigen were examined by hematoxylin-eosin staining. DNA extracted from VZV antigen-positive slides were analyzed by PCR for VZV DNA. Consistent with prior studies of the association with VZV in GCA-positive and GCA-negative TAs [2, 14], a cumulative study of GCA-positive, GCA-negative, and normal TAs revealed VZV antigen in 73/104 (70 %) GCA-positive and 58/100 (58 %) GCA-negative TAs compared to 11/61 (18 %) normal TAs ($p < 0.0001$) [4]. In GCA-positive and GCA-negative TAs, viral antigen was predominantly in the adventitia and present in skip areas. Of 58 GCA-positive, VZV antigen-positive TAs examined, all contained cellular DNA and 23 (40 %) contained VZV DNA. Of 58 GCA-negative, VZV antigen-positive TAs examined, 51 contained cellular DNA, 9 (18 %) of which contained VZV DNA. Nine of 11 normal VZV antigen-positive TAs contained cellular DNA, of which 3 (33 %) contained VZV DNA. Adventitial inflammation was seen adjacent to viral antigen in 26 (52 %) of 58 GCA-negative subjects whose TAs contained VZV antigen; no inflammation was seen in normal TAs containing VZV antigen. VZ virions were also found in a GCA-positive TA that contained VZV antigen (Fig. 7.1), indicating productive virus infection [2]. The finding of VZV antigen predominantly in adventitia of GCA-positive TAs and the presence of VZV with inflammation in the adventitia of GCA-negative TAs support the notion that after VZV reactivation from

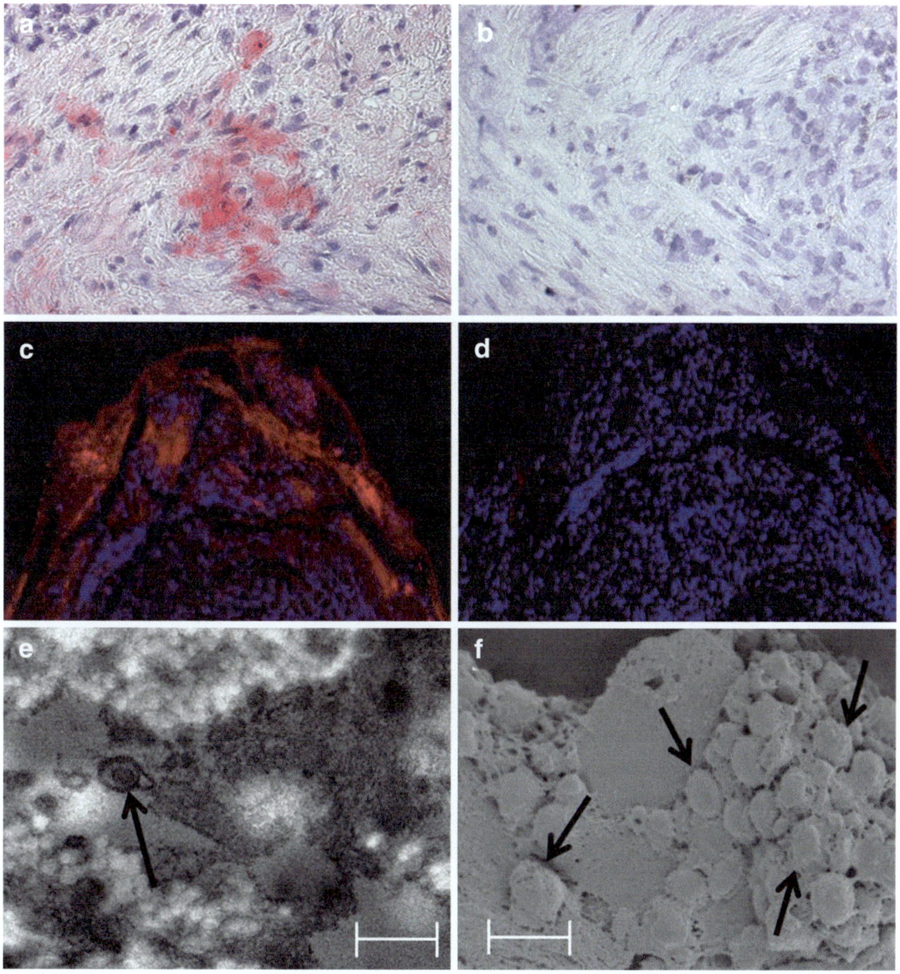

Fig. 7.1 Immunofluorescence staining and ultrastructural imaging of VZV-infected temporal artery. Immunohistochemical staining with rabbit anti-VZV IE63 antibody revealed VZV antigen in the media (**a**, *pink color*), but not after staining with rabbit anti-HSV-1 antibody (**b**). Immunofluorescence staining with a mouse anti-VZV IgG antibody with different specificity from that used for immunohistochemistry revealed VZV antigen in the adventitia (**c**, *red color*), but not when primary antibody was omitted (**d**). Sections adjacent to those containing VZV antigen revealed an enveloped virus particle upon examination by transmission electron microscopy (**e**, *arrow*) and a cluster of virus particles in the adventitia egressing through an outer cell wall (**f**, *arrows*) by scanning electron microscopy. Viral particles appear slightly larger than 200 nm because they were sputter-coated with a gold alloy. Scale bars in **e** and **f** = 300 nm (From Gilden et al. [2]; reprinted with permission from Wolters Kluwer Health)

ganglia, virus spreads transaxonally to the arterial adventitia followed by transmural spread with accompanying inflammation. The tempo and evolution of GCA (transmural inflammation and necrosis with giant and/or epithelioid cells) after virus infection of the adventitia and adventitial inflammation remains to be determined.

7.4 Granulomatous Aortitis

Finally, because granulomatous arteritis characterizes the pathology not only of intracerebral VZV vasculopathy and GCA but also of granulomatous aortitis, and because intracerebral VZV vasculopathy and GCA are strongly associated with productive VZV infection in cerebral and temporal arteries, respectively, we studied human aortas for VZV antigen and VZV DNA. Using three different anti-VZV antibodies, we found VZV antigen in all of 11 aortas with pathologically verified granulomatous aortitis, in one case of nongranulomatous aortitis, and in 5/18 (28 %) control aortas obtained at autopsy [5]. The presence of VZV antigen in granulomatous aortitis was highly significant ($P = 0.0001$) as compared to control aortas, in which VZV antigen was never associated with pathology, indicating subclinical reactivation. VZV DNA was found in most aortas containing VZV antigen. The frequent clinical, radiological, and pathological involvement of the aorta and other large vessels in GCA patients correlates with the significant detection of VZV in granulomatous aortitis.

Granulomatous aortitis and other cardiovascular diseases due to VZV are not surprising since virus can reactivate from dorsal root and autonomic ganglia and travel transaxonally to cardiac vessels and the myocardium. In one case report, a 60-year-old man who abused corticosteroids developed zoster and died 5 months later; at autopsy, VZV antigen was widespread in arteries and organs, including the coronary arteries, aorta, and the bundle of His. More generally, a self-controlled case series study showed a 2.4-fold increased risk of ischemic stroke and 1.7-fold increased risk of myocardial infarction within 2 weeks of zoster that gradually resolved over 6 months [7].

7.5 Treatment of GCA

As in the pathology of GCA, VZV antigen and associated inflammation are usually patchy and detected in noncontiguous skip areas. The research-focused evaluation of 50 sections per TA biopsy is not practical for routine diagnostic work-up, and, while immunohistochemical evaluation is worthwhile, a negative result does not rule out GCA or VZV reactivation. Rather, the clinician must consider the patient's clinical and laboratory presentation.

As for treatment of GCA, no trials have yet been conducted to determine whether antivirals and steroids confer additional benefit to steroids alone. Although many GCA patients improve with steroids, there are numerous reports of GCA patients who not only fail steroids but also develop more disseminated VZV vasculopathy and die [3, 15]. If VZV triggers the immunopathology of GCA, antiviral treatment is likely to confer additional benefit to corticosteroid-treated GCA patients and may contribute to shortened corticosteroid treatment and its concurrent adverse effects, although the optimal antiviral regimen remains to be determined. Indeed, in one case report of a patient with GCA and Takayasu arteritis who had failed corticosteroids, treatment with intravenous acyclovir led to dramatic clinical improvement consistent with a causative role for VZV. We currently treat GCA with prednisone, 1 mg/kg, along with valacyclovir, 1 gm three times daily. If the patient improves after 4–6 weeks, we recommend tapering prednisone while continuing antiviral agents for another 4–6 weeks. Long-term antiviral drugs are far less risky than long-term steroids.

7.6 Conclusion

Virological analysis of TAs from patients with clinically suspected GCA whose TA biopsies were pathologically positive or negative revealed the presence of VZV in most TA biopsies, particularly in skip areas that correlate with adjacent GCA pathology or adventitial inflammation, supporting the notion that the majority, if not all cases of GCA, are due to VZV. The presence of VZV in GCA-positive and GCA-negative TAs reflects the possible role of VZV in triggering the immunopathology of GCA and indicates the need for treatment in both groups of patients with clinically suspected GCA with antiviral drugs in addition to corticosteroids. Whether oral antiviral agents and steroids are as effective as intravenous acyclovir and steroids, as well as the dosage and duration of treatment, remains to be determined.

Acknowledgments Supported in part by National Institutes of Health grants NS093716 (Gilden), AG032958 (Gilden and Nagel), and NS094758 (Nagel). The authors thank Marina Hoffman for editorial review and Cathy Allen for manuscript preparation.

References

1. Gilden DH, Kleinschmidt-DeMasters BK, Wellish BS et al (1996) Varicella zoster virus, a cause of waxing and waning vasculitis: The New England Journal of Medicine case 5-1995 revisited. Neurology 47:1441–1446
2. Gilden D, White T, Khmeleva N et al (2015) Prevalence and distribution of VZV in temporal arteries of patients with giant cell arteritis. Neurology 84:1948–1955
3. Gilden D, White T, Galetta SL et al (2015) Widespread arterial infection by varicella zoster virus explains refractory giant cell arteritis. Neurol Neuroimmunol Neuroinflamm 2:e125. doi:10.1212/NXI.0000000000000125
4. Gilden D, White T, Khmeleva N, Boyer PJ, Nagel MA (2016) VZV in biopsy-positive and -negative giant cell arteritis: analysis of 100+ temporal arteries. Neurol Neuroimmunol Neuroinflamm 3:e216. doi:10.1212/NXI.0000000000000216
5. Gilden D, White T, Boyer PJ et al (2016) Varicella zoster virus infection in granulomatous arteritis of the aorta. J Infect Dis 213(12):1866–1871 Pil: jiw101
6. Mathias M, Nagel MA, Khmeleva N et al (2013) VZV multifocal vasculopathy with ischemic optic neuropathy, acute retinal necrosis and temporal artery infection in the absence of zoster rash. J Neurol Sci 325:180–182

7. Minassian C, Thomas SL, Smeeth L, Douglas I, Brauer R, Langan SM (2015) Acute cardiovascular events after herpes zoster: a self-controlled case series analysis in vaccinated and unvaccinated older residents of the United States. PLoS Med 12(12):e1001919. doi:10.1371/journal.pmed.1001919
8. Nagel MA, Cohrs RJ, Mahalingam R et al (2008) The varicella zoster virus vasculopathies: clinical, CSF, imaging, and virologic features. Neurology 70:853–860
9. Nagel MA, Traktinskiy I, Azarkh Y et al (2011) Varicella zoster virus vasculopathy: analysis of virus-infected arteries. Neurology 77:364–370
10. Nagel MA, Traktinskiy I, Stenmark KR, Frid MG, Choe A, Gilden D (2013) Varicella zoster virus vasculopathy: immune characteristics of virus-infected arteries. Neurology 80:62–68
11. Nagel MA, Russman AN, Feit H et al (2013) VZV ischemic optic neuropathy and subclinical temporal artery infection without rash. Neurology 80:220–222
12. Nagel MA, Bennett JL, Khmeleva N et al (2013) Multifocal VZV vasculopathy with temporal artery infection mimics giant cell arteritis. Neurology 80:2017–2021
13. Nagel MA, Khmeleva N, Boyer PJ et al (2013) Varicella zoster virus in the temporal artery of a patient with giant cell arteritis. J Neurol Sci 335:228–230
14. Nagel MA, White T, Khmeleva N et al (2015) Analysis of varicella zoster virus in temporal arteries biopsy-positive and -negative for giant cell arteritis. JAMA Neurol 72:1281–1287
15. Nagel MA, Lenggenhager D, White T et al (2015) Disseminated VZV infection and asymptomatic VZV vasculopathy after steroid abuse. J Clin Virol 66:72–75
16. Salazar R, Russman AN, Nagel MA et al (2011) Varicella zoster virus ischemic optic neuropathy and subclinical temporal artery involvement. Arch Neurol 68:517–520

Chapter 8
Herpes Zoster and Vascular Risk

Charlotte Warren-Gash and Judith Breuer

Abbreviations

AIS	Acute ischaemic stroke
CI	Confidence interval
CPRD	Clinical Practice Research Datalink
EHR	Electronic health record
HR	Hazard ratio
HZ	Herpes zoster
HZO	Herpes zoster ophthalmicus
IR	Incidence ratio
IRR	Incidence rate ratio
MI	Myocardial infarction
PVD	Peripheral vascular disease
SAH	Subarachnoid haemorrhage
SVV	Simian varicella virus
THIN	The Health Improvement Network
TIA	Transient ischaemic attack
VZV	Varicella zoster virus

C. Warren-Gash (✉)
Institute of Health Informatics, University College London,
222 Euston Road, London NW1 2DA, UK

Faculty of Epidemiology and Population Health, London School of Hygiene
and Tropical Medicine, Keppel Street, London WC1E 7HT, UK
e-mail: c.warren-gash@ucl.ac.uk

J. Breuer
Division of Infection & Immunity, University College London,
Cruciform Building, London WC1E 6BT, UK
e-mail: j.breuer@ucl.ac.uk

C.P.N. Watson et al. (eds.), *Herpes Zoster: Postherpetic Neuralgia and Other Complications*, DOI 10.1007/978-3-319-44348-5_8

8.1 Introduction

Herpes zoster is a substantial global public health problem: around 95 % of people have been infected with varicella zoster virus (90 % by the age of 15 years [1]). After primary infection the virus establishes lifelong latency, mainly in trigeminal and dorsal root ganglia [2]. Reactivation results in herpes zoster – a painful blistering condition – which occurs in 25–35 % of people, rising to nearly half of those aged over 80 years. In the United States, there are an estimated one million cases per year, such that herpes zoster is the single most common infection of the nervous system [3]. In England and Wales, there are an estimated 88,650 (95 % C.I. 65,000–113,000) cases annually in immunocompetent people aged ≥60 years, of whom 18,200 (95 % C.I. 13,500–23,300) experience ongoing pain at 3 months and 1750 (95 % C.I. 1300–2200) are hospitalised [4].

8.1.1 Complications of Herpes Zoster

Although the commonest complication is post-herpetic neuralgia, herpes zoster is associated with a range of inflammatory neurological and ocular complications. There is also increasing epidemiological evidence that herpes zoster may trigger acute vascular events among certain populations. These events may be local to the site affected by VZV reactivation, e.g. stroke following herpes zoster ophthalmicus, or may comprise more distant systemic complications such as acute coronary syndrome. This is in keeping with a growing body of literature highlighting the role of infections as vascular triggers, e.g. acute respiratory infections such as influenza and urinary tract infections are associated with increased myocardial infarction risk, especially among the elderly [5, 6]. Such insights raise the intriguing possibility that intervening to prevent or treat infections may reduce the risk of vascular complications. This is discussed further in Sect. 7.

8.2 Study Designs

Before reviewing the evidence, we first highlight key features of the major observational study designs used so far to investigate the relationship between VZV reactivation and vascular events. We consider the advantages and disadvantages of the different study types especially in the context of research using electronic health records (EHRs).

(i) Self-controlled case series studies

Originally developed from the cohort design to model the effect of transient exposures such as vaccines on short-term health outcomes such as risk of adverse effects in time periods following vaccination, the self-controlled case series is based upon within-person comparisons [7]. It generates a measure of effect known as an incidence ratio through conditional Poisson regression. The incidence ratio relates the incidence of outcome events such as stroke occurring in time periods immediately following exposure, e.g. to herpes zoster to the incidence of outcomes occurring in baseline time periods for each individual (Fig. 8.1).

As each person acts as their own control, this implicitly controls for the effect of fixed confounders such as gender and genetic factors. This is a major advantage for EHR studies: in routinely collected health data, many potential confounders may be unmeasured or poorly quantified, leading to difficulty controlling for these factors during analysis. With self-controlled case series, it is still possible to control for time-varying confounders in models. The use of within-person comparisons also helps to reduce selection biases associated with an inappropriate choice of control group that may distort the relationship between exposure and outcome in the study population compared to the general population. Although self-controlled case series is relatively statistically efficient compared to the cohort design, the statistical power of the study to detect an effect should be considered when interpreting results. As with any other observational study design, care must be taken to avoid bias due to misclassification of either outcome or exposure.

(ii) Cohort studies

Cohort studies may be used to investigate the effect of an exposure such as herpes zoster on an outcome such as stroke. In these studies, groups of subjects exposed and unexposed to herpes zoster are followed over time and incidence rates of vascular events compared between groups using regression modelling. Prospective cohort studies follow exposed and unexposed subjects who are initially disease-free at

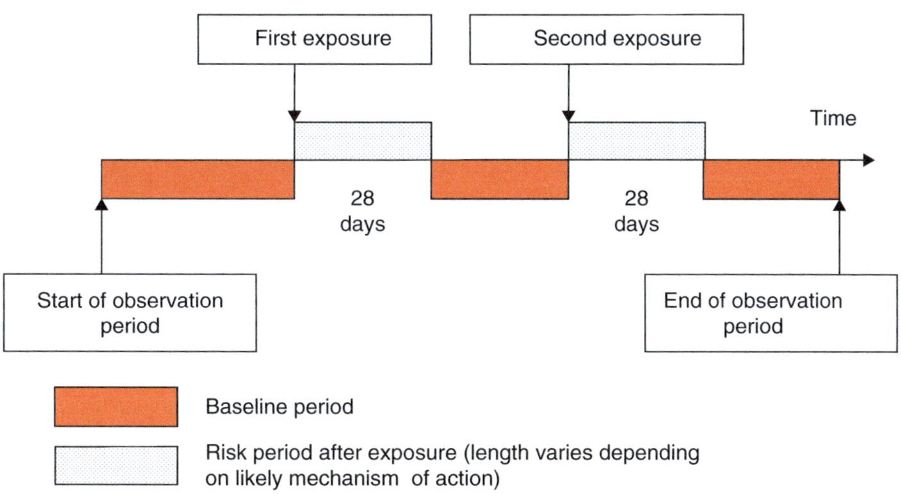

Fig. 8.1 Timeline of one individual in a self-controlled case series showing risk periods and baseline periods of follow-up

baseline into the future, whereas historical cohort studies rely on the availability of adequate records from the past to identify outcome events occurring in exposed and unexposed groups (Fig. 8.2).

Cohort studies are useful to investigate the effect of an exposure on multiple outcomes, for establishing the temporal relationship between an exposure and an outcome (here ensuring that herpes zoster occurred before the vascular complication and not vice versa), providing a direct measure of outcome incidence and, to some extent, minimising biases compared to other designs such as the case control study. There are limitations, particularly with cohort studies that use routinely collected EHR data: it may be difficult to control for all desired confounders because of missing data, information bias due to misclassification of exposure or outcome due to inaccurate recording could also distort the effect seen and differential loss to follow-up between exposed and unexposed groups in a cohort could introduce selection bias.

(iii) Case control studies

Case control studies have a retrospective design, in which a comparison is performed between 'cases', who are subjects with the disease or outcome of interest, and 'controls' or subjects who do not have the outcome under study. In a case control study, the presence or absence of previous exposure is compared between cases and controls, typically using logistic regression analysis (Fig. 8.3).

Some advantages of case control studies are that they allow investigation of rare outcomes and are relatively quick and cheap to conduct. They may enable investigation of exposures with long latent periods. Disadvantages of case control studies

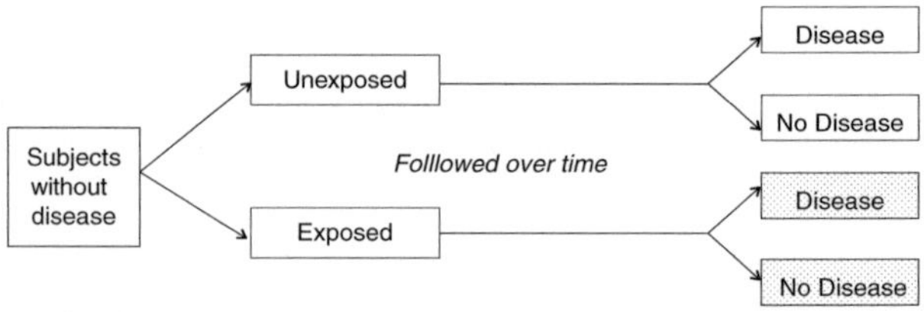

Fig. 8.2 Cohort study timeline

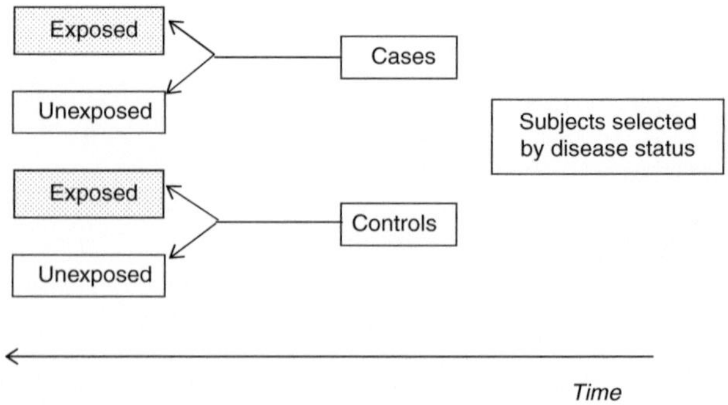

Fig. 8.3 Case control study timeline

include the potential for bias, the difficulty investigating rare exposures and the risk of reverse causality in which it is not possible to establish whether the exposure truly preceded the outcome.

8.3 Epidemiology of Shingles and Stroke

The majority of epidemiological evidence for a link between shingles and acute cerebrovascular events such as stroke or transient ischaemic attack (TIA) comes from observational studies conducted in large databases of electronic health records. A summary of these studies is given below.

(i) Self-controlled case series studies

In the UK, Langan and colleagues conducted a self-controlled case series study using primary care records from the UK Clinical Practice Research Datalink (CPRD) to investigate the age-adjusted relative incidence of stroke in time periods following herpes zoster compared to other time periods. Data from 6584 individuals showed a 63 % increase in stroke incidence after herpes zoster in weeks 1–4 (adjusted IR 1.63, 95 % C.I. 1.32–2.02), which gradually fell over time to baseline after week 26. A stronger effect was observed with HZO with the highest adjusted IR seen 5–12 weeks after zoster (IR 3.38, 95 % C.I. 2.18–5.24). The use of within-person comparisons implicitly controlled for the effect of fixed confounders in this analysis, which was an advantage compared to other studies [8]. Although studies using EHRs risk possible misclassification of outcome and exposure, this is likely to be random and therefore to dilute the size of any effects seen rather than to generate 'false positive' findings.

Recently the same group conducted a similar analysis of ischaemic stroke after herpes zoster using administrative claims data from Medicare – a health insurance plan that covers around 15 % of the US population, mainly those aged ≥ 65 years. Self-controlled case series analysis based on 42,954 individuals showed a marked transient increase in the relative incidence of stroke after herpes zoster – adjusted IR 2.37 (95 % C.I. 2.17–2.59) in week one, followed by a gradual resolution over 6 months [9]. Analysing data for cases of HZO separately gave similar results: adjusted IR 2.73 (95 % C.I. 2.22–3.35) for week one. While the self-controlled case series design has the advantages described above, even in a large database, there was limited power for some secondary analyses, e.g. differentiating the effect of HZ on haemorrhagic versus ischaemic strokes.

Another self-controlled case series study focussing on UK primary care data from four different databases (CPRD, THIN, QResearch and IMS Disease Analyser Mediplus) investigated the relative incidence of stroke after chickenpox in children and adults. Stroke incidence was increased in the 60 children included in the analysis, IR 4.07 (95 % C.I. 1.96–8.45) for months 1–6, and to a lesser extent for the 500 adults, IR 2.13 (95 % C.I. 1.05–4.36). No effect was seen on TIA and there was no longer-term effect past 6 months [10]. As in the HZ studies above, this study was not limited by between-person confounding, and although the sample size was modest, a clear effect was demonstrated.

In summary, a transient increase in risk of stroke was shown consistently in the 6 months following chickenpox and herpes zoster in self-controlled case series studies using both UK and US primary care data. It was unclear whether HZO is associated with a greater stroke risk than HZ as the two studies comparing the effects of HZ and HZO on stroke gave conflicting results. There is no evidence from these studies that stroke risk persisted beyond 6 months.

(ii) Cohort studies

In Taiwan a historical cohort of 7760 patients who had received treatment for acute herpes zoster between 1997 and 2001 was matched with 23,280 randomly selected subjects without a record of herpes zoster treatment using data from a national insurance database. In a period of 1 year following shingles, the adjusted hazard ratio for stroke after herpes zoster affecting any site was 1.31 (95 % C.I. 1.06–1.60). The corresponding adjusted HR for stroke after HZO was 4.28 (95 % C.I. 2.01–9.03) [11]. In a similar study by the same group using the Taiwan National Health Insurance Research Database, 658 patients with a diagnosis of HZO in 2003 and 2004 were compared to a cohort of 1974 randomly selected patients matched on age and sex. After adjusting for demographic characteristics, some comorbidities and medications, the HR for stroke after HZO was 4.52 (95 % C.I. 2.45–8.33) [12]. These studies were limited by a lack of adjustment for other important confounders such as obesity. They did not provide information on the timing of strokes in relation to HZ.

In the UK, a historical cohort of 106,601 HZ patients and 213,202 non-HZ patients matched for age, sex and general practice was identified in primary care data from the Health Improvement Network (THIN) and followed over 24 years. The adjusted HR for TIA was increased in all patients with herpes zoster (HR 1.15 (95 % C.I. 1.09–1.21)) but it was not increased for stroke 1.02 (95 % C.I. 0.98–1.07). Restricting to those with HZ aged under 40 years revealed a significant increase in the HR for stroke (1.74 (95 % C.I. 1.13–2.66)) and a higher HR for TIA (2.42 (95 % C.I. 1.34–4.36)) [13]. This study is not directly comparable to other studies described so far as it focussed on long-term rather than short-term effects of HZ. It is also difficult to exclude the potential for residual confounding in cohort studies based on routine clinical data.

In Denmark, a population cohort was devised of 4,620,980 adults resident in Denmark between 1995 and 2008. 117,926 patients with HZ were identified through prescriptions for antiviral drugs using the Danish National Register of Medicinal Product Statistics. Stroke/TIA (a composite outcome) was identified from hospital discharge diagnosis codes. The IRR for stroke after HZ was 2.27 (95 % C.I. 1.83–2.82) for the time period up to 14 days. For 14 days to 1 year, the IRR was 1.17 (95 % C.I. 1.09–1.24), falling to 1.05 (95 % C.I. 1.02–1.09) for >1 year. As with the previous study, the IRR was highest in the youngest age group <40 years (IRR 5.52 (95 % C.I. 1.38–22.1)) compared to the oldest group (IRR 2.06 (95 % C.I. 1.62–2.62)) for the time period <14 days [14]. As this study did not assess HZ diagnoses directly, instead using prescription for antiviral drugs as a proxy, it assessed the effect of treated zoster on stroke. Risk of misclassification of exposure was likely to be significant in this study: the exposed group included individuals with any other conditions treated with the same antiviral agents, such as herpes simplex virus, and the unexposed could have included those with untreated HZ. As for the other cohort studies, residual confounding may have influenced results.

In Sweden, a population-based study used linked data from a primary healthcare register and Swedish patient register (for inpatient and outpatient hospital visits) covering the second largest county in Sweden to identify 13,296 patients with a HZ clinical code between 2008 and 2010. Age- and sex-adjusted stroke incidence rates were calculated for these patients and compared to rates in the general population, giving an overall IRR of 1.34 (95 % C.I. 1.12–1.62). When restricted to patients aged under 40 years, the IRR rose to 10.3 (95 % C.I. 3.87–27.6) [15]. In a nested validation study, 86 % of 112 randomly sampled HZ cases defined clinically could be verified through records review. Using a general population comparison group hindered the ability to control adequately for confounders and may have introduced selection bias.

Overall, most cohort studies showed an increased short-term risk of stroke after HZ, with some evidence that HZO had a greater effect on stroke than HZ at other sites. In three of the studies, the effect of HZ on stroke was greater in younger individuals, especially among those aged under 40 years. From these studies, the longer-term risks of stroke and the effect of HZ on TIA were less clear.

(iii) Case control or case series studies

One case control study nested within a village-based Ugandan population cohort focussed on serum antibody titres rather than clinical VZV reactivation. It has long been recognised that repeated episodes of, often subclinical, VZV reactivation result in ‘boosts’ to VZV serum IgG antibody titres [16]. IgM antibodies can also be detected after herpes zoster, although it is not clear how long this class of antibodies persists in serum following virus reactivation [17]. This study compared titres of IgM and IgG antibodies to VZV in stored serum samples in 31 cases of clinically confirmed stroke with 132 controls matched on sex, age and village. No difference was seen between titres in cases and controls in repeated serum samples taken in the 15-year period prior to diagnosis [18]. This pilot study was limited by its small sample size and lack of control for confounding. Although using antibody titre as a measure of VZV reactivation might help to increase the sensitivity of exposure measurement through capturing subclinical virus reactivation, results are not directly comparable to those from studies with clinical measures of VZV exposure.

Another small case control study in a French paediatric clinic population of 11 children with arterial ischaemic stroke and 44 controls matched for age, gender and site of residence showed that 64 % of cases had had chickenpox in the 9 months prior to stroke compared to 9 % of controls – a frequency corresponding to the expected varicella incidence in France [19]. This study is limited by very small numbers, lack of information on the choice of control subjects and a lack of control for confounding. It is also unclear whether the 9-month cut-off was prespecified.

Finally a series of 70 children aged 6 months to 10 years presenting consecutively with arterial ischaemic stroke to two Canadian hospitals showed that 22 (31 %) reported varicella in the preceding 12 months, compared to an expected frequency of 9 % based on the annual varicella incidence in Canadian children [20]. Children with a varicella history were more likely to have hemiparesis, basal ganglia infarcts, anterior circulation infarcts and stenosis of proximal portions of major cerebral arteries compared with AIS occurring in children without varicella. This exploratory study was based on relatively small numbers and used a population comparison group so it was not possible to control for confounding.

These case control or case series studies provided limited evidence for an increase in reported chickenpox in time periods preceding stroke in children compared to the expected population incidence. The one small study investigating VZV antibody titres did not show an association with stroke risk.

8.4 Epidemiology of Shingles and Other Acute Vascular Complications

Although stroke was the most commonly investigated acute vascular outcome after HZ, several of these studies also investigated the effect of HZ on cardiac events. Two studies focussed specifically on myocardial infarction (MI). In the US self-controlled case series study by Minassian et al., a transient increase in MI risk was also seen which gradually returned to baseline by 3–6 months. In week one after HZ, the IR for MI was 1.68 (95 % C.I. 1.47–1.92). For HZO there was also an increase in MI risk – IR 2.06 (95 % C.I. 1.52–2.79) [9]. In the UK cohort study in THIN investigating long-term risks of events after shingles, the adjusted HR for MI was 1.10 (95 % C.I. 1.05–1.16) in all subjects and 1.49 (95 % C.I. 1.04–2.15) when restricting to patients aged <40 years [13].

Several other studies investigated the effect of HZ on broader cardiac endpoints. The Swedish cohort study described earlier investigated the incidence of cardiovascular disease excluding stroke after HZ and showed an elevated risk in the year following HZ diagnosis – IRR 1.82 (95 % C.I. 1.42–2.33) [15]. In Taiwan, a cohort study using the Taiwan National Insurance Research Database showed that the incidence of acute coronary syndrome was 1.24-fold higher in 57,956 HZ patients than 231,832 patients without HZ: HR 1.15 (95 % C.I. 1.07–1.24) after adjustment for age, sex and comorbidities [21]. Another cohort study in the same Taiwanese database showed small increases in the incidence rate of arrhythmias (adjusted HR 1.22 (95 % C.I. 1.12–1.34)) and coronary artery disease (adjusted HR 1.24 (95 % C.I. 1.02–1.28)) in 19,483 HZ patients compared to 77,932 non-HZ patients [22]. These studies only adjusted for a limited number of confounding factors.

Many other vascular manifestations of either primary or reactivated varicella have also been reported in the literature including venous sinus thrombosis [23, 24], spinal cord thrombosis [25] and peripheral arterial [26] and venous thrombosis [27].

The evidence for these is limited to case reports, and subjects frequently have coexisting coagulation abnormalities, which make it difficult to establish the relative contribution of VZV.

In general, these studies suggest a pattern of transient increased risk of acute cardiac events following HZ which is similar to that seen for stroke. The effect sizes for cardiac events were generally more modest than those for stroke. At present there is insufficient data to assess the risk of noncardiac non-stroke vascular events after HZ.

8.5 Herpes Zoster Vaccination and Antivirals

Systematic reviews of randomised controlled trials of herpes zoster vaccine have confirmed that vaccine is effective for preventing herpes zoster disease, safe and generally well tolerated [28, 29]. None of the included trials has published data on vascular endpoints. At the time of writing, we are not aware of any trials of zoster vaccine to mitigate vascular risk.

Most of the observational studies described above did not investigate the effects of zoster vaccine, partly due to its relatively recent introduction, e.g. the vaccine was first offered in England and Wales in 2013. Even in the future, the narrow target population recommended to receive vaccine – in England and Wales, this is currently people aged 70 years with a catch-up campaign for individuals aged 78 years – as well as the low vaccine uptake (52.8 % for people aged 70 years in 2014/15) [30] will limit the use of routine vaccine data for research.

In the US study based on Medicare data, it was possible to identify people who had been vaccinated, but zoster vaccination did not appear to modify the association between herpes zoster and either MI or stroke. This study's self-controlled case series design meant that only individuals with HZ and stroke were included, so the proportion of participants who had received vaccine was extremely low (n=2.8 % vaccinated before zoster who had a stroke and 2.3 % vaccinated before zoster who had an MI) [9].

Similarly, most included studies did not investigate the effect of antiviral treatment after diagnosis of herpes zoster on stroke or cardiac risk. Only one self-controlled case series presented data on antiviral effects. In this study oral antivirals were given to 55 % of individuals: IRs for stroke were lower among those receiving antivirals compared with those not treated, suggesting a protective effect. For example, in the risk period 1–4 weeks after HZ, the IR was 1.23 (95 % C.I. 0.79–1.71) for the 3647 people prescribed antivirals, whereas for the 2937 cases not prescribed oral antivirals, the IR was higher at 2.14 (95 % C.I. 1.62–2.84) [8]. In two other studies, an antiviral prescription was used in the diagnosis of herpes zoster so it was not possible to quantify the effect of antivirals on vascular risk [9, 14].

In conclusion, the effect of vaccination against herpes zoster on the risk of subsequent cardiovascular events is not known. Although limited data suggest that antiviral drugs prescribed after HZ may reduce the risk of stroke, further definitive evidence is needed.

8.6 Mechanisms

Using EHR data alone, it is not possible to elucidate precise mechanisms for the relationship between VZV reactivation and vascular risk. The only animal model that approximates the natural history of VZV infection is the macaque simian varicella virus (SVV). While there are presently no data on vascular complications following SVV reactivation in non-human primates [31], SVV, like VZV, has been associated with pathological vascular changes [32]. In humans we are able to generate some mechanistic understanding through (a) considering parallels with other studies of interactions between infections and chronic disease and (b) using data from small clinic-pathological and neuroimaging studies of patients with vascular complications of VZV.

It is likely that a combination of factors contributes to the triggering of stroke or MI following herpes zoster. First, in common with other acute infections such as influenza, inflammation associated with systemic infection may result in endothelial dysfunction and a state of hypercoagulability [33]. On a background of arteriopathy, this may be sufficient to disrupt atheromatous plaques, causing a transient increase in the risk of acute vascular events. As well as effects on inflammatory and coagulation pathways, haemodynamic effects of infections such as increased sympathetic activity and vasoconstriction might also contribute to the disruption of atherosclerotic plaques [34]. These indirect effects are common across a range of infections.

The neurotropic nature of VZV also predisposes to a particular phenomenon – VZV vasculopathy – in which VZV virus spreads directly along nerve fibres to blood vessels, leading to pathological vascular remodelling. VZV vasculopathy is diagnosed by the detection of VZV DNA in CSF, the presence of anti-VZV IgG or anti-VZV IgM antibody in CSF or the presence of anti-VZV IgM antibody in serum [35] in combination with a range of vascular changes demonstrable on neuroimaging. In around one third of cases, changes consistent with VZV vasculopathy may be seen without rash, suggesting that significant neurovascular pathology may occur with subclinical VZV reactivation [36]. Small clinico-pathological studies suggest that early VZV vasculopathy primarily affects the adventitia, followed by virus spreading transmurally and later to the vessel lumen [37]. VZV vasculopathy plausibly contributes to both haemorrhagic and ischaemic strokes: disruption of smooth muscle cells in the media affects the risk of aneurysm formation, while intimal thickening contributes to vascular occlusion and ischaemia. The suggestion that HZO has a greater effect on stroke than herpes zoster in general supports a role for VZV vasculopathy in stroke pathogenesis. Other vascular complications associated with VZV vasculopathy include giant cell arteritis, which is discussed in detail by Gilden and Nagel in Chap. 7.

8.7 What Is Not Known/Future Directions

There are many unanswered questions about the nature of vascular risk following herpes zoster. Developing sufficient longitudinal linked data sources including, for example, EHR data linked to data on imaging, serology and genomics will enable more sophisticated studies to be conducted. These could also assess the role of subclinical virus reactivation – a potential window for early intervention.

Understanding which population groups are affected and during which time periods will be key to enabling effective targeting of personalised preventive or therapeutic options. Developing the evidence base around the effect of vaccine and antiviral agents on vascular outcomes through robust clinical trials is likely to be an important step. There are also gaps in current understanding of the mechanisms governing vascular links with infections. Improving mechanistic understanding could aid drug development and targeting.

In conclusion the evidence supporting a role for herpes zoster in vascular risk continues to grow. Our challenge is to harness these new insights into effective, personalised strategies to benefit population health.

References

1. Guess H, Broughton D, Melton L et al (1986) Population-based studies of varicella complications. Pediatrics 78(4 pt 2):723–727
2. Rovnak J, Kennedy PGE, Badani H et al (2015) A comparison of herpes simplex virus type 1 and varicella-zoster virus latency and reactivation. J Gen Virol 96(7):1581–1602
3. Insinga DRP, Itzler RF, Pellissier JM et al (2005) The incidence of herpes zoster in a United States administrative database. J Gen Intern Med 20(8):748–753
4. van Hoek AJ, Gay N, Melegaro A et al (2009) Estimating the cost-effectiveness of vaccination against herpes zoster in England and Wales. Vaccine 27(9):1454–1467
5. Smeeth L, Thomas SL, Hall AJ et al (2004) Risk of myocardial infarction and stroke after acute infection or vaccination. N Engl J Med 351(25):2611–2618
6. Warren-Gash C, Hayward AC, Hemingway H et al (2012) Influenza infection and risk of acute myocardial infarction in England and Wales: a CALIBER self-controlled case series study. J Infect Dis 206(11):1652–1659
7. Whitaker HJ, Farrington CP, Spiessens B et al (2006) Tutorial in biostatistics: the self-controlled case series method. Stat Med 25(10):1768–1797
8. Langan SM, Minassian C, Smeeth L et al (2014) Risk of stroke following herpes zoster: a self-controlled case-series study. Clin Infect Dis 58(11):1497–1503
9. Minassian C, Thomas SL, Smeeth L et al (2015) Acute cardiovascular events after herpes zoster: a self-controlled case series analysis in vaccinated and unvaccinated older residents of the United States. PLoS Med 12(12):e1001919
10. Thomas SL, Minassian C, Ganesan V et al (2014) Chickenpox and risk of stroke: a self-controlled case series analysis. Clin Infect Dis 58(1):61–68
11. Kang J-H, Ho J-D, Chen Y-H et al (2009) Increased risk of stroke after a herpes zoster attack, a population-based follow-up study. Stroke 40(11):3443–3448
12. Lin H-C, Chien C-W, Ho J-D (2010) Herpes zoster ophthalmicus and the risk of stroke: a population-based follow-up study. Neurology 74(10):792–797
13. Breuer J, Pacou M, Gauthier A et al (2014) Herpes zoster as a risk factor for stroke and TIA a retrospective cohort study in the UK. Neurology 82(3):206–212
14. Sreenivasan N, Basit S, Wohlfahrt J et al (2013) The short- and long-term risk of stroke after herpes zoster - a nationwide population-based cohort study. PLoS One 8(7):e69156
15. Sundström K, Weibull CE, Söderberg-Löfdal K et al (2015) Incidence of herpes zoster and associated events including stroke—a population-based cohort study. BMC Infect Dis 15(1):488
16. Hope-Simpson RE (1965) The nature of herpes zoster: a long-term study and a new hypothesis. Proc R Soc Med 58(1):9–20

17. Kangro HO, Ward A, Argent S et al (1988) Detection of specific IgM in varicella and herpes zoster by antibody-capture radioimmunoassay. Epidemiol Infect 101(1):187–195
18. Asiki G, Stockdale L, Kasamba I et al (2015) Pilot study of antibodies against varicella zoster virus and human immunodeficiency virus in relation to the risk of developing stroke, nested within a rural cohort in Uganda. Trop Med Int Health 20(10):1306–1310
19. Sébire G, Meyer L, Chabrier S (1999) Varicella as a risk factor for cerebral infarction in childhood: a case-control study. Ann Neurol 45(5):679–680
20. Askalan R, Laughlin S, Mayank S et al (2001) Chickenpox and stroke in childhood: a study of frequency and causation. Stroke J Cereb Circ 32(6):1257–1262
21. Wang C-C, Lin C-L, Chang Y-J et al (2014) Herpes zoster infection associated with acute coronary syndrome: a population-based retrospective cohort study. Br J Dermatol 170(5):1122–1129
22. Wu P, Lin C-L, Sung F-C et al (2014) Increased risk of cardiovascular events in patients with herpes zoster: a population-based study. J Med Virol 86(5):772–777
23. Siddiqi SA, Nishat S, Kanwar D et al (2012) Cerebral venous sinus thrombosis: association with primary varicella zoster virus infection. J Stroke Cerebrovasc Dis 21(8):917.e1–917.e4
24. Chan J, Bergstrom RT, Lanza DC et al (2004) Lateral sinus thrombosis associated with zoster sine herpete. Am J Otolaryngol 25(5):357–360
25. Orme HT, Smith AG, Nagel MA et al (2007) VZV spinal cord infarction identified by diffusion-weighted MRI (DWI). Neurology 69(4):398–400
26. Massano J, Ferreira D, Toledo T et al (2008) Stroke and multiple peripheral thrombotic events in an adult with varicella. Eur J Neurol 15(10):e90–e91
27. Rabah F, El-Banna N, Abdel-Baki M et al (2012) Post varicella thrombosis-report of two cases and literature review. Pediatr Infect Dis J 31(9):985–987
28. Gagliardi AMZ, Gomes Silva BN, Torloni MR et al. (2012) Vaccines for preventing herpes zoster in older adults. Cochrane Database Syst Rev (10):CD008858.
29. Sanford M, Keating GM (2010) Zoster vaccine (Zostavax): a review of its use in preventing herpes zoster and postherpetic neuralgia in older adults. Drugs Aging 27(2):159–176
30. Public Health England. Shingles vaccine coverage report, England, Sept 2014 to May 2015 [Cited 2016 Jan 28]. Available from: https://www.gov.uk/government/uploads/system/uploads/attachment_data/file/448407/hpr2615_shngls.pdf
31. Ouwendijk WJ, Verjans GM (2015) Pathogenesis of varicelloviruses in primates. J Pathol 235(2):298–311
32. Roberts ED, Baskin GB, Soike K et al (1984) Pathological changes of experimental simian varicella (Delta herpesvirus) infection in African green monkeys (Cercopithecus aethiops). Am J Vet Res 45(3):523–530
33. Arbab-Zadeh A, Nakano M, Virmani R et al (2012) Acute Coronary Events. Circulation 125(9):1147–1156
34. Corrales-Medina VF, Madjid M, Musher DM (2010) Role of acute infection in triggering acute coronary syndromes. Lancet Infect Dis 10(2):83–92
35. Nagel MA, Gilden D (2016) Developments in varicella zoster virus vasculopathy. Curr Neurol Neurosci Rep 16(2):12. doi:10.1007/s11910-015-0614-5
36. Nagel MA, Cohrs RJ, Mahalingam R et al (2008) The varicella zoster virus vasculopathies Clinical, CSF, imaging, and virologic features. Neurology 70(11):853–860
37. Nagel MA, Traktinskiy I, Azarkh Y et al (2011) Varicella zoster virus vasculopathy: analysis of virus-infected arteries. Neurology 77(4):364–370

Chapter 9
Antiviral Therapy and Local Treatment for Herpes Zoster

Anne A. Gershon

In the twenty-first century, there is great interest in prevention of zoster, both in healthy and immunocompromised patients. Successful prevention of clinical reactivation of *varicella zoster virus* (VZV) is obviously preferable to treatment after VZV has reactivated and caused disease. Despite the development of effective zoster vaccines [1, 2], however, there will continue to be a need for antiviral therapy for zoster in patients who have not received immunization or in whom a vaccine has not been entirely effective. Today, most individuals who develop zoster, particularly if they are immunocompromised or over 50 years old, are recommended to receive specific therapy mainly to decrease morbidity from VZV infection; death from zoster is uncommon or rare [3]. Therapy of zoster, even if given early after onset, unfortunately, probably does not prevent development of postherpetic neuralgia (PHN), as discussed below. On the other hand, relief of pain and promotion of healing of the rash of zoster are obviously greatly welcomed by patients, and antiviral therapy is given to most zoster patients.

9.1 Local Treatment for Zoster

Patients should be told that treatment of the skin rash involves keeping it clean with soap and warm water, religiously avoiding scratching, allowing the weeping areas to air dry, and covering the rash with loose sterile gauze bandages if that proves comfortable. Some physicians prescribe local therapy such as calamine lotion, Burrow's solution, capsaicin, or lidocaine to decrease pain and itching. Patients should be reminded that they can spread chicken pox to contacts who are susceptible to that

A.A. Gershon
Professor of Pediatrics, Columbia University College of Physicians and Surgeons,
630 West 168th Street, New York, 10032 NY, USA
e-mail: aag1@columbia.edu

C.P.N. Watson et al. (eds.), *Herpes Zoster: Postherpetic Neuralgia and Other Complications*, DOI 10.1007/978-3-319-44348-5_9

disease, although they will not transmit zoster to individuals who are immune to chicken pox.

9.2 Development of Antivirals

The nucleoside acyclovir was the first antiviral medication found to be effective and safe for treatment of infections due to *herpes simplex virus* (HSV) and VZV, and today it is widely used. Acyclovir was approved by the Food and Drug Association (FDA) in 1982, and since that time, it has played a major role in treatment of infections caused by VZV, particularly in elderly and/or immunocompromised patients.

One of acyclovir's most important attributes is that it is remarkably safe and well tolerated, compared with antiviral drugs that had been developed earlier. Its safety and efficacy are due to its being taken up preferentially by cells that are infected with HSV or VZV, compared to uptake into uninfected cells. In infected cells acyclovir is triply phosphorylated by viral thymidine kinase and host cell enzymes; in cells infected with HSV, acyclovir triphosphate concentrations are 40–100 times higher than in uninfected cells [4]. Acyclovir acts as an antiviral in two ways. First, acyclovir triphosphate prevents viral DNA synthesis by inhibiting viral DNA polymerase. In addition, it is a DNA terminator and prevents elongation of viral DNA. There is little incorporation of acyclovir into cellular DNA. Acyclovir penetrates well into cerebrospinal fluid (CSF), achieving about half the levels of drug present in plasma. Acyclovir is thus an extremely effective, safe, and well-tolerated antiviral drug. Both HSV and VZV are sensitive to acyclovir, although VZV is less so and requires higher doses of acyclovir than does HSV [5].

Therapy with acyclovir, especially if it is given intravenously, can be associated with serious adverse reactions including nephrotoxicity [6]. Nephrotoxicity can be avoided by decreasing dosage if there is renal compromise, assuring good hydration of the patient, and avoiding rapid administration. Renal toxicity is due to crystallization of acyclovir in the kidney. Renal toxicity occurs more frequently in patients with comorbidities. Other important adverse effects of acyclovir include neurotoxicity and gastrointestinal manifestations. Neurotoxicity is more frequent in patients with comorbidities and includes a wide variety of neurological symptoms such as lethargy, confusion, hallucinations, and tremors. These symptoms usually remit after acyclovir is discontinued. Gastrointestinal adverse events include nausea, vomiting, and diarrhea, which also disappear after acyclovir is stopped.

9.3 Newer Antivirals Against VZV

Valacyclovir was developed some years after acyclovir, from which it was derived. Valacyclovir is especially useful because it has significantly increased oral bioavailability in comparison to acyclovir. Valacyclovir is rapidly converted to acyclovir in the body and is concentrated in infected cells so that it too avoids cellular toxicity.

Valacyclovir is highly effective against VZV, achieves three to five times the bioavailability of acyclovir, and has a similar clinical antiviral spectrum. In practical terms, acyclovir administered orally must be given five times a day, whereas valacyclovir is given only three times a day. In addition, orally administered acyclovir is variably absorbed, compared to valacyclovir. Data on absorption in older patients for acyclovir, moreover, are lacking.

A similar drug, famciclovir, is converted into penciclovir after oral administration; penciclovir, which is not absorbed orally, behaves similarly to acyclovir against VZV when administered as famciclovir. Today adults with VZV infections therefore are often treated with valacyclovir or famciclovir for the reasons described above. Depending on how ill they are, immunocompromised patients may first be given intravenous acyclovir; after healing has clearly commenced, the treatment can be changed to oral valacyclovir or famciclovir [5]. Because these newer antivirals are administered orally, they may be less likely to result in toxic reactions than intravenous acyclovir.

Application of antiviral drugs to the rash of zoster is of no therapeutic use. Patients should also be counseled about the importance of remembering to take all of their antiviral medications every day; skipping doses is not a good idea.

9.4 Impact of Antiviral Therapy on PHN

Today it is widely accepted that orally administered valacyclovir and famciclovir are useful to speed the healing of zoster in both children over age 2 years and adults. The controversy comes in when one is considering whether early administration of antiviral therapy has an impact on PHN. In an early study, oral antiviral therapy with valacyclovir (1000 mgm three times a day) or acyclovir 800 mg five times a day, for 7–14 days in immunocompetent individuals, did not prevent PHN although it shortened its duration [7]. A number of other studies also failed to show that acyclovir reduced the duration of the incidence of PHN [8]. In 1996, however, a meta-analysis was published which involved 691 older patients with zoster who had participated in one of four double-blind studies of acyclovir (800 mg five times daily) or placebo for HZ. Resolution of zoster-associated pain PHN at 3–6 months after HZ was evaluated. Acyclovir appeared to accelerate resolution of acute pain, especially in persons over 50 years old. The incidence of PHN was lower by a factor of two in those who received acyclovir compared to those who had received placebo [9]. Subsequently, therefore, double-blind, randomized, multicenter studies comparing the safety and efficacy of valacyclovir and famciclovir within 72 h of zoster onset were conducted. In one study, results with valacyclovir and famciclovir were similar. In addition, it was recognized that the longer half-life of famciclovir compared to valacyclovir did not seem to result in a better outcome with famciclovir. Valacyclovir was more cost-effective than famciclovir [10]. In another study, immunocompetent patients who were over age 50 were treated for 7 days and followed up for 24 weeks. Safety and efficacy were similar for both drugs. In one study it appeared that treatment decreased the incidence and duration of PHN [11]. In another study, however, there seemed to be an effect on PHN duration although

PHN was not prevented [10]. In still another study, when prednisone was added to acyclovir, and studied in a randomized fashion, there was no effect on rapidity of healing of zoster or resolution of pain in the 3–6 months after zoster, although addition of prednisone seemed to improve the quality of life during the healing process [12]. Which drug regimen was best did not emerge from these studies, except that it seemed that oral acyclovir was probably inferior to the newer antivirals.

A randomized, double-blind trial of famciclovir vs acyclovir for treatment of localized zoster in immunocompromised adolescents and adults involving 148 patients who were given acyclovir (800 mg five times daily) or oral famciclovir (500 mg three times a day) showed similar results with regard to healing of rash and loss of acute pain [13].

Some of these studies and additional ones were reviewed in a 2014 Cochrane report analyzing six clinical trials (five of acyclovir and one of famciclovir) involving 1319 patients [14]. It was concluded that oral acyclovir for treatment of zoster did not reduce the incidence of PHN and that the evidence concerning the efficacy of valacyclovir and famciclovir for prevention of PHN was inconclusive. In this meta-analysis, the incidence of PHN in placebo groups was 11–60 %, and in treatment groups, it was 12–58 %. It was proposed that additional trials of famciclovir or other new antivirals be carried out in larger groups of patients, with more attention being paid to the severity of pain and quality of life, and that immunocompromised patients be studied as well [14]. Whether such studies will be carried out at this point is questionable, although it is clear that oral acyclovir would no longer be appropriate as a comparator drug.

At this stage, it seems that the best general approach to patients with zoster is to treat immunocompetent and stable immunocompromised patients over age 50 with oral valacyclovir or famciclovir for 7–14 days, depending on the severity of the infection and the condition of the patient. Administration of an antiviral as soon as possible after onset of rash, before 72 h has elapsed, is vastly preferred. It seems appropriate to tell patients that early therapy with valacyclovir or famciclovir would be expected to reduce the acute pain of zoster and might shorten the duration of PHN if it occurs in the 3–6 months following the zoster rash. There is general agreement that antivirals are of no use to treat established PHN [8, 15] (also see Chap. 23).

9.5 Future Directions

Now that it is recognized that zoster can present without a skin rash, it would be appropriate for new studies of efficacy of valacyclovir or famciclovir be carried out in patients with what might be described as occult zoster. One example is giant cell arteritis, including temporal arteritis and aortitis, as discussed in Chap. 7. Gilden, Nagel, and their colleagues have identified reactivated VZV by PCR and immunofluorescence arterial walls in their studies of these vascular diseases, which can lead to aneurysms and strokes in some patients [16–18]. Often these patients have no skin rash because their zoster results from reactivation of VZV latent in autonomic ganglia, the axons of which do not project to the skin but to blood vessels. These diseases seem to be further complicated by immune responses to VZV, which produce some of the manifestations of vasculitis. If VZV infection of arteries could be diagnosed early, the antiviral treatment might be helpful to these patients. A potential screening test for arterial infection by VZV is testing saliva for VZV DNA. This assay has been used to identify symptomatic and asymptomatic individuals with VZV reactivation with or without rash [19–24]. Open-label studies of valacyclovir suggest that some forms of chronic vascular involvement with VZV may respond to acyclovir [25, 26] .

Another form of occult VZV infection that might be amenable to antiviral therapy that calls out for therapeutic study is infection of the gastrointestinal tract with or without VZV rash [24]. Gastrointestinal VZV occurs due to reactivation of the virus from latency in the enteric nervous system (ENS), one of the three branches of the autonomic nervous system. This form of reactivation may result in abdominal pain, perforation of the bowel, and pseudoobstruction [24, 27]. It can be diagnosed in some patients by the transient presence of VZV DNA in saliva [24]. There are data to suggest that valacyclovir may benefit these patients if they are diagnosed promptly [24]. Additional study of visceral zoster is warranted.

9.6 Conclusions

Every effort should be made to treat patients with zoster promptly, within 72 h of rash onset, particularly those over age 50 and/or who are immunocompromised. It is recommended to begin oral therapy with oral valacyclovir or famciclovir rather than acyclovir (unless the patient is quite ill) while awaiting diagnostic testing to confirm the clinical diagnosis. Diagnosis is best made by performing polymerase chain reaction (PCR) on skin scabs or vesicular fluid collected with a sterile swab and held at room temperature in a sterile container. Most large hospital laboratories have the capability to perform PCR for VZV; PCR is more sensitive and available than viral culture. Because valacyclovir and famciclovir are so well tolerated and safe, treatment can even be carried out if diagnostic testing is not available and clinical suspicion of zoster is strong (unilateral, neuropathic, burning, allodynia, etc.). Early therapy speeds healing and may even shorten the duration of PHN should it occur.

It is fervently hoped that with newer highly effective subunit vaccines against zoster, the incidence of this disease will finally begin to fall and PHN will become a rarity.

Acknowledgments Grant support: R01DK093094; Biken Institute of Osaka University.

References

1. Oxman MN, Levin MJ, Johnson GR et al (2005) A vaccine to prevent herpes zoster and postherpetic neuralgia in older adults. N Engl J Med 352:2271–2284
2. Lal H, Cunningham AL, Godeaux O et al (2015) Efficacy of an adjuvanted herpes zoster subunit vaccine in older adults. N Engl J Med 372:2087–2096 .2137-21
3. Mahamud A et al (2012) Herpes Zoster-Related Deaths in the United States: Validity of Death Certificates and Mortality Rates 1979–2007. Clin Infect Dis 55:960–966

4. Elion GB (1982) Mechanism of action and selectivity of acyclovir. Am J Med 73:7–13
5. YIn M, Brust J, Tieu H, Hammer S (2009) Antiherpesvirus, anti-hepatitis virus, and anti-respiratory virus agents. In: Richman D, Whitley R, Hayden F (eds) Clinical virology, 3rd edn. ASM Press, Washington, DC, pp. 217–264
6. Tilson HH (1988) Monitoring the safety of antivirals. The example of the acyclovir experience. Am J Med 85:116–122
7. Beutner KR, Friedman DJ, Forszpaniak C, Andersen PL, Wood MJ (1995) Valaciclovir compared with acyclovir for improved therapy for herpes zoster in immunocompetent adults. Antimicrob Agents Chemother 39:1546–1553
8. Kost RG, Straus SE (1996) Postherpetic neuralgia--pathogenesis, treatment, and prevention. N Engl J Med 335:32–42
9. Wood MJ, Kay R, Dworkin RH, Soong S-J, Whitley RJ (1996) Oral acyclovir therapy accelerates pain resolution in patients with herpes zoster: a meta analysis of placebo controlled trials. Clin Infect Dis 22:341–347
10. Tyring SK, Beutner KR, Tucker BA, Anderson WC, Crooks RJ (2000) Antiviral therapy for herpes zoster: randomized, controlled clinical trial of valacyclovir and famciclovir therapy in immunocompetent patients 50 years and older. Arch Fam Med 9:863–869
11. Alper BS, Lewis PR (2000) Does treatment of acute herpes zoster prevent or shorten postherpetic neuralgia? J Fam Pract 49:255–264
12. Whitley RJ, Weiss H, Gnann JW Jr et al (1996) Acyclovir with and without prednisone for the treatment of herpes zoster. A randomized, placebo-controlled trial. The National Institute of Allergy and Infectious Diseases Collaborative Antiviral Study Group. Ann Intern Med 125:376–383
13. Tyring S, Belanger R, Bezwoda W, Ljungman P, Boon R, Saltzman R (2001) A randomized, double-blind trial of famciclovir versus acyclovir for the treatment of localized dermatomal herpes zoster in immunocompromised patients. Cancer Invest 19:13–22
14. Chen N, Li Q, Yang J, Zhou M, Zhou D He L (2014) Antiviral treatment for preventing postherpetic neuralgia. Cochrane Database Syst Rev (2):CD006866
15. Dworkin RH, Johnson RW, Breuer J et al (2007) Recommendations for the management of herpes zoster. Clin Infect Dis 44(Suppl 1):S1–S26
16. Nagel MA, Gilden D (2016) Developments in varicella zoster virus vasculopathy. Curr Neurol Neurosci Rep 16:12
17. Gilden D, White T, Boyer PJ et al (2016) Varicella zoster virus infection in granulomatous arteritis of the aorta. J Infect Dis 213(12):1866–1871
18. Gilden D, Nagel MA (2016) Varicella zoster virus and giant cell arteritis. Curr Opin Infect Dis 29(3):275–279
19. Mehta SK, Tyring SK, Gilden DH et al (2008) Varicella-zoster virus in the saliva of patients with herpes zoster. J Infect Dis 197:654–657
20. Mehta SK, Cohrs RJ, Forghani B, Zerbe G, Gilden DH, Pierson DL (2004) Stress-induced subclinical reactivation of varicella zoster virus in astronauts. J Med Virol 72:174–179
21. Cohrs RJ, Mehta SK, Schmid DS, Gilden DH, Pierson DL (2008) Asymptomatic reactivation and shed of infectious VZV in astronauts. J Med Virol 80:1116–1122
22. Birlea M, Cohrs RJ, Bos N, Mehta SK, Pierson DL, Gilden D (2014) Search for VZV DNA in saliva of healthy individuals aged 20-59 years. J Med Virol 86:360–362
23. Papaevangelou V, Quinlivan M, Lockwood J et al (2013) Subclinical VZV reactivation in immunocompetent children hospitalized in the ICU associated with prolonged fever duration. Clin Microbiol Infect 5:E245–E251
24. Gershon AA, Chen J, Gershon MD (2015) Use of saliva to identify varicella-zoster virus (VZV) infection of the gut. Clin Infect Dis 61:536–544
25. Gilden D, White TM, Nagae L, Gurdin WH, Boyer PJ, Nagel MA (2015) Successful antiviral treatment of giant cell arteritis and takayasu arteritis. JAMA Neurol 72:943–946
26. Cheng-Ching E, Jones S, Hui FK et al (2015) High-resolution MRI vessel wall imaging in varicella zoster virus vasculopathy. J Neurol Sci 351:168–173
27. Edelman DA, Antaki F, Basson MD, Salwen WA, Gruber SA, Losanoff JE (2009) Ogilvie syndrome and herpes zoster: case report and review of the literature. J Emerg Med 39:696–700

Chapter 10
Dermatologic Manifestations of Herpes Zoster

Ranjani Somayaji, James A. Elliott, and R. Gary Sibbald

10.1 Introduction

Herpes zoster (HZ), also called 'shingles', is a unique neurocutaneous entity that has been recognized since historic times. HZ represents a recurrent infection state of varicella-zoster virus (VZV) and commonly occurs in elderly [46] and immunocompromised hosts [3, 13, 63]. This chapter will focus on dermatological issues regarding diagnosis, pathology, cutaneous complications and management of HZ.

The name herpes is derived from the Greek word meaning 'to creep or crawl' [57], a description that depicts the spreading nature of herpetic vesicular lesions of the skin. The term zoster also has a Greek origin meaning 'girdle', and both the words zoster and shingles describe the 'wrapping around' distribution of HZ that is commonly noted on the trunk. HZ lesions are classically a unilateral rash limited to one or two dermatomes consisting of erythematous, maculopapular lesions that evolve into vesicles that later crust as they resolve. HZ is commonly preceded by pain 48–72 h prior to rash onset, with new lesions appearing for up to 5 days. Total duration of infection ranges from 10 to 15 days [63].

Despite its characteristic appearance, shingles was often historically confused with the lesions of smallpox [28]. However, in 1875, VZV transmission was

R. Somayaji
Division of Infectious Disease, Department of Medicine, University of Calgary, Calgary, AB, Canada

J.A. Elliott
Toronto Regional Wound Healing Clinic, Mississauga, Canada

R.G. Sibbald (✉)
Dalla Lana Faculty of Public Health and Department of Medicine, Faculty of Medicine, University of Toronto, Toronto, Canada

Trillium Health and Partners, Mississauga, Canada
e-mail: rgarysibbald@gmail.com

C.P.N. Watson et al. (eds.), *Herpes Zoster: Postherpetic Neuralgia and Other Complications*, DOI 10.1007/978-3-319-44348-5_10

demonstrated by inoculation of vesicular fluid from an individual infected with chickenpox to healthy ones [53]. This was further studied, and replicated with use of vesicular fluid from HZ lesions, where inoculation into healthy individuals resulted in chickenpox infections [7, 35]. These findings led to the understanding that HZ was a result of latent VZV reactivation and that the two viruses were one and the same [23]. In 1958, biologic definition of VZV (and HZ) virus was made possible with isolation of the virus from patients with chickenpox and shingles infections [62].

10.2 Clinical Features

HZ has a unilateral dermatomal distribution and classically is a vesicular eruption. Commonly, the first manifestation of zoster preceding the rash is pain, which can be severe. It may be sharply localized or diffuse and is often described as burning, aching or lancinating. Forty percent of patients experience pain more than 4 days prior to the skin eruption, and 35 % experience it less than 48 h prior to the skin condition [67]. In occasional cases, the pain is reported to have started more than 100 days prior to the development of the distinctive skin rash. This phenomenon is referred to as 'preherpetic neuralgia' [26]. Based on the distribution of the pain, it may mimic migraine, acute glaucoma, myocardial infarction, pleurisy, appendicitis or duodenal ulcer. Fever, malaise, myalgia, paraesthesia or pruritus may accompany the pain.

Early on with HZ infection, a maculopapular rash with surrounding erythema occurs, which then develops into fluid-filled vesicles within 12–24 h [63]. Characteristically, it starts in a localized part of one or two dermatomes, and rarely more contiguous dermatomes, and spreads in a linear fashion over the next 3–5 days, stopping abruptly or slightly overlapping the midline. Fixed urticarial erythema precedes closely grouped vesicles that tend to enlarge, umbilicate and then become pustular as leukocytes invade the vesicular fluid. As the haemorrhagic component (local bleeding of the underlying friable tissue) of the vesicles develops, the surrounding urticarial erythema dissipates. This haemorrhagic blistering, often in grouped grape-like clusters, is so distinctive that its presence in a linear distribution must be thought of as HZ until proven otherwise. Individual lesions may also become necrotic and ulcerate. The lymph nodes draining the affected area are often enlarged acutely and may be tender.

Mild infections run a shorter course and can occur with macules and papules that regress in 7–10 days without progressing to vesicles [8]. In typical infections, crusting of the lesions occurs in 7–10 days, after which the patient is no longer considered contagious. Re-epithelialization then follows and is usually complete in 2–3 weeks in children [6, 47] and young adults and in 3–4 weeks in older individuals. The eruption usually resolves with no residual rash, although occasionally post-inflammatory hyper- and hypopigmentation may be seen. If severe ulceration occurs or healing is delayed, permanent scarring of the skin may result. This is most commonly seen in elderly, undernourished or immunocompromised patients [13, 46].

Any dermatome may be affected, but the dermatomes are not affected with equal frequency [46]. The distribution of the rash of HZ tends to mirror the areas most heavily affected by the primary VZV infections. As chickenpox tends to be centripetally distributed on the trunk primarily, thoracic and lumbar dermatomes are most commonly affected, accounting for approximately half of all cases. This distribution may be due to contiguous spread from infected skin to the ganglia [25] at the time of primary infection or alternatively due to reactivation stimuli that may occur more commonly in this area. HZ is less common in children under 10, but when it occurs it is more common in the cervical and sacral areas [48].

HZ infections involving the eyelid are also relatively common and characteristically involve the first (ophthalmic) branch of the trigeminal nerve (termed herpes zoster ophthalmicus). This occurs in 10–20 % of affected individuals [46] and has a particular propensity for elderly patients. HZ ophthalmicus with involvement of the nasociliary branch of the trigeminal nerve is a sight-threatening condition requiring close monitoring, referral to ophthalmology and aggressive treatment. Lesions on the nose tip (termed Hutchinson's sign) can herald corneal involvement in HZ, but absent nasal lesions do not rule it out. Corneal inflammation, or keratitis, may subsequently lead to iridocyclitis (iris and ciliary body) or secondary glaucoma [63]. Additionally, the sequela of neuroparalytic keratitis has also been described where the interruption of sensory input to the cornea occurs following HZ and increases risk of further corneal injury (i.e. ulceration) or infection leading to visual loss [63]. One rare complication following zoster ophthalmicus is granulomatous cerebral angiitis [63].

In addition to the truncal HZ manifestations and HZ ophthalmicus, other unusual manifestations include involvement of the 2nd or 3rd branches of the trigeminal nerve, which results in herpetic intraoral lesions. Lesion locations can include the palate, uvula and tonsillar fossa with maxillary branch involvement and the mouth floor, buccal mucous membrane or tongue with mandibular branch involvement. For intraoral HZ infections, toothache may be a presenting sign. If the geniculate ganglion is involved, this is termed the Ramsay Hunt syndrome and is characterized by a triad of HZ lesions in the ear, affecting the tympanum, ear, tonsillar fauces, anterior pillar, pinna and the external auditory canal, as well as ipsilateral facial palsy and loss of taste in the anterior 2nd/3rd of the tongue [30]. However, the involvement of the geniculate ganglion remains to be confirmed with pathological studies.

Occasionally, HZ may be manifested by pain without skin involvement [15, 41]. This is referred to as 'zoster sine herpete' and is presumably due to an abortive eruption. Conversely, some of these cases may represent a very prolonged interval between the onset of the preherpetic neuralgia and the characteristic rash of HZ [26]. However, in others, the rash does not manifest. The diagnosis in this situation is very difficult and is often made on a post hoc basis by demonstration of rising antibody titres to the virus (i.e. a fourfold rise over 10 days).

Although most cases of HZ are viewed as sharply localized to one or more dermatomes, careful observation may often reveal a few sparse papules and vesicles outside of the affected dermatome(s) [32]. Tyring [59] documented this finding in approximately 33 % of immunocompetent patients. Resembling the lesions of varicella, these are considered to represent haematogenous spread and an inadequate host immune defence system to contain the lesions [13]. The dissemination is thought to occur within circulating leukocytes and particularly in monocytes.

Cutaneous dissemination, arbitrarily defined as more than 20 lesions outside the affected dermatome(s), is infrequent, occurring in only 2 % of HZ cases [46]. An immunosuppressed state is associated with an increased risk of disseminated HZ infections and can occur in 15–30 % of cases [50, 52]. Systemic involvement, resulting in pneumonia, hepatitis, retinitis or acute retinal necrosis, meningitis and encephalitis, [10, 34] can be associated with cutaneous dissemination and may be fatal [27].

Although HZ is classically an acute infection, chronic HZ may also occur in immunocompromised hosts, especially with HIV infection. New HZ lesions may form without healing of pre-existing lesions, and these infections may be quite severe and debilitating. These persistent infections have been associated with increased drug-resistant (i.e. acyclovir) isolates.

10.3 Complications

10.3.1 Pain

Almost all patients have pain associated with acute HZ. The most common and intractable sequel of HZ is postherpetic neuralgia (PHN) [33]. The true frequency of PHN depends on the definition used and the population studied. One definition of PHN is pain persisting more than 4 weeks after the onset of the HZ infection [31]. The reported incidence of PHN varies between 5 and 57 %, according to different studies [51]. More information about PHN and other neurological complication is detailed elsewhere in this book in Chap. 6 [10, 11–24, 34, 37].

10.3.2 Cutaneous Complications

10.3.2.1 Bacterial Infection

Infection of cutaneous lesions was reported in 2.3 % of cases in a 60-day follow-up study of complications of HZ [22]. The most common secondary infective complication is impetigo, although erysipelas or cellulitis may also occur.

10.3.2.2 Scarring

Individual lesions can scar as a consequence of necrotic zoster lesions, dermal ulceration or secondary infection. Sarcoid-like lesions [5] and granuloma formation [68] have also been reported within HZ scars.

10.3.2.3 Post-inflammatory Hyperpigmentation

Significant thick crusts are often noted, particularly following secondary bacterial infection or ulceration. These crusts resolve leaving local dusky erythematous macules that may then become pigmented.

10.3.2.4 Gangrene

Gangrenous skin lesions rarely (usually in association with a significant immunocompromised state) result from deep necrosis and ulceration [8].

10.3.2.5 Wolf's Isotopic Response

This response describes the occurrence of a new skin disorder occurring at the site of a previous or another unrelated and healed skin disease. The original disease is most commonly HZ [65]. The mechanism is unknown but may have genetic and environmental factors along with a decreased local immunity from the primary disease process.

10.3.3 Ocular Complications

Eye complications occur in approximately 2 % of patients with HZ [22, 46] and in 67 % of patients with ophthalmic nerve involvement. In short, HZ may complicate any aspect of the ophthalmic or periophthalmic structures. For a more in-depth review of these complications, the reader is referred to Chap. 5.

10.4 Histopathology

The cutaneous lesions of HZ are histologically indistinguishable from varicella or herpes simplex infection. In all, the most common feature is an intraepidermal blister formed by 'ballooning degeneration' of infected epidermal cells in the lower to middle epidermis, resulting in marked acantholysis. Ballooning degeneration causes marked swelling of epidermal cells. Balloon cells have a homogeneous, eosinophilic cytoplasm, with possible eosinophilic inclusion bodies. These are separated by clear zones from the nuclear membrane and are known as Cowdry type A inclusions. As balloon cells lose their intercellular bridges, the acantholysis occurs, leading to dissolution of the epidermis, i.e. reticular degeneration. Reticular degeneration is due to the membrane left behind by the cells forming a netlike structure. This is a process in which the epidermal cells become greatly distended by intracellular oedema, so that many of the cells burst. This occurs mainly at the periphery of viral vesicles. Reticular degeneration is not specific for herpetic or other vesicular viral infections and can also occur in nutritional deficiencies. Multinuclear giant cells are characteristic of herpetic lesions [61]. These occur due to fusion of adjacent infected cells, in or near the blisters. Smears taken from the floor of an early, freshly opened vesicle and stained with a Giemsa or toluidine blue stain (available in many hospital laboratories) show this characteristic feature and can be used to aid in the diagnosis of VZV infections [38].

The host inflammatory reaction promoted by HZ infection consists of an early response of lymphocytes and monocytes, and the later appearance of neutrophils. This may progress to a leukocyclastic vasculitis, with neutrophilic cells infiltrating and destroying blood vessel walls and associated cell fragments due to release of lysosomal enzymes and cell dissolution. This destruction is often associated with haemorrhage and purpura due to leaky blood vessel walls. The purpura associated with leukocytoclastic vasculitis is often palpable due to the inflammatory infiltrate and can be distinguished from the flat purpura associated with fragile vessel walls of the elderly or associated with bleeding disorders. In sensory ganglia these histopathological features of inflammation, haemorrhage and necrosis are very pronounced [4, 25].

10.5 Diagnosis of HZ

As HZ commonly has a very characteristic presentation, the diagnosis is usually readily made by history and physical examination. However, because the pain may precede the development of vesicles by days, HZ must also be considered in the differential diagnosis of any unexplained localized acute pain syndrome.

Problems in diagnosis may arise early in the course of the disease. Localized dermatoses such as crawling insect bites, allergic and irritant contact dermatitis and localized infections such as viral exanthems (i.e. enteroviruses, group A coxsackieviruses), impetigo and folliculitis may mimic the disease [63]. Historically, smallpox was also confused with VZV infection [28]. However, the progression of HZ to its characteristic painful, grouped lesions in a dermatomal distribution usually distinguishes it from these entities.

Classically in HZ the lesions are at differing stages of evolution. The distribution pattern may help, as zosteriform HSV infections occur most commonly in the maxillary distribution of the trigeminal nerve and in the sacral areas, in contrast to HZ, recurrent lesions of the buttocks are almost always herpes simplex and not zoster. This site is the third most common for HSV after the lips and genitals and is due to front-to-back body contact between individuals ('cupping'). Pain is also more pronounced with HZ than with HSV infections. Furthermore, zosteriform HSV tends to recur, whereas HZ usually occurs only once and is usually not bilateral, as can sometimes occur with herpes simplex in this region. Notably, when patients experience an HZ-like eruption and a local recurrence, these tend to more commonly be caused by HSV. Despite this, the distinction between HSV and HZ can still pose difficulties in some cases. In one randomized controlled trial, 4.5 % of clinically suspected cases of HZ were in fact revealed to be HSV when lesions were cultured [58].

The diagnosis of HZ can be confirmed through viral cultures by isolating the virus in tissue or through acute and convalescent serologic tests to demonstrate conversion or rising titres [63]. For viral culture sampling, any crust if present should be lifted to expose the basal epidermal cells. When viral cultures are sent, HSV takes 24–48 h, and HZ requires approximately 5–7 days to grow. A Tzanck smear is a diagnostic test which is performed by scraping an HZ lesion base and will demonstrate multinucleated giant cells. However, the sensitivity of the Tzanck smear is only 60 % and does not allow for differentiation between HSV and VZV infection. Commercially available direct fluorescent antibody (DFA) staining may be used of smears obtained from vesicular lesion scrapings and can distinguish between HSV and VZV infections [39]. However, if lesions are atypical, DFA testing is relatively insensitive and lacks the specificity to guide therapeutic decisions.

Increasingly, polymerase chain reaction (PCR)-based tests have become the test of choice for diagnosis of HZ infections due to high sensitivity, high specificity, quick processing time and stability of specimens [39]. PCR tests can also be utilized for cerebrospinal fluid (CSF) to detect VZV DNA and can be helpful in diagnosis in cases when classic HZ is not present. However, use of PCR has yet to be widespread as testing has increased associated costs, in addition to non-uniform performance standards between laboratories and health regions. Other antibody tests that may be used for diagnosis include fluorescence antibody to membrane antigen (FAMA), immune adherence haemagglutination and enzyme-linked immunosorbent assays (ELISA) [19].

10.6 Therapy of HZ

Prior to the 1960s, a number of therapies were reported as efficacious but were never tested in a controlled manner. These included injections of pituitary extract, vitamin B1 and B12, cobra venom, quinine, proteolytic enzymes, autohemotherapy, sodium iodine and ergotamine [12, 45]. The goals of HZ therapy are the following: to relieve the symptoms of acute pain; to decrease the duration, dissemination and infectivity of the skin rash; and to prevent or shorten the duration of complications, including postherpetic neuralgia. Since the latter is covered extensively elsewhere in this book, only the treatment of acute HZ is described here.

10.6.1 General Measures

Although HZ is self-limited, many patients will recover completely without specific therapy within a month. Topical measures may be used for symptom relief. One such measure is the use of wet dressings or soaks that are moistened gauze applied to the skin for hydrating the skin locally. These serve three functions. First, the application of fluid to the eroded (loss of surface epidermis) or ulcerated skin (complete loss of epidermis with a dermal or deeper base) restores a relatively physiological environment to the exposed nerve endings that are responsible for transmitting sensations such as pain and itching. Second, by softening and dissolving crust with compresses of saline or water-moistened gauze and then wringing out of the excess fluid may cause the net movement from the skin to the cloth. This method provides an autolytic debridement function to remove solidified protein that would otherwise serve as a nidus for infection. Third, removal of the crusts and vesicles reduces the damaging effect of fluid entrapment and therefore reduces maceration.

The fluids used for compressing can be normal saline, tap water, Burrow's solution (aluminium acetate) and colloidal oatmeal (Aveeno®). These compresses are usually applied for 10–15 min four times daily. If the skin lesions are impetiginized, topical povidone iodine solution or a topical antibiotic cream including mupirocin, fusidic acid or polymyxin/gramicidin combination can be applied thereafter. Neomycin should be avoided alone or in combination products because of its allergic sensitizing potential and cross sensitivity with aminoglycosides. The patient is contagious by direct contact until the lesions have crusted. Topical antipruritic creams or lotions can be administered in the form of pramoxine or menthol, preferably without topical steroids. Storing these preparations in the refrigerator may provide an additive cooling, anti-itch effect.

Antipyretic agents may also be administered. Acetaminophen is preferable to acetylsalicylic acid in children because of the association of the latter with Reye's syndrome. Controlling pain, especially at night so the patient can rest, is essential. Over-the-counter analgesics, nonsteroidal anti-inflammatory agents, tricyclic antidepressants, gabapentin and opioid analgesics may all be considered based on the severity of the pain. The use of antidepressants, and anticonvulsants (gabapentin, pregabalin), and other agents is discussed in detail in another chapter (Chap. 19). Zoster with pregnancy raises particular issues [17, 18].

With HZ infections in the healthcare setting, infection control measures must be implemented. These measures will depend on the immune status of the host and the extent of infection. For localized (dermatomal) disease, standard precautions are followed, and the rash area should be kept completely covered (i.e. clothing or dressing) [9]. For either localized HZ infections in immunocompromised hosts or disseminated infection, standard precautions with additional airborne and contact precautions must be followed. Once disseminated infection is ruled out or the lesions are crusted, then these measures may be discontinued [9]. If healthcare professionals are exposed to VZV or HZ, they require post-exposure monitoring for 8 to 21 days or vaccination within 3 to 5 days of exposure based on their vaccination status [9]. In cases of immunosuppression or VZV vaccine contraindications, administer VZ immune globulin following exposure.

10.6.2 Systemic Antibacterial Agents

The role of oral and parenteral antibacterial agents is controversial. Due to the intensity of HZ eruptions, many physicians are tempted to prevent secondary bacterial infections. It is often assumed the progressive cloudiness of the vesicular eruption that becomes pustular is bacterial infection. This is, however, uncommon and such therapy is usually not needed. The presence of frank purulence, honey-coloured crusting and deep ulceration may be more suggestive of superimposed impetiginization. Since the most common offending organisms is *Staphylococcus aureus*, a penicillinase-resistant semisynthetic penicillin, a cephalosporin and a broad-spectrum antibiotic aimed at this organism are the most appropriate choices. More specifically, cloxacillin, cephalexin, amoxicillin clavulanate, trimethoprim and sulfamethoxazole are all appropriate antibiotics depending on local epidemiology of methicillin-resistant *S. aureus* [54].

10.6.3 Glucocorticosteroids

One of the most controversial therapies for HZ is the use of oral glucocorticosteroid drugs. The use of systemic corticosteroid therapy has been suggested as a viable alternative (adrenocorticotrophic hormone) in the treatment of HZ infections [44, 55]. A number of subsequent investigations followed in which corticosteroid derivatives were used in the management of HZ [2, 16, 20, 24, 49, 56]. However, these early investigations failed to utilize proper experimental methodology, making their conclusions suspect. A survey of 73 practising dermatologists, 56 of whom responded, showed 81 % routinely used corticosteroids in otherwise healthy patients older than age 60 who had HZ [40]. Critics have argued that side effects including an increased risk of herpes dissemination [1, 16] and avascular necrosis of the femoral head [42] are reasons against their use. More recent studies demonstrated an acute reduction in pain with no change in PHN risk [64, 66]. Further, only the combination of corticosteroids and antivirals (acyclovir) has been studied and only in immunocompetent host. There is no evidence for corticosteroids in isolation with greater concern for immunocompromised patients. Thus, with the availability of effective antiviral therapies, corticosteroids now have little evidence to recommend their usage.

10.6.4 Antiviral Agents

Antiviral agents have been employed in the treatment of HZ for over 30 years. Although this is elaborated elsewhere in this book, a few points warrant mention. Idoxuridine, cytosine arabinoside, vidarabine and, more recently, acyclovir, valacyclovir and famciclovir have all been used in the acute management of HZ. The antiviral drugs acyclovir, valacyclovir and famciclovir (as well as other available formulations) are the mainstay of treatment for HSV and VZV infections and act following thymidine kinase activation to inhibit viral DNA polymerase. This group of drugs are generally considered to promote faster resolution of vesicles and to decrease the amount of acute pain experienced [67]. Additionally, they have been demonstrated to reduce the duration of acute zoster-associated postherpetic pain (ZAP) [58, 64, 66]. Antivirals have not been proven to prevent severe postherpetic neuralgia [11]. For maximal benefit, oral antivirals should be started within the 72 h of rash onset. They are recommended for immunocompromised patients and in those patients at higher risk of developing postherpetic neuralgia (i.e. patients over 50 years of age, involvement of greater skin surface area, severe pain at presentation and those with a coexisting illness) [14, 29, 43, 66]. In one study [58], famciclovir 500 mg three times daily for 7 days decreased duration of postherpetic pain by 60 days but not severe postherpetic neuralgia [11]. Comparable benefits are probably obtained with valacyclovir 1000 mg three times daily. The newer second-generation agents, valacyclovir (57 %) and famciclovir (77 %), have a higher oral absorption rate than acyclovir (20 %), allowing a decrease in dosage. There was no increase in side effects with those agents when compared to placebo in the controlled studies [58, 60].

10.7 Conclusion

Despite the fact that HZ has been recognized for very many years, its treatment still presents a challenge in both the immunocompetent and even more so in the immunocompromised patient. This chapter has focused on dermatological aspects of HZ but briefly reviews other issues which are dealt with in more detail in other chapters. The judicious use of pain relievers may ameliorate the acute phase of the condition. In view of the higher risk of potentially life-threatening complications of HZ in immunocompromised patients and the high morbidity associated with postherpetic neuralgia, antivirals (i.e. valacyclovir or famciclovir) are recommended for both immunocompromised patients and those at high risk of developing PHN (i.e. individuals with ophthalmic zoster or over age of 50 years). The benefits of the newer-generation antivirals in all patients have been documented. As HZ and its complications may be severe, prevention is critical. Vaccination against varicella virus is now widely available. A zoster prevention vaccine is currently available to decrease the risk of herpes zoster and postherpetic neuralgia in immunocompetent individuals over 50 years of age. A new vaccine is on the horizon which is more efficacious and more broadly applicable to the immunosuppressed (Chap. 24, [36]).

References

1. Anderson DJ, Janoff EN (1987) Herpes zoster infection in a patient on methotrexate given prednisone to prevent post-herpetic neuralgia. Ann Intern Med 107:783
2. Appelman D (1955) Treatment of herpes zoster with ACTH. N Engl J Med 20:693–695
3. Arvin A, Pollard R, Rasmussen L, Merigan T (1980) Cellular and humoral immunity in the pathogenesis of recurrent herpes viral infections in patients with lymphoma. J Clin Invest 65:869–878
4. Bastian FO, Rabson AS, Yee CL (1974) Herpesvirus varicellae isolated from human dorsal root ganglia. Arch Pathol 97:331–333
5. Bisaccia E, Scarborough DA, Carr RD (1983) Cutaneous sarcoid granuloma formation in herpes zoster scars. Arch Dermatol 119:788–789
6. Brunell PA, Kotchmar GSJ (1981) Zoster in infancy: failure to maintain virus latency following intrauterine infection. J Pediatr 98:71–73

7. Brunsgaard E (1932) The mutual relation between zoster and varicella. Br J Dermatol Syph 44:1–21
8. Burgoon CJ, Burgoon J, Baldridge G (1957) The natural history of herpes zoster. JAMA 164:265–269
9. CDC (2013) Preventing varicella in health care settings [WWW Document]. Cent. Dis. Control Prev. URL www.cdc.gov/chickenpox/hcp/healthcare-setting.html. Accessed 6.1.16
10. Chang C, Woo E, Yu Y, Huang C, Chin D (1987) Herpes zoster and its neurological complications. Postgrad Med J 63:85–89
11. Chen N, Li Q, Yang T et al. (2014) Antiviral therapy for the treatment of postherpetic neuralgia. Cochrane Database Syst Rev (2): CD006866
12. De Moragas JM, Kierland RR (1957) The outcome of patients with herpes zoster. AMA Arch Derm 75:193–196
13. Dolin R, Reichman RC, Mazur MH, Whitley RJ (1978) Herpes zoster-varicella infections in immunosuppressed patients. Ann Intern Med 89:375–388
14. Dworkin RH, Boon RJ, Griffin DR, Phung D (1998) Postherpetic neuralgia: impact of famciclovir, age, rash severity, and acute pain in herpes zoster patients. J Infect Dis:S76–S80
15. Easton H (1970) Zoster sine herpete causing acute trigeminal neuralgia. Lancet 2:1065–1066
16. Elliott FA (1964) Treatment of herpes zoster with high doses of prednisone. Lancet 2:610–611
17. Enders G (1984) Varicella-zoster virus infection in pregnancy. Prog Med Virol 29:166–196
18. Eyal A, Friedman M, Peretz B, Paldi E (1983) Pregnancy complicated by herpes zoster. A report of two cases and literature review. J Reprod Med 28:600–603
19. Forghani B, Schmidt N, Dennis J (1978) Antibody assays for varicella-zoster virus: comparison of enzyme immunoassay with neutralization, immune adherence hemagglutination, and complement fixation. J Clin Microbiol 8:545–552
20. Frank L, Lysiak R (1953) Herpetic and post-herpetic pain treated with cortisone and ACTH. N Y State J Med 53:2379
21. Fugelso PD, Reed WB, Newman SB, Beamer JE (1973) Herpes zoster of the anogenital area affecting urination and defaecation. Br J Dermatol 89:285–288
22. Galil K, Choo P, Donahue JG, Platt R (1997) The sequelae of herpes zoster. Arch Intern Med 157:1209–1213
23. Garland J (1943) Varicella following exposure to herpes zoster. N Engl J Med 228:336–337
24. Gelfand ML (1954) Treatment of herpes zoster with cortisone. JAMA 154:911–912
25. Ghatak N, Zimmerman H (1973) Spinal ganglion in herpes zoster. A light and electron microscopic study. Arch Pathol 95:411–415
26. Gilden DH, Dueland AN, Cohrs R, Martin JR, Kleinschmidt-DeMasters BK, Mahalingam R (1991) Preherpetic neuralgia. Neurology 41:1215–1218
27. Gnann J, Whitley RJ (1991) Natural history and treatment of varicella-zoster in high-risk populations. J Hosp Infect 18:317–329
28. Gordon J, Meader FM (1929) The period of infectivity and serum prevention of chickenpox. JAMA 93:2013–2015
29. Harrison RA, Soong S, Weiss HL, Gnann JWJ, Whitley RJ (1999) A mixed model for factors predictive of pain in AIDS patients with herpes zoster. J Pain Symptom Manage 17: 410–417
30. Hato N, Kisaki H, Honda N, Gyo K, Murakami S, Yanagihara N (2000) Ramsay Hunt syndrome in children. Ann Neurol 48:254–256
31. Hope-Simpson RE (1965) The nature of herpes zoster: a long-term study and a new hypothesis. Proc R Soc Med 58:9–20
32. Huff JC (1988) Antiviral treatment in chickenpox and herpes zoster. J Am Acad Dermatol 18:204–206
33. Huff JC, Bean B, Balfour HHJ, Laskin OL, Connor JD, Corey L, Bryson YJ (1988) Therapy of herpes zoster with oral acyclovir. Am J Med 85:84–89

34. Jemsek J, Greenberg SB, Taber L, Harvey D, Gershon A, Couch RB (1983) Herpes zoster-associated encephalitis: clinicopathologic report of 12 cases and review of the literature. Medicine Baltimore 62:81–97
35. Kundratitz K (1925) Experimentelle Übertragungen von Herpes zoster auf Menschen und die Bezichungen von Herpes zoster zu Varicellen. Z Für Kinderheilkd 39:379–387
36. Lal H, Cunningham AL, Godeaux O et al (2015) Efficacy of an adjuvanted herpes zoster subunit vaccine in older adults. N Engl J Med 372(22):2087–2296
37. Landthaler M, Heuser M (1979) Paralytic abdominal hernia in zoster. Hautarzt 30:432–433
38. Leinweber B, Kerl H, Cerroni L (2006) Histopathologic features of cutaneous herpes virus infections (herpes simplex, herpes varicella/zoster): a broad spectrum of presentations with common pseudolymphomatous aspects. Am J Surg Pathol 30:50–58
39. Leung J, Harpaz R, Baughman AL, Heath K, Loparev V, Vázquez M, Watson BM, Schmid DS (2010) Evaluation of laboratory methods for diagnosis of varicella. Clin Infect Dis 51:23–32
40. Levinson W, Shaw JC (1985) Treatment of herpes zoster with corticosteroids--fact or faith? West J Med 142:117–118
41. Lewis GW (1958) Zoster sine herpete. Br Med J 2:418–421
42. Mills K (1986) Herpes zoster and corticosteroid therapy. Med J Aust 145:60
43. Nagasako EM, Johnson RW, Griffin DR, Dworkin RH (2002) Rash severity in herpes zoster: correlates and relationship to postherpetic neuralgia. J Am Acad Dermatol 46:834–839
44. Nickel WR (1951) Herpes Zoster Treated with ACTH. AMA Arch Derm Syphilol 64:372
45. Portenoy RK, Duma C, Foley KM (1986) Acute herpetic and postherpetic neuralgia: clinical review and current management. Ann Neurol 20:651–664
46. Ragozzino MW, Melton LJ 3rd, Kurland LT, Chu CP, Perry HO (1982) Population-based study of herpes zoster and its sequelae. Medicine Baltimore 61:310–316
47. Rogers RS 3rd, Tindall JP (1972) Herpes zoster in children. Arch Dermatol 106:204–207
48. Roizenblatt S, Rosa NS (2014) Herpes zoster involving the S1 dermatome. N Engl J Med 370:2031–2031. doi:10.1056/NEJMicm1311499
49. Sauer GC (1955) Herpes zoster; treatment of postherpetic neuralgia with cortisone, corticotropin, and placebos. AMA Arch Derm 71:488–491
50. Schimpff S, Serpick A, Stoler B, Rumack B, Mellin H, Joseph JM, Block J (1972) Varicella-zoster infection in patients with cancer. Ann Intern Med 76:241–254
51. Schmader KE, Studenski S (1989) Are current therapies useful for the prevention of postherpetic neuralgia? A critical analysis of the literature. J Gen Intern Med 4:83–89
52. Sokal JE, Firat D (1965) Varicella-zoster infection in Hodgkin's disease: clinical and epidemiological aspects. Am J Med:452–463
53. Steiner P (1875) Zur inokulation der varicellen. Wien Med Wochenschr 25:306
54. Stevens DL, Bisno AL, Chambers HF, Dellinger EP, Goldstein EJ, Gorbach SL, Hirschmann JV, Kaplan SL, Montoya JG, Wade JC (2014) Practice guidelines for the diagnosis and management of skin and soft tissue infections: 2014 update by the infectious diseases society of America. Clin Infect Dis 59:147–159
55. Sulzberger MB, Sauer GC, Herrmann F, Baer RL, Milberg IL (1951) Effects of ACTH and cortisone on certain diseases and physiologic functions of the skin: I Effects of ACTH. J Invest Dermatol 16:323–337
56. Sutton G (1984) Steroidstoss therapy in the treatment of herpes zoster. Br J Clin Pract 38:21–24
57. Taylor-Robinson D, Caunt AE (1972) Varicella virus. Springer, Vienna
58. Tyring S, Barbarash RA, Nahlik JE, Cunningham A, Marley J, Heng M, Jones T, Rea T, Boon R, Saltzman R (1995) Famciclovir for the treatment of acute herpes zoster: effects on acute disease and postherpetic neuralgia. A randomized, double-blind, placebo-controlled trial. Collaborative Famciclovir Herpes Zoster Study Group. Ann Intern Med 123:89–96
59. Tyring SK (1992) Natural history of varicella zoster virus. Semin Dermatol 11:211–217
60. Tyring SK, Beutner K, Tucker BA, Anderson WC, Crooks RJ (2000) Antiviral therapy for herpes zoster: randomized, controlled clinical trial of valacyclovir and famciclovir therapy in immunocompetent patients 50 years and older. Arch Fam Med 9:863–869

61. Tyzzer EE (1906) The histology of the skin lesions in varicella. J Med Res 14:361–392
62. Weller TH, Witton HM (1958) The etiologic agents of varicella and herpes zoster; serologic studies with the viruses as propagated in vitro. J Exp Med 108:869–890
63. Whitley RJ (2015) Chickenpox and herpes zoster (Varicella-Zoster Virus). In: Bennett JE, Dolin R, Blaser MJ (eds) Mandell, Douglas, and Bennett's principles and practice of infectious diseases. Elsevier Saunders, Philadelphia, pp 1731–1737
64. Whitley RJ, Weiss H, Gnann JWJ, Tyring S, Mertz GJ, Pappas PG, Schleupner CJ, Hayden F, Wolf J, Soong SJ (1996) Acyclovir with and without prednisone for the treatment of herpes zoster. A randomized, placebo-controlled trial. The National Institute of Allergy and Infectious Diseases Collaborative Antiviral Study Group. Ann Intern Med 125:376–383
65. Wolf R, Wolf D, Ruocco E, Brunetti G, Roucco V (2011) Wolf's isotopic response. Clin Dermatol 29:237–240
66. Wood MJ, Johnson RW, McKendrick MW, Taylor J, Mandal BK, Crooks J (1994) A randomized trial of acyclovir for 7 days or 21 days with and without prednisolone for treatment of acute herpes zoster. N Engl J Med 330:896–900
67. Wood MJ, Ogan PH, McKendrick MW, Care CD, McGill JI, Webb EM (1988) Efficacy of oral acyclovir treatment of acute herpes zoster. Am J Med 85:79–83
68. Wright AL, Cotton DW, Winfield DA, Messenger AG (1989) Granuloma formation in herpes zoster scars. Dermatologica 179:46–46

Part III
Postherpetic Neuralgia: Assessment, Pathology, Pathophysiology

Chapter 11
The Effect of Herpes Zoster and Postherpetic Neuralgia on Health-Related Quality of Life, Function, Employment-Related Productivity, and the Cost-Effectiveness of the Vaccine

Melanie Drolet

11.1 Introduction

Although some progress has been made in the treatment of herpes zoster and postherpetic neuralgia, about 20–35 % of individuals will experience herpes zoster in their life [1–3]. Among individuals with herpes zoster, around 10–30 % will develop postherpetic neuralgia, one of the common, debilitating, and challenging complications of herpes zoster [4–10]. Both conditions, particularly postherpetic neuralgia, are known to have a considerable negative impact on health-related quality of life (HRQoL), and, unfortunately, available therapeutic options are only partially effective.

Given recent evidence that a varicella zoster virus (VZV) vaccine is effective at preventing herpes zoster and postherpetic neuralgia and associated burden [8], clinicians and decision makers are being asked to make recommendations regarding the use and funding of the VZV vaccine. Information regarding the burden of herpes zoster and postherpetic neuralgia, their impact of HRQoL, and the potential cost-effectiveness (costs compared to health benefits) of vaccination are needed to inform evidence-based policy decision and to guide clinicians in their recommendations. The objective of this chapter is to summarize the most recent evidence regarding the (1) impact of herpes zoster and postherpetic neuralgia on HRQoL and (2) cost-effectiveness of vaccination against herpes zoster.

M. Drolet, PhD
Centre de recherche du CHU de Québec-Université Laval, Québec, QC, Canada
e-mail: Melanie.Drolet@crchudequebec.ulaval.ca

C.P.N. Watson et al. (eds.), *Herpes Zoster: Postherpetic Neuralgia and Other Complications*, DOI 10.1007/978-3-319-44348-5_11

11.2 Health-Related Quality of Life (HRQoL)

11.2.1 General Concept and Definitions

Although there is still controversy regarding the exact definition of HRQoL and how it should be measured, it is generally recognized as a broad subjective multidimensional construct encompassing general health; physical, emotional, social, cognitive, and role functioning; as well as physical symptoms [11–13]. The World Health Organization defines quality of life as "individuals' perception of their position in life in the context of the culture and value systems in which they love and in relation to their goals, expectations, standards and concerns" [13]. Hence, HRQoL is a broad, complex, and multidimensional concept reflecting the overall sense of well-being.

In the past decades, HRQoL has become an important aspect of clinical research. Indeed, HRQoL is now considered as one of the most important outcome in many clinical trials assessing the efficacy of interventions. HRQoL is also frequently used to document and monitor the burden associated with several diseases. Moreover, HRQoL represents a meaningful tool to assess the impact of interventions among the elderly, particularly when increasing life expectancy is sometimes achieved at the detriment of quality of life [14].

11.2.2 HRQoL Measures

Since HRQoL is a subjective construct, the patients themselves should be questioned about their perceptions of HRQoL [11]. HRQoL instruments aim to assess, using the most rigorous, valid, and standardized methods, how individuals perceive the impact of various situations (e.g., diseases, treatments, screening tests) on different aspects of their life. A large number of patient-reported HRQoL instruments have been developed, and these instruments can be divided into two broad categories: generic and disease-specific instruments [11]. Generic instruments are designed to assess the impact of wide range of conditions or diseases on HRQoL. Given that norms from the general population are usually available, it is possible to compare the HRQoL of patients to the HRQoL of the general population. On the other hand, because these instruments cover general HRQoL domains, they may not be sensitive enough to capture significant changes related to a specific disease (e.g., improvements following treatments, worsening of heath condition). Disease-specific instruments have been developed to overcome this limitation of generic instruments.

11.2.3 Generic and Disease-Specific Instruments Commonly Used to Examine the Impact of Herpes Zoster and Postherpetic Neuralgia on HRQoL

Several generic instruments have been used to examine the different HRQoL domains affected by herpes zoster and postherpetic neuralgia. The most commonly used instruments are the EuroQol or EQ-5D [15–17] and the SF-36 (or the shortened version, the SF-12) [18–21]. These instruments have been extensively validated and used among diverse populations from several countries. The EQ-5D is comprised of five HRQoL dimensions: mobility, self-care, usual activities, pain or discomfort, and anxiety or depression. Individuals are asked to indicate their level of problems in each dimension (no problem, some problems, or severe problems). The level of problems for each domain are then translated into a single EQ-5D score varying from 0 to 1, with higher scores indicating higher HRQoL. These overall EQ-5D scores can be compared to age- and sex-adjusted population norms [22, 23] or to scores from a control group not affected by the disease to estimate the HRQoL loss because of the disease. The SF-36, one of the most widely used HRQoL instrument in epidemiological studies, is comprised of 36 questions grouped into eight scales: physical functioning, role physical, bodily pain, general health, vitality, social functioning, role emotional, and mental health. Two scores, physical health and mental health, are derived from these eight scales. The SF-12 is a shortened version of the SF-36, also including eight scales that can be summarized into the physical health and mental health summary scores. Scores for the eight scales and the two summary measures are usually calculated using norm-based scoring algorithms to obtain a mean of 50 and a standard deviation of 10 (norm-based scoring), with higher scores indicating higher HRQoL. This standardization makes it possible to meaningfully compare scores of the different scales and the physical and mental summary scores and to identify HRQoL domains with greater detriments. Similarly to the EQ-5D, summary scores can also be compared to age- and sex-adjusted population norms [22, 24, 25] or to a control group to estimate HRQoL loss associated with the disease.

Two instruments have been specifically designed to assess the impact of herpes zoster and postherpetic neuralgia-associated pain/discomfort on activities of daily living and function. The Zoster Brief Pain Inventory (ZBPI) has been developed by Coplan et al. [26] to quantify, on an 11-point Likert scale (0 = no interference to 10 = complete interference), the interference of pain/discomfort from herpes zoster and postherpetic neuralgia with seven activities of daily living: general activity, mood, walking ability, work, relation with others, sleep, and enjoyment of life. This instrument was used in the Shingles Prevention Study [8] and has shown good reliability and validity [26]. The Zoster Impact Questionnaire (ZIQ) was developed for the purposes of herpes zoster vaccine trials and measures interference of pain with the following activities of daily living: ability to clothe oneself, bathe, groom oneself, eat, travel, shop, leave the house, concentrate, do housework, prepare meals, do leisure activities, and be sexually active [26]. However, this instrument has not been formally validated against other HRQoL questionnaires [26].

11.3 Impact of Herpes Zoster on HRQoL

The acute phase of herpes zoster is characterized by a unilateral, dermatomal, and vesicular rash, associated with dermatomal pain or discomfort [27, 28]. Although the rash and pain associated with herpes zoster usually disappear within 1 month, acute herpes zoster has a significant impact on HRQoL and functional status. Several studies, conducted in North America [29–31], Europe [32–35], or Asia [36–38], have consistently reported an important impact of acute herpes zoster on all HRQoL domains. Table 11.1 summarizes the impact of acute herpes zoster, as measured at the beginning of the episode, by either generic instruments (EQ-5D, SF-12/36) or the herpes zoster-specific instrument (ZBPI-ADL). Studies generally recruited individuals with newly diagnosed herpes zoster (i.e., within 7 or 14 days after rash onset or with a visible rash) and aged ≥50–60 years old. At recruitment, the mean pain severity reported by patients from the various studies ranged from 5.0 to 6.4/10.0 (equivalent to moderate pain). Although definitions and cutoff used to define severe pain varied between studies, the proportion of newly diagnosed patients reporting severe pain was around 40 %.

11.3.1 Impact of Herpes Zoster on HRQoL: Generic Instruments

Five studies examined the impact of acute herpes zoster on HRQoL using the EQ-5D [29, 30, 34, 37, 38] (Table 11.1). Unsurprisingly, these studies consistently reported that pain/discomfort was the health domain most frequently affected by herpes zoster, with more than 80 % of patients reporting this problem [30, 34, 38]. Problems in performing usual activities (35–55 %) and symptoms of anxiety/depression (34–65 %) were also frequently reported. More specifically, in MASTER (Monitoring and Assessing Shingles Through Education and Research), a pan-Canadian prospective observational study [30], significant detriments in the five HRQoL domains (mobility, self-care, usual activities, pain/discomfort, anxiety/depression) were observed at recruitment compared to the pre-herpes zoster status (Fig. 11.1). At recruitment, 94 %, 55 %, and 46 % of the 261 patients recruited within 14 days of rash onset reported pain/discomfort, problems in performing usual activities, and symptoms of anxiety/depression compared to 36 %, 13 %, and 23 % prior to herpes zoster, respectively. However, these proportions returned to pre-herpes zoster values after pain has stopped.

Table 11.1 Impact of acute herpes zoster on HRQoL at recruitment as measured using the EQ-5D, SF-36/SF-12, or ZBPI-ADL

	North America		Europe				Asia	
	Schmader et al. 2007 [29]	Drolet et al. 2010 [30]	Chidiac et al. 2001 [32]	Bouhassira et al. 2012 [33]	Gater et al. 2014 [34]	Bricourt et al. 2014 [35]	Song et al. 2014 [37]	Tsai et al. 2014 [38]
Description								
Country	USA	Canada	France	France	UK	Italy	South Korea	Taiwan
Population								
N	102	261	8103	1354	229	413	46 incident cases	150
Age	≥60 years old	≥50 years old	No restriction (mean = 55 years old)	≥50 years old	≥50 years old	≥50 years old	≥50 years old	≥50 years old
Recruitment	Within 14 days after rash onset	Within 14 days after rash onset	Acute herpes zoster	Within 7 days after rash onset	With visible rash	New diagnosis	Within 7 days after rash onset	107 cases within 14 days of rash onset
Results								
Pain								
Pain measure	ZBPI worst pain	ZBPI worst pain	SF-36 bodily pain	ZBPI worst pain	ZBPI worst pain	VAS	ZBPI worst pain	ZBPI worst pain
Pain severity								
None	0: 10 %	0: 5 %	–		0: 11 %			0: 5 %
Mild	1–3: 25 %	1–2: 11 %	16 %		1–4: 30 %			1–2: 9 %
Moderate	4–7: 47 %	3–6: 40 %	43 %		5–6: 22 %			3–6: 39 %
Severe	8–10: 18 %	7–10: 44 %	41 %		7–10: 37 %			7–10: 47 %
Mean		5.7		5.3	5.0	5.8	6.4[a]	
Impact on HRQoL								

(continued)

Table 11.1 (continued)

	North America		Europe				Asia	
	Schmader et al. 2007 [29]	Drolet et al. 2010 [30]	Chidiac et al. 2001 [32]	Bouhassira et al. 2012 [33]	Gater et al. 2014 [34]	Bricourt et al. 2014 [35]	Song et al. 2014 [37]	Tsai et al. 2014 [38]
EQ-5D mean score	0.67[b]	0.59			0.65		0.67	0.67
EQ-5D % with problems								
Pain/discomfort		93 %			80 %			85 %
Usual activities		55 %			48 %			35 %
Anxiety/depression		46 %			34 %			65 %
Mobility		31 %			39 %			29 %
Self-care		17 %			14 %			26 %
SF-36/SF-12[c]								
SF-12 physical mean score SF-12 mental mean score	39[b] 44[b]			44[b] 41[b]		39 42		
SF-36 physical mean score SF-36 mental mean score					42 43			
Domains mean score HZ patients vs. norm[d]								
Physical functioning			70 vs. 85[a, e]		43 vs. 44			
Role physical			65 vs. 83		40 vs. 45			
Bodily pain			65 vs. 75		39 vs. 47			
General health			63 vs. 70		46 vs. 48			
Vitality			52 vs. 60		41 vs. 51			
Social functioning			72 vs. 83		39 vs. 49			

Role emotional			67 vs. 84		43 vs. 48			
Mental health			60 vs. 70		43 vs. 52			
ZBPI-ADL measure	% with interference ≥ 5	% with interference ≥ 5		Mean	Mean			% with interference ≥5
ZBPI-ADL domains							NA for incident cases only	
Sleep	43 %	64 %		4.3[a]	4.5			53 %
Enjoyment of life	51 %	58 %		2.4	4.0			49 %
General activities	51 %	53 %		3.8	3.8			43 %
Mood	41 %	46 %		3.5	3.4			53 %
Normal work	41 %	45 %		3.0	3.3			41 %
Relations with others	26 %	31 %		NA	2.1			36 %
Walking abilities	22 %	29 %		NA	1.7			29 %
Overall				3.1	3.3			

[a]Estimated from figures
[b]Data were available by pain categories. The weighted mean scores (score for each pain category multiplied by the proportion of individual in each pain category) were recalculated to obtain an overall score for the total population of the study
[c]Unless specified, SF-12 and SF-36 scores were standardized using norm-based scoring algorithm to obtain a mean = 50 and standard deviation = 10
[d]Scores of individuals with herpes zoster are compared to norms from the general population
[e]Scores have not been standardized

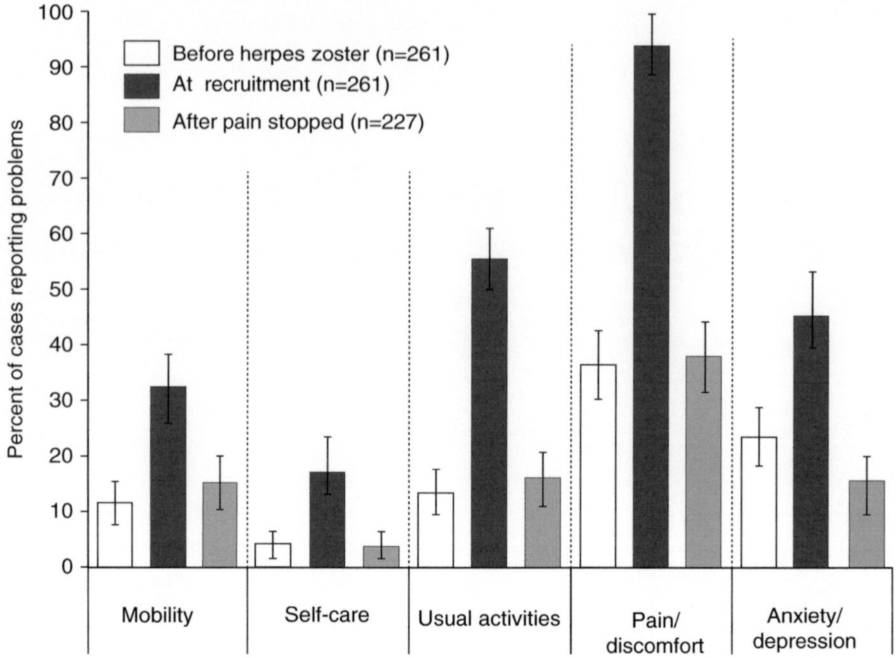

Fig. 11.1 Impact of herpes zoster on health-related quality of life: EQ-5D domains affected over time for 261 individuals newly diagnosed with herpes zoster (MASTER)

A significant impact of herpes zoster on physical and mental health was also observed when using the SF-36 or SF-12 generic instruments, and this negative impact was strongly associated with the severity of pain [29]. More specifically, significant differences with population norms were observed for the following domains: role physical, bodily pain, vitality, social functioning, role emotional, and mental health [34].

11.3.2 Impact of Herpes Zoster on HRQoL: Disease-Specific Instrument

Herpes zoster also has a considerable impact in activities of daily living, as measured by the ZBPI- ADL [29, 30, 33, 34, 38]. The four activities of daily living more frequently affected by herpes zoster are sleep (43–64 % of individuals reported interference of pain ≥5), enjoyment of life (49–58 %), general activities (43–53 %), and mood (41–53 %) [29, 30, 38]. For example, in MASTER, 64 %, 58 %, 53 %, and 46 % of individuals newly diagnosed with herpes zoster reported interference of pain ≥5/10 with sleep, enjoyment of life, general activities, and mood, respectively [30] (Fig. 11.2). Once again, very few individuals (<4 %) reported interference of pain in activities of daily living after pain cessation. These data, as well as

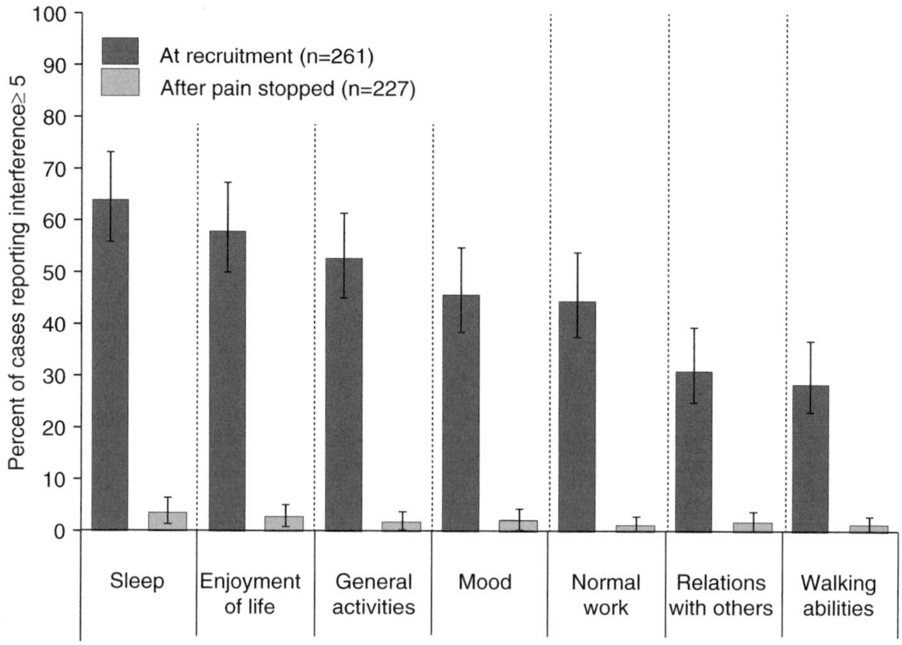

Fig. 11.2 Impact of herpes zoster on activities of daily living: interference of pain ≥5 for 261 individuals newly diagnosed with herpes zoster (MASTER)

those from other studies [29, 33, 36, 37], clearly indicate that the impact of herpes zoster on HRQoL and functional status is closely associated with pain.

11.4 Impact of Postherpetic Neuralgia on HRQoL

Although there is still some controversy regarding the exact definition of postherpetic neuralgia, it is generally recognized as the most common complication of herpes zoster and one of the most challenging to treat. When defined as pain persisting for more than 90 days after rash onset, between 8 and 27 % of individuals with herpes zoster develop postherpetic neuralgia [4–10]. Older age, greater number of lesions, and greater acute pain have been consistently reported as key risk factors for postherpetic neuralgia [5, 10, 39–42]. It has also been suggested that premorbid functional status, possibly a marker of poor health status, could potentially increase the risk of postherpetic neuralgia [10].

Several studies have examined the impact of postherpetic neuralgia on HRQoL and functional status in North America and Europe [30, 32, 33, 35, 43–45] (Table 11.2). Some studies prospectively followed individuals with herpes zoster to identify those who developed postherpetic neuralgia (i.e., incident cases) [30, 33, 35], whereas other studies recruited individuals experiencing postherpetic neuralgia for different periods of time (i.e., prevalent cases) [32, 43–45]. The pain severity was generally higher

Table 11.2 Impact of postherpetic neuralgia on HRQoL as measured using the EQ-5D, SF-36/SF-12, or ZBPI-ADL

	North America		Europe				
	Oster et al. 2007 [43]	Drolet et al. 2010 [30]	Chidiac et al. 2001 [32]	Bouhassira et al. 2012 [33]	Serpell et al. 2014 [44]	Bricourt et al. 2014 [35]	Laurent et al. 2014 [45]
Description							
Country	USA	Canada	France	France	UK	Italy	France
Population							
N with PHN	385	63	935	127	152	73	108
Age	≥65 years old	≥50 years old	No restriction, mean = 69 years old	≥50 years old	≥50 years old	≥50 years old	No restriction, mean = 73 years old
PHN definition	Pain persisting more than 90 days after rash onset	Pain ≥3 persisting more than 90 days after rash onset	Pain persisting after clearance of lesions	Pain persisting more than 90 days after rash onset	Pain persisting more than 90 days after rash onset	Pain persisting more than 90 days after rash onset	Pain persisting more than 90 days after rash onset
Timing of pain and HRQoL measures	Mean = 3.3 years after rash onset	90 days after rash onset	NA	90 days after rash onset	Mean = 3.5 years after rash onset	90 days after rash onset	Mean = 2 years after pain onset
Results							
Pain							
Pain measure	ZBPI worst pain	ZBPI worst pain	SF-36 bodily pain	ZBPI worst pain	ZBPI worst pain	VAS	ZBPI worst pain
Pain severity							
None					0: 10 %		
Mild	0–3: 20 %[a]	1–2: 16 %	9 %	NA	1–4: 29 %		13 %
Moderate	4–7: 35 %	3–6: 9 %	39 %		5–6: 17 %		30 %
Severe	8–10: 45 %	7–10: 15 %	52 %		7–10: 44 %		57 %
Mean	6.0	3.8			5.4	3.7	6.5

Impact on HRQoL							
EQ-5D mean score	0.61	0.67			0.56[b]		
EQ-5D % with problems							
Pain/discomfort	97%[b]	88 %			90 %		
Usual activities	61 %	43 %			66 %		
Anxiety/depression	52 %	56 %			46 %		
Mobility	47 %	42 %			55 %		
Self-care	12 %	29 %			21 %		
SF-36/SF-12[c]							
SF-12 physical mean score				40[a]		40	34
SF-12 mental mean score				41		40	56
SF-36 physical mean score					38[b]		
SF-36 mental mean score					43		
Domains mean score HZ patients vs. norm[d]							
Physical functioning			60 vs. 85[a,e]		39 vs. 44		45
Role physical			44 vs. 83		38 vs. 45		46
Bodily pain			46 vs. 75		39 vs. 47		45
General health			47 vs. 70		42 vs. 48		45
Vitality			42 vs. 60		42 vs. 51		30
Social functioning			50 vs. 83		41 vs. 49		60
Role emotional			46 vs. 84		40 vs. 48		55
Mental health			50 vs. 70		44 vs. 52		52
ZBPI-ADL measure	Mean	% with interference ≥ 5		Mean	Mean		Mean

(continued)

Table 11.2 (continued)

	North America		Europe				
	Oster et al. 2007 [43]	Drolet et al. 2010 [30]	Chidiac et al. 2001 [32]	Bouhassira et al. 2012 [33]	Serpell et al. 2014 [44]	Bricourt et al. 2014 [35]	Laurent et al. 2014 [45]
ZBPI-ADL domains							
Sleep	3.8	29 %		3.0[a]	3.4		3.4
Enjoyment of life	4.5	31 %		2.5	3.8		NA
General activities	3.7	25 %		4.0	3.1		4.5
Mood	4.3	30 %		3.3	3.4		4.4
Normal work	NA	22 %		2.9	2.9		4.0
Relations with others	3.0	19 %		NA	NA		3.0
Walking abilities	NA	25 %		NA	NA		2.7
Overall	NA			3.1	2.9		3.7

[a]Estimated from figures
[b]Data were available by pain categories. The weighted mean scores (score for each pain category multiplied by the proportion of individual in each pain category) were recalculated to obtain an overall score for the total population of the study
[c]Unless specified, SF-12 and SF-36 scores were standardized using norm-based scoring algorithm to obtain a mean = 50 and standard deviation = 10
[d]Scores of individuals with postherpetic neuralgia are compared to norms from the general population
[e]Scores have not been standardized

among prevalent cases (5.4–6.5/10) [43–45] compared to incident cases (3.8 /10) [30, 35], and the impact of postherpetic neuralgia on HRQoL is likely to be higher in studies assessing prevalent cases. Therefore, results from these studies must be interpreted with caution as individuals experiencing pain for long periods of time may only represent a subset of all individuals confronted with postherpetic neuralgia.

11.4.1 Impact of Postherpetic Neuralgia on HRQoL: Generic Instruments

Studies recruiting prevalent or incident cases of postherpetic neuralgia provide complementary information. The former type of study provides important information about the long-term burden of postherpetic neuralgia. For example, Serpell et al. [44] showed that, on average of 3.5 years after the onset of herpes zoster, individuals with postherpetic neuralgia still had significant detriments in all SF-36 domains compared to age-matched UK population norms; the most important negative impact was observed for vitality, social function, and mental health. Persistent problems were also observed when using the EQ-5D: 90 % reported pain/discomfort, 55 % mobility problems, 21 % problems with self-care, 66 % problems in performing usual activities, and 46 % symptoms of anxiety/depression.

In comparison, studies prospectively following individuals with herpes zoster until the development of postherpetic neuralgia provide information on the natural evolution of the disease over time as well as the evolution of the disease's impact on HRQoL. In MASTER [30], 24 % (63/261) of individuals with herpes zoster developed

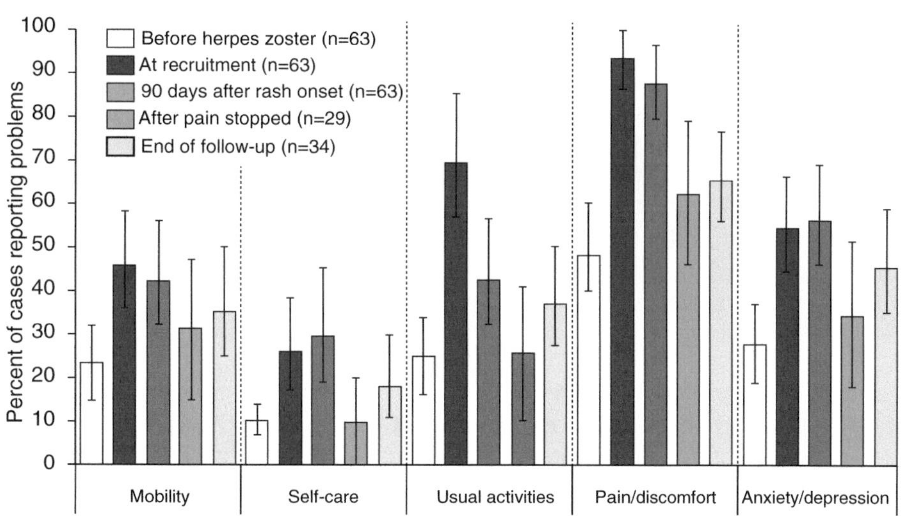

Fig. 11.3 Impact of postherpetic neuralgia on health-related quality of life: EQ-5D domains affected over time for 63 individuals who developed postherpetic neuralgia (MASTER)

postherpetic neuralgia. At the beginning of this phase (i.e., 90 days after rash onset), significant detriments in the five quality of life domains were still reported, compared to the pre-herpes zoster status (Fig. 11.3): 88 % reported pain/discomfort, 42 % mobility problems, 29 % problems with self-care, 43 % problems in performing usual activities, and 56 % anxiety/depression. However, almost half of patients with postherpetic neuralgia (46 %: 29/63) had pain cessation during the 6-month follow-up. Among these patients, the proportion reporting problems for the different domains returned to pre-herpes zoster values after pain stopped. For the other half of patients (54 %, 34/63) still affected by postherpetic neuralgia at the end of the 6-month follow-up, considerable proportions of individuals still reported pain/discomfort (65 %) and symptoms of anxiety/depression (45 %) compared to their situation prior to herpes zoster, although these differences were not statistically significant. Another interesting observation with these data is that the proportions of individuals who reported problems prior to herpes zoster is more important among individuals who developed postherpetic neuralgia compared to the overall group of subjects (Figs. 11.1 and 11.3). This observation supports the hypothesis that the premorbid functional status could influence the risk of developing postherpetic neuralgia.

11.4.2 Impact of Postherpetic Neuralgia on HRQoL: Disease-Specific Instrument

As expected, postherpetic neuralgia also interferes with activities of daily living. The four activities of daily living more frequently affected by postherpetic neuralgia are sleep, enjoyment of life, general activities, and mood [30, 33, 43–45]. More specifically, in MASTER [30], about 30 % of individuals reported interference of pain in these activities at the beginning of the postherpetic neuralgia phase. Among patients who had pain cessation during the follow-up, very few still reported pain interference with activities of daily living after pain cessation (proportions not significantly different from 0). However, nearly 15 % of patients still experiencing pain at the end of follow-up reported interference of pain ≥5 with all activities of daily living examined, except enjoyment of life. Of note, these proportions are much lower compared to the beginning of the postherpetic neuralgia phase, suggesting that for the majority of patients with postherpetic neuralgia, pain decreases over time as well as its detrimental impact on HRQoL (Fig. 11.4).

11.5 Impact of Herpes Zoster and Postherpetic Neuralgia on Employment-Related Productivity

A better understanding of the impact of herpes zoster and postherpetic neuralgia on employment-related productivity is important for several reasons. First, work is a crucial aspect of the overall well-being of individuals, and maintaining work after a

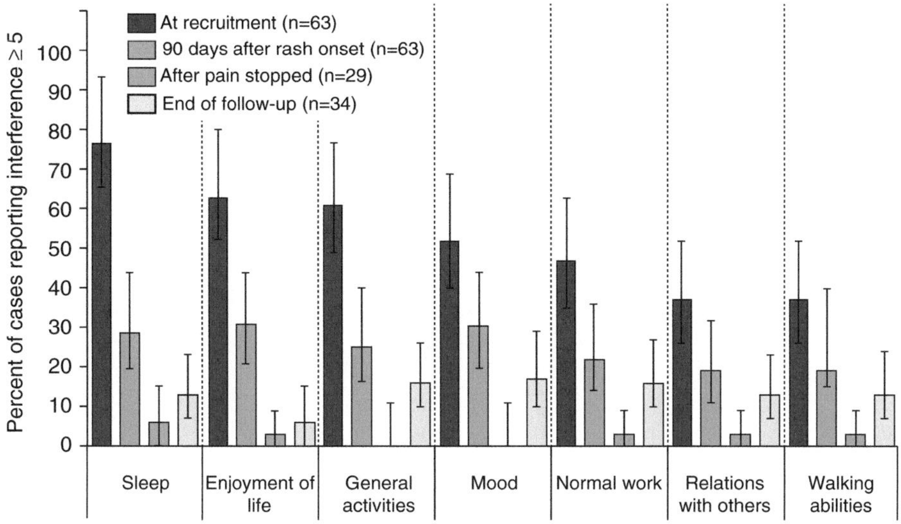

Fig. 11.4 Impact of postherpetic neuralgia on activities of daily living: interference of pain ≥5 for 63 individuals who developed postherpetic neuralgia in MASTER

serious disease is important for the overall HRQoL, including physical and mental health [46, 47]. In addition, earnings from work are necessary to meet basic needs and maintain usual life activities. Second, although the incidence of herpes zoster and postherpetic neuralgia increases with age, many individuals diagnosed with these conditions are still of working age (i.e., <65–70 years old) [48]. Employment-related productivity problems are thus likely to affect a great proportion of individuals confronted with herpes zoster and postherpetic neuralgia. Third, productivity loss associated with different diseases is an important aspect considered in economic analysis of different interventions [49]. For all these reasons, employment-related productivity is becoming an integral part of the overall assessment of the impact of different diseases on HRQoL.

11.5.1 General Concept and Measures

Absenteeism (absences from work because of illness) was traditionally used as the main outcome when examining productivity loss associated with a disease. However, recent evidence suggests that presenteeism (decreased effectiveness at work because of illness) might have a greater impact on the overall productivity loss [50–52]. Absenteeism represents the total time off work because of illness, while presenteeism is estimated by the time at work, weighted by the decreased effectiveness because of illness. For example, individuals are asked to rate, from 0 % (not effective at all) to 100 % (completely effective/able to work like before the disease), their effectiveness at work for a given day or period [53]. The time at work is then multiplied by the percent reduction in effectiveness at work. Total employment-related productivity loss can then be obtained by summing the number of hours or days lost due to absenteeism and/or presenteeism.

11.5.2 Productivity Loss Associated with Herpes Zoster and Postherpetic Neuralgia

A few studies have indicated that herpes zoster and postherpetic have a negative impact on employment-related productivity [34, 35, 53–56]. The majority of employed individuals (50–65 %) reported taking time off work because of herpes zoster, for an average of 26–37 h of absenteeism per employee [34, 53–56]. Moreover, a greater proportion of employed individuals (51–76 %) reported decreased productivity at work because of herpes zoster pain, for an average of 34–84 h of presenteeism [53, 54]. Pain severity and duration were associated with greater employment-related productivity loss [53, 54]. For example, in MASTER, 80 % and 100 % of individuals who developed postherpetic neuralgia reported absenteeism and presenteeism, respectively, for an average of 45 h of absenteeism and 122 h of presenteeism per employee [53]. These studies suggest that presenteeism potentially has a greater impact on productivity loss associated with herpes zoster and postherpetic neuralgia [53, 54]. Overall, the average total employment-related productivity loss associated with herpes zoster and postherpetic neuralgia ranged from 62 to 116 h per employee [53, 54].

11.6 Economic Analyses of the Herpes Zoster Vaccine

11.6.1 General Concept and Definitions

The number of health-care interventions largely exceeds the society's capacity to pay for them. Economic evaluations provide a formal framework to examine the social desirability of an intervention compared with other uses of the same scarce resources [49]. Questions such as "Is vaccination worthwhile compared to alternative use of the same resources?" and "If vaccination is to be implemented, who should be vaccinated and at what age?" can be examined using economic evaluations. In these analyses, both the costs and consequences of an intervention are considered. The costs include (1) the costs for the health-care system (e.g., cost of implementing the intervention and health-care costs saved), (2) the costs for the patients and their family (e.g., out-of-pocket cost for medications, treatments), and (3) employment-related productivity loss (e.g., productivity loss from the patients and caregivers). The consequences of an intervention refer to the potential health benefits that can be gained from the intervention as well as the adverse effects of the intervention, if any. When health benefits are expressed in terms of effectiveness (e.g., life years gained, number of cases averted), wc get cost-effectiveness analysis. When health benefits are expressed in terms of quality-adjusted-life-years gained (QALYs), a measure combining health gains from reduced mortality (quantity of life) and from reduced morbidity (quality of life), we get cost-utility analysis [49]. QALYS are generally estimated using multi-attribute instruments, such as the EQ-5D or the SF-36/12 (through the SF-6D, a utility instrument derived from the SF-36/12 questions [49, 57]). Therefore, the information presented previously about the impact of herpes zoster and postherpetic neuralgia on HRQoL is translated into QALYs to constitute the main outcome of health benefits used in economic analysis.

Although there is no clear recommendation regarding the maximal amount that policy makers are willing to pay for each additional QALY gained, a commonly used rule of thumb is that interventions are considered cost-effective if their cost/QALY gained is below the per capita GPD [58, 59]. Using this cutoff, interventions are generally considered cost-effective in Canada if their cost-utility ratio is less than 40,000$/QALYs gained [60].

11.6.2 Economic Evaluations of Vaccination against Herpes Zoster

Many studies have performed economic evaluation of vaccination against herpes zoster [2, 61–73], and several reviews (systematic and non-systematic) [74–79] have synthetized these results. The great majority of economic analysis suggested that vaccination against herpes zoster could be a cost-effective intervention, under three main conditions: (1) age at vaccination is around 60–70 years old, (2) the vaccine duration of protection is longer than 10 years, and (3) the vaccine reduces the risk of developing postherpetic neuralgia by about 70 % across all age groups, as estimated in the Shingles Prevention Study [8]. The incidence of herpes zoster and postherpetic neuralgia increases with age. On the other hand, vaccine efficacy against herpes zoster significantly decreases with older age at vaccination, and the duration of vaccine protection is still uncertain. Consequently, the optimal age at vaccination is the one where vaccine efficacy is high (i.e., among younger individuals) while offering protection for a sufficient period of time to cover ages at increased risk of herpes zoster and postherpetic neuralgia (i.e., older individuals). For these reasons, most economic analyses suggested that the optimal age at vaccination is around 65 years old. Vaccination before 65 years old offers the highest vaccine efficacy but remains less cost-effective as a result of uncertainties regarding the vaccine duration of protection. Vaccination beyond 65 years is also less cost-effective as a result of decreased vaccine efficacy against herpes zoster. However, recent data suggesting waning of vaccine protection against herpes zoster and postherpetic neuralgia over time [80, 81] could have an important impact on these results and reinforce the need to target optimal ages at vaccination.

11.7 Conclusion

Herpes zoster is a frequent condition, particularly among the elderly. Although the rash and pain associated with herpes zoster usually disappear within 1 month, studies conducted in North America, Europe, and Asia consistently reported that herpes zoster has an important impact on several HRQoL domains. Postherpetic neuralgia is the most common complication of herpes zoster and one of the most challenging to treat. Once again, studies conducted in North America and Europe clearly indicated that postherpetic neuralgia can seriously affect all HRQoL domains, and this negative impact persists as long as pain persists. It is also important to point out that these severe detriments in HRQoL were observed among populations receiving treatments against herpes zoster or postherpetic neuralgia. The main treatment for herpes zoster is antiviral medication, and, to reach optimal efficacy, antivirals have to be started within 72 h of rash onset. Unfortunately, this short time window is difficult to achieve in clinic as many patients may wait for several days before seeking care. Treatments for postherpetic neuralgia are only partially effective and are often associated with severe adverse effects. Thus, available therapeutic options are largely suboptimal, and the treatment of postherpetic neuralgia is too often unsuccessful. These data reinforce the need for supplementary preventive strategies. The safety and efficacy of the vaccine have been demonstrated in large placebo-controlled trials, and several economic analyses have concluded that vaccination against herpes zoster could be cost-effective if the vaccine is given at 60–70 years old and confers protection for at least 10 years. In light of this evidence, vaccination against herpes zoster represents a promising strategy to reduce the overall burden of herpes zoster and postherpetic neuralgia. A new vaccine is on the horizon which appears more effective and more broadly applicable to immunosuppressed patients [82].

References

1. Brisson M, Edmunds WJ, Law B et al (2001) Epidemiology of varicella zoster virus infection in Canada and the United Kingdom. Epidemiol Infect 127(2):305–314
2. Edmunds WJ, Brisson M, Rose JD (2001) The epidemiology of herpes zoster and potential cost-effectiveness of vaccination in England and Wales. Vaccine 19(23–24):3076–3090
3. Insinga RP, Itzler RF, Pellissier JM, Saddier P, Nikas AA (2005) The incidence of herpes zoster in a United States administrative database. J Gen Intern Med 20(8):748–753
4. Oxman MN (2000) Clinical manifestation of herpes zoster. In: Arvin AM, Gerson AA (eds) Varicella zoster virus: virology and clinical management. Cambridge University Press, Cambridge, pp. 246–275
5. Opstelten W, Zuithoff NP, van Essen GA et al (2007) Predicting postherpetic neuralgia in elderly primary care patients with herpes zoster: prospective prognostic study. Pain 132(Suppl 1):S52–S59
6. Schmader KE (2002) Epidemiology and impact on quality of life of postherpetic neuralgia and painful diabetic neuropathy. Clin J Pain 18(6):350–354
7. Dworkin R, Schmader K (2001) The epidemiology and natural history of herpes zoster and postherpetic neuralgia. Herpes zoster and postherpetic neuralgia, 2nd edn. Elsevier, Amsterdam, pp. 39–65

8. Oxman MN, Levin MJ, Johnson GR et al (2005) A vaccine to prevent herpes zoster and postherpetic neuralgia in older adults. N Engl J Med 352(22):2271–2284
9. Scott FT, Leedham-Green ME, Barrett-Muir WY et al (2003) A study of shingles and the development of postherpetic neuralgia in East London. J Med Virol 70(Suppl 1):S24–S30
10. Drolet M, Brisson M, Schmader K et al (2010) Predictors of postherpetic neuralgia among patients with herpes zoster: a prospective study. J Pain 11(11):1211–1221
11. Fayers PM, Machin D (2000) Quality of life – assessment, analysis & interpretation
12. International society for quality of life research. What is health-related quality of life research? Available at http://www.isoqol.org/about-isoqol/what-is-health-related-quality-of-life-research. Accessed 27 Apr 2016.
13. World Health Organization Quality of Life. WHOQOL Measuring quality of life. 1997. Available at http://www.who.int/mental_health/media/68.pdf. Accessed 27 Apr 2016.
14. Centers for Disease control and Prevention. Helath-related Quality of life (HRQOL). Available at http://www.cdc.gov/hrqol/concept.htm. Accessed 27 Apr 2016.
15. Brazier J, Jones N, Kind P (1993) Testing the validity of the Euroqol and comparing it with the SF-36 health survey questionnaire. Qual Life Res 2(3):169–180
16. Huang IC, Willke RJ, Atkinson MJ, Lenderking WR, Frangakis C, Wu AW (2007) US and UK versions of the EQ-5D preference weights: does choice of preference weights make a difference? Qual Life Res 16(6):1065–1072
17. Euroqol. What is EQ-5D. Available at http://www.euroqol.org/. Accessed 27 Apr 2016.
18. Ware J Jr, Kosinski M, Keller SD (1996) A 12-Item Short-Form Health Survey: construction of scales and preliminary tests of reliability and validity. Med Care 34(3):220–233
19. Gandek B, Ware JE, Aaronson NK et al (1998) Cross-validation of item selection and scoring for the SF-12 Health Survey in nine countries: results from the IQOLA Project. International Quality of Life Assessment. J Clin Epidemiol 51(11):1171–1178
20. McHorney CA, Ware JE Jr, AE R (1993) The MOS 36-Item Short-Form Health Survey (SF-36): II. Psychometric and clinical tests of validity in measuring physical and mental health constructs. Med Care 31(3):247–263
21. McHorney CA, Ware JE, Jr, Lu JF, Sherbourne CD (1994). The MOS 36-item Short-Form Health Survey (SF-36): III. Tests of data quality, scaling assumptions, and reliability across diverse patient groups. Med Care; 32(1): 40–66.
22. Fryback DG, Dunham NC, Palta M et al (2007) US norms for six generic health-related quality-of-life indexes from the National Health Measurement study. Med Care 45(12):1162–1170
23. Johnson JA, Pickard AS (2000) Comparison of the EQ-5D and SF-12 health surveys in a general population survey in Alberta. Canada Med Care 38(1):115–121
24. Ware JE, Kosinski M (2001) SF-36 physical and mental health summary scales: a manual for users of version 1, 2nd edn. Lincoln, QualityMetric Incorporated
25. Hopman WM, Towheed T, Anastassiades T et al (2000) Canadian normative data for the SF-36 health survey. Canadian Multicentre Osteoporosis Study Research Group. CMAJ 163(3):265–271
26. Coplan PM, Schmader K, Nikas A et al (2004) Development of a measure of the burden of pain due to herpes zoster and postherpetic neuralgia for prevention trials: adaptation of the brief pain inventory. J Pain 5(6):344–356
27. Gnann JW Jr, Whitley RJ (2002) Clinical practice. Herpes zoster. N Engl J Med 347(5):340–346
28. Head H, Campbell AW, Kennedy PG (1997) The pathology of Herpes Zoster and its bearing on sensory localisation. Rev Med Virol 7(3):131–143
29. Schmader KE, Sloane R, Pieper C et al (2007) The impact of acute herpes zoster pain and discomfort on functional status and quality of life in older adults. Clin J Pain 23(6):490–496
30. Drolet M, Brisson M, Schmader KE et al (2010) The impact of herpes zoster and postherpetic neuralgia on health-related quality of life: a prospective study. CMAJ 182(16):1731–1736
31. Katz J, Cooper EM, Walther RR, Sweeney EW, Dworkin RH (2004) Acute pain in herpes zoster and its impact on health-related quality of life. Clin Infect Dis 39(3):342–348

32. Chidiac C, Bruxelle J, Daures JP et al (2001) Characteristics of patients with herpes zoster on presentation to practitioners in France. Clin Infect Dis 33(1):62–69
33. Bouhassira D, Chassany O, Gaillat J et al (2012) Patient perspective on herpes zoster and its complications: an observational prospective study in patients aged over 50 years in general practice. Pain 153(2):342–349
34. Gater A, Abetz-Webb L, Carroll S, Mannan A, Serpell M, Johnson R (2014) Burden of herpes zoster in the UK: findings from the zoster quality of life (ZQOL) study. BMC Infect Dis 14:402
35. Bricout H, Perinetti E, Marchettini P et al (2014) Burden of herpes zoster-associated chronic pain in Italian patients aged 50 years and over (2009-2010): a GP-based prospective cohort study. BMC Infect Dis 14:637
36. Aunhachoke K, Bussaratid V, Chirachanakul P et al (2011) Measuring herpes zoster, zoster-associated pain, post-herpetic neuralgia-associated loss of quality of life, and healthcare utilization and costs in Thailand. Int J Dermatol 50(4):428–435
37. Song H, Lee J, Lee M et al (2014) Burden of illness, quality of life, and healthcare utilization among patients with herpes zoster in South Korea: a prospective clinical-epidemiological study. Int J Infect Dis 20:23–30
38. Tsai TF, Yao CA, Yu HS et al (2015) Herpes zoster-associated severity and duration of pain, health-related quality of life, and healthcare utilization in Taiwan: a prospective observational study. Int J Dermatol 54(5):529–536
39. Opstelten W, Mauritz JW, de Wit NJ, van Wijck AJ, Stalman WA, van Essen GA (2002) Herpes zoster and postherpetic neuralgia: incidence and risk indicators using a general practice research database. Fam Pract 19(5):471–475
40. Whitley RJ, Weiss HL, Soong SJ, Gnann JW (1999) Herpes zoster: risk categories for persistent pain. J Infect Dis 179(1):9–15
41. Coen PG, Scott F, Leedham-Green M et al (2006) Predicting and preventing post-herpetic neuralgia: are current risk factors useful in clinical practice? Eur J Pain 10(8):695–700
42. Helgason S, Petursson G, Gudmundsson S, Sigurdsson JA (2000) Prevalence of postherpetic neuralgia after a first episode of herpes zoster: prospective study with long term follow up. BMJ 321(7264):794–796
43. Oster G, Harding G, Dukes E, Edelsberg J, Cleary PD (2005) Pain, medication use, and health-related quality of life in older persons with postherpetic neuralgia: results from a population-based survey. J Pain 6(6):356–363
44. Serpell M, Gater A, Carroll S, Abetz-Webb L, Mannan A, Johnson R (2014) Burden of post-herpetic neuralgia in a sample of UK residents aged 50 years or older: findings from the Zoster Quality of Life (ZQOL) study. Health Qual Life Outcomes 12:92
45. Laurent B, Vicaut E, Leplege A, Bloch K, Leutenegger E (2014) Prevalence and impact on quality of life of post-herpetic neuralgia in French medical centers specialized in chronic pain management: the ZOCAD study. Med Mal Infect 44(11–12):515–524
46. Anderson NB, Armstead CA (1995) Toward understanding the association of socioeconomic status and health: a new challenge for the biopsychosocial approach. Psychosom Med 57(3):213–225
47. Federal provincial and territorial advisory committee on population health. Strategies for population health: investing in the health of Canadians (1994) Canada: Minister of Supply and Services
48. Yawn BP, Saddier P, Wollan PC, St Sauver JL, Kurland MJ, Sy LS (2007) A population-based study of the incidence and complication rates of herpes zoster before zoster vaccine introduction. Mayo Clin Proc 82(11):1341–1349
49. Drummond MF, Sculpher MJ, Torrance GW, O'Brien B, Stoddart GL (2005) Methods for the economic evaluation of health care programmes, 3rd edn. Oxford Medical Publication, Oxford
50. Aronsson G, Gustafsson K, Dallner M (2000) Sick but yet at work. An empirical study of sickness presenteeism. J Epidemiol Community Health 54(7):502–509
51. Prochaska JO, Evers KE, Johnson JL et al (2011) The well-being assessment for productivity: a well-being approach to presenteeism. J Occup Environ Med 53(7):735–742

52. Stewart WF, Ricci JA, Chee E, Morganstein D (2003) Lost productive work time costs from health conditions in the United States: results from the American Productivity Audit. J Occup Environ Med 45(12):1234–1246
53. Drolet M, Levin MJ, Schmader KE et al (2012) Employment related productivity loss associated with herpes zoster and postherpetic neuralgia: a 6-month prospective study. Vaccine 30(12):2047–2050
54. Singhal PK, Makin C, Pellissier J, Sy L, White R, Saddier P (2011) Work and productivity loss related to herpes zoster. J Med Econ 14(5):639–645
55. Scott FT, Johnson RW, Leedham-Green M, Davies E, Edmunds WJ, Breuer J (2006) The burden of Herpes Zoster: a prospective population based study. Vaccine 24(9):1308–1314
56. White RR, Lenhart G, Singhal PK et al (2009) Incremental 1-year medical resource utilization and costs for patients with herpes zoster from a set of US health plans. Pharmacoeconomics 27(9):781–792
57. Brazier JE, Roberts J (2004) The estimation of a preference-based measure of health from the SF-12. Med Care 42(9):851–859
58. World Health Organization (2001) Investing in health for economic development. Report of the Commission on Macroeconomics and Health. Geneva
59. World Health Organization. WHO-CHOICE: World Health Organization statistical information system: choosing interventions that are cost effective. Available: http://www.who.int/choice/en. Accessed 30 May 2016.
60. Baldwin JR, Brown M, Maynard JP (2005) Insights on the Canadian Economy: interprovincial differences in GDP per capita, labour productivity and work intensity: 1990–2003. StatCan Catalogue no. 11–624-MIE-No.011
61. Hornberger J, Robertus K (2006) Cost-effectiveness of a vaccine to prevent herpes zoster and postherpetic neuralgia in older adults. Ann Intern Med 145(5):317–325
62. Rothberg MB, Virapongse A, Smith KJ (2007) Cost-effectiveness of a vaccine to prevent herpes zoster and postherpetic neuralgia in older adults. Clin Infect Dis 44(10):1280–1288
63. Pellissier JM, Brisson M, Levin MJ (2007) Evaluation of the cost-effectiveness in the United States of a vaccine to prevent herpes zoster and postherpetic neuralgia in older adults. Vaccine 25(49):8326–8337
64. Brisson M, Pellissier JM, Camden S, Quach C, De Wals P (2008) The potential cost-effectiveness of vaccination against herpes zoster and post-herpetic neuralgia. Hum Vaccin 4(3):238–245
65. van Hoek AJ, Gay N, Melegaro A, Opstelten W, Edmunds WJ (2009) Estimating the cost-effectiveness of vaccination against herpes zoster in England and Wales. Vaccine 27(9):1454–1467
66. Najafzadeh M, Marra CA, Galanis E, Patrick DM (2009) Cost effectiveness of herpes zoster vaccine in Canada. Pharmacoeconomics 27(12):991–1004
67. Moore L, Remy V, Martin M, Beillat M, McGuire A (2010) A health economic model for evaluating a vaccine for the prevention of herpes zoster and post-herpetic neuralgia in the UK. Cost Eff Resour Alloc 8:7
68. Annemans L, Bresse X, Gobbo C, Papageorgiou M (2010) Health economic evaluation of a vaccine for the prevention of herpes zoster (shingles) and post-herpetic neuralgia in adults in Belgium. J Med Econ 13(3):537–551
69. van Hoek AJ, Melegaro A, Gay N, Bilcke J, Edmunds WJ (2012) The cost-effectiveness of varicella and combined varicella and herpes zoster vaccination programmes in the United Kingdom. Vaccine 30(6):1225–1234
70. van Lier A, van Hoek AJ, Opstelten W, Boot HJ, de Melker HE (2010) Assessing the potential effects and cost-effectiveness of programmatic herpes zoster vaccination of elderly in the Netherlands. BMC Health Serv Res 10:237
71. Bilcke J, Marais C, Ogunjimi B, Willem L, Hens N, Beutels P (2012) Cost-effectiveness of vaccination against herpes zoster in adults aged over 60 years in Belgium. Vaccine 30(3):675–684

72. Bresse X, Annemans L, Preaud E, Bloch K, Duru G, Gauthier A (2013) Vaccination against herpes zoster and postherpetic neuralgia in France: a cost-effectiveness analysis. Expert Rev Pharmacoecon Outcomes Res 13(3):393–406
73. Le P, Rothberg MB (2015) Cost-effectiveness of Herpes Zoster vaccine for persons aged 50 years. Ann Intern Med 163(7):489–497
74. Damm O, Ultsch B, Horn J, Mikolajczyk RT, Greiner W, Wichmann O (2015) Systematic review of models assessing the economic value of routine varicella and herpes zoster vaccination in high-income countries. BMC Public Health 15:533
75. de Boer PT, Wilschut JC, Postma MJ (2014) Cost-effectiveness of vaccination against herpes zoster. Hum Vaccin Immunother 10(7):2048–2061
76. Kawai K, Preaud E, Baron-Papillon F, Largeron N, Acosta CJ (2014) Cost-effectiveness of vaccination against herpes zoster and postherpetic neuralgia: a critical review. Vaccine 32(15):1645–1653
77. Peden AD, Strobel SB, Forget EL (2014) Is herpes zoster vaccination likely to be cost-effective in Canada? Can J Public Health 105(4):e287–95
78. Szucs TD, Pfeil AM (2013) A systematic review of the cost effectiveness of herpes zoster vaccination. Pharmacoeconomics 31(2):125–136
79. Drolet M, Oxman MN, Levin MJ et al (2013) Vaccination against herpes zoster in developed countries: state of the evidence. Hum Vaccin Immunother 9(5):1177–1184
80. Schmader KE, Oxman MN, Levin MJ et al (2012) Persistence of the efficacy of zoster vaccine in the Shingles Prevention Study and the Short-Term Persistence Substudy. Clin Infect Dis 55(10):1320–1328
81. Morrison VA, Johnson GR, Schmader KE et al (2015) Long-term persistence of zoster vaccine efficacy. Clin Infect Dis 60(6):900–909
82. Lal H et al (2015) Efficacy of an adjuvanted herpes zoster subunit vaccine in older adults. N Engl J Med 372:2087–2096

Chapter 12
The Pathology of Postherpetic Neuralgia and Postherpetic Itch

Anne Louise Oaklander

12.1 Introduction

Postherpetic neuralgia (PHN) is the most studied focal neuropathic pain condition. In addition to being common, zoster affects all demographic groups and is unusually easy to diagnose and neuroanatomically localize from its characteristic dermatomal rash and scars. The day of onset of the rash is usually remembered or indicated in the medical records. These characteristics make it ideal for research. Studies and trials of PHN have informed our knowledge of neuropathic pain and how to treat it and brought neuropathic itch into awareness. This chapter's focus is on the pathological findings shingles causes in the sensory ganglia, peripheral nerves, and spinal cord. Advanced imaging studies including anatomical and functional magnetic resonance imaging and analyses of brain perfusion are beginning to reveal zoster's secondary and tertiary effects on the central nervous system [13, 25, 56].

12.2 Early Pathological Studies

Reports about the pathological consequences of shingles extend from the nineteenth century. Bright [4] associated shingles with the nervous system as well as the skin, and Friedrich Wilhelm Felix von Bärensprung [47, 48] first linked herpes zoster to hemorrhagic lesions of the spinal dorsal root ganglion (DRG), the nidus of

A.L. Oaklander, MD, PhD (✉)
Department of Neurology, Massachusetts General Hospital and Harvard Medical School, Boston, MA, USA

Department of Pathology (Neuropathology), Massachusetts General Hospital, 275 Charles St. /Warren Bldg. 806, Boston, MA 02114, USA
e-mail: aloaklander@mgh.harvard.edu

C.P.N. Watson et al. (eds.), *Herpes Zoster: Postherpetic Neuralgia and Other Complications*, DOI 10.1007/978-3-319-44348-5_12

varicella-zoster virus (VZV) latency and reactivation. Later nineteenth-century reports included two cases of zoster with hemorrhage and inflammation in the gasserian ganglion [42, 53] and four cases with hemorrhage, inflammation, and destruction of ganglion cells in thoracic and cervical ganglia [8, 23]. Chandelux [5] described one chronic case with ganglion cell and connective tissue loss.

12.3 Head and Campbell [17]

The acute and chronic pathology of zoster was comprehensively described in 1900 in a 170-page landmark monograph "The pathology of Herpes Zoster and its bearing on sensory localization" written by neurologists Sir Henry Head (1861–1940) (Fig. 12.1) and A. W. Campbell and published in the journal *Brain*. As discussed in Oaklander [32], this masterpiece remains not only the primary reference about zoster pathology, but their meticulous drawings of the cutaneous locations of zoster rashes in almost 500 patients (Fig. 12.2a) enabled them to map the locations of the sensory dermatomes (Fig. 12.2b), information used every day in medical practice. It also included meticulous drawings of the gross and microscopic anatomic pathology from 21 shingles patients who later died for various

Fig. 12.1 Sir Henry Head (1861–1940) was a pioneer in academic medicine who published widely in neurology. He was also an accomplished medical educator and editor, poet, and literary critic. He coauthored the definitive work on the clinical and pathological features of zoster (Head and Campbell [17]). His academic awards included the Royal Medal in 1908, knighthood in 1927, the Gold Medal of the Royal Society of Medicine in 1929, and Fellowship in the Royal Society (Image credit: Wellcome Library, London)

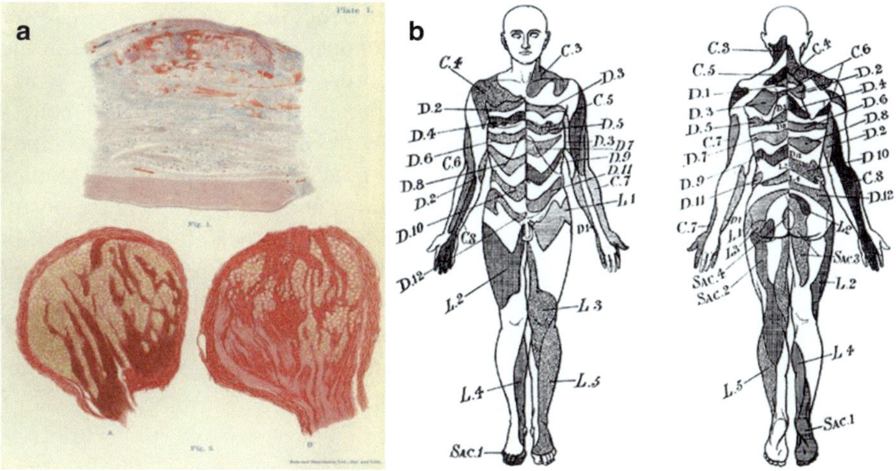

Fig. 12.2 Head and Campbell correlated their meticulous of mapping the location of zoster rashes on the skin of patients under their care at the Rainhill Asylum with later pathologic localization of the damage that person's shingles had caused to their nervous system (From Head and Campbell [17]). (**a**) (*Plate 1*). The image at the top depicts a longitudinal section of the right seventh dorsal ganglion of the side in Case 2, James T., age 35, who died from syphilitic heart failure 8 days after the onset of his zoster eruption. At this early stage, the corresponding dorsal roots had not yet begun to degenerate (not shown). Stained with hematoxylin and eosin. The images at the bottom depict Head and Campbell's colored drawing of both seventh dorsal ganglia from Case 9, Richard K, aged 36, who died from syphilitic dementia 153 days after the onset of his zoster eruption. The unaffected contralateral left ganglion is at the left, and the right zoster-affected seventh dorsal ganglion is at the lower right. Stained by Marchi's method and counterstained by van Gieson's picro hematoxylin method. (**b**) This monolithic project enabled them to map the classic cutaneous dermatomes

reasons, often tuberculosis or syphilitic dementia. These included 8 patients with acute zoster (death occurring 4–30 days after the rash) and 13 who died 57–790 days after onset. Only one was clearly noted to have lingering pain at the time of death. Head and Campbell corroborated earlier reports of ganglionic changes – acute infiltration of small round inflammatory cells, extravasation of blood, destruction of sensory neurons, axons, and satellite cells (Fig. 12.2a) – and also described the end stage of scarring in the affected part of the ganglion and thickening of its sheath. They established that only one ganglion is affected by zoster (in immunocompetent patients), that lesions include hemorrhage and inflammation, and that the dorsal part of the ganglion is most often affected. They reported that ganglia can appear normal after mild eruptions and mapped the somatotopic organization of the dorsal columns and where they terminate in the brain. Head and Campbell added new information about the time, course, and nature of degeneration and subsequent scarring in sensory roots and peripheral nerves. They established these as secondary to the ganglionic lesion in most cases, although in two cases they found inflammation and hemorrhage in the peripheral nerve as well as the ganglion. They described in detail the location, onset, slower resolution, and complete disappearance of secondary degenerative change in the ipsilateral posterior column of the spinal cord. Degenerating central axons appeared as early as day 9 after rash onset and persisted as long as 5 and 9 months after the rash. By tracing degenerating tracts with Marchi staining, they helped establish somatosensory anatomy. Of particular note in light of subsequent studies of patients with PHN [50] is their lack of documentation of changes in the posterior horn of the spinal cord, although they studied this area carefully. Presumably the vast majority of their patients did not have PHN. Head labored in ignorance of the cause of shingles, since viruses had not yet been discovered, but we still cannot answer his questions about why shingles prefers the thoracic and V_1 ophthalmic ganglia, why only one ganglion is affected, and why the dorsal part of the ganglion is preferentially afflicted.

12.4 Later Pathological Studies of Zoster-Affected Ganglia and Central Nervous Tissue

Denny-Brown et al. [7] reported the pathology of two patients with acute thoracic zoster (one with and one without pain) and one with the rare Ramsay-Hunt syndrome (facial nerve paralysis from cranial zoster). They corroborated the presence of hemorrhagic ganglionic inflammation with one ganglion containing a cyst. All three patients had inflammatory central nervous system changes, findings first reported in the French [11, 24] and German literature [52]. They described "poliomyelitis" (inflammation in the gray matter) of the anterior and dorsal horns accompanied by leptomeningitis and pleocytosis in the cerebrospinal fluid. They also found peripheral mononeuritis in nerves distal to the ganglion and in the anterior root. They used this to explain the facial paralysis in their Ramsay-Hunt case since the geniculate ganglion was completely spared, with a ganglionitis in this case involving the second cervical ganglion. In that same year Raymond D. Adams [1] described another case of direct cellular infiltration of the facial nerve during herpetic infection.

The advent of immunofluorescence and electron microscopy yielded further insights. Esiri and Tomlinson [10] demonstrated the presence of VZV within the trigeminal nerve and ganglion of a patient with ophthalmic zoster. Ghatak and Zimmerman [14] visualized VZV particles in the affected spinal ganglia of a case of HZ associated with Hodgkin's disease, plus the intranuclear inclusions in ganglionic neurons described earlier by Cheatham [6] and McCormick et al. [27]. Ghatak also found multinucleated giant cells and alterations suggestive of mitosis within neurons in the sensory ganglia and speculated that these might reflect zoster superimposed on Hodgkin's disease. Imaging studies have shown that PHN is more frequent in shingles patients with involvement of the central nervous system [54].

12.5 Pathological Studies of Nerves Extending from Zoster-Affected Ganglia

Head and Campbell [17] examined the nerves extending from zoster-affected ganglia and demonstrated that changes develop later than in the affected ganglia. Noordenbos [30] first analyzed the size distribution of axons in four intercostal nerve biopsies from patients with PHN. He found fewer large myelinated fibers and more smaller myelinated axons and postulated that zoster preferentially damages larger DRG cells with consequent loss of their thicker myelinated fibers and that "remaining sensibility was subserved by fibers of the slowest conduction rate." This led him to the idea that interactions between different fiber types might influence pain, and the concept that "fast blocks slow." His theories helped stimulate Ronald Melzack and Patrick Wall to propose the gate control theory of pain [28], but his interpretation that small-caliber neurons were spared by zoster was wrong. The autopsy studies had already made it clear that no type of neuron is preferentially spared. More likely, the increased numbers of thin axons in zoster-affected nerves reflected regenerating axons, which are not distinguishable from preserved small-diameter axons by light microscopy. All types of peripheral neurons extend thin unmyelinated axon sprouts after injury.

Next, Zacks et al. [55] studied four peripheral nerve biopsies from patients with prior HZ, two with and two without pain. They found evidence for Wallerian degeneration, which they presumed was secondary to perikaryal damage and direct inflammatory changes within associated proximal nerves. They found that small fibers were present early but eventually disappeared, which is the natural history of axonal sprouting and regrowth. Later, collagen proliferated and the nerves became masses of fibrous tissue containing rare small myelinated fibers. They were not able to correlate the presence and severity of fibrosis with postherpetic neuralgia. Although they suggested that the paucity of surviving fibers implicated central mechanisms in PHN, it was obvious from their patients' symptoms (most had allodynia, hyperesthesia and, dysesthesia) that peripheral mechanisms also contributed.

12.6 Autopsy Attempts to Discover the Pathologic Signature of PHN

Very few early autopsied cases were documented as having PHN close to the time of death (Head and Campbell case 7, Denny-Brown case 2, Zacks cases 2 and 3, the 4 Noordenbos biopsies). Given shorter life expectancies in the eighteenth century, and the fact Head and Campbell's subjects were incarcerated in an asylum, many with infectious encephalitis, few among the early autopsied cases were likely to have PHN, given that older age is the biggest risk factor for developing PHN after shingles [19].

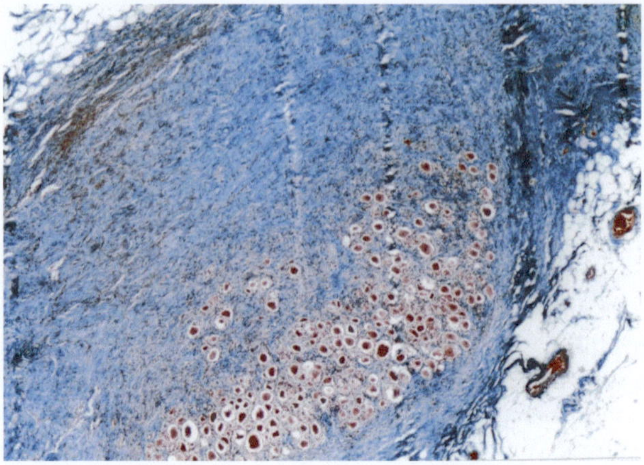

Fig. 12.3 This right T8 dorsal root ganglion from a patient with 5 years of PHN shows a large area of pan-cellular necrosis and fibrosis (scarring) in only this one DRG (*upper left*). Residual normal-appearing ganglion is at lower right (Masson trichrome × 10) (From: Watson et al. [50])

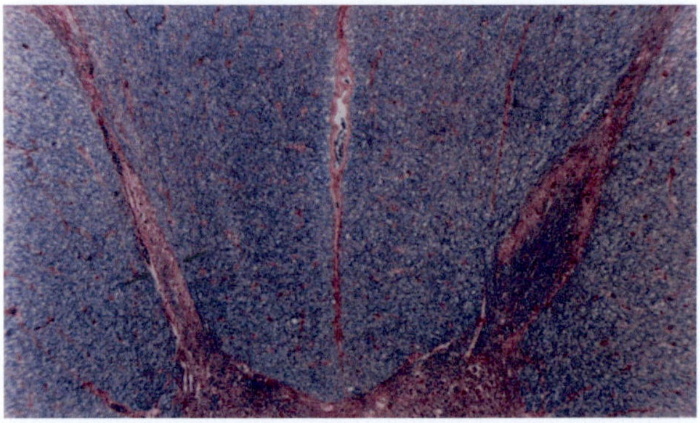

Fig. 12.4 Atrophy of the dorsal horn of the spinal cord on the shingles-affected side (*left side*). This involves loss of myelin as evidenced by the reduced darker staining of the central dorsal horn compared with the contralesional control side) (MBP × 2.5). In contrast to the single DRG affected, dorsal horn atrophy and cell, axon, and myelin loss with DRG fibrosis were found only in patients with persistent PHN (From: Watson et al. [50])

Noordenbos [30] and Zacks et al. [55] first focused on PHN and attempted to discover its pathogenesis. Then Watson et al. [49] described autopsy findings from a patient with 5 years of severe thoracic PHN. Despite involvement of only one ganglion (Fig. 12.3), the dorsal-horn atrophy (Fig. 12.4) sometimes affected several adjacent spinal-cord segments. This finding offers a pathological explanation for why PHN usually symptoms typically affect several contiguous dermatomes (Fig. 12.5) even though zoster rashes and ganglion damage affect only one ganglion in most immunocompetent patients. The cause is that primary afferents branch superficially in Lissauer's tract and send collaterals a few segments rostrally and caudally before penetrating the dorsal horn, so more than one dorsal-

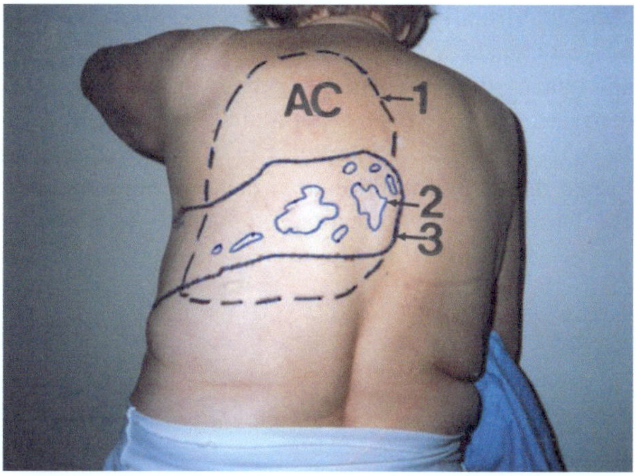

Fig. 12.5 Patient AC: (*1*) margin of area of allodynia to skin stroking; (*2*) postherpetic scarring; (*3*) margin of area of reduced sensation (From: Herpes Zoster and Postherpetic Neuralgia. 1st ed. 1993 Elsevier page 156)

horn segment becomes deafferented and vulnerable to transsynaptic degeneration after VZV reactivation in a single sensory ganglion. Further autopsies comparing three patients with severe PHN with two without PHN after their shingles [50] confirmed ganglionic changes, dorsal-horn atrophy, and persistent dorsal-horn labeling for substance P and CGRP. Dorsal horn atrophy and cell, axon, and myelin loss with fibrosis in the sensory ganglion were found only in the patients with PHN, whereas loss of myelin and axons in the nerve and/or sensory root was found in cases with and without pain. Morphometric study of the peripheral nerves corroborated the finding of fewer large myelinated fibers and more thin ones, consistent with sprouting of remaining DRG neurons.

12.7 The Pathology of Facial (Trigeminal) Zoster and PHN

Although the ophthalmic division of the trigeminal ganglion (V_1) is the second most common location for zoster after the torso, there are fewer pathological studies of trigeminal zoster [17, 42, 53]. Two of the Head and Campbell [17] cases were trigeminal, one with rash in the mandibular V_3 division 30 day premortem and the other with the rash in V_1 6 months before death. They described similar pathology in the gasserian (trigeminal) ganglia as in DRG. Secondary central degeneration was detected only in the V_3 case, in the trigeminal root, and spinal tract of V. Reske-Nielsen et al. [38] autopsied three V_1 cases with death 22–117 days after the eruption – pain status unmentioned – and found lesions in the ophthalmic part of the gasserian ganglion and in the mesencephalic trigeminal nucleus. In contrast, autopsy study of a case of severe V_1 PHN after herpes zoster ophthalmicus (HZO) revealed demyelination, loss of myelinated fibers, fibrosis, and axon sprouting in the ophthalmic and supraorbital nerves, but no major central pathology [51]. Hamrah et al.'s [16] study of 27 patients with V_1 zoster used the newest method – in vivo corneal confocal microscopy (IVCM) – to characterize the neuropathology within patients' corneas. Virtually all corneal innervation is nociceptive, explaining why even the lightest contact with the cornea is painful. IVCM permits noninvasive visualization and quantitation of the nerve endings in the cornea of living patients. This method demonstrated that degeneration and pathology of the subbasal nerve plexus (Fig. 12.6) is correlated with the severity of loss of corneal sensation. It additionally confirmed the presence of bilateral distal neurite loss in cases of clinically unilateral HZO as described on the torso (vide infra).

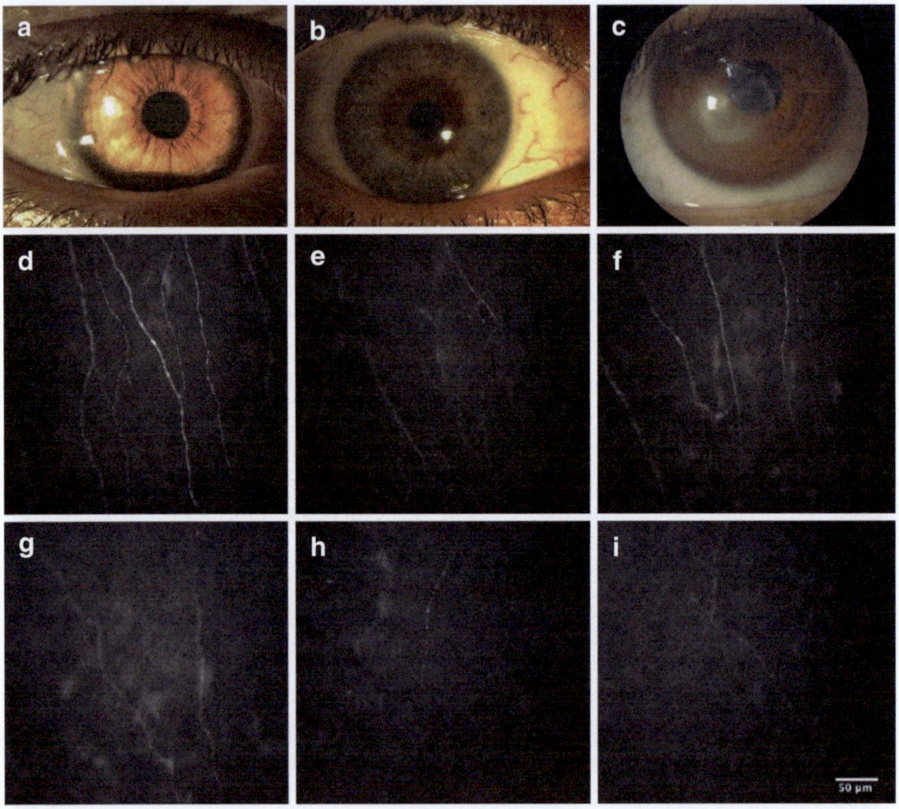

Fig. 12.6 Ocular pathology in herpes zoster ophthalmicus (HZO). (**a**) Slit lamp photo of normal cornea. (**b**) Slit lamp photo of contralateral clinically unaffected eye of patient with unilateral HZO. (**c**) Slit lamp photo of eye affected by herpes zoster ophthalmicus demonstrating corneal inflammation and clouding. (**d**) In vivo corneal confocal microscopy (IVCM) image of corneal subbasal nerve plexus within a normal eye. (**e**) IVCM image of subbasal nerve plexus in clinically unaffected contralateral eye of HZO patient reveals fewer and shorter fibers (bilateral pathology). (**f**) IVCM image of subbasal nerve plexus in affected eye of patient with HZO and no corneal sensory loss demonstrates relatively preserved subbasal fibers with shorter neurites. (**g**) IVCM image of subbasal nerve plexus in affected eye of patient with HZO and mild corneal sensory loss demonstrates fewer and shorter subbasal neurites. (**h, i**) IVCM image of subbasal nerve plexus in HZO-affected eyes with severe corneal sensory loss demonstrates far fewer, thin subbasal neurites (Reprinted from: Hamrah et al. [16])

12.8 Skin-Biopsy Studies of Living Patients with Prior Shingles

Immunohistochemically labeling skin biopsies with the pan-axonal marker anti-PGP9.5 permits neuropathological data to be gathered from living patients. For the first time, it enabled enough PHN patients were studied to discover a pathological correlate for their pain. Earlier studies of sensory function in patients previously affected by zoster had yielded murky results [3, 34, 40], perhaps because psychophysical techniques rely on subjective perceptions in contrast to pathology.

Unlike markers of specific peptides, transmitters, or receptors, PGP9.5 appears to label all types of neural processes coursing through non-neural tissue and make them visible using light microscopy [21, 46]. The neural identity of PGP9.5-immunopositive cutaneous fibers has been confirmed by electron microscopy [18]; the antigen is neuronal cytoplasmic protein gene product 9.5 (PGP9.5)/ubiquitin-C-terminal hydrolase 1 (UCHL-1), a protease involved in processing of ubiquitin precursors and ubiquinated proteins. In the epidermis, the axon bundles in the dermis spread into distinct terminal branches (neurites). These come almost exclusively from the nociceptive C- and A-delta nerve endings that transduce and transmit signals of pain and itch [44], hence the relevance to PHN and postherpetic itch (PHI).

Ebert [9] performed the first PGP9.5 studies of previously zoster-affected skin and reported that segmental axonal losses extended to the skin. Rowbotham et al. [41] studied allodynic PHN-affected skin and found no correlation between density of epidermal innervation and severity of allodynia, perhaps because the large myelinated "touch fibers" end in the dermis and are not captured in measurements of epidermal innervation. The next skin-biopsy study that was of 34 patients with or without PHN after shingles at least 3 months earlier [31] reported strong correlations between the presence of PHN and the severity of loss of epidermal innervation. Subjects with PHN averaged 339 ± 97 neurites/mm^2 of skin surface area in their previously shingles-affected dermatome as compared with 1661 ± 262 neurites/mm^2 in subjects without PHN pain after shingles [31]. A related biopsy study of previously shingles-affected skin from 38 subjects ([33]; Fig. 12.7) identified a threshold of 650 neurites/mm^2 that predicted who did or did not have PHN, implicating peripheral deafferentation as a mechanism of PHN pain. The presence of a threshold (step-function relationship) (Fig. 12.8) suggested that neuropathic pain did not occur despite degeneration of nociceptors as long as a minimum number survived, and neuropathic pain (PHN) ensured abruptly when neuronal densities dropped slightly below the threshold. It was proposed that PHN was in part a somatosensory hallucination, analogous to tinnitus after hearing loss and the Charles Bonnet syndrome after loss of primary afferent visual neurons [20, 43]. Studies of risk factors for developing PHN after zoster support this hypothesis, since rash severity and duration (surrogate markers for ganglion cell loss) add risk for PHN [12].

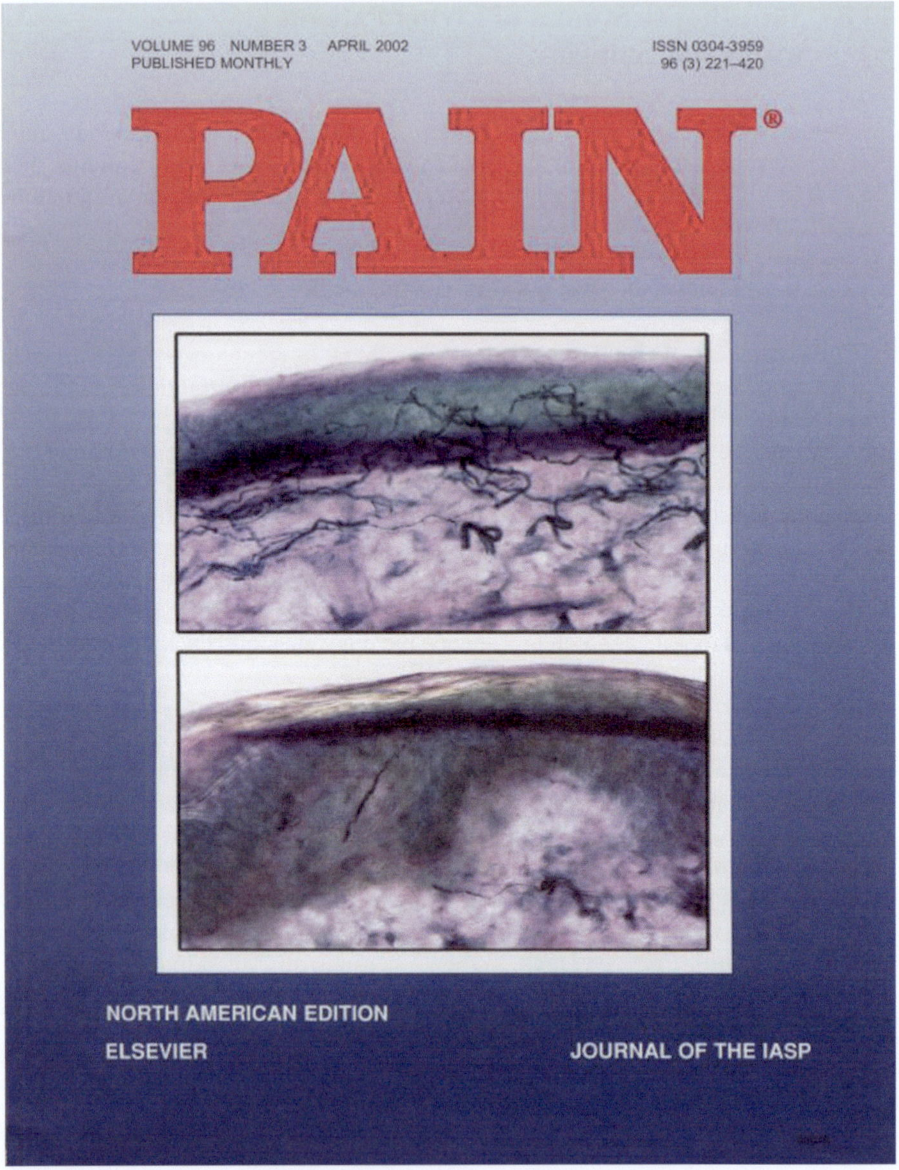

Fig. 12.7 PGP9.5 immunolabeling of nerve fibers within vertical sections of human punch skin biopsies. The stratum corneum is uppermost, with the epidermis and dermis beneath. Both biopsies are from previously shingles-affected skin. The subject whose biopsy is on top is a 75-year-old woman with no PHN after shingles on her back. She had 1184 neurites/mm^2 skin surface area. The subject whose biopsy is below is a 72-year-old woman with PHN after shingles on her back and shoulder. She had 145 neurites/mm^2 skin surface area (Reprinted from: Oaklander, AL. *PAIN* Vol 96, number 3, April 2002)

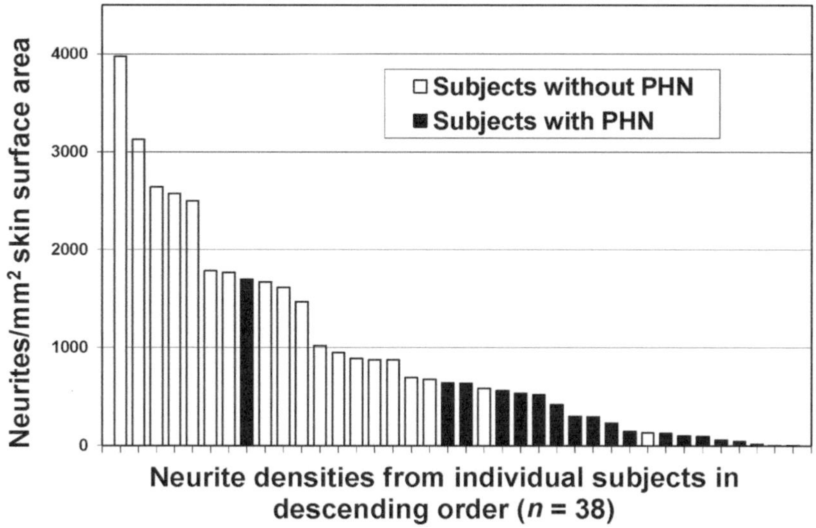

Fig. 12.8 Analysis of densities of epidermal nerve endings from the torso of patients previously affected by shingles at least 3 months prior revealed that a threshold of approximately 650 neurites/mm^2 was 95 % sensitive and 89 % specific in distinguishing between subjects with and without PHN (From: Oaklander [33])

12.9 Bilateral Neuropathology after Unilateral Shingles

These skin-biopsy studies also confirmed earlier isolated reports that unilateral shingles also damages segmental "mirror-image" neurons directly contralateral from a zoster-affected ganglion. This was discovered when Oaklander et al. [31] used as a control not only skin mirror-image to the shingles but also shin from a different dermatome on the torso. Contralateral damage occurred despite the lack of contralateral shingles eruptions or pain and did not extend to distant contralateral dermatomes. In light of this report, other authors appreciated that their earlier data also included evidence of bilateral effects of unilateral shingles [2, 50]. An autopsied case of clinically unilateral ophthalmic PHN [51] had clear damage in the contralateral ophthalmic and supraorbital nerves. Previous spinal cases also demonstrated that the spectrum of myelinated fibers from the contralateral non-affected side differed from that of age-matched controls (Cases 1 and 2: Watson et al. [50]). Another previous case (Case 5: Watson [50]) had contralateral inflammatory changes in nerves of the affected segment and bilaterally including in segments above and below the zoster-affected segment.

Head and Campbell [17] reported no evidence of VZV reactivation in mirror-image contralateral ganglia (e.g., Fig. 12.2a), demonstrating that loss of contralesional axons does not merely reflect contralateral spread of VZV. Others hypothesized that this might reflect occult zoster myelitis [15, 50]. However, when Oaklander and Brown [37] demonstrated similar segmental contralesional axonopathy after one-sided distal nerve injuries in rats, this confirmed that the contralesional loss of skin innervation after unilateral zoster is a transcellular effect, a phenomenon often reported but often unappreciated and little understood [22]. Surprisingly, a significant correlation was reported between severity of PHN pain and severity of loss of epidermal neurites in mirror-image skin [31]. Perhaps this is a surrogate marker for occult spinal-cord changes. Application of IVCM to both eyes of 27 patients affected by unilateral HZO confirmed that contralesional corneas had significant reductions in total nerve length, number of nerves, and main nerve trunks as compared with control eyes (Fig. 12.6; [16]).

12.10 The Pathology of Postherpetic Itch (PHI) and Associated Self-Injury

Varicella eruptions are notoriously itchy, and so is zoster. Dermatomal itch is a common early sign of zoster, and in mild cases it can be the only sensory symptom. After shingles some patients are left with severe disabling PHI, either accompanying PHN or alone. Patients with PHI can injure and scar their skin by repeated scratching. Rarely they scratch deeply and cause serious self-injury. The well-known case of a woman who painlessly scratched through her skull and into her brain after V_1 zoster [35] raised awareness of PHI. Skin biopsies from her scalp demonstrated that her PHI arose in nearly total denervation of her previously zoster-affected skin (Fig. 12.9). Her self-injury was caused by the conjunction of severe PHI and loss of the protective nociceptive sensations that normally limit scratching by triggering pain when the skin is lacerated. This paper hypothesized that the pathological substrate of PHI is abnormal spontaneous firing of the few remaining itch-transmitting C-fibers plus the lack of enough nearby sensory fibers to generate normal the inhibition in the dorsal horn that spatially and temporally limits itch.

Itch had not been a subject of neurological or pathological study, and it was not included among the endpoints in most clinical trials studying PHN, and there has been no systematic study of treatments. Clinical experience and case reports [35] show that most patients' itch is relieved by local administration of local anesthetics, meaning that peripheral axonal signaling is required. The historical HZ literature does not mention "itch" or "pruritus," but sometimes refers to "distressing sensory symptoms" or "self-injurious scratching." Another early term "*pruritus sine materia*" was used for dermatologically unexplained regional itch syndromes including PHI. This was, and still often is, attributed to psychogenic causes. "Trigeminal tropic syndrome" is a term from the early twentieth century that described self-injurious facial scratching after trigeminal damage. The most common cause then was neurosurgical transaction of trigeminal branches during early (ineffective) attempts to relieve trigeminal neuralgia, but now it is PHI.

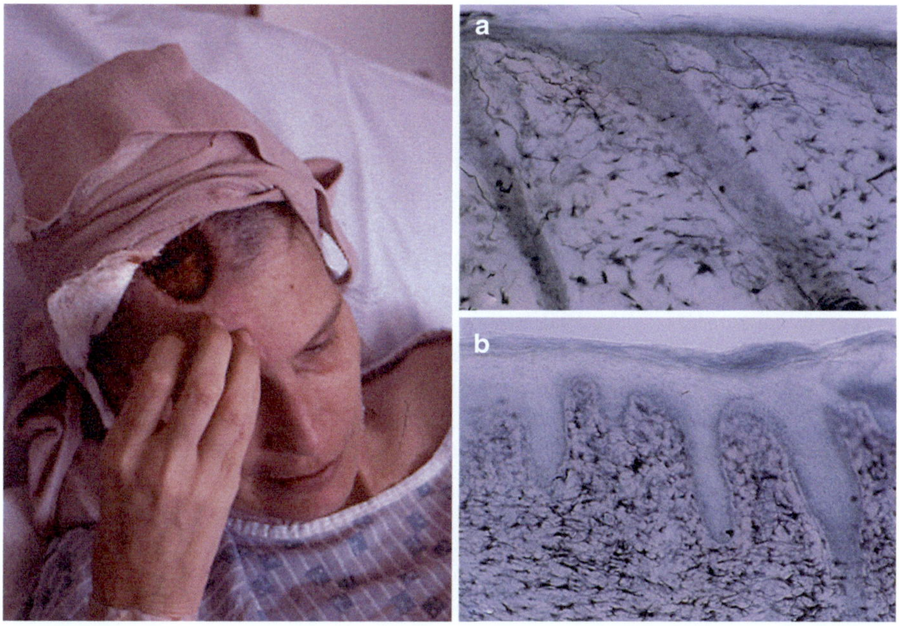

Fig. 12.9 At left, image of a woman with severe postherpetic itch but no postherpetic neuralgia, after ophthalmic zoster. Within a year, she had painlessly scratched through her frontal skull into her brain causing severe brain injury. Sensory testing revealed loss of pain but not itch sensation in the affected dermatome. Skin biopsies from normal (Panel **a**) and itchy scalp (Panel **b**) demonstrated loss of 96 % of epidermal innervation in the itchy area (From: Oaklander et al. [35])

The first systematic study of PHI pooled data from 586 adults with shingles or PHN [36]. It revealed that PHI was usually mild or moderate in severity and that itch was a common in both acute zoster and PHN. It confirmed clinical impressions that PHI was more likely after HZ eruptions on the head, face, and neck than on the torso. Perhaps itch was evolutionarily selected to be more highly represented on the face to help protect animals from insects drawn to facial mucous membranes.

PHI has been modeled in animals including mice, rats, and monkeys by surgical extirpation of individual dorsal root ganglia. The resulting scratching and self-injury centered on the denervated dermatome is referred to as "overgrooming" or "autonomy" in the literature. Initially interpreted as a sign of pain, neuropathic overgrooming/autotomy was later recognized to be modeling neuropathic itch. As in PHI, self-injurious scratching develops in some but not all rodents after experimental lesions throughout the PNS and spinal cord that damage pain/itch pathways [54]. The animal models reproduce the rostro-caudal gradient of PHI, with overgrooming developing after rhizotomy of cervical and thoracic, but not lumbosacral dorsal roots [26]. Genetic [29] and dietary [45] factors influence whether or not rats scratch after nerve injuries and thus they might contribute to PHI as well. Neuropathic itch in lesioned rats appears maintained by spontaneous activity in second-order, lamina I, itch-projection neurons. As postulated in PHI, this is driven by the conjunction of spontaneously firing peripheral itch fibers plus loss of dorsal-horn inhibitory circuits [39].

12.11 Summary

Pathology studies have demonstrated that shingles is caused by VZV reactivation, usually in only one sensory ganglion. This causes hemorrhagic inflammation that spreads distally along peripheral and central axons. The central axons of affected primary afferent neurons also degenerate, affecting the spinothalamic tracts and dorsal columns and sometimes other areas of the spinal cord, meninges, and brain. After zoster subsides, it usually leaves pathological scars in one ganglion and the corresponding skin, nerves and nerve roots, and spinal cord. In some patients, these processes cause PHN and/or PHI.

These pathological findings have revealed information about the causes of neuropathic pain and itch, not only caused by zoster, but by extension to other neuropathies as well. As in other neurodegenerative illnesses, complex multi-cell circuits are involved. Both peripheral and central mechanisms appear to trigger PHN and PHI. Stimulus-evoked pain symptoms such as allodynia or hyperalgesia require abnormal signaling from remaining damaged peripheral pain neurons. For itch, evidence from PHI patients and rodent models confirms the participation of both peripheral and central mechanisms. More is known about the neuropathology of PHN than any other focal neuropathic sensory syndrome. Because it is caused by a single monophonic disease and lesions can be precisely localized from the skin rash and scarring, zoster has been ideal for studying pain and itch mechanisms. Head and Campbell's eighteenth-century large-scale autopsy studies of well-characterized patients with prior shingles should be resumed in the twenty-first century. Antemortem internet capture of autopsy consents and clinical details ("deep phenotyping") in large patient registries that include photographs ("selfies") localizing the shingles lesions make this feasible.

Acknowledgments Funded in part by the Public Health Service. NINDS R01 AG42866.

Conflict of Interest Statement None to report.

References

1. Adams RD (1944) Herpes zoster: Pathologic features. Bull N Engl Med Center 6:12
2. Baron R, Saguer M (1994) Axon-reflex reactions in affected and homologous contralateral skin after unilateral peripheral injury of thoracic segmental nerves in humans. Neurosci Lett 165:97–100
3. Baron R, Saguer M (1995) Mechanical allodynia in postherpetic neuralgia: evidence for central mechanisms depending on nociceptive C-fiber degeneration. Neurology 45(12):S63–S65
4. Bright R (1831) Reports of medical cases. London 2(Part 1):383
5. Chandelux (1879) Observation pour servir a l'histoire des lesions nerveuses dans le zona. Archives de Physiologie XI:674
6. Cheatham WJ (1953) The relation of heretofore unreported lesions to pathogenesis of herps zoster. Am J Pathol 29:401–412
7. Denny-Brown D, Adams RD, Fitzgerald PJ (1944) Pathologic features of herpes zoster: a note on 'geniculate herpes'. Arch Neurol Psychiatr (Chicago) 57:216

8. Dubler (1884) Ueber Neuritis boi Herpes Zoster. Virchows Arch, XCVI:195
9. Ebert MH (1949) Histologic changes in sensory nerves of the skin in herpes zoster. Arch Dermatol Syphil 60:641–648
10. Esiri MM, Tomlinson AH (1972) Herpes zoster: demonstration of virus in trigeminal nerve and ganglion by immunofluorescence and electron microscopy. J Neurol Sci 14:35–48
11. Faure-Beaulieu M, Lhermitte J (1929) Les lesions medullaires du zona idiopathique: La myelite zosterienne. Rev Neurol 1:1250–1258
12. Forbes HJ, Thomas SL, Smeeth L, Clayton T, Farmer R, Bhaskaran K, Langan SM (2016) A systematic review and meta-analysis of risk factors for postherpetic neuralgia. Pain 157(1):30–54
13. Geha PY, Baliki MN, Wang X, Harden RN, Paice JA, Apkarian AV (2008) Brain dynamics for perception of tactile allodynia (touch-induced pain) in postherpetic neuralgia. Pain 138(3):641–656
14. Ghatak NR, Zimmerman HM (1973) Spinal ganglion in herpes zoster: a light and electron microscopic study. Arch Pathol 95:411–415
15. Haanpää M, Dastidar P, Weinberg A (1998) CSF and MRI findings in patients with acute herpes zoster. Neurology 51:1045–1411
16. Hamrah P, Cruzat A, Dastjerdi MH, Pruss H, Zheng L, Shahatit BM, Bayhan HA, Dana R, Pavan-Langston D (2013) Unilateral herpes zoster ophthalmicus results in bilateral corneal nerve alteration: an in vivo confocal microscopy study. Ophthalmology 120:40–47
17. Head H, Campbell AW (1900) The pathology of herpes zoster and its bearing on sensory localization. Brain 23:353–523
18. Hilliges M, Wang L, Johansson O (1995) Ultrastructural evidence for nerve fibers within all vital layers of the human epidermis. J Invest Dermatol 104:134–137
19. Hope-Simpson RE (1965) The nature of herpes zoster: a long-term study and a new hypothesis. Proc R Soc Med 58:9–20
20. Jastreboff PJ (1990) Phantom auditory perception (tinnitus): mechanisms of generation and perception. Neurosci Res 8:221–254
21. Karanth SS, Springall DR, Kuhn DM, Levene MM, Black JM (1991) An immunocytochemical study of cutaneous innervation and the distribution of neuropeptides and protein gene product 9.5 in man and animals. Am J Anat 191:368–383
22. Koltzenburg M, Wall PD, McMahon SB (1999) Does the right side know what the left side is doing? TINS 22:122–127
23. Lesser E (1881) Beitrage zur Lehre vom Herpes Zoster. Virchows Arch XXXVI:391
24. Lhermitte J, Nicolas (1924) Les lesions spinales du zona: La myelite zosterienne. Rev Neurol 1:361–364
25. Liu J, Hao Y, Du M, Wang X, Zhang J, Manor B et al (2013) Quantitative cerebral blood flow mapping and functional connectivity of postherpetic neuralgia pain: a perfusion fMRI study. Pain 154(1):110–118
26. Lombard MC, Nashold BS Jr, Albe-Fessard D, Salman N, Sakr C (1979) Deafferentation hypersensitivity in the rat after dorsal rhizotomy: a possible animal model of chronic pain. Pain 6(2):163–174
27. McCormick WF, Rodnitzky RL, Schochet SS et al (1969) Varicella-zoster encephalomyelitis: a morphologic and virologic study. Arch Neurol 21:559–570
28. Melzack R, Wall PW (1965) Pain mechanisms: a new theory. Science 150:971–979
29. Mogil JS, Wilson SG, Bon K, Lee SE, Chung K, Raber P et al (1999) Heritability of nociception II. Types of nociception revealed by genetic correlation analysis. Pain 80:83–93
30. Noordenbos W (1959) Pain: problems pertaining to the transmission of nerve impulses which give rise to pain. Amsterdam, Elsevier. Chapter 1 pp 4–10; ch 10 pp 68–80
31. Oaklander AL, Romans K, Horasek S, Stocks A, Hauer P, Meyer RA (1998) Unilateral postherpetic neuralgia is associated with bilateral sensory neuron damage. Ann Neurol 44:789–795
32. Oaklander AL (1999) The pathology of shingles; Head and Campbell's 1900 monograph. Arch Neurol 56:1292–1294

33. Oaklander AL (2001) The density of remaining nerve endings in human skin with and without postherpetic neuralgia after shingles. Pain 92:139–145
34. Pappagallo M, Oaklander AL, Quatrano-Piacentini AL, Clark MR, Raja SN (2000) Heterogenous patterns of sensory dysfunction in postherpetic neuralgia suggest multiple pathophysiologic mechanisms. Anesthesiology 92:691–698
35. Oaklander AL, Cohen SP, Raju SVY (2002) Intractable postherpetic itch and cutaneous deafferentation after facial shingles. Pain 96:9–12
36. Oaklander AL, Bowsher D, Galer BS, Haanpää ML, Jensen MP (2003) Herpes zoster itch: preliminary epidemiologic data. J Pain 4:338–343
37. Oaklander AL, Brown JM (2004) Unilateral nerve injury produces bilateral loss of distal innervation. Ann Neurol 55:639–644
38. Reske-Nielsen E, Oster S, Pedersen B (1986) Herpes zoster ophthalmicus and the mesencephalic nucleus. Acta Pathol Microbiol Immunol Scand 94:263–269
39. Ross SE, Mardinly AR, McCord AE, Zurawski J, Cohen S, Jung C et al (2010) Loss of inhibitory interneurons in the dorsal spinal cord and elevated itch in Bhlhb5 mutant mice. Neuron 65:886–898
40. Rowbotham MC (1996) Relationship of pain, allodynia and thermal sensation in postherpetic neuralgia. Brain 119:347–354
41. Rowbotham MC, Yosipovitch G, Connolly MK, Finlay D, Forde G, Fields HL (1996) Cutaneous innervation density in the allodynic form of postherpetic neuralgia. Neurobiol Dis 3:205–214
42. Sattler (1875) Ueber das Wesen des Herpes ophthalmicus. Anzelger der K. K. Gesellschaft der Herzte in Wien. Protocoll der sitzung (This will be found at the end of the Medizinische Jahrbuch von der K.K. Gesellschaft der Aertze)
43. Schultz G, Melzack R (1991) The Charles Bonnet syndrome: 'phantom visual images'. Perception 20:809–825
44. Simone DA, Noland M, Johnson T, Wendelschafer-Crabb G, Kennedy WR (1998) Intradermal injection of capsaicin in humans produces degeneration and subsequent reinnervation of epidermal nerve fibers. J Neurosci 18:8947–8959
45. Shir Y, Ratner A, Seltzer Z (1997) Diet can modify autotomy behavior in rats following peripheral neurectomy. Neurosci Lett 236:71–74
46. Thompson RJ, Doran JF, Jackson P, Dhillon AP, Rode J (1983) PGP 9.5 – a new marker for vertebrate neurons and neuroendocrine cells. Brain Res 278:224–228
47. von Bärensprung FGF (1861) Die Gürtelkrankheit. Ann Char-Krankenh Zu Berlin 9:40–238
48. von Bärensprung FGF (1862) Beiträge zur Kenntnis des Zoster. Ann Char-Krankenh Zu Berlin 10:96–104
49. Watson CPN, Morshead C, Van der Kooy D, Deck JH, Evans RJ (1988) Postherpetic neuralgia: post-mortem analysis of a case. Pain 34:129–138
50. Watson CPN, Deck JH, Morshead C, Van der Kooy D, Evans RJ (1991) Postherpetic neuralgia: further post-mortem studies of cases with and without pain. Pain 44:105–117
51. Watson CPN, Midha R, Devor M, Nag S, Munro C, Dostrovsky JO (2000) Trigeminal postherpetic neuralgia postmortem: clinically unilateral, pathologically bilateral. In: Devor M, Rowbotham MC, Wiesenfeld-Hallin Z (eds) Proceedings of the 9th world congress on pain. Progress in pain research and management, IASP Press, Seattle
52. Wohlwill F (1924) Zur pathologischen Anatomie des Nervensystems vom Herpes Zoster. Z ges Neurol Psychiatr 89:170–212
53. Wyss O (1871) Beitrag zur Kentniss des Herpes Zoster. Arch Heilk XVI:261
54. Yezierski RP, Liu S, Ruenes GL, Kajander KJ, Brewer KL (1998) Excitotoxic spinal cord injury: behavioral and morphological characteristics of a central pain model. Pain 75:141–155
55. Zacks SL, Langfitt TW, Elliott FA (1964) Herpetic neuritis: a light and electron microscopic study. Neurology 14:744–750
56. Zhang Y, Liu J, Li L, Du M, Fang W, Wang D et al (2014) A study on small-world brain functional networks altered by postherpetic neuralgia. Magn Reson Imaging 32:359–365

Chapter 13
Neural Basis of Pain in Herpes Zoster and Postherpetic Neuralgia: The Ectopic Pacemaker Hypothesis

Marshall Devor

13.1 Introduction

Herpes zoster (HZ, "shingles") and postherpetic neuralgia (PHN) are caused by viral infection of primary sensory neurons. They are generally considered to be two phases, early and late, of the same condition. In both phases the principle cause of morbidity is pain. However, pain is not a direct effect of the virus. Rather, pain results from specific forms of neural activity impacting a conscious brain. This chapter considers the causal chain that begins with varicella zoster virus (VZV) infection and ends with the experience of pain. In the long run, the solution to HZ and PHN will likely be prevention by efficient immunization. In the meanwhile, understanding VZV-associated pain mechanisms can provide better approaches to clinical management and the alleviation of suffering. It can also yield insights into the cause and treatments of other painful conditions triggered by disease or injury affecting the nervous system, neuropathic pain.

13.2 Spontaneous and Evoked Neuropathic Pain

PHN is a Neuropathic Pain Condition Pain in HZ is usually attributed to the herpetic rash and associated inflammation, although pathological changes in nerve fibers are also present. For PHN there is general consensus about its classification as a neuropathic pain condition, along with painful diabetic polyneuropathy, phantom limb pain, erythromelalgia, and many others. Some pain professionals express

M. Devor
Department of Cell and Developmental Biology, Institute of Life Sciences and Center for Research on Pain, The Hebrew University of Jerusalem, Jerusalem 91904, Israel
e-mail: marshlu@mail.huji.ac.il

C.P.N. Watson et al. (eds.), *Herpes Zoster: Postherpetic Neuralgia and Other Complications*, DOI 10.1007/978-3-319-44348-5_13

skepticism about the logic of placing this grab bag of clinically different conditions under a single header. After all, they are precipitated by very different factors (viral infection, metabolic disorder, trauma, genetic mutation, etc.) and have very different clinical presentations and natural histories. But along with others, I hold that despite the differences, the underlying causes of pain in neuropathic conditions are the same. This is why a diagnosis of neuropathic can be made with considerable specificity based on answers to four questions [7] and based on why the fundamental mechanisms of neuropathic pain can be laid out at a reasonable level of resolution in a single book chapter [23]. It is also why a small and quirky selection of drugs drawn from very different pharmacological categories (certain antidepressants, certain anticonvulsants, and local anesthetics) are used in the treatment of neuropathic pain across the diagnostic spectrum [35].

The common denominator of painful neuropathic conditions is faulty regulation of neuronal excitability. Specifically, they feature *hyperexcitability* at one or more locations in the neural circuitry that processes pain signals. The basic mechanisms of neuronal "excitability" (the ability of neurons to generate and propagate electrical nerve impulses) have been known for a long time, although we still lack a fundamental understanding of the homeostatic factors that regulate excitability in healthy individuals. Moreover, we generally understand the cellular events that lead to hyperexcitability and have at our disposal a variety of membrane-stabilizing drugs and synaptic blockers that effectively suppress hyperexcitability. Treating neuropathic pain, including HZ and PHN, is largely a practical problem: how to deliver appropriate suppressive drugs to the key location(s) at a reasonable personal and financial cost, without causing intolerable side effects. Tackling this problem is not trivial. But as far as one can tell, it does not involve any insurmountable conceptual problems. The first question that needs to be asked is "where precisely are the impulses coming from that are causing this patient's pain?"

Positive and Negative Symptoms Until fairly recently neuropathic pain was considered paradoxical. Damaging a nerve was expected to cause "silence," much like cutting a telephone line. It should render innervated skin (or other tissue) numb and unresponsive to applied stimuli ("negative symptoms"). Why, then does it often cause (1) spontaneous (ongoing) pain and (2) hyper sensitivity to stimuli, including in areas of skin that are partly denervated ("positive symptoms"), the key symptoms also in HZ and PHN? Explanations of these two anomalies have emerged with the discovery of two processes, both attributable in large part to Patrick Wall (and his students), to whom this volume is dedicated. These are (1) ectopic hyperexcitability in injured nerves and (2) "central sensitization." The present discussion of HZ and PHN rests on these two processes.

HZ and PHN present with spontaneous pain in the skin, usually described as burning or stinging in quality, but sometimes also including itch and pain paroxysms [16, 123]. Frequently there is also extreme tenderness to light touch ("tactile allodynia") and "hyperpathia." "Hyperpathia" refers to the explosive appearance of pain as stimulus intensity in increased stepwise into the noxious range, often with aftersensation. These positive symptoms are often accompanied by sensory deficits (negative signs) especially to warming and cooling of the skin. There is a good deal of interindividual variability in the symptomatology as described in this volume and elsewhere [85, 89, 123]. Because spontaneous and evoked pain are quite different in terms of mechanism, they will be discussed here separately. Some authors have the bad habit of using the word "pain" in place of "spontaneous pain," indicating the type only when referring to evoked pain. Here pain type, spontaneous or evoked, is specified unless the meaning is both.

PNS vs. CNS and the Idea of Pain Centralization For the most part, spontaneous pain and other ongoing sensations in the territory served by a damaged nerve is due to electrical impulses that arise in the peripheral nervous system (PNS). Ongoing pain following central nervous system (CNS) injury ("central pain") is supposed to be due to spontaneous firing arising in the CNS, although here, too, activity in peripheral sensory neurons might be a key contributor [130]. That having been said, sensory experience obviously requires a conscious brain, and pain-provoking signals originating in the PNS can clearly be enhanced and diminished by CNS processes. Central modulation, however, is very different from the popular notion that peripheral neuropathic pain may actually be driven by signals originating in the CNS.

The evidence for a PNS origin of most (or all) peripheral neuropathic pain is its elimination when the pain driver is silenced or when impulses generated in the PNS are prevented from reaching the CNS by nerve or spinal block [41, 49, 118]. The block, of course, needs to include the appropriate nerve fibers and to be (verifiably) effective/complete. Also, pain relief cannot be expected to outlast the duration of action of the blocking agent, although for reasons that remain uncertain it sometimes does. With currently available local anesthetic agents, the duration of block is measured in hours unless the drug is continuously replenished using repeated application or a depot or pumping device. It is irrational to declare a block, or a series of blocks, a failure if pain initially subsides but later returns. Likewise, the block needs to be made proximal to the location at which the spontaneous impulse discharge is generated. Occasional reports of pain relief following distal block failed to control for systemic effects of anesthetics, placebo, and/or spontaneous remission (e.g., [58]).

Many in the clinical community believe that prolonged noxious input in patients with neuropathy can cause pain drivers to migrate from the PNS to the CNS and become intractable. This idea, "centralization," is a popular explanation of transition to chronicity and connects with thinking about the transition of HZ to PHN as reflecting progression toward a CNS process. The reality of centralization, however, is equivocal. Usually lacking is evidence that nerve or spinal block indeed lose their prior effectiveness [49]. Centralization also doesn't sit well when one considers situations where the source of the peripheral drive is known and can be definitively removed, like childbirth, the passage of a kidney stone, treatment of a painful tooth, or replacement of an arthritic hip joint. Pain fades quickly even if the peripheral drive had been intense and prolonged. Even the prototypical centralized pain, phantom limb pain in amputees, is unlikely to be driven by neural activity originating in the cortex [73, 118].

PNS Generators of Spontaneous and Evoked Discharge Normal sensory nerve fibers are designed to generate impulses when adequate stimuli, mechanical, thermal, or chemical, are applied to their specialized transducer endings (Fig. 13.1).

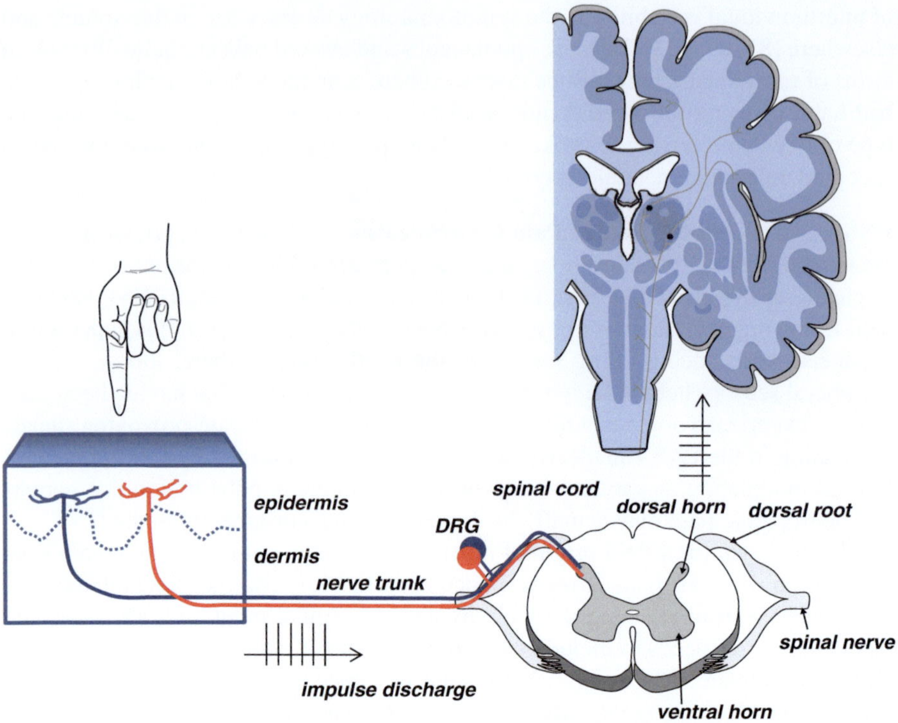

Fig. 13.1 Sketch of the pain system. Impulses are generated in the cutaneous sensory endings of primary sensory neurons whose cell bodies (somata) reside in the segmental dorsal root ganglia (DRGs). The sensory impulses propagate along axons in the peripheral nerve, pass the DRG, proceed along the dorsal root, and then enter the spinal cord (some also send an ascending branch direct to the lower brainstem). After integration and processing in the spinal cord, the sensory signal proceeds to higher levels of the CNS

Pressing on healthy axons at mid-nerve does not evoke impulses or corresponding sensation. However, if there has been a local nerve injury, the mid-nerve axon changes its properties and locally acquires the ability to respond to stimuli and even to fire spontaneously. Examples are percussion on the wrist in cases of carpal tunnel entrapment, or over stump neuromas in amputees, which induces a sudden shock-like sensation, the Tinel sign. The anatomical structures associated with these ectopic sources of impulse generation ("electrogenesis") are swollen end bulbs, axon sprouts, and patches of demyelination. A Tinel-like response can sometimes be evoked over long expanses of nerve, as in painful diabetic polyneuropathy. This suggests the presence of mechanosensitive pacemaker sites scattered along the length of the nerve, probably in the form of regenerating sprouts or dying-back axon ends.

Animal studies have identified a second major generator of ectopic spontaneous and evoked afferent discharge, the dorsal root ganglion (DRG) [23, 119]. Indeed, DRGs may be more important than neuromas as generators of spontaneous neuropathic pain [72] (Fig. 13.2). There is also (indirect) evidence of this in humans. For example, during straight-leg lifting, mechanical force is applied to the ganglion by tensioning the sciatic nerve. In patients with radicular low back pain, this gives rise to ectopic firing and shooting leg pain [18, 68, 84]. The DRG is located proximally in the PNS, in the intervertebral foramen. Thus, if the DRG contributes a significant fraction of the impulses responsible for ongoing pain in a particular individual, nerve block (distal to the ganglion) will not stop the pain. Block centrally (spinal block) or at the level of the DRG itself (intraforaminal block) will. Impulses generated ectopically in the DRG propagate into the CNS, causing a sensation felt in (i.e., referred to) the corresponding dermatome. Impulses also propagate antidromically from the DRG out into the skin where they release peptide neuromodulators resulting in so-called neurogenic inflammation. This adds to whatever inflammatory process might already be present in the skin [121].

Ectopic pacemakers, both at nerve injury sites and DRGs, are remarkably sensitive to local anesthetics and other membrane-stabilizing drugs. For example, whereas 2 % lidocaine is used to block impulse conduction in nerves, the ectopic *generation* of impulses is suppressed by concentrations orders of magnitude lower.

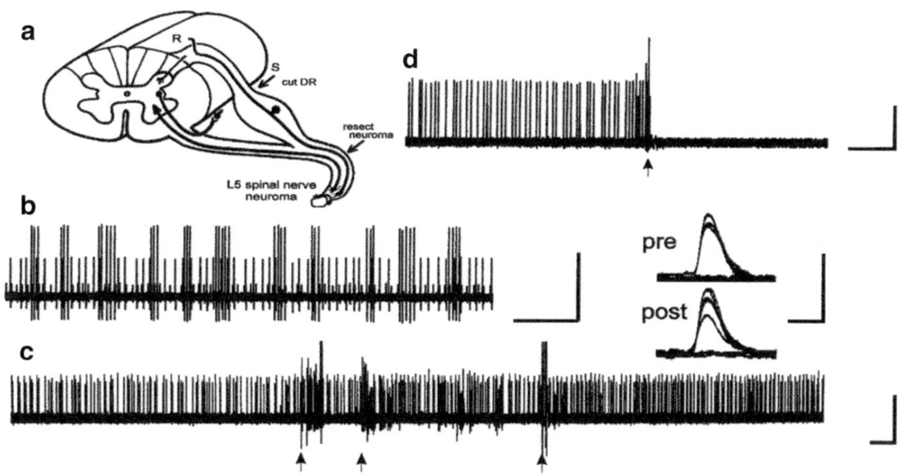

Fig. 13.2 Ectopic discharge in injured nerves originates at the nerve injury site and in the corresponding DRG(s). (**a**) Sketch of the experimental setup showing the proximal cut end of the spinal nerve (neuroma), DRG, dorsal root (from which impulse activity in fine axon bundles is recorded (*R*)), and the spinal cord. (**b**) Spontaneous discharge in three axons, recorded from a fine axon bundle in the dorsal root. Note the all-or-none spike amplitude of each axon. (**c**) This second axon bundle originally contained two spontaneously active axons (inset labeled "pre"). A priori it was not clear whether this activity originates in the neuroma, the DRG, or both. However, it persisted following stepwise transection of the nerve just central to the neuroma indicating an origin in the DRG. The timing of transection is indicated by three *vertical arrows* below the spike trace and the location by "resect neuroma" in (**a**). Note the brief episodes of acute injury discharge following each cut. A third axon began to fire during the resection (inset labeled "post"). (**d**) Activity in a third axon bundle was silenced by severing the dorsal root (arrow marked "cut DR" in (**a**)). Calibrations: (**b–d**) 1.0 s/250 μV; inset, 0.2 ms/250 μV (From Liu et al. [72])

This is reflected in experimental preparations, and also in clinical practice [28]. For example, systemic lidocaine at serum levels of 2.5 μg/ml is sufficient to significantly reduce ongoing pain in most neuropathic conditions [120]. This compares to 20 mg/ml (2 %) used for nerve block, nearly 10,000 times more. Systemic lidocaine does not work by blocking nerve conduction. That would be lethal. It works by suppressing ectopic electrogenesis. The same holds for other membrane-stabilizing agents administered systemically [14, 23].

Mechanisms of Electrogenesis The cellular and molecular processes responsible for ectopic electrogenesis and hyperexcitability in neuropathic pain conditions are rapidly coming into focus. They include (1) altered gene expression in the cell soma of sensory neurons, (2) dysregulation of the trafficking of molecules-of-excitability by axoplasmic transport from the neuronal cell body in the DRG to the axonal membrane (especially ion channels and receptors), and (3) changes in the response properties of these molecules themselves [23]. The quality of the spontaneous pain felt (burning, cramping, tingling, paroxysmal, electric shock-like) is a reflection of the specific type(s) of afferents contributing to the ectopic barrage and their firing rate, pattern, and degree of synchrony (esp. for pain paroxysms). The term "ectopic" refers to the *location* of electrogenesis, not to the abnormal firing itself or whether it is spontaneous or evoked. The normal location of impulse generation is the specialized sensory ending in the skin (or other tissues). Any other location is ectopic. How do these general principles of electrogenesis in the event of neuropathy apply to HZ and PHN?

13.3 Spontaneous Pain in HZ and PHN

13.3.1 VZV Causes Varicella, HZ, and PHN

Herpes zoster is believed to result from the resurgence of latent VZV infection [69, 99, 133]. The original infection usually occurs in childhood and causes the symptoms of chickenpox (varicella). This disease, signaled by widespread mild skin eruptions and (usually painless) itch, is rapidly suppressed by the immune system. VZV virus particles are neurotropic and highly selective for cutaneous primary sensory neurons of the DRG and trigeminal ganglion (TRG), avoiding motor neurons and afferents that serve deep tissues. As such they preferentially enter the terminals of sensory axons in the skin and migrate with the retrograde axoplasmic flow back to the sensory cell soma in the DRG. There they hide in a latent form for the remainder of the individual's life, evading the immune system. In addition to neuronal somata (of all types), VZV has also been detected in the satellite glial cells that surround the cell soma. There are significant neuron-glial interactions in the sensory ganglia that might play a role in HZ/PHN pain [47].

In unfortunate individuals, decades after the initial chickenpox infection, the latent VZV is reactivated causing a second VZV-related disease, HZ/PHN. The reasons for reactivation remain elusive. Whatever they are, reactivation is mostly limited to a single ganglion or a single division in the case of the TRG. This is based on a survey of infected ganglia in relation to rash location [50, 122]. VZV load in the reactivated ganglion skyrockets with many of the new virus particles being transported back from the cell soma to the axon end in the skin where they once again cause a rash. This is herpes zoster (HZ). Unlike chickenpox, the rash in HZ is primarily described as burning/stinging pain, it is usually accompanied by tactile allodynia, and it is focal, located in the dermatome of the infected DRG. Reactivation in additional ganglia is presumed to be prevented by an efficient immune response. But alternatively, it could be that the reactivation process itself is a rare, random event, and that most people don't live long enough to suffer a second hit. Either way, it is uncommon for immune-competent individuals to develop a second VZV-related rash in the same or a different dermatome.

HZ usually runs its course in a few weeks. Pain fades as the virus is cleared from the skin and ganglion, and the skin heals. Residual scarring is not uncommon, but simple recovery is the rule. However, again unlike chickenpox, in some individuals the dermatomal skin where the rash used to be remains painful. This residual pain is postherpetic neuralgia (PHN). Pain in PHN also tends to fade over time, but in ~20 % of cases, it lasts for >1 year and in some it persists for a lifetime. The likelihood of HZ transitioning into PHN increases with age at the time of HZ, severity of the HZ, and other risk factors [37, 53]. Interestingly, this transition is apparently seamless, marked only by drying out of the rash. The continuity of the pain in terms of dermatomal location and quality of sensation suggests that the underlying pain mechanism is the same in HZ and PHN. In this sense PHN is not so much a "complication" of HZ as a continuation; it is not a third VZV-related disease.

HZ has been known since antiquity, and from the beginning, the cause of pain was referred to the rash-inflamed skin. The seminal nineteenth-century discovery of inflammation and cell loss in the DRG of the painful dermatome added an explanation for PHN, deafferentation of the spinal cord [50]. These two mechanisms, inflammation and deafferentation, remain the undisputed explanations of HZ and PHN to this day [33]. In the spirit of Patrick Wall, I will now revisit this received wisdom.

13.3.2 Do the Impulses That Cause Spontaneous Pain in HZ and PHN Originate in the Skin?

Irritable Nociceptor Endings in the HZ Rash As the virus that causes chickenpox and HZ/PHN is the same, reasons for the major differences in symptomatology and natural history need to be sought elsewhere. VZV is a "syncytial" virus which contributes to rash by inducing fusion of neighboring skin cells causing vesicles, pustules, and secondarily an immune response in the skin [17]. Scratching may lead to bacterial infection and exacerbation of the inflammation. The whole body rash in chickenpox and accompanying myalgia, malaise, and low-grade fever reflect the fact that this is a systemic disease. The feeling of being ill is thought to be due to circulating cytokines which can sensitize the pain system at many locations [131]. As topical remedies relieve the itch transiently, and the itch always resolves when the rash and inflammation resolve, there is little doubt that the major cause of sensory symptoms is inflammatory sensitization and activation of the transducer endings of cutaneous sensory axons. This mechanism, "peripheral sensitization," is also the normal explanation of ongoing pain in HZ, with some extending it also to PHN. However, there are reasons to question this conclusion.

In chickenpox virus particles are delivered to the skin by the circulation, and the rash is widespread. In HZ the VZV particles reach the skin via axonal transport from the infected ganglion, and hence the rash is dermatomal. Skin inflammation is present in both conditions, but the sensory experience is very different: itch in one case and burning pain in the other. Peripheral sensitization reflects the action of inflammatory mediator molecules on the process of impulse initiation. A mix of mediators is present ("inflammatory soup"), derived from local mast cells, invasive immune cells, cellular breakdown products, products of local enzymatic action, blood-borne agents, and mediators released from axon endings [55]. These affect impulse discharge mostly by binding to specific receptor molecules on the axonal membrane. Different types of sensory axons express different receptors, and terminate differentially in the spinal cord accounting for differences in the quality of the sensation felt.

The itch of chickenpox suggests that the inflammatory soup contains primarily puritic mediators such as histamine which activate itch-specific sensory endings. Itch can also occur in HZ [87]. The most prominent sensation, however, is ongoing burning pain suggesting a mix of mediators that selectively activate heat-sensitive nociceptors (e.g., bradykinin, PGE_2). One wonders why the same virus (VZV) in the same environment (skin) would yield a very different mix of mediators. In the pain field, we teach that inflammatory mediators activate nociceptors. But itch is a more common dermatological symptom than pain, and in some conditions (e.g., leishmaniasis) weepy inflamed lesions are reported to be painless. A peculiarity of HZ is that not infrequently patients report paresthesias and pain beginning 2–3 days *before* the rash appears. Indeed, in some a rash never appears (zoster sine herpete). Another peculiarity that dissociates the pain from the rash is the observation that while the pain usually clears along with the rash (HZ), in PHN the rash clears but the pain remains. The transition from HZ to PHN does not even entail a marked change in the sensory quality of pain, something that might be expected of a shift from an inflammatory to a neuropathic cause. I acknowledge, however, that the claimed seamlessness of the transition reflects the impression of experienced clinicians I consulted, not studies that have actually tracked the transition using quantitative methods.

Uncertainty about the primacy of inflammation as a cause of pain in HZ (or PHN) also comes from the observation that systemic and topical NSAIDs and corticosteroids are only moderately effective in HZ and largely ineffective in PHN [1, 15, 30, 39, 77, 82, 92]. Given the disruption of the skin barriers due to the rash, topical drugs are expected to reach epidermal nerve endings and provide relief. Anti-inflammatory creams dispensed by compounding pharmacists may help, but such remedies usually contain local anesthetics and/or other membrane stabilizers such as amitriptyline and ketamine which are known to act on ectopic pacemaker sites [23, 103].

Finally, when the firing threshold of heat-sensitive nociceptors in skin inflamed experimentally drops below body temperature, firing becomes spontaneous [64]. This causes an ongoing burning sensation which can be relieved by cooling the skin (consider sunburn and CRPS). However, cooling is not notably effective at relieving ongoing burning pain in HZ or PHN. Indeed, patients who live in cold climates often report that it *exacerbates* pain [85]. This may be due to the fact that ectopia in C-fiber pacemakers in experimental neuromas is exacerbated by cold [76]. To summarize, counterintuitive as it may seem, we need to take seriously the thought that inflammation-sensitized nociceptors are *not* key drivers of spontaneous pain in HZ or PHN. These arguments also hold for tactile allodynia as discussed below.

Nonetheless, there are solid reasons to believe that ongoing pain in HZ/PHN is often driven from the skin. Most important are the observations that infiltrating the painful skin with local anesthetics often provides dramatic pain relief, and likewise diagnostic block of somatic nerves that serve the painful area [86, 97, 100]. This, of course, is incompatible with a CNS origin. A caveat is that these veteran studies lacked placebo controls and control for systemic effects of the anesthetics used. Along similar lines surgical excision of the painful skin may provide striking short-term pain relief although months later pain frequently returns [8, 94]. Interestingly, after freezing (intercostal) nerves, a process that also severs axons but facilitates the formation of regenerating sprouts, pain returns much sooner [54]. But if the skin innervation is indeed a major source of pain-provoking impulses in HZ and PHN, and the cause is not irritable, inflammation-sensitized nociceptors, what is it?

13.3.3 Ectopic Pacemakers in the Skin

Epidermal Fiber Loss: Degeneration vs. Dying-Back Histopathological studies of the skin in patients with HZ and PHN, especially recent studies based on punch biopsy samples taken at the time that pain was experienced, provide a likely answer. Specifically, these studies reveal substantial small-fiber loss in the epidermis where most nociceptors end (Oaklander's Chap. 12 in this volume and [9, 93, 114]. An additional indicator of C-fiber loss is reduced cutaneous blood flow response to the focal application of histamine [121]. Pathological changes are also seen in axon bundles in deeper dermal layers and proximally along nerve trunks, notably demyelination [122, 124, 132]. This suggests that damage to myelinated axons may also contribute to pain. The loss of epidermal endings tends to vary with the severity of the pain and in advanced PHN can apparently reach a stage of near-total denervation. As expected, severe loss blunts evoked sensation, but this hypoesthesia is often accompanied by particularly severe ongoing pain. These observations substantiate the long-standing conclusion that PHN is a neuropathic pain condition and extend it to HZ.

Epidermal fiber loss in HZ and PHN is usually attributed to VZV-induced destruction of cell bodies in the infected DRG followed by anterograde (Wallerian) degeneration of axon endings in the skin. But how can fiber *loss* explain ongoing pain, tactile allodynia, or hyperpathia? Denervation, partial or complete, is expected to cause sensory loss, i.e., negative signs and symptoms, not pain! The likely explanation is the formation of ectopic pacemakers and the emergence of ectopic electrogenesis akin to that seen in experimental neuromas. It needs to be made clear that epidermal fiber loss and the other structural abnormalities in the skin and nerves does *not* necessarily mean that the parent neuron in the DRG has died. Correspondingly, even gross severance of major nerves causes retrograde loss of only a minority of the axotomized DRG somata, and over an extended period of time (unless the lesion occurs at a very early age) [25, 112]. If all axotomized neurons died, there could never be a Tinel sign in amputees. I hypothesize that rather than reflecting cell death in the DRG, the loss of epidermal endings represents "dying-back" of the distal end of the axon (due to metabolic stress in the infected DRG neurons) with preservation of more proximal parts of the neuron including the cell soma and its central connectivity. The new retracted axon end is the ectopic pacemaker.

Ectopic Pacemakers and Pain in HZ/PHN If this is the correct interpretation of the histopathological findings, one would expect to find the distal end of C-fiber axons in the form of end bulbs and/or sprout(s) proximal to the epidermis, i.e., as microneuromas in the dermal plexus and the distal nerve trunk. In and among these structures, one might also expect to find end bulbs of large diameter fibers and axons that have shed myelin segments near the axon end and probably also regenerating sprouts and axons undergoing remyelination. Indeed, such structures, most notably myelin disorders, have been documented in pathology samples taken from patients with HZ and PHN [122, 132]. In other forms of experimental and clinical neuropathy, neuroma endings and patches of demyelinization tend to be marked with accumulations of Na^+ channels visualized using appropriate antibodies [32](Fig. 13.3). The mid-nerve accumulation of Na^+ channels at these ectopic locations is believed to underlie ectopic hyperexcitability and hence both spontaneous discharge and ectopic responsiveness to applied stimuli (Tinel sign [23]). In HZ/PHN such "hotspots" (ectopic pacemakers) are probably scattered in dermal fiber bundles and possibly further proximally along the course of the nerve trunk, although I am not aware of any studies that have attempted to visualize them in specimens taken from HZ/PHN patients. I am also unaware of whether the Tinel sign can or cannot be evoked by tapping along the proximal course of the relevant nerve (e.g., along an intercostal nerve in patients with distal allodynia).

Ectopic pacemakers in experimental neuromas are sensitive to specific inflammatory mediators, and this is also to be expected of hotspots in dermal and subdermal axon branches in HZ/PHN [29, 125]. Mediators released from keratinocytes are particularly powerful drivers of spontaneous discharge in neuroma endings [52, 98]. Finally, a growing literature based on microneurographic recordings from single nerve fibers has shown that in a variety of neuropathic pain conditions, spontaneous impulse discharge associates with ongoing pain, including in skin that has been partly denervated [13, 61, 104]. To the best of my knowledge, such recordings have not yet been made in patients with HZ or PHN.

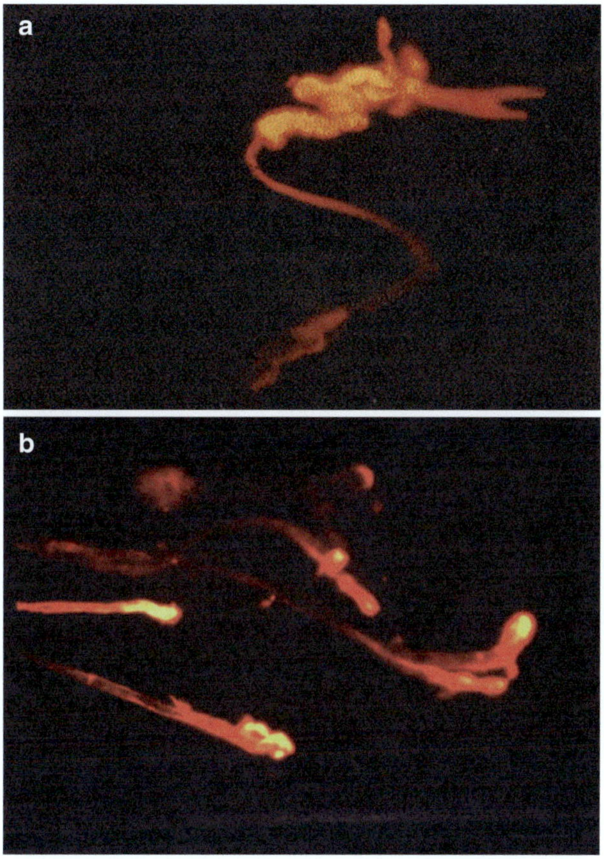

Fig. 13.3 Na^+ channel accumulation at ectopic pacemaker sites. (**a**) Photomicrograph of a neuroma end bulb with sprouts. Myelin is present until about 250 μm from the axon end (lower part of image). The *red-orange* glow indicates immunolabeling with an antibody that recognizes voltage-sensitive Na^+ channels. (**b**) A cluster of neuroma end bulbs immunolabeled for Na^+ channels (From Devor et al. [27])

All of these observations form the basis for the hypothesis that ongoing pain in HZ and PHN is driven by ongoing activity in disseminated ectopic pacemakers, in dermal fiber bundles, and perhaps at more proximal locations (see below). This accounts for the resolution of pain with the clearing of the rash in most HZ patients and its persistence, without prominent change in sensory quality, when HZ becomes PHN. Specifically, PHN patients are the minority in whom dermal pacemakers continue to fire spontaneously even without a boost by inflammatory mediators. Since the ongoing pain is driven by ectopia in the same sensory neurons in the early and the late phases, the sensation is similar. The disease was neuropathic from the beginning. Inflammation is only an exacerbating factor, contributing mostly in the early stage. How does this hypothesis fit with effective therapeutic modalities targeted to the skin?

Pain Relief by Suppression of Ectopic Pacemaker Activity in the Skin The observation that infiltrating the painful skin with local anesthetics, or proximal nerve block, transiently suppresses ongoing pain and allodynia [86, 97, 100] is consistent with the hypothesis that firing and pain originate in fibers that are dying-back (Fig. 13.4). There is an oddly persistent idea that effects of nerve blocks, single or multiple, might long outlast the action of the blocking drug and even prevent the transition of HZ to PHN. Only marginally plausible a priori, this idea continues to lack firm empirical support. But there are other approaches to sustaining pain relief. With the intent of extending the duration of analgesia, transdermal lidocaine patches have been introduced. These provide slow release of the anesthetic for an extended period of time. Although the patch depot contains 5 % lidocaine, skin penetration is poor, and the concentration in the skin is much lower, far too low to block impulses propagating along axons in deep dermal bundles or in major nerve trunks at a distance. But as discussed above, very low concentrations suffice to suppress ectopic *electrogenesis* at pacemaker sites within these bundles. Indeed, the transdermal lidocaine patch has proven efficacy in HZ/PHN [20, 38, 80]. Lidocaine at low concentrations may also have other analgesic effects, for example, by acting on keratinocytes and immune cells [102].

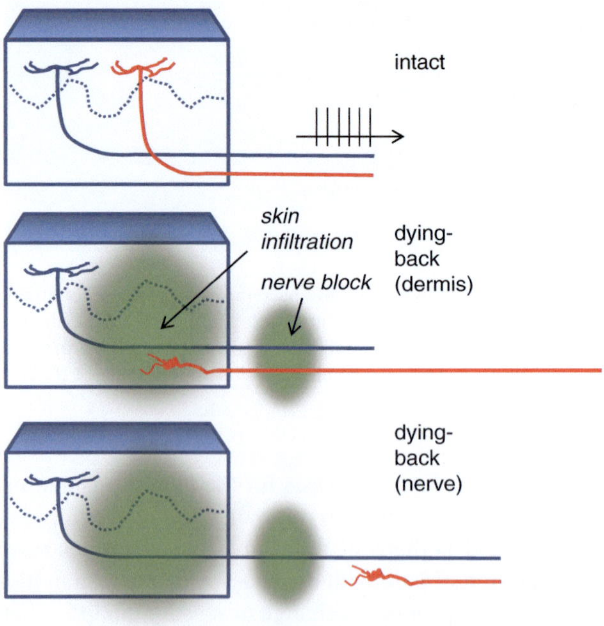

Fig. 13.4 Afferent innervation of the skin illustrating dying-back. Sketches, from top to bottom, show intact innervation by two afferent axons, dying-back of one axon into the dermal plexus, and dying-back into the cutaneous nerve. If spontaneous pain is due to ectopic electrogenesis in the dying-back end bulbs, skin infiltration or distal nerve block with local anesthetic (*large and small green clouds*) will yield transient relief in the second case, but not in the third. In both cases, the skin will lose responsiveness to applied stimuli

A strategy for further extending the duration of analgesia is topical capsaicin (8 %), also best applied via a slow-release transdermal patch. Capsaicin is a direct agonist of the Na^+ and Ca^{2+}-permeant TRPV1 receptor that is expressed by many nociceptors. Activating TRPV1 initiates (painful) impulse discharge and introduces high levels of Ca^{2+} into the axon. This disaggregates microtubules, blocks axoplasmic transport, and causes anterograde degeneration of the axon, destroying distal pacemaker sites and depleting pain-related neurotransmitters in the dorsal horn [42, 107]. This can result in partial pain suppression lasting for several weeks until regeneration restores pacemaker activity. The clinical effect, however, is only partial, and benefit is accrued in only a small minority of patients (NNT > 8) [11, 19]. The more complete effect of lidocaine, despite poorer penetration, suggests that much of the pain in HZ/PHN may be driven by afferents that do not express TRPV1, perhaps including touch-sensitive Aβ-fibers (see below). Infiltrating the skin with botulinum toxin also appears to have efficacy [105]. The toxin is thought to act by suppressing synaptic release at central endings of peripheral afferents that drive pain [34, 75]. This is yet another indication that the pain originates in primary afferent neurons that still have central terminals.

Finally, there was once strong support for the idea that activity in the sympathetic nervous system enhances pain in HZ/PHN and therefore that sympathectomy might arrest HZ pain or even prevent its transition to chronicity. Indeed, experimental studies confirmed that sympathetic efferent activity can exacerbate ectopia in experimental neuromas and DRGs [26, 79]. However, the clinical evidence that sympathetic block or sympatholysis effectively attenuate HZ/PHN pain, or prevent PHN, is weak [86, 129].

13.3.4 "Deafferentation" and the Spinal Cord as a Pain Driver in HZ and PHN

Deafferentation Classical postmortem studies dating back more than a century laid the foundation for the current belief that in PHN (and HZ?), the impulses that drive ongoing pain originate in the deafferented spinal cord or brain. Notable among them is the landmark opus of Head and Campbell [50] which reported in 21 cases examined severe hemorrhagic and inflammatory damage in one, and only one, DRG along with degenerative changes in associated nerves. However, as pointed out by Oaklander in Chap. 12 in this volume (also see [122]), caveats apply. These include the likelihood that most or all of the patients had HZ, not PHN, the likelihood that most also had other diseases that may have compromised the DRGs (especially tertiary syphilis) and uncertainty about the very presence of pain. Modern DRG specimens are not described as being bloody. And while demyelination and nerve fiber loss is corroborated, its extent, certainly in HZ, is not consistent with the heavy Wallerian degeneration that would be expected from the near-total necrosis of the ganglion as described by Head and Campbell. The 1900 study also does not mention ipsilateral atrophy of the dorsal horn, one of the most prominent features in long-standing PHN specimens (2–18 years) in contemporary studies [122].

Modern studies report cell loss in the dermatomal DRG in PHN, but little if any in HZ. However, none to date has applied the quantitative methods considered essential for cell counting in DRGs in experimental studies, including serial sectioning and unbiased sampling [111]. While neuronal loss undoubtedly occurs in chronic cases (>2 years after HZ), the published images show that even then the effect is far from total (see Fig. 12.3 in Chap. 12 of this volume [123]). In addition there is little loss of afferent peptidergic axons (SP and CGRP) in the ipsilateral dorsal horn as occurs after dorsal rhizotomy in experimental animals [43, 122]. True, major loss of immunostaining for SP and CGRP was reported in the DRG, but this does not necessarily indicate cell death. Nociceptive afferents in the DRG are known to decrease the synthesis of these peptides after nerve injury (with de novo synthesis in some Aβ touch afferents) as part of a major change in the pattern of gene expression that occurs in neuropathy [81, 83]. All of these considerations should moderate our confidence about the accuracy of the 100-year-old conclusion that HZ/PHN causes rapid and massive necrosis of one DRG.

Anesthesia Dolorosa But even if one DRG were destroyed completely, this would still not form a sound basis for the hypothesis that PHN (and certainly not HZ) is due to spinal deafferentation, with the pain-causing electrogenesis occurring in the CNS. In the past dorsal rhizotomy and ganglionectomy were used as a therapeutic strategy for treating chronic pain including that following peripheral nerve injury. Since afferents in most peripheral nerves enter the spinal cord through several adjacent segments, therapeutic rhizotomy typically included several adjacent dorsal roots. Such surgeries generally provided short-term relief proving that the pain did not arise in the CNS. The approach was ultimately abandoned, however, because of the appearance of a new pain problem, "anesthesia dolorosa" [126]. Anesthesia dolorosa is an intractable chronic pain state believed to be triggered by massive loss of presynaptic afferent terminals in the dorsal horn. The loss is presumed to initiate a compensatory increase in "amplification" in the circuitry of the deafferented spinal cord and trigeminal brainstem, akin to automatic gain control in electronic circuitry. This results in a "false" sensory signal, generated in the CNS, causing ongoing pain referred to the otherwise numb, insensate body part. The concept of a CNS origin is consistent with the fact that little short of destruction of the dorsal horn itself by DREZ surgery (dorsal root entry zone ablation) is effective [56, 96]. Note that in contrast to (multiple root) dorsal rhizotomy, peripheral nerve injury *denervates* peripheral tissue but does not significantly *deafferent* the spinal cord. Most sensory neurons in the DRG survive axotomy. In fact, as we have seen, nerve injury *per se* often increases afferent input to the spinal cord in the form of ectopic discharge.

Anesthesia dolorosa qualifies as deafferentation pain, but neither HZ nor PHN are likely to involve deafferentation-induced anesthesia dolorosa for a variety of reasons. First, in the large majority of patients with HZ, and many with PHN, the pain has come and gone before anesthesia dolorosa, a late-onset phenomenon, would be expected to appear. Second, single-level rhizotomy or ganglionectomy almost never triggers anesthesia dolorosa. For example, lumbosacral rhizotomy is carried out to relieve dystonia and spasticity in patients with cerebral palsy, and C2 rhizotomy and/or ganglionectomy is done to relieve chronic headaches and occipital neuralgia and even to facilitate procedural screw placement [40, 90, 91, 108]. Anesthesia dolorosa is very rare following these and other single ganglion procedures, and it does not appear rapidly. Third, major, albeit incomplete destruction of the trigeminal ganglion is carried out routinely in the treatment of trigeminal neuralgia using balloon compression, radiofrequency lesioning, or gamma-knife ablation. Here too, anesthesia dolorosa is a rare and delayed complication [10]. Finally, while ipsilateral atrophy of the dorsal horn is a reliable finding in long-term PHN, its relation to pain is unproven. It is not present in HZ, unlikely in early-stage PHN, and it may well be present in long-surviving PHN patients in whom pain eventually resolved.

Effects of Dorsal Rhizotomy and Ganglionectomy on Ongoing Pain in HZ and PHN If the impulses driving spontaneous pain in HZ or PHN originated in the spinal dorsal horn or in the brain, then this pain would be unaffected by rhizotomy or ganglionectomy. Both procedures have been carried out in patients with chronic, severe PHN usually after affirming pain relief by diagnostic segmental block. This surgery usually produces immediate and definitive elimination of the pain, but the effect is not durable. Therefore, in light of the risk of triggering anesthesia dolorosa, the procedure has been abandoned for PHN as it has for most other conditions [88, 109, 126]. Failure as a therapeutic modality, however, has no bearing on the conclusion that drivers of the original pain resided in the PNS. In summary, as for inflammation, the evidence base suggesting that HZ and/or PHN are due to spinal deafferentation is not entirely secure. Results of spinal block, which are more difficult to interpret, are discussed next.

The DRG as a Driver of Pain in HZ and PHN The overall impression from the material just reviewed is that the principal driver of spontaneous pain in HZ and PHN is peripheral. More specifically, if the reports cited above on skin infiltration and distal nerve block are to be taken at face value, the signal in most patients originates in the skin. However, the high rate of (diagnostic) success reported by Nurmikko et al. [86] and the others may not be representative. For example Zacks et al. [132], excised nerves serving painful skin in two cases of PHN without affecting ongoing pain in either. I failed to find further publications on this, but a colleague (Christoph Maier, Bochum) volunteered relevant results from his unpublished PhD thesis as follows: "I performed intercostal blocks in 28 patients with acute zoster (<12 weeks) and 18 patients with PHN. Immediate effects were: 25% of the acute cases, and 11% of the chronic were pain free for 6-8 hr. But in 32% (acute) and 50% (chronic) there was no effect." (Quoted with permission.). What might be driving pain in cases in which skin infiltration and nerve block are ineffective, or only partly effective?

A prime candidate is the DRG (Figs. 13.2 and 13.5). The DRG is the initial location of the causative VZV infection. Likewise, as noted above, it appears to be a major pain source in other neuropathic pain conditions. DRG neurons are progressively lost over time with massive loss seen mostly in late stage disease. For most patients one should therefore imagine a process where only a fraction of DRG neurons are compromised due to the toxic inflammatory environment in the ganglion. Several factors might cause "sick" neurons to become electrically hyperexcitability and a source of ongoing ectopic discharge and pain. These range from generalized metabolic stress resulting in membrane depolarization and enhancement of spike-triggering subthreshold oscillations [3] to more direct effects of inflammatory mediators as in experimental neuromas. The VZV might even play a direct role by altering the expression of molecules of excitability in the DRG, especially upregulation of Na^+ channels [46, 57]. Interestingly, although neuron-to-neuron cell fusion has not been described in VZV-infected DRGs, this does occur following infection with the related herpes simplex virus (HSV), with hyperexcitability as a consequence [78].

The DRG has no tissue-tissue or tissue-blood barriers and is readily accessible to targeted therapeutic intervention [21]. In animal models and human amputees, transient silencing of DRG ectopia yields transient pain relief [110, 118]. Nor has the DRG been overlooked as a potential target for pain control in HZ and PHN. Going as far back as the 1930s [101], there have been reports of intraganglionic, epidural, and intrathecal injection of lidocaine, procaine, or bupivacaine, usually together with depot-form (particulate) corticosteroids. Thoracic zoster is the usual target. Since the thoracic dorsal roots are short, the virus-infected DRG would also have been exposed to the local anesthetic in these studies (Fig. 13.5). Variable therapeutic benefit has been reported in randomized controlled trials [15, 66, 116]. But to know if the DRG is an important pain source, we want to know to what extent pain is suppressed *acutely*, during the first hours after spinal injection of lidocaine. Unfortunately, this information was not reported. In a personal communication with the first author of one large study, it was explained that the study objective was therapeutic efficacy; information on acute effects was probably noted but was not reported and is not currently available. Interestingly, adding steroids to the local anesthetic provides little additional benefit suggesting that as in the skin, inflammation is not the major driver of ectopia in the DRG.

Most valuable would be to know if there is a difference in the acute effect of depositing lidocaine on the spinal nerve distal to the DRG vs. on the surface of the DRG itself, within the intervertebral foramen. If the DRG is a major pain generator of spontaneous pain, intraforaminal injection ought to be far more effective (Fig. 13.5). In the absence of published data on this, I queried a number of physicians who use epidural injection in their daily practice. Only one reply, from Dr. Shane Brogan (Salt Lake City), was illuminating although tentative and lacking placebo control. I quote with permission: “Anecdotally, I have done selective nerve root block (and presumably blocking the DRG) on PHN patients and the response is always interesting. Some patients describe pain relief (seldom with long-term benefit) and others describe an anesthesia dolorosa situation where they are anesthetic to exam but their pain is somewhat disturbingly unchanged. But I have not seen any papers quantifying this better.” The obvious implication of this observation is that block on or central to the DRG eliminates both evoked and spontaneous pain, while block just distal to the DRG eliminates sensation evoked by stimulating the skin, but not ongoing pain (hence anesthesia dolorosa). This is precisely the outcome expected if a major portion of the spontaneous pain signal is generated within the DRG (Fig. 13.5).

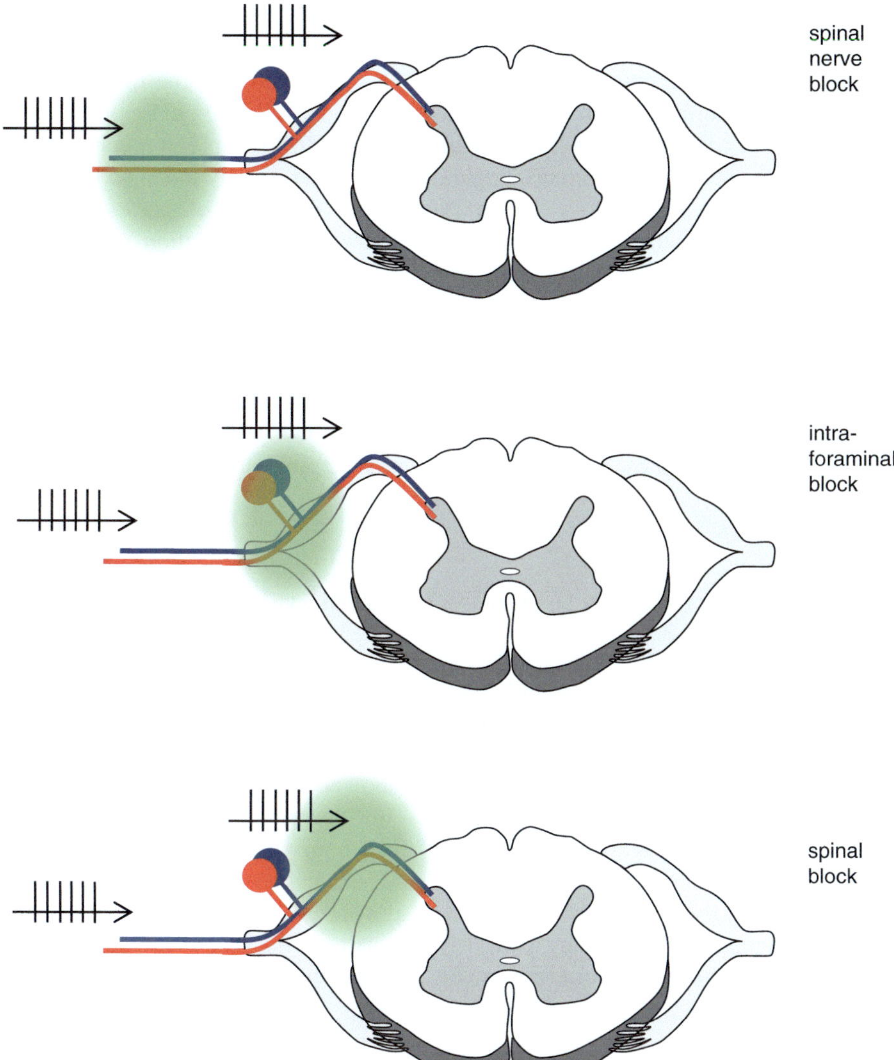

Fig. 13.5 Anticipated effects of diagnostic paraspinal blocks with local anesthetics. In each of the three sketches, there is afferent input (spontaneous and/or evoked) from the periphery (note impulse trains). Spinal nerve, intraforaminal, and spinal blocks (*green clouds*) will all suppress this signal assuming a lidocaine concentration adequate to block impulse propagation in axons (2 %) is used. However, if the DRG is a source of (spontaneous) discharge and pain, this signal will be suppressed by intraforaminal block using high or low lidocaine concentrations, but not by (spinal) nerve block. The effect of spinal block on signals originating in the DRG depends on whether dorsal root axons are exposed to blocking concentrations of lidocaine (2 %), or alternatively whether after the drug has spread to ectopic pacemaker sites within the ganglion the concentration remains high enough to suppress local impulse initiation

Remission of Pain in HZ and PHN Punch biopsy shows a rough association between the amount of epidermal fiber loss and pain report. This is expected by the ectopic pacemaker hypothesis as the more fibers that die-back, the more ectopic pacemakers are formed. But there is very little recovery of epidermal innervation by regenerative or collateral sprouting even after years of follow-up, nor much reversal of sensory loss [93]. Why, then does pain in acute zoster usually fade in a matter of weeks, and PHN pain within a year in most cases? A number of factors probably contribute. These include reduced inflammation with the resolution of the rash, decline in viral load in the DRG, depletion of pacemakers with the gradual loss of neuronal somata and consequent Wallerian degeneration of the distal axon, and the natural tendency for ectopic firing to subside over weeks, at least in neuroma endings of A-fibers (see below) [23]. In the few remaining cases in which PHN persists indefinitely a population of active DRG neurons probably persists indefinitely. Another possibility is that CNS hyperexcitability amplifies afferent input from adjacent segments (Fig. 13.6). The significance of overlapping innervation in the skin is discussed below.

13.4 Hypersensibility to Applied Stimuli and Tactile Allodynia

The main message of the previous section is that impulses that drive *spontaneous* pain in HZ and PHN arise in the PNS, alternatively in afferent fibers in the skin, at ectopic pacemaker sites disseminated along peripheral nerves, and in the affected segmental DRG. Based on published results on skin infiltration and nerve block, dermal generators appear to be the major contributors, but it is not clear how general this conclusion is. A priori it would appear likely that the relative contribution of each source might vary among patients and over time. For PHN, at least, diagnostic blocks should probably be a part of the usual workup (Fig. 13.7). In contrast, with regard to tactile allodynia, location is known; activity obviously originates in the skin where the stimulus is applied. Here the relevant question is why light-touch stimuli evoke pain?

13.4.1 Tactile Allodynia: Irritable Nociceptors or Central Sensitization?

Sensitive Nociceptor Endings and Dermal Pacemakers It is appealing to believe that tenderness to the touch (tactile allodynia) in HZ and PHN is due to irritable mechano-nociceptor endings in the skin. This is the most widely quoted explanation [33]. But as discussed above, a number of observations are inconsistent with this explanation. For example, tactile allodynia may appear before or without zoster rash, and like ongoing pain it apparently continues unchanged following the clearance of the rash and the transition to PHN. A variant of this hypothesis, more in line with neuropathic mechanisms, is that tactile allodynia reflects exaggerated mechanosensitivity of ectopic pacemakers at dying-back ends of C-fibers. However, this is also unlikely. Among the most compelling reasons are [4–6, 12, 62, 63, 95, 106, 113]:

1. In animal models of tactile allodynia one does not find C-fiber endings that respond to the weak forces that evoke tactile allodynia. This contrasts with the marked sensitization of C-fiber afferents to warm stimuli. In some HZ/PHN patients, the exquisite sensitivity to the very lightest touch goes well beyond the allodynia typical of the inflamed skin.
2. In humans it takes ~1–2 s for impulses carried in C-fiber nociceptors to reach the spinal cord from the hand or foot and even longer to reach levels of conscious perception. C-fibers conduct at ~1 m/s. And yet, everyday experience shows that pain on touching mildly burned skin, or areas affected by HZ or PHN, is felt almost immediately upon touch, certainly not after a delay of seconds. Therefore, C-fibers, which constitute the large majority of mechano-nociceptors, could not be carrying this signal. Aδ nociceptors are fast enough, but they are rare in comparison to C-fibers. Furthermore, a brushstroke to allodynic skin does not induce both first and second pain (Aδ- and C-fiber mediated), and I am not aware of evidence that sensitization processes differ between the two fiber types.
3. Selective block of touch-responsive Aβ-fibers eliminates tactile allodynia without affecting heat allodynia, while C-fiber block attenuates heat allodynia but not tactile allodynia. This and other evidence has convinced most investigators that the impulses that cause tactile allodynia are carried by normal, fast-conducting, touch-responsive Aβ-fibers, not C-nociceptors. What can cause impulses in touch-sensitive A-fibers to evoke pain sensation?

Central Sensitization Numerous specific neurophysiological mechanisms have been identified in the CNS that can alter the normal match between the type of afferent fiber activated and the sensation evoked. Each of these can render light touch painful. The mechanisms include long-term synaptic potentiation (LTP), loss of spinal inhibition, de novo expression of CGRP in Aβ neurons, reversal of the chloride pump, and activation of spinal microglia [128]. The umbrella term used for the entire collection of such processes is "central sensitization." Central sensitization is not rare or pathological; it is an everyday event that protects injured tissue. For example, Aβ input accounts for the ordinary tenderness of abrasions and sunburn. The idea that nociceptors do not have a monopoly on pain, that Aβ touch afferents are often primary pain drivers ("Aβ-pain"), constitutes a revolution in the way we now understand the pain system [22, 24, 48, 127].

Central sensitization is a labile state that can turn on and off in a matter of minutes or hours. It is usually induced by nociceptive input from the periphery. C-fiber impulses from a mild burn or from an active neuroma, for example, initiate central sensitization. The resulting tactile allodynia then lasts for as long as the maintaining nociceptive drive persists. In patients with PHN, this can be years. But as soon as the maintaining drive subsides, the central sensitization and the allodynia fade. This is presumably why spontaneous pain and allodynia decline together.

Allodynia can also be stopped intentionally by cooling the skin (when sensitized heat nociceptors are the active drivers), or in neuropathy by suppressing ectopic pacemaker firing with lidocaine. Once the effect of the lidocaine fades, however, the allodynia returns [44, 65]. Importantly, just as Aβ activity evoked by light brushing is felt as pain when central sensitization is present, Aβ activity arising spontaneously from neuroma endings, sites of demyelination, or active DRG neurons will be felt as spontaneous (neuropathic) pain. Ectopic firing in Aβ afferents may thus contribute as much to ongoing *and* evoked pain in patients with HZ and PHN as ectopia in C-nociceptors. Suppressing peripheral ectopic discharge has two benefits: (1) it reduces the direct drive of spontaneous pain due to spontaneous activity in C- and A-fiber afferents and (2) it suppresses central sensitization, eliminating tactile allodynia. Note that while the terms "central sensitization" and "pain centralization" sound the same, they have very different meanings and clinical consequences.

13.4.2 The Significance of Dermatomal Overlap

Viral resurgence in one DRG yields the single-dermatome distribution of pain in HZ/PHN. However, due to segmental overlap in skin innervation [45, 60] any given patch of skin also contains sensory endings of afferent neurons that reside in at least the two adjacent DRGs (Fig. 13.6). By the ectopic pacemaker hypothesis, the major source of spontaneous pain is electrogenesis in dying-back axon endings of infected DRG neurons and in cell somata resident in the DRG. There is evidence that after injury to the spinal nerve of one ganglion, "uninjured" (i.e., non-axotomized) afferents in the immediately adjacent ganglia may also begin to fire spontaneously [2]. It is not known whether something similar occurs following VZV infection of a single ganglion. But if it did, then in principle the neighboring DRGs could also contribute to pain in HZ and PHN.

Ongoing activity in infected afferents is presumed to be responsible for maintaining central sensitization. However, in the presence of central sensitization, Aβ touch afferents whose cell soma resides in adjacent non-infected (intact) ganglia are also expected to contribute to tactile allodynia. In advanced PHN, where there is considerable loss of neurons in the infected DRG and corresponding loss of ectopic pacemaker sites in the skin and nerve due to Wallerian degeneration, spontaneous firing in the adjacent ganglia might sustain ongoing pain. Moreover, if this activity were able to maintain central sensitization [22, 83], the adjacent ganglia could also support tactile allodynia. Such contributions by adjacent DRGs are speculative at the moment. In the absence of a satisfactory animal model of HZ/PHN, it will be necessary to examine these possibilities in dedicated clinical studies in which diagnostic blocks are made to the individual ganglia in turn (Fig. 13.6).

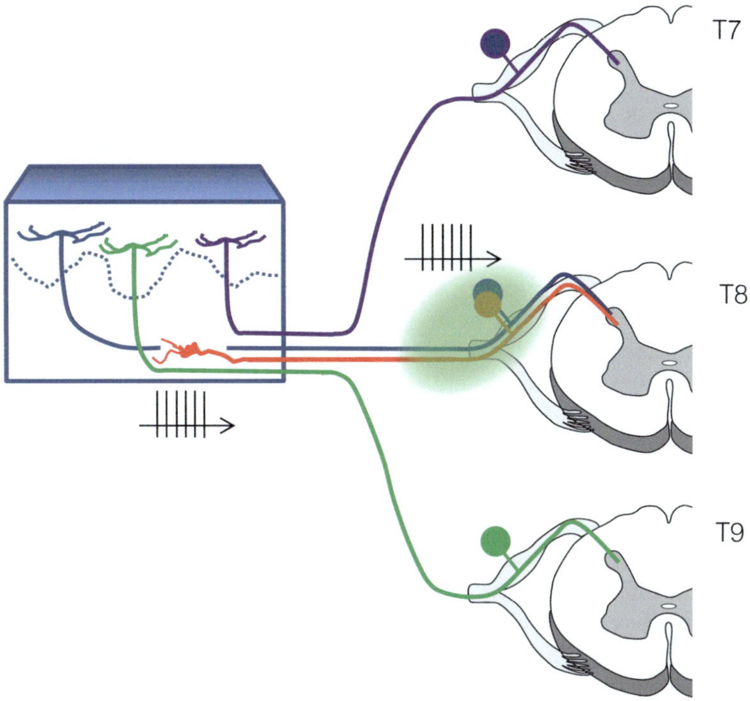

Fig. 13.6 Several adjacent DRGs contribute afferent axons to any given patch of the skin. In this sketch there has been reactivation of VZV infection in the T8 DRG which has caused dying-back of some T8 afferents into the dermal plexus where ectopic discharge is generated. Ectopic discharge is also generated in the DRG (spike trains). Blocking the T8 afferent pathway central to the point at which T7 and T9 afferents peel off of the cutaneous nerve (*green cloud*) will leave a degree of functionality in the partly denervated skin due to afferents from the T7 and T9 segments

13.5 Summary and Perspective

The Ectopic Pacemaker Hypothesis of Pain in HZ and PHN Despite the remarkable progress in our understanding of the biology of neuropathic pain, there has been disappointingly little translation into the clinic thus far. In this chapter I have considered the causes of pain in HZ and PHN in light of the new knowledge. After pointing out significant shortcomings in the traditional explanations, inflammation, and deafferentation, I proposed an alternative, the ectopic pacemaker hypothesis of pain in HZ and PHN. *This hypothesis holds that (1) spontaneous pain is due to spontaneous impulse discharge arising at ectopic locations in the PNS and (2) tactile allodynia results from intensification of the sensory effects of normal Aβ touch afferents by central sensitization maintained by the ectopia.* The hypothesis is rooted in a single pathophysiological process, hyperexcitability. Normally, the excitability of afferent neurons is regulated within narrow bounds (stabilized) by homeostatic processes that remain poorly understood. When nerve injury or disease interferes, hyperexcitability, ectopic firing, and positive symptoms (pain) can result. Negative symptoms in HZ/PHN (hypesthesia, hypalgesia) are a simple consequence of fiber damage or loss. I suspect that this general framework, applied here to HZ/PHN, applies no less to other painful medical conditions properly attributed to neuropathy. Prominent among these are conditions that involve dying-back, such as painful diabetic neuropathy and perhaps fibromyalgia [104, 115].

Medical Management in Light of the Ectopic Pacemaker Hypothesis: Rationale and Avenues for Improvement An understanding of mechanism can illuminate the mode of action of effective treatments and guide future developments. None of the therapeutic agents available today for HZ and PHN are considered adequate. But some have measurable efficacy, and when this is added to the inevitable (and ultimately desirable) effects of context and placebo, they are the best therapeutic modalities we have. A quick review of the agents for which there is a consensus on efficacy is therefore likely to be informative [16, 31, 35, 51, 117]. As topical treatments were discussed above, I will focus here on orally available drugs.

First, opiates are effective presumably due to activation of central pain-control pathways. Side effects and abuse potential are widely recognized. Anti-inflammatory drugs, including systemic corticosteroids, seem to have modest efficacy at best, even in HZ. This is odd for a condition classically thought to be caused by inflammation. Indeed, this should probably have raised misgivings long ago. Systemic local anesthetics are effective presumably because they suppress ectopia at very low plasma concentrations. But at present they have no practical orally available counterpart. The remaining drugs with recognized efficacy include *particular* agents from two very different pharmacological categories: antidepressants and anticonvulsants. Among the antidepressants, tricyclics and SNRIs tend to be effective (and at doses too low to be antidepressant), while SSRIs are not. Among the anticonvulsants, carbamazepine and gabapentin are somewhat effective while barbiturates are not. The striking factor that differentiates the effective from the ineffective drugs in both categories is that the effective ones tend to have membrane stabilizing effects in addition to the actions for which they are marketed. That is, their pharmacological profile includes a local anesthetic-like action, and/or they have been shown experimentally to suppress ectopic hyperexcitability in injured sensory neurons [14, 23]. If HZ and PHN are due primarily to hyperexcitable afferents, then it is expected that agents which suppress excitability should be analgesic. This also accounts for the fact that the same quirky list of drugs used for HZ and PHN are used for neuropathic pain states across the diagnostic spectrum.

But an important caveat needs to be kept in mind with respect to the use of these drugs. Since all neural functions, and also muscle and some endocrine functions, depend on membrane excitability, suppressing excitability with a systemic drug is likely to have unwanted, generally suppressive, side effects. These tend to limit the dose that can be used, limiting analgesic efficacy in parallel. Since excitability involves a variety of ion channels (Na^+, K^+, Ca^{2+}, HCN, etc.) and receptors, each with a variety of subtypes and isoforms, different drugs are expected to show different efficacy and side effect profiles. For this reason, the current practice of titrating drug dose in a given patient, and switching among drugs if need be, makes clinical and pharmacological sense. Likewise, subtype-selective membrane stabilizers now in development may have advantages. But in the end, most of the drugs used currently share the same side effect profile: drowsiness, reduced cognitive focus, dizziness, and nausea. These are all CNS effects. If it is so that the desired effect (suppression of ectopic pacemakers) plays out in the PNS, the usefulness of our standard drugs might be improved by preventing their central action. Cardiology exploited this strategy long ago by peripheralizing β-blockers.

Another rational approach is to specifically target the relevant pain generator. Thus, if pain in a particular HZ/PHN patient is due to impulses generated in the skin, a transdermal approach makes sense. On the other hand, if the DRG is the primary pain driver in this patient, then topical lidocaine or capsaicin is unlikely to work (Figs. 13.4, 13.5, and 13.7). Rather, one should target the DRG directly using systemic membrane-stabilizing drugs or better by delivering such agents direct to the intervertebral foramen via a port, or a catheter attached to an implantable pump system [118]. As noted above, ectopic electrogenesis is silenced by concentrations of membrane stabilizers much lower than is required to block impulse propagation. Thus, if accurately targeted, a very low concentration of lidocaine, amitriptyline, etc. should be sufficient. In principle, the pump reservoir could contain a high concentration of the drug, permitting very slow pumping rates and an extended duration between refills. Since only local electrogenesis will be affected and not conduction

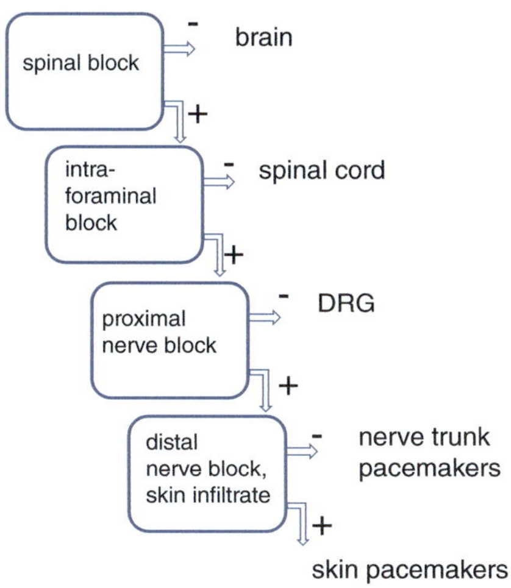

Fig. 13.7 Flow chart indicates a diagnostic routine for locating sources of ectopic spontaneous electrogenesis (the driver of ongoing pain) using local anesthetic blocks at the level of the spinal cord, intervertebral foramen, proximal nerve, distal nerve, or skin. For each type of block, if the spontaneous pain persists (i.e., if the block was ineffective), the likely pain source is indicated by the arrow marked (-). As a practical matter blocks would usually begin at the skin or distal nerve or distal nerve and proceed proximally

along *en passage* axons, sensation from the periphery should remain normal as should motor function.

The use of electrical stimulation delivered to the DRG has been advocated recently [70]. It is hard to imagine this providing relief in HZ/PHN patients with tactile allodynia, as such stimulation preferentially activates Aβ afferents and should provoke pain. In patients without allodynia, however, DRG stimulation might work by closing spinal gates. Alternatively, some non obvious mechanisms might cause stimulation to suppress ectopic discharge [71].

Lacunae, and a Call for Observational Evidence from the Clinic The ectopic pacemaker hypothesis laid out here arose in an intellectual exercise: fitting clinical observations about HZ/PHN into a theoretical framework on neuropathic pain that has been evolving for the past few decades. The evidence was gleaned more from incidental observations than focused research. As pointed out at several junctures, it was sometimes necessary to draw conclusions from reports lacking essential controls, from unpublished observations and from clinical impressions (I hope that I did not miss any seminal publications). For this reason, the hypothesis presented is tentative. Dedicated research would help firm up some of the soft spots. But even without such efforts, it might be possible to answer some questions if authors of therapeutic studies were simply to report immediate effects of interventions such as lidocaine blocks. Specifically, I urge clinicians who see patients with HZ and PHN to collect and share information on the following issues.

1. Is the transition from HZ to PHN indeed seamless in terms of the patient's sensory experience?
2. Does cooling the skin, or infiltration with wide-spectrum anti-inflammatory agents, in fact fail to provide relief?
3. What is the short-term effect across patients of infiltrating painful skin with lidocaine, or blocking relevant nerves? A clear distinction should be made between spontaneous and evoked pain, and control for placebo and systemic effects of the drug needs to be kept in mind (Fig. 13.2).
4. Persistence of pain following definitive spinal block would establish whether, and to what extent, the spontaneous pain signal is generated in the brain. Intraforaminal block, best preceded by block of the plexus or the spinal nerve distal to the foramen, could determine if the generator is in the spinal cord, the DRG or further peripherally (Figs. 13.6, 13.7).
5. Can a Tinel sign be evoked by percussion along the course of relevant nerves?
6. It is time for microneurography to be applied to HZ/PHN. Cases with lumbosacral distribution of allodynia might permit easier access to a suitable nerve. Adding proximal and distal nerve blocks would reveal the likely origin of spontaneous firing.
7. Punch biopsy of the skin has become a common diagnostic tool. With regard to pain, the status of dermal and subdermal fibers, including large myelinated fibers, is of interest in addition to C-nociceptors in the epidermis. Is there ectopic accumulation of Na^+ channels or other molecules-of-excitability using appropri-

ate immunolabels? Examination of biopsied or postmortem nerve samples in this way could also be informative.

HZ and PHN are relatively common conditions, and they take an enormous toll in human suffering. Answers to questions that bear on pain mechanism are a high priority. One might have hoped that answers could be obtained using animal models. Indeed, around the time of the publication of the previous edition of this book, the first such model appeared [36]. Additional papers followed, many demonstrating tactile allodynia and some noting evidence of hyperexcitability, including in Aβ touch afferents [41, 67]. However, while VZV infection can be induced in vitro, there have been difficulties demonstrating viral reactivation in vivo, or obtaining titers of virus in DRGs that are high enough to convincingly mimic the human disease. Rodents are highly resistant to VZV infection [59, 74, 99]. Until these issues are resolved, progress on pain mechanism in HZ and PHN will probably be based mostly on observations in vitro and in human patients.

The ectopic pacemaker hypothesis of pain in HZ and PHN stresses three principles: (1) diagnostic identification, in the individual patient, of where the pain-provoking impulses are coming from, (2) targeting the primary source(s), and (3) focusing on suppression of ectopic electrogenesis (using membrane-stabilizing drugs and strategies) in preference to anti-inflammatory approaches. Incorporating these principles could benefit clinical outcome.

Acknowledgments I wish to thank the following individuals for their helpful responses to queries about unpublished observations concerning herpes zoster and postherpetic neuralgia, and for their guidance on published material: Ralf Baron, Shane Brogan, Perry Fine, Ron Goldstein, Michaela Kress, John Loeser, Florella Magora, Christof Maier, Turo Nurmikko, Anne Louise Oaklander, Gabor Racz, Srinivasa Raja, Zvi Harry Rappaport, Andrew Rice, Michael Rowbotham, Ze'ev Seltzer, Jordi Serra, Claudia Sommer, Maarten van Kleef, and Bart van Wijck. A special note of thanks is due to Peter Watson for his feedback on numerous factual and conceptual issues related to the text. The authors' research on neuropathic pain mechanisms is funded by the Israel Science Foundation, the Seymour and Cecile Alpert Chair in Pain Research, and the Hebrew University Center for Research on Pain.

References

1. Ahmed SU, Zhang Y, Chen L, Cohen A, St Hillary K, Vo T, Houghton M, Mao J (2015) Effect of 1.5 % topical diclofenac on clinical neuropathic pain. Anesthesiology 123:191–198
2. Ali Z, Ringkamp M, Hartke TV, Chien HF, Flavahan NA, Campbell JN, Meyer R (1999) Uninjured cutaneous C-fiber nociceptors develop spontaneous activity and alpha adrenergic sensitivity following L6 spinal nerve ligation in the monkey. J Neurophysiol 81:455–466
3. Amir R, Michaelis M, Devor M (2002) Burst discharge in primary sensory neurons: triggered by subthreshold oscillations, maintained by depolarizing afterpotentials. J Neurosci 22:1187–1198
4. Andrew D, Greenspan JD (1999) Mechanical and heat sensitization of cutaneous nociceptors after peripheral inflammation in the rat. J Neurophysiol 82:2649–2656
5. Banik RK, Brennan TJ (2004) Spontaneous discharge and increased heat sensitivity of rat C-fiber nociceptors are present in vitro after plantar incision. Pain 112:204–213

6. Banik RK, Brennan TJ (2008) Sensitization of primary afferents to mechanical and heat stimuli after incision in a novel in vitro mouse glabrous skin-nerve preparation. Pain 138:380–391
7. Bouhassira D, Attal N, Alchaar H, Boureau F, Brochet B, Bruxelle J, Cunin G, Fermanian J, Ginies P, Grun-Overdyking A, Jafari-Schluep H, Lanteri-Minet M, Laurent B, Mick G, Serrie A, Valade D, Vicaut E (2005) Comparison of pain syndromes associated with nervous or somatic lesions and development of a new neuropathic pain diagnostic questionnaire (DN4). Pain 114:29–36
8. Browder J, de Veer JA (1949) Herpes zoster: a surgical procedure for the treatment of postherpetic neuralgia. Ann Surg 130:622–635
9. Buonocore M, Gatti AM, Amato G, Aloisi AM, Bonezzi C (2012) Allodynic skin in postherpetic neuralgia: histological correlates. J Cell Physiol 227:934–938
10. Burchiel KJ (2015) Surgical management of pain, 2nd edn. Thieme, New York, p. 631
11. Campbell CM, Diamond E, Schmidt WK, Kelly M, Allen R, Houghton W, Brady KL, Campbell JN (2016) A randomized, double-blind, placebo-controlled trial of injected capsaicin for pain in Morton's neuroma. Pain 157:1297–1304
12. Campbell JN, Raja SN, Meyer RA, MacKinnon SE (1988) Myelinated afferents signal the hyperalgesia associated with nerve injury. Pain 32:89–94
13. Campero M, Serra J, Marchettini P, Ochoa JL (1998) Ectopic impulse generation and autoexcitation in single myelinated afferent fibers in patients with peripheral neuropathy and positive sensory symptoms. Muscle Nerve 21:1661–1667
14. Catterall WA (1987) Common modes of drug action on Na+ channels: local anaesthetics, antiarrhythmics and anticonvulsants. Trends Pharmacol Sci 8:57–65
15. Chen N, Yang M, He L, Zhang D, Zhou M, Zhu C (2010) Corticosteroids for preventing postherpetic neuralgia. Cochrane Database Syst Rev 3:CD005582
16. Cohen JI (2013) Herpes zoster. N Engl J Med 369:1766–1767
17. Cole NL, Grose C (2003) Membrane fusion mediated by herpesvirus glycoproteins: the paradigm of varicella-zoster virus. Rev Med Virol 13:207–222
18. Defrin R, Devor M, Brill S (2014) Tactile allodynia in patients with lumbar radicular pain (sciatica). Pain 155(12):2551–2559
19. Derry S, Sven-Rice A, Cole P, Tan T, Moore RA (2013) Topical capsaicin (high concentration) for chronic neuropathic pain in adults. Cochrane Database Syst Rev 2:CD007393
20. Derry S, Wiffen PJ, Moore RA, Quinlan J (2014) Topical lidocaine for neuropathic pain in adults. Cochrane Database Syst Rev 7:CD010958
21. Devor M (1999) Unexplained peculiarities of the dorsal root ganglion. Pain suppl.6:S27–S35
22. Devor M (2009) Ectopic discharge in Abeta afferents as a source of neuropathic pain. Exp Brain Res 196:115–128
23. Devor M (2013) Neuropathic pain: pathophysiological response of nerves to injury. Chapter 61. In: McMahon SL, Koltzenburg M, Tracey I, Turk DC (eds) Wall and Melzack's textbook of pain, 6th edn. Churchill Livingstone, London, pp. 861–888
24. Devor M, Basbaum A, Bennett G, Blumberg H, Campbell J, Dembowsky K, Guilbaud G, Janig W, Koltzenberg M, Levine J, Otten U, Portenoy R (1991) Mechanisms of neuropathic pain following peripheral injury. In: Basbaum A, Besson J-M (eds) Towards a new pharmacology of pain. Dahlem Konferenzen. Wiley, Chichester, pp. 417–440
25. Devor M, Govrin-Lippmann R (1991) Neurogenesis in adult rat dorsal root ganglia: on counting and the count. Somatosens Motor Res 8:9–12
26. Devor M, Janig W (1981) Activation of myelinated afferents ending in a neuroma by stimulation of the sympathetic supply in the rat. Neurosci Lett 24:43–47
27. Devor M, Keller CH, Deerinck T, Levinson SR, Ellisman MH (1989) Na+ channel accumulation on axolemma of afferents in nerve end neuromas in Apteronotus. Neurosci Lett 102:149–154
28. Devor M, Wall PD, Catalan N (1992a) Systemic lidocaine silences ectopic neuroma and DRG discharge without blocking nerve conduction. Pain 48:261–268

29. Devor M, White DM, Goetzl EJ, Levine JD (1992b) Eicosanoids, but not tachykinins, excite C-fiber endings in rat sciatic nerve-end neuromas. Neuroreport 3:21–24
30. Dworkin RH, Johnson RW, Breuer J, Gnann JW, Levin MJ, Backonja M, Betts RF, Gershon AA, Haanpaa ML, McKendrick MW, Nurmikko TJ, Oaklander AL, Oxman MN, Pavan-Langston D, Petersen KL, Rowbotham MC, Schmader KE, Stacey BR, Tyring SK, van Wijck AJ, Wallace MS, Wassilew SW, Whitley RJ (2007) Recommendations for the management of herpes zoster. Clin Infect Dis 44(Suppl 1):S1–26
31. Dworkin RH, O'Connor AB, Kent J, Mackey SC, Raja SN, Stacey BR, Levy RM, Backonja M, Baron R, Harke H, Loeser JD, Treede RD, Turk DC, Wells CD (2013) Interventional management of neuropathic pain: NeuPSIG recommendations. Pain 154:2249–2261
32. England JD, Happel LT, Kline DG, Gamboni F, Thouron CL, Liu ZP, Levinson SR (1996) Sodium channel accumulation in humans with painful neuromas. Neurology 47:272–276
33. Fields HL, Rowbotham M, Baron R (1998) Postherpetic neuralgia: irritable nociceptors and deafferentation. Neurobiol Dis 5:209–227
34. Filipovic B, Matak I, Bach-Rojecky L, Lackovic Z (2012) Central action of peripherally applied botulinum toxin type A on pain and dural protein extravasation in rat model of trigeminal neuropathy. PLoS One 7:e29803
35. Finnerup NB, Attal N, Haroutounian S, McNicol E, Baron R, Dworkin RH, Gilron I, Haanpaa M, Hansson P, Jensen TS, Kamerman PR, Lund K, Moore A, Raja SN, Rice AS, Rowbotham M, Sena E, Siddall P, Smith BH, Wallace M (2015) Pharmacotherapy for neuropathic pain in adults: a systematic review and meta-analysis. Lancet Neurol 14:162–173
36. Fleetwood-Walker SM, Quinn JP, Wallace C, Blackburn-Munro G, Kelly BG, Fiskerstrand CE, Nash AA, Dalziel RG (1999) Behavioural changes in the rat following infection with varicella-zoster virus. J Gen Virol 80(Pt 9):2433–2436
37. Forbes HJ, Thomas SL, Smeeth L, Clayton T, Farmer R, Bhaskaran K, Langan SM (2016) A systematic review and meta-analysis of risk factors for postherpetic neuralgia. Pain 157:30–54
38. Galer BS, Jensen MP, Ma T, Davies PS, Rowbotham MC (2002) The lidocaine patch 5 % effectively treats all neuropathic pain qualities: results of a randomized, double-blind, vehicle-controlled, 3-week efficacy study with use of the neuropathic pain scale. Clin J Pain 18:297–301
39. Galluzzi KE (2009) Managing herpes zoster and postherpetic neuralgia. J Am Osteopath Assoc 109:S7–12
40. Gande AV, Chivukula S, Moossy JJ, Rothfus W, Agarwal V, Horowitz MB, Gardner PA (2016) Long-term outcomes of intradural cervical dorsal root rhizotomy for refractory occipital neuralgia. J Neurosurg 125:102–110
41. Garry EM, Delaney A, Anderson HA, Sirinathsinghji EC, Clapp RH, Martin WJ, Kinchington PR, Krah DL, Abbadie C, Fleetwood-Walker SM (2005) Varicella zoster virus induces neuropathic changes in rat dorsal root ganglia and behavioral reflex sensitisation that is attenuated by gabapentin or sodium channel blocking drugs. Pain 118:97–111
42. Gibson SJ, McGregor G, Bloom SR, Polak JM, Wall PD (1982) Local application of capsaicin to one sciatic nerve of the adult rat induces a marked depletion in the peptide content of the lumbar dorsal horn. Neuroscience 7:3153–3162
43. Gibson SJ, Polak JM, Bloom SR, Sabate IM, Mulderry PM, Ghatei MA, McGregor GP, Morrison JF, Kelly JS, Evans RM et al (1984) Calcitonin gene-related peptide immunoreactivity in the spinal cord of man and of eight other species. J Neurosci 4:3101–3111
44. Gracely R, Lynch S, Bennett G (1992) Painful neuropathy: altered central processing, maintained dynamically by peripheral input. Pain 51:175–194
45. Greenberg SA (2003) The history of dermatome mapping. Arch Neurol 60:126–131
46. Guedon JM, Yee MB, Zhang M, Harvey SA, Goins WF, Kinchington PR (2015) Neuronal changes induced by Varicella Zoster Virus in a rat model of postherpetic neuralgia. Virology 482:167–180
47. Hanani M (2012) Intercellular communication in sensory ganglia by purinergic receptors and gap junctions: implications for chronic pain. Brain Res 1487:183–191

48. Hardy JD, Wolf HG, Goodell H (1952) Pain sensations and reactions. William and Wilkins, New York
49. Haroutounian S, Nikolajsen L, Bendtsen TF, Finnerup NB, Kristensen AD, Hasselstrom JB, Jensen TS (2014) Primary afferent input critical for maintaining spontaneous pain in peripheral neuropathy. Pain 155:1272–1279
50. Head H, Campbell AW (1900) The pathology of herpes zoster and its bearing on sensory localization. Brain 23:353–523
51. Hempenstall K, Nurmikko TJ, Johnson RW, A'Hern RP, Rice AS (2005) Analgesic therapy in postherpetic neuralgia: a quantitative systematic review. PLoS Med 2:e164
52. Hou Q, Barr T, Gee L, Vickers J, Wymer J, Borsani E, Rodella L, Getsios S, Burdo T, Eisenberg E, Guha U, Lavker R, Kessler J, Chittur S, Fiorino D, Rice F, Albrecht P (2011) Keratinocyte expression of calcitonin gene-related peptide beta: implications for neuropathic and inflammatory pain mechanisms. Pain 152:2036–2051
53. Johnson RW (2007) Zoster-associated pain: what is known, who is at risk and how can it be managed? Herpes 14(Suppl 2):30–34
54. Jones MJ, Murrin KR (1987) Intercostal block with cryotherapy. Ann R Coll Surg Engl 69:261–262
55. Julius D, Basbaum AI (2001) Molecular mechanisms of nociception. Nature 413:203–210
56. Kanpolat Y, Tuna H, Bozkurt M, Elhan AH (2008) Spinal and nucleus caudalis dorsal root entry zone operations for chronic pain. Neurosurgery 62:235–242 discussion 242-234
57. Kennedy PG, Montague P, Scott F, Grinfeld E, Ashrafi GH, Breuer J, Rowan EG (2013) Varicella-zoster viruses associated with post-herpetic neuralgia induce sodium current density increases in the ND7-23 Nav-1.8 neuroblastoma cell line. PLoS One 8:e51570
58. Kibler RF, Nathan PW (1960) Relief of pain and paraesthesiae by nerve block distal to a lesion. J Neurol Neurosurg Psychiatry 23:91–98
59. Kinchington PR, Goins WF (2011) Varicella zoster virus-induced pain and post-herpetic neuralgia in the human host and in rodent animal models. J Neurovirol 17:590–599
60. Kirk EJ, Denny-Brown D (1970) Functional variation in dermatomes in the macaque monkey following dorsal root lesions. J Comp Neurol 139:307–320
61. Kleggetveit IP, Namer B, Schmidt R, Helas T, Ruckel M, Orstavik K, Schmelz M, Jorum E (2012) High spontaneous activity of C-nociceptors in painful polyneuropathy. Pain 153:2040–2047
62. Kocher L, Anton F, Reeh PW, Handwerker HO (1987) The effect of carrageenan-induced inflammation on the sensitivity of unmyelinated skin nociceptors in the rat. Pain 29:363–373
63. Koltzenburg M, Kees S, Budweiser S, Ochs G, Toyka KV (1994a) The properties of unmyelinated nociceptive afferents change in a painful chronic constriction neuropathy. In: Gebhart G, Hammond D, Jensen T (eds) Progress in pain research and management, vol 2. IASP Press, Seattle, pp. 511–521
64. Koltzenburg M, Kress M, Reeh PW (1992) The nociceptor sensitization by bradykinin does not depend on sympathetic neurons. Neuroscience 46:465–473
65. Koltzenburg M, Torebjork H, Wahren L (1994b) Nociceptor modulated central sensitization causes mechanical hyperalgesia in acute chemogenic and chronic neuropathic pain. Brain 117:579–591
66. Kotani N, Kushikata T, Hashimoto H, Kimura F, Muraoka M, Yodono M, Asai M, Matsuki A (2000) Intrathecal methylprednisolone for intractable postherpetic neuralgia. N Engl J Med 343:1514–1519
67. Kress M, Fickenscher H (2001) Infection by human varicella-zoster virus confers norepinephrine sensitivity to sensory neurons from rat dorsal root ganglia. FASEB J 15:1037–1043
68. Kuslich S, Ulstro C, Michael C (1991) The tissue origin of low back pain and sciatica. Orthop Clin North Am 22:181–187
69. Levin MJ, Cai GY, Manchak MD, Pizer LI (2003) Varicella-zoster virus DNA in cells isolated from human trigeminal ganglia. J Virol 77:6979–6987
70. Liem L (2015) Stimulation of the dorsal root ganglion. Prog Neurol Surg 29:213–224

71. Lisney SJW, Devor M (1987) Afterdischarge and interactions among fibers in damaged peripheral nerve in the rat. Brain Rcs 415:122–136
72. Liu C-N, Wall PD, Ben-Dor E, Michaelis M, Amir R, Devor M (2000) Tactile allodynia in the absence of C-fiber activation: altered firing properties of DRG neurons following spinal nerve injury. Pain 85:503–521
73. Makin TR, Scholz J, Filippini N, Henderson Slater D, Tracey I, Johansen-Berg H (2013) Phantom pain is associated with preserved structure and function in the former hand area. Nat Commun 4:1570
74. Markus A, Lebenthal-Loinger I, Yang IH, Kinchington PR, Goldstein RS (2015) An in vitro model of latency and reactivation of varicella zoster virus in human stem cell-derived neurons. PLoS Pathog 11:e1004885
75. Matak I, Bach-Rojecky L, Filipovic B, Lackovic Z (2011) Behavioral and immunohistochemical evidence for central antinociceptive activity of botulinum toxin A. Neuroscience 186:201–207
76. Matzner O, Devor M (1987) Contrasting thermal sensitivity of spontaneously active A- and C-fibers in experimental nerve-end neuromas. Pain 30:373–384
77. Max MB, Schafer SC, Culnane M, Dubner R, Gracely RH (1988) Association of pain relief with drug side effects in postherpetic neuralgia: a single-dose study of clonidine, codeine, ibuprofen, and placebo. Clin Pharmacol Ther 43:363–371
78. Mayer ML, James MH, Russell RJ, Kelly JS, Pasternak CA (1986) Changes in excitability induced by herpes simplex viruses in rat dorsal root ganglion neurons. J Neurosci 6:391–402
79. McLachlan E, Janig W, Devor M, Michaelis M (1993) Peripheral nerve injury triggers noradrenergic sprouting within dorsal root ganglia. Nature 363:543–546
80. Meier T, Wasner G, Faust M, Kuntzer T, Ochsner F, Hueppe M, Bogousslavsky J, Baron R (2003) Efficacy of lidocaine patch 5 % in the treatment of focal peripheral neuropathic pain syndromes: a randomized, double-blind, placebo-controlled study. Pain 106:151–158
81. Miki K, Fukuoka T, Tokunaga A, Noguchi K (1998) Calcitonin gene-related peptide increase in the rat spinal dorsal horn and dorsal column nucleus following peripheral nerve injury: up-regulation in a subpopulation of primary afferent sensory neurons. Neuroscience 82:1243–1252
82. Moore RA, Chi CC, Wiffen PJ, Derry S, Rice AS (2015) Oral nonsteroidal anti-inflammatory drugs for neuropathic pain. Cochrane Database Syst Rev 10:CD010902
83. Nitzan-Luques A, Devor M, Tal M (2011) Genotype-selective phenotypic switch in primary afferent neurons contributes to neuropathic pain. Pain 152:2413–2426
84. Nordin M, Nystrom B, Wallin U, Hagbarth K-E (1984) Ectopic sensory discharges and paresthesiae in patients with disorders of peripheral nerves, dorsal roots and dorsal columns. Pain 20:231–245
85. Nurmikko T, Bowsher D (1990) Somatosensory findings in postherpetic neuralgia. J Neurol Neurosurg Psychiatry 53:135–141
86. Nurmikko T, Wells C, Bowsher D (1991) Pain and allodynia in postherpetic neuralgia: role of somatic and sympathetic nervous systems. Acta Neurol Scand 84:146–152
87. Oaklander AL, Bowsher D, Galer B, Haanpaa M, Jensen MP (2003) Herpes zoster itch: preliminary epidemiologic data. J Pain 4:338–343
88. Onofrio BM, Campa HK (1972) Evaluation of rhizotomy: review of 12 years' experience. J Neurosurg 36:751–755
89. Pappagallo M, Oaklander AL, Quatrano-Piacentini AL, Clark MR, Raja SN (2000) Heterogenous patterns of sensory dysfunction in postherpetic neuralgia suggest multiple pathophysiologic mechanisms. Anesthesiology 92:691–698
90. Park TS, Johnston JM (2006) Surgical techniques of selective dorsal rhizotomy for spastic cerebral palsy. Technical note. Neurosurg Focus 21:e7
91. Patel AJ, Gressot LV, Boatey J, Hwang SW, Brayton A, Jea A (2013) Routine sectioning of the C2 nerve root and ganglion for C1 lateral mass screw placement in children: surgical and functional outcomes. Childs Nerv Syst 29:93–97

92. Peppin JF, Albrecht PJ, Argoff C, Gustorff B, Pappagallo M, Rice FL, Wallace MS (2015) Skin matters: a review of topical treatments for chronic pain. Part two: treatments and applications. Pain Ther 4:33–50
93. Petersen KL, Rice FL, Farhadi M, Reda H, Rowbotham MC (2010) Natural history of cutaneous innervation following herpes zoster. Pain 150:75–82
94. Petersen KL, Rowbotham MC (2007) Relief of post-herpetic neuralgia by surgical removal of painful skin: 5 years later. Pain 131:214–218
95. Pitcher GM, Henry JL (2008) Governing role of primary afferent drive in increased excitation of spinal nociceptive neurons in a model of sciatic neuropathy. Exp Neurol 214:219–228
96. Prestor B (2001) Microsurgical junctional DREZ coagulation for treatment of deafferentation pain syndromes. Surg Neurol 56:259–265
97. Puri N (2011) Modified Jaipur block for the treatment of post-herpetic neuralgia. Int J Dermatol 50:1417–1420
98. Radtke C, Vogt PM, Devor M, Kocsis JD (2010) Keratinocytes acting on injured afferents induce extreme neuronal hyperexcitability and chronic pain. Pain 148:94–102
99. Reichelt M, Zerboni L, Arvin AM (2008) Mechanisms of varicella-zoster virus neuropathogenesis in human dorsal root ganglia. J Virol 82:3971–3983
100. Riopelle JM, Naraghi M, Grush KP (1984) Chronic neuralgia incidence following local anesthetic therapy for herpes zoster. Arch Dermatol 120:747–750
101. Rosenak S (1938) Procaine injection treatment of herpes zoster. Lancet:1056–1058
102. Sawynok J (2014) Topical analgesics for neuropathic pain: preclinical exploration, clinical validation, future development. Eur J Pain 18:465–481
103. Sawynok J, Zinger C (2016) Topical amitriptyline and ketamine for post-herpetic neuralgia and other forms of neuropathic pain. Expert Opin Pharmacother 17:601–609
104. Serra J, Collado A, Sola R, Antonelli F, Torres X, Salgueiro M, Quiles C, Bostock H (2014) Hyperexcitable C nociceptors in fibromyalgia. Ann Neurol 75:196–208
105. Shackleton T, Ram S, Black M, Ryder J, Clark GT, Enciso R (2016) The efficacy of botulinum toxin for the treatment of trigeminal and postherpetic neuralgia: a systematic review with meta-analyses. Oral Surg Oral Med Oral Pathol Oral Radiol 122:61–71
106. Shim B, Kim DW, Kim BH, Nam TS, Leem JW, Chung JM (2005) Mechanical and heat sensitization of cutaneous nociceptors in rats with experimental peripheral neuropathy. Neuroscience 132:193–201
107. Simone DA, Nolano M, Johnson T, Wendelschafer-Crabb G, Kennedy WR (1998) Intradermal injection of capsaicin in humans produces degeneration and subsequent reinnervation of epidermal nerve fibers: correlation with sensory function. J Neurosci 18:8947–8959
108. Sindou M, Mifsud JJ, Boisson D, Goutelle A (1986) Selective posterior rhizotomy in the dorsal root entry zone for treatment of hyperspasticity and pain in the hemiplegic upper limb. Neurosurgery 18:587–595
109. Sugar O, Bucy PC (1951) Postherpetic trigeminal neuralgia. AMA Arch Neurol Psychiatry 65:131–145
110. Sukhotinsky I, Ben-Dor E, Raber P, Devor M (2004) Key role of the dorsal root ganglion in neuropathic tactile hypersensibility. Eur J Pain 8:135–143
111. Tandrup T (2004) Unbiased estimates of number and size of rat dorsal root ganglion cells in studies of structure and cell survival. J Neurocytol 33:173–192
112. Tandrup T, Woolf CJ, Coggeshall RE (2000) Delayed loss of small dorsal root ganglion cells after transection of the rat sciatic nerve. J Comp Neurol 422:172–180
113. Tsuboi Y, Takeda M, Tanimoto T, Ikeda M, Matsumoto S, Kitagawa J, Teramoto K, Simizu K, Yamazaki Y, Shima A, Ren K, Iwata K (2004) Alteration of the second branch of the trigeminal nerve activity following inferior alveolar nerve transection in rats. Pain 111:323–334
114. Uceyler N, Valet M, Kafke W, Tolle TR, Sommer C (2014) Local and systemic cytokine expression in patients with postherpetic neuralgia. PLoS One 9:e105269

115. Uceyler N, Zeller D, Kahn AK, Kewenig S, Kittel-Schneider S, Schmid A, Casanova-Molla J, Reiners K, Sommer C (2013) Small fibre pathology in patients with fibromyalgia syndrome. Brain 136:1857–1867
116. van Wijck AJ, Opstelten W, Moons KG, van Essen GA, Stolker RJ, Kalkman CJ, Verheij TJ (2006) The PINE study of epidural steroids and local anaesthetics to prevent postherpetic neuralgia: a randomised controlled trial. Lancet 367:219–224
117. van Wijck AJ, Wallace M, Mekhail N, van Kleef M (2011) Evidence-based interventional pain medicine according to clinical diagnoses. 17. Herpes zoster and post-herpetic neuralgia. Pain Pract 11:88–97
118. Vaso A, Adahan HM, Gjika A, Zahaj S, Zhurda T, Vyshka G, Devor M (2014) Peripheral nervous system origin of phantom limb pain. Pain 155:1384–1391
119. Wall PD, Devor M (1983) Sensory afferent impulses originate from dorsal root ganglia as well as from the periphery in normal and nerve injured rats. Pain 17:321–339
120. Wallace MS, Dyck JB, Rossi SS, Yaksh TL (1996) Computer-controlled lidocaine infusion for the evaluation of neuropathic pain after peripheral nerve injury. Pain 66:69–77
121. Wasner G, Kleinert A, Binder A, Schattschneider J, Baron R (2005) Postherpetic neuralgia: topical lidocaine is effective in nociceptor-deprived skin. J Neurol 252:677–686
122. Watson CP, Deck JH, Morshead C, Van der Kooy D, Evans RJ (1991) Post-herpetic neuralgia: further post-mortem studies of cases with and without pain. Pain 44:105–117
123. Watson CP1, Morshead C, Van der Kooy D, Deck J, Evans RJ (1988) Post-herpetic neuralgia: post-mortem analysis of a case. Pain 34:129–38
124. Watson CPN, Midha R, Devor M, Nag S, Munro C, Dostrovsky JO (2000) Trigeminal postherpetic neuralgia postmortem: clinically unilateral, pathologically bilateral. In: Devor M, Rowbotham M, Wiesenfeld-Hallin Z (eds) Proceedings of the 9th world congress on pain, progress in pain research and management, vol 17. IASP Press, Seattle, pp. 733–739
125. White DM, Zimmermann M (1988) Changes in the content and release of substance P and calcitonin-gene related peptide in rat cutaneous nerve neuroma. In: Dubner R, Gebhart G, Bond M (eds) Proceedings of the fifth world congress on pain. Elsevier, Amsterdam, pp. 109–113
126. White J, Sweet W (1969) Pain and the neurosurgeon. Thomas, Springfield
127. Woolf CJ (1983) Evidence for a central component of postinjury pain hypersensitivity. Nature 306:686–688
128. Woolf CJ (2011) Central sensitization: implications for the diagnosis and treatment of pain. Pain 152:S2–15
129. Wu CL, Marsh A, Dworkin RH (2000) The role of sympathetic nerve blocks in herpes zoster and postherpetic neuralgia. Pain 87:121–129
130. Yang Q, Wu Z, Hadden JK, Odem MA, Zuo Y, Crook RJ, Frost JA, Walters ET (2014) Persistent pain after spinal cord injury is maintained by primary afferent activity. J Neurosci 34:10765–10769
131. Yirmiya R, Pollak Y, Morag M, Reichenberg A, Barak O, Avitsur R, Shavit Y, Ovadia H, Weidenfeld J, Morag A, Newman ME, Pollmacher T (2000) Illness, cytokines, and depression. Ann N Y Acad Sci 917:478–487
132. Zacks SI, Elliott FA, Langfitt TW (1964) Herpetic neuritis: a light and electron microscopic study. Neurology 14:744–750
133. Zerboni L, Arvin A (2015) Neuronal subtype and satellite cell tropism are determinants of varicella-zoster virus virulence in human dorsal root ganglia xenografts in vivo. PLoS Pathog 11:e1004989

Chapter 14
Persistent VZV Ganglionitis May Be the Cause of Postherpetic Neuralgia

Don Gilden and Maria A. Nagel

Dr. Donald Harvey Gilden (1937–2016) We were most saddened to hear of the passing of Dr. Donald Harvey Gilden, contributor to very important chapters for this book (Chaps. 7 and 14). These chapters provide new, much-needed insight into the broader impact of the varicella zoster virus regarding vascular diseases and, also, a new idea about the pathophysiology of postherpetic neuralgia.

His co-author, Dr. Maria Nagel, a colleague and co-author of these chapters, states: "Donald was a wonderful mentor. He had an ability to identify projects that would ultimately improve human health, bring collaborators together, and successfully drive science forward. He had profound respect for academic medicine and his partners in science. Even in hospice care, he thoughtfully planned the future direction of the lab and career development of his beloved colleagues."

Donald Gilden was a National Institutes of Health-funded researcher who published more than 420 scientific papers and book chapters. He also edited several books on virus infections of the central nervous system. In 1994, an editorial by Dr. Gilden, published in the *New England Journal of Medicine,* promoted increased attention to developing shingles vaccines.

In the latter part of his career, Dr. Gilden's lab demonstrated that varicella zoster virus, which causes chickenpox and shingles, can also cause strokes and aneurysms. In 2015, his research group published a paper on varicella zoster virus as a cause of giant cell arteritis, a disease that can cause headaches and vision loss in the elderly.

Personally, this editor (CPNW) was in awe of his contributions to varicella zoster virus research, and the preparation of these two chapters for this book during his final illness is a great tribute to his courage and dedication to this important work.

Ave atque vale,
C. Peter N. Watson

D. Gilden, MD
Departments of Neurology and Immunology and Microbiology, University of Colorado School of Medicine, 12700 E. 19th Avenue, Mail Stop B182, Aurora, Colorado 80045, USA
e-mail: don.gilden@ucdenver.edu

M.A. Nagel, MD (✉)
Department of Neurology, University of Colorado School of Medicine, 12700 E. 19th Avenue, Mail Stop B182, Aurora, Colorado 80045, USA
e-mail: maria.nagel@ucdenver.edu

C.P.N. Watson et al. (eds.), *Herpes Zoster: Postherpetic Neuralgia and Other Complications*, DOI 10.1007/978-3-319-44348-5_14

14.1 Zoster and Postherpetic Neuralgia

Varicella zoster virus (VZV) is a ubiquitous, neurotropic human herpesvirus. Primary infection usually causes chickenpox (varicella), after which virus becomes latent in cranial nerve ganglia, dorsal root ganglia, and autonomic ganglia along the entire neuraxis. A natural decline in VZV-specific cell-mediated immunity with advancing age or immunosuppression results in virus reactivation, usually manifesting as zoster (shingles), dermatomal-distribution pain, and rash. A light touch to the affected area of skin commonly produces pain (allodynia). Pain that persists for 3 months or more after zoster rash is operationally defined as postherpetic neuralgia (PHN). Age is the most important factor that predicts the development of PHN [3, 20]. Although PHN is rare before age 50, the incidence is 43–47.5 % after age 50 and slightly greater in women and after trigeminal-distribution zoster [4, 13]. About one million Americans develop zoster annually, resulting in 200,000–400,000 new cases of PHN yearly [14].

14.2 Postherpetic Neuralgia, Preherpetic Neuralgia, and Zoster Sine Herpete

Although not nearly as common as PHN, two other qualitatively identical, closely related forms of radicular pain without rash are associated with VZV infection. The first is prolonged radicular pain preceding zoster rash (preherpetic neuralgia). Generally, zoster pain and rash occur within hours or a couple of days of each other. Uncommonly, pain precedes rash by weeks, even as much as 100 days, so-called "preherpetic" neuralgia [8]. The severe, burning radicular nature of the pain, followed weeks to months later by zoster rash, likely indicates that the pain resulted from a smoldering ganglionitis. The second chronic radicular pain syndrome associated with VZV infection without rash is zoster sine herpete. The concept of zoster sine herpete (shingles without rash) is nearly 100 years old, a notion that received credence when Lewis [15] described numerous zoster patients who, days later, also developed pain without rash in a different dermatome distribution, often on the opposite side. The first serologic evidence of zoster sine herpete occurred in a physician who developed acute trigeminal distribution pain associated with a fourfold rise in complement-fixing antibody to VZV but not to HSV [6]. Virologic confirmation of zoster sine herpete did not come until the analysis of two men with thoracic-distribution radicular pain that had lasted for months to years revealed PCR-amplifiable VZV DNA but not HSV DNA in their CSF and blood MNCs [8]. After diagnosis, both men were treated successfully with intravenous acyclovir. While PHN is the most common form of chronic radicular pain, the quality of "preherpetic" neuralgia and "zoster sine herpete" is no different. Importantly the latter two conditions are produced by VZV. Overall, it is possible, if not likely, that PHN, preherpetic neuralgia, and zoster sine herpete are all manifestations of productive ganglionitis. Why PHN is so much more common remains unknown.

14.3 The Possible Viral Cause of Postherpetic Neuralgia (PHN)

The cause and pathogenesis of PHN are unknown. Two non-mutually exclusive theories are (1) that the excitability of ganglionic or even spinal cord neurons is altered or (2) that there is persistent virus infection (not latency) in ganglia. Analysis of blood mononuclear cells (MNCs), cerebrospinal fluid (CSF), and ganglia from patients with PHN or patients with zoster sine herpete (chronic radicular pain without rash) suggests that this may be the case (detailed below). The finding that PHN may be due to chronic active infection by the same virus, which decades earlier had produced only a relatively benign childhood exanthema, represents one of the most exciting medical developments that has resulted from the application of molecular virologic strategies and techniques to analysis of human tissue.

There have been precious few instances in which ganglia from patients who had PHN before death have been analyzed pathologically. Analysis of ganglia from an early case of PHN of 2.5-month duration revealed diffuse and focal infiltration by chronic inflammatory cells (Fig. 14.1; [11]). The inflammatory response in the ganglion of this subject raised the possibility of prolonged viral infection.

Although virologic analyses of ganglia from PHN patients are wanting, VZV DNA was detected in blood MNCs of PHN patients 1–8 years after zoster [5]. In contrast, in zoster patients who did not develop PHN, VZV DNA was found in blood MNCs only up to 38 days after zoster and not at all after disappearance of zoster pain [7]. A more extensive study detected VZV DNA in MNCs up to 8 years after zoster in 11/51 patients with PHN, but not in MNCs of 19 zoster patients without PHN who were analyzed 1–31 years after zoster, or in any of 11 elderly age- and gender-matched subjects with no history of zoster [17].

How are these virological findings best explained? The most rational explanation is that MNCs, particularly antigen-presenting cells, acquire virus while trafficking through ganglia and ultimately digest virus. This would explain the detection of

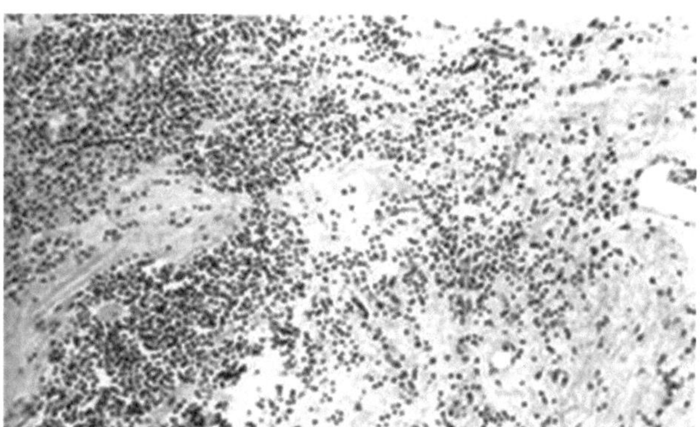

Fig. 14.1 Hematoxylin and eosin (H & E)-stained sections of dorsal root ganglia from patients with postherpetic neuralgia reveal diffuse and focal infiltration by chronic inflammatory cells (Figure from Gilden et al. [11]; reprinted with permission from Wolters Kluwer Health, Inc)

some, but not all, regions of the VZV genome in circulating MNCs [16], compared to the presence of the entire virus genome in latently infected ganglia. Digestion of viral DNA by MNCs also helps to explain why VZV DNA is found randomly in MNCs from only 20 % of patients with PHN since there would only be a chance occurrence that fragments of viral DNA will be present in blood samples.

The above findings were later supported by a correlative clinical-molecular virological study conducted over an 11-year period in an immunocompetent elderly woman with PHN [10]. Initially, blood MNCs contained VZV DNA on two consecutive occasions. After treatment with oral famciclovir, her pain resolved and blood MNCs were negative for VZV DNA. However, over the years, the patient voluntarily stopped antiviral treatment five times, pain always recurred within 1 week, and blood MNCs contained many, but not all, regions of the VZV genome on all five occasions. Chronic VZV ganglionitis-induced PHN best explains both a gratifying clinical response to famciclovir and the recurrence of VZV DNA in MNCs whenever famciclovir was discontinued.

More evidence that long-standing radicular pain could be due to a chronic ganglionitis comes from studies of patients with zoster sine herpete (pain without rash). Lewis [15] originally suggested zoster sine herpete in patients with zoster who had dermatomal distribution pain in areas distinct from zoster. A clinical link between PHN and zoster sine herpete was later provided by a report of four patients who, years after the resolution of pain from trigeminal zoster, developed zoster sine herpete in the same distribution of the trigeminal nerve. Facial surgery in three patients and a tooth abscess in one patient precipitated zoster sine herpete. Unfortunately, none of these patients was studied virologically [21].

Importantly, of the first two patients with zoster sine herpete to be studied virologically, both had VZV DNA in blood MNCs and CSF and were cured of pain after treatment with intravenous acyclovir [9, 22]. Virological analysis of another patient with lumbar distribution zoster that was followed by chronic radicular sacral distribution zoster sine herpete revealed that the persistent pain was produced by active VZV infection [18]. Another patient with ipsilateral truncal sensory deficit and ophthalmic zoster sine herpete, who had reduced serum/CSF ratios of VZV antibody indicative of active virus infection, was treated with intravenous acyclovir; within 7 days, the sensory-deficit progression stopped, although mild neuralgia on the left face persisted [25]. The most compelling evidence that persistent radicular pain without rash can be caused by a chronic active VZV ganglionitis came from analysis of a trigeminal ganglionic mass removed from an immunocompetent adult who had experienced relentless trigeminal distribution pain for more than a year. Pathological and virological studies revealed that the patient's zoster sine herpete was caused by an active VZV ganglionitis [12]. Finally, a recent report described a patient with chronic active VZV ganglionitis whose MRI revealed inflammation in ganglia and nerve roots corresponding to persistent pain [2].

14.4 Perspective on Antiviral Therapy for Postherpetic Neuralgia (PHN)

Current treatment for PHN includes neuroleptic drugs and various analgesics, including opiates, to alleviate pain. PHN may be caused by damage to ganglionic neurons induced by viral replication during acute zoster; another possibility is that an excessive barrage of nerve impulses generated during acute zoster produces long-term damage [24]. Thus, if persistent productive ganglionic infection continues, long-term pain would not be surprising, and there would be the potential to reduce PHN by antiviral treatment.

The two reports summarized below are those that treated PHN with antivirals, not studies which used antivirals at the onset of zoster to prevent PHN. Acosta and Balfour [1] described improvement in one of six PHN patients who received intravenous acyclovir followed by oral acyclovir. Finally, our treatment of 15 patients with moderate to severe PHN with intravenous acyclovir followed by oral valacyclovir led to clinical improvement in 8 (53 %), warranting further investigation of antiviral therapy in a larger, randomized, double-blind, placebo-controlled trial [19]. Further studies are needed to determine levels of antiviral drug in CSF after oral administration of acyclovir, famciclovir, or valacyclovir compared to intravenous administration of acyclovir, the only drug approved for intravenous use. Valacyclovir has already been shown to compare favorably with famciclovir in speeding the resolution of pain in zoster [23].

14.5 Conclusion

Virological analyses of ganglia corresponding to the area of pain in patients with PHN are wanting. Inflammation has been found in ganglia from two PHN subjects who were studied postmortem. Besides PHN, two other qualitatively identical closely related forms of radicular pain without rash are associated with VZV infection. The first is prolonged radicular pain preceding zoster rash. The second chronic radicular pain syndrome associated with VZV infection is zoster sine herpete. In patients with zoster sine herpete, correlative clinical-virological analyses indicate a productive VZV ganglionitis. Finally, the presence of VZV DNA and proteins in peripheral blood MNCs of many patients with PHN combined with a favorable response of some patients with zoster sine herpete and PHN to antiviral treatment provide further evidence that PHN is caused by chronic active VZV infection in ganglia. Because only a few studies have used antiviral therapy to treat PHN and with conflicting results, larger, double-blinded studies, which give antiviral therapy intravenously, are needed.

Acknowledgments Supported in part by the National Institutes of Health Grants NS093716 (Gilden), AG032958 (Gilden and Nagel), and NS094758 (Nagel). The authors thank Marina Hoffman for editorial review and Cathy Allen for manuscript preparation.

Editorial Comment

C. Peter N. Watson

The concept of at least some cases of postherpetic neuralgia (PHN) (and zoster sine herpete) being due to a smoldering VZV ganglionitis as ventured here is an exciting one with this important research being in an early phase. This raises the possibility of treatment of PHN and other conditions (other neuropathic facial and enigmatic neuropathic pain elsewhere) with antiviral drugs orally (famciclovir, valacyclovir) and intravenously (aciclovir).

The following two autopsied cases [26] support the opinion expressed in this chapter. The figure shows acute inflammatory cells in a patient (case 2) with chronic postherpetic neuralgia of 22-month duration indicating recent acute inflammation. Case 4 showed a significant amount of collagen and marked axon and myelin loss of the affected second thoracic intercostal nerve compatible with residual scarring from remote inflammation in a 72-year-old man with an acute case of herpes zoster of 5-week duration indicating preexisting activity of long standing. Both these are supportive of ongoing disease activity long before and long after the clinical presentation.

Figure: Hematoxylin and eosin (H & E) × 25-stained section of the affected right tenth thoracic nerve from a patient with postherpetic neuralgia of 22-month duration revealing diffuse and focal infiltration by inflammatory cells. There are many prominent collections of lymphocytes in the right tenth nerve and ganglion (case 2 from Watson et al. [26]).

References

1. Acosta EP, Balfour HH Jr (2001) Acyclovir for treatment for postherpetic neuralgia: efficacy and pharmacokinetics. Antimicrob Agents Chemother 45:2771–2774
2. Birlea M, Nagel MA, Khmeleva N et al (2014) Varicella zoster virus trigeminal ganglioneuritis without rash. Neurology 82:90–92
3. Brown GR (1976) Herpes zoster: correlation of age, sex, distribution, neuralgia and associated disorders. South Med J 69:576–578
4. DeMoragas JM, Kierland RR (1957) The outcome of patients with herpes zoster. Arch Dermatol 75:193–196
5. Devlin ME, Gilden DH, Mahalingam R, Dueland AN, Cohrs R (1992) Peripheral blood mononuclear cells of the elderly contain varicella-zoster virus DNA. J Infect Dis 165:619–622
6. Easton HG (1970) Zoster sine herpete causing acute trigeminal neuralgia. Lancet 2:1065–1066
7. Gilden DH, Devlin M, Wellish M et al (1988) Persistence of varicella-zoster virus DNA in blood MNCs of patients with varicella or zoster. Virus Genes 2:299–305
8. Gilden DH, Dueland AN, Cohrs R, Martin JR, Kleinschmidt-DeMasters BK, Mahalingam R (1991) Preherpetic neuralgia. Neurology 41:1215–1218
9. Gilden DH, Wright RR, Schneck SA, Swaltney JM Jr, Mahalingam R (1994) Zoster sine herpete, a clinical variant. Ann Neurol 35:530–533
10. Gilden DH, Cohrs RJ, Hayward AR, Wellish M, Mahalingam R (2003) Chronic varicella zoster virus ganglionitis – a possible cause of postherpetic neuralgia. J Neurovirol 9:404–407

11. Gilden DH, Cohrs RJ, Mahalingam R (2005) VZV vasculopathy and postherpetic neuralgia: progress and perspective on antiviral therapy. Neurology 64:21–25
12. Hevner RF, Vilela MD, Rostomily RC et al (2003) An unusual cause of trigeminal distribution pain and tumour. Lancet Neurol 2:567–572
13. Hope-Simpson RE (1975) Postherpetic neuralgia. J R Col Gen Pract 25:571–575
14. Insinga RP, Itzler RF, Pellissier JM, Saddier P, Nikas AA (2005) The incidence of herpes zoster in a United States administrative database. J Gen Intern Med 20:748–753
15. Lewis GW (1958) Zoster sine herpete. Br Med J 2:418–421
16. Mahalingam R, Wellish M, Wolf W et al (1990) Latent varicella zoster virus DNA in human trigeminal and thoracic ganglia. N Engl J Med 323:627–631
17. Mahalingam R, Wellish M, Brucklier J, Gilden DH (1995) Persistence of varicella-zoster virus DNA in elderly patients with postherpetic neuralgia. J Neurovirol 1:130–133
18. Morita Y, Osaki Y, Doi Y, Forghani B, Gilden DH (2003) Chronic active VZV infection manifesting as zoster sine herpete, zoster paresis and myelopathy. J Neurol Sci 212:7–9
19. Quan D, Hammack BN, Kittelson J, Gilden DH (2006) Improvement of postherpetic neuralgia after treatment with intravenous acyclovir followed by oral valacyclovir. Arch Neurol 63:940–942
20. Ragozzino MW, Melton LJ III, Kurland LT, Chu CP, Perry HO (1982) Population-based study of herpes zoster and its sequelae. Medicine 61:310–316
21. Schott GD (1998) Triggering of delayed-onset postherpetic neuralgia. Lancet 351:419–420
22. Terada K, Niizuma T, Kawano S, Kataoka N, Akisada T, Orita Y (1998) Detection of varicella-zoster virus DNA in peripheral mononuclear cells from patients with Ramsay Hunt syndrome or zoster sine herpete. J Med Virol 56:359–363
23. Tyring SK, Beutner KR, Tucker BA et al (2000) Antiviral therapy for herpes zoster: randomized, controlled clinical trial of valacyclovir and famciclovir therapy in immunocompetent patients 50 years and older. Arch Fam Med 9:863–869
24. Wall PD (1993) An essay on the mechanisms which may contribute to the state of postherpetic neuralgia. In: Watson CPN (ed) Pain research and clinical management, vol 8, herpes zoster and postherpetic neuralgia. Elsevier Science Publishers, Amsterdam, pp. 123–138
25. Yamada S, Atsuta N, Tokunaga S, Motegi Y (2003) Ipsilateral truncal sensory deficit in a patient with ophthalmic zoster sine herpete. Neurology 60:1049–1050
26. Watson CPN, Deck JH, Morshead C et al (1991) Post-herpetic neuralgia postmortem: more cases with and without pain. Pain 44:105–117

Chapter 15
A Comparison of Clinical Features and Mechanisms of Trigeminal Postherpetic Neuralgia and Trigeminal Neuralgia

Barry J. Sessle and C. Peter N. Watson

When the wind is southerly I know a hawk from a handsaw.

- Hamlet

Acronyms

AMPA	Alpha-amino-3-hydroxy-5-isoxazde-proprionic acid
CGRB	Calcitonin gene-related peptide
GABA	Gamma amino butyric acid
NMDA	N-methyl-D-aspartate
CNS	Central nervous system
DRG	Dorsal root ganglion
HZ	Herpes zoster
MVD	Microvascular decompression
NS	Nociceptive specific
PHN	Postherpetic neuralgia
SNRIs	Serotonin reuptake inhibitors
TCAs	Tricyclic antidepressants
TN	Trigeminal neuralgia
TPHN	Trigeminal postherpetic neuralgia
VBSNC	Trigeminal brainstem sensory nuclear complex
WDR	Wide dynamic range
VZV	Varicella zoster virus

B.J. Sessle (✉)
Faculty of Dentistry, Department of Physiology, Faculty of Medicine, and Centre for the Study of Pain, University of Toronto, Toronto, ON, M5G 1G6, Canada
e-mail: barry.sessle@utoronto.ca

C.P.N. Watson
University of Toronto, Department of Medicine, Division of Neurology, 99 Avenue Rd, #908, Toronto, ON M5R 2G5, Canada
e-mail: peter.watson@utoronto.ca

C.P.N. Watson et al. (eds.), *Herpes Zoster: Postherpetic Neuralgia and Other Complications*, DOI 10.1007/978-3-319-44348-5_15

15.1 Introduction

Trigeminal postherpetic neuralgia (TPHN) refers to postherpetic neuralgia (PHN) that is manifested in the orofacial region. TPHN and trigeminal neuralgia (TN) are both examples of neuropathic pain (similar birds), and although some of their features appear similar, the two pain states are almost as a different as a hawk and heron (handsaw in Shakespearean parlance). Although both occur in the orofacial region, TPHN and TN have little else in common other than the term "neuralgia," which is an older term applied rather loosely when there is electric shocklike or stabbing neuropathic pain in any neuropathic condition. Both are accompanied by allodynia (pain from touch), but this is different in nature (see below). It is unknown why TN-like pain, which has unique trigger zones and is responsive to carbamazepine, does not appear to occur in dermatomes below the head, areas commonly affected by PHN.

A failure to understand the clinical, pathological, and therapeutic differences (see Table 15.1) will result in therapeutic failure if TN is subjected to the guidelines

Table 15.1 A comparison of TPHN and TN

	Trigeminal postherpetic neuralgia (TPHN)	Trigeminal neuralgia (TN)
Epidemiology Natural history Etiology	Increases with age, natural history, improvement in most	Exacerbations and remissions initially, incidence increases with age, associations: vascular anomalies V nerve root, multiple sclerosis
Genetic	Overall 10 % have PHN at 1 month, but at age 60, this is 50 %	Familial cases
Clinical	Steady burning pain, non-triggered shocks, dynamic mechanical allodynia, unilateral	Electric shocks triggered from localized areas (allodynia) (e.g., nasolabial area) unilateral
Pathology	Major injury to one dorsal root ganglion, nerve, nerve root, spinal dorsal horn	Minor injury (vascular loop) compresses trigeminal nerve root
Treatment: medical	Antidepressants (TCAs, SNRIs), gabapentinoids (gabapentin, pregabalin), opioids (Chap. 19)	Carbamazepine, oxcarbazepine, gabapentinoids, no response to antidepressants or opioids
Treatment: surgical	A number of case reports of different procedures of variable quality and follow-up (Chap. 20)	Temporary: a variety of minor non-curative procedures with good initial response (gamma knife, intracavernous glycerol, balloon compression, and radiofrequency lesions of Gasserian ganglion) curative: microvascular decompression (MVD) of trigeminal root

for neuropathic pain in general (Chap. 19), as the treatment, both medically and surgically, is quite different.

Since previous editions of this book, more and moderately effective therapeutic options have resulted with the introduction of new drugs such as the gabapentinoids (gabapentin and pregabalin), tricyclic antidepressants (TCAs), serotonin norepinephrine reuptake inhibitors (SNRIs) (duloxetine), lidocaine and capsaicin patches, and zoster prevention vaccines.

TPHN is much like PHN in other locations and other neuropathic pain conditions, but TN appears to be a condition unique to the trigeminal system, suggesting a different pathophysiology. However, there is a possibility that these differences are more apparent than real.

This chapter will address the epidemiological, clinical, pathological, and therapeutic differences between TN and TPHN, their few similarities, the implications of the comparative features, and the attempt to explain them on pathophysiological grounds. The two case studies that follow illustrate the typical features of these different neuropathic orofacial pain disorders.

15.2 Case Studies

Case 1: TPHN

An 83-year-old woman presented with pain on the right forehead, pain appearing particularly later in the day and keeping her awake at night. She was otherwise in good health. This followed an attack of herpes zoster (HZ) in that area 6 months previously. She described three main types of pain on the right side of her forehead: (1) a constant, steady, burning pain; (2) pain on touching the skin (dynamic mechanical allodynia when she was clinically examined), combing her hair, and even the wind blowing on her face; and (3) several times a day, she would have brief non-triggered electric shocklike pains. She rated all the facets of her pain at 8–9/10, particularly later in the day and on retiring for the night. The pain caused severe insomnia. She had previously been treated with gabapentin (600 mg daily) that was not effective and made her feel off balance and caused swelling of her feet and weight gain. She had failed to have relief from acetaminophen (325 mg)/codeine (30 mg), which made her constipated. She had also been treated with carbamazepine to no avail. On examination, pale scarring was evident over the right forehead. There was a blunting of sensation to localized or nonmoving tactile stimuli (which typically activate large sensory fibers) and pin and cold stimuli (typically activating small sensory fibers) over the area of the ophthalmic division (V1) of the right trigeminal nerve. There was dynamic mechanical allodynia (pain on stroking with cotton) on the right side of the forehead. The right corneal reflex was reduced. The gabapentin dose was reduced from 200 mg three times a day to 100 mg given later in the day at noon and with her evening meal. Because of her age, it was decided (because of potential side effects) not to use a TCA or SNRI antidepressant, which are possible first choices of treatment in neuropathic pain guidelines, and to instead introduce a smaller dose of gabapentin (100 mg bid) and the combination pill of oxycodone (5 mg)/ acetaminophen (325 mg) at night starting with ½ to one pill. The dose was chosen because of her age, the timing because of the late occurrence of the severe pain. With this, her pain became well controlled, her insomnia was relieved, and she has remained on this regimen, unchanged, for 4 years.

Comment: This is a typical case of TPHN in an aged individual that caused a major disruption in the patient's life as the day progressed with severe insomnia and was controlled by low doses of two drugs. The use of a low dose of the opioid combined with gabapentin was vital and has not been problematic after screening the patient for a history of drug abuse and psychiatric illness despite the contentious role of opioids and concern about the potential risk of abuse (Chap. 18).

Case 2: TN

A 52-year-old man presented with a 6-year history of pain in the mandibular division (V3) of the right trigeminal nerve, predominantly inside the mouth in the lower jaw. There was triggering of this pain by very localized light touch from a small area in the right nasolabial area that produced a severe (9/10) electric shock-like pain lasting 5 to 30 s many times a day. Washing the face, brushing the teeth, shaving, speaking, and chewing all triggered these severe shocks. The longest he would be free of this pain was approximately 30 minutes. There was no steady pain. The course of the pain over the years had been one of being initially episodic, lasting weeks and poorly controlled with low-dose (400 mg/day) short-acting carbamazepine. However, he had periods of long remission of up to 1 year. The present history was of unremitting severe pain for 3 months. He had been correctly diagnosed as having TN type I. This pain was progressively more severe and frequent and was now interfering with his life, particularly his work as a salesman that involved extensive traveling and driving. He was switched to long-acting carbamazepine (100 mg every 8 h) with rescue short-acting carbamazepine, the dose was titrated to 200 mg every 8 h, and his pain came under control and continued at 12 months follow-up.

Comment: This is a case of typical or classical TN (therefore, type I) with unilateral electric shocklike, triggered pain in the lower face and was diagnosed as TN without atypical features such as steady constant pain. It responded to an appropriate dose and dosing intervals of long-acting carbamazepine.

15.3 Historical

The roots of PHN and TN extend back many years. In the case of PHN, the detailed description of the pathological features goes back to the remarkable, extensive Head and Campbell article in the journal *Brain* (1900). These pathological studies are showing among other findings severe damage to the dorsal root ganglion (DRG) (Fig. 15.1, [35]). Edgar Hope-Simpson's meticulous studies of the epidemiology and natural history of varicella and HZ in the 1960s (Fig. 15.2, [36]) established that

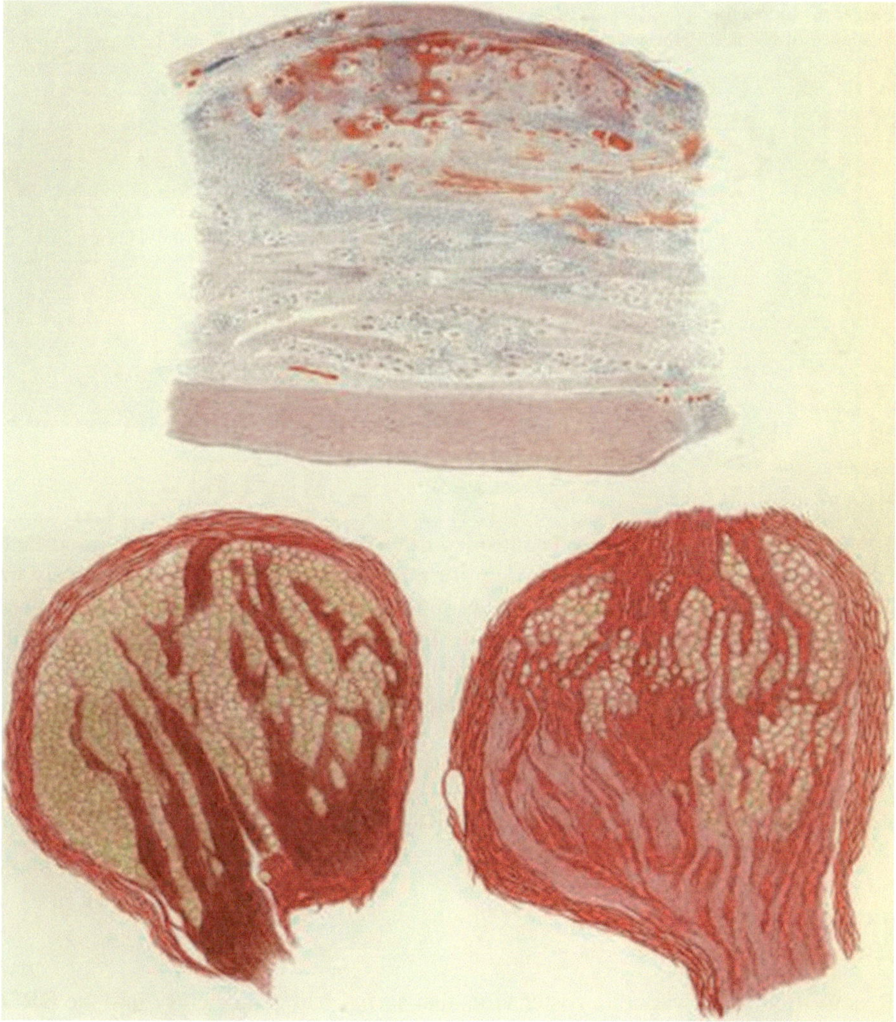

Fig. 15.1 Lower right image fibrosis in the DRG in long-standing PHN patient [35]. Upper image is acute zoster with hemorrhagic inflammation in the dorsal aspect of DRG. Lower left is the control ganglion from the chronic case

HZ and varicella were the same virus and that the natural history of HZ-related pain was one of improvement and resolution in most cases but less so with increasing age of the zoster victims.

In the case of TN, this neuropathic pain state was first described fully by John Fothergill in 1723 [25]. In the nineteenth century, Armand Trousseau made the observation that the attacks of TN were much like epilepsy [71]. This led, over the years, to the exploration of anticonvulsant therapy for neuropathic pain in general and the discoveries of phenytoin for TN in the 1940s [6] and carbamazepine in the

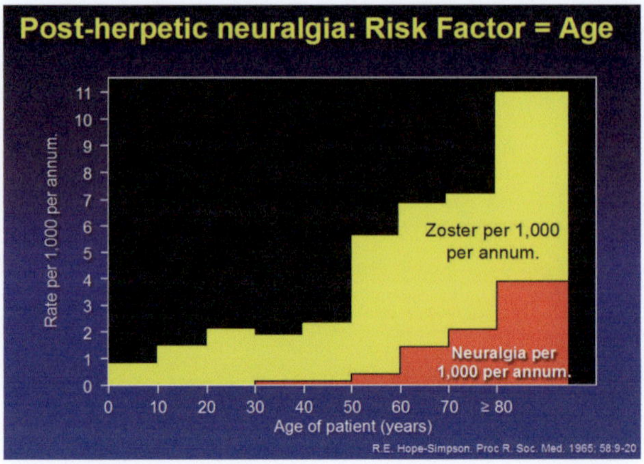

Fig. 15.2 Increasing incidence of HZ and PHN with age [36]

1960s [7] involving randomized controlled trials that established the most important current and effective treatment for any form of neuropathic pain but selectively in TN. Trousseau's epilepsy concept of TN spilled over into the investigation of treatment of other neuropathic pain conditions by various anticonvulsants and led to the current enthusiasm for the gabapentinoids and others. Unfortunately carbamazepine is rarely effective for neuropathic pain conditions other than TN.

15.4 Epidemiology, Incidence, and Natural History of TN and TPHN

15.4.1 Clinical Features

Reactivation of the varicella zoster virus that earlier had gained entry into the DRG or trigeminal (Gasserian) ganglion is of etiological importance in the development of PHN. HZ and PHN are unilateral and affect preferentially the ophthalmic and thoracic regions, a localization which is unexplained. In contrast, other childhood exanthems such as the measles rash are generalized over the body, and smallpox rash affects the distal limbs in children and adults.

PHN has three major facets to the pain. There is a steady and often burning or hot pain, a usually non-triggered electric shocklike component (if patients are questioned closely), and pain on touch (allodynia) that is (unlike TN) dynamic mechanical allodynia (pain on moving touch such as stroking with cotton). Usually there are sensory changes on clinical examination (unlike TN) such as a reduction of touch, cold, and pinprick stimuli over the affected area and over the scarring. The cornea which is supplied by the ophthalmic division [V1]) may be affected with loss of vision or of the eye itself (Chap. 5).

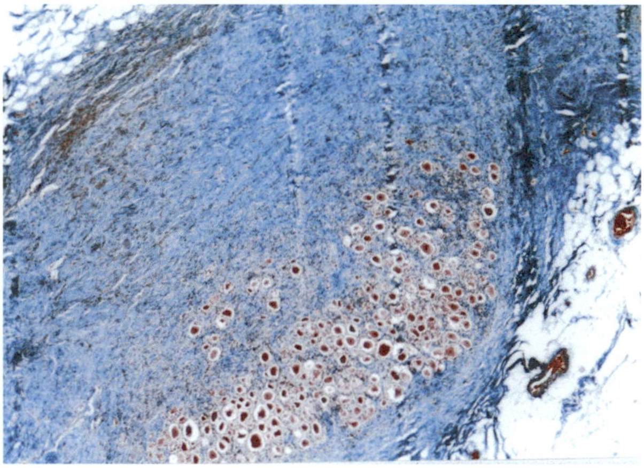

Fig. 15.3 Fibrosis of the DRG in PHN seen in upper image with survival of ganglion cells in lower image (Masson trichrome ×12) [75]

Typically TN has a variable natural history with episodes of remissions and exacerbations for no apparent reason which are relieved well or modified significantly by carbamazepine. Although the natural history is not well known (see [24, 27, 42]), a proportion of patients do appear to follow a progressive course requiring increasing doses of medication and ultimately the necessity for surgery (CPNW personal observations).

Clinically, TN is usually idiopathic and has a rare association with multiple sclerosis, injuries to the trigeminal nerve and tumors but is associated with vascular loops pressing on the trigeminal root in many patients [42]. Differences between the clinical features of TN and TPHN are striking. TN, unlike TPHN, usually manifests as unilateral electric shocklike pain triggered by a low-intensity stimulus (e.g., tactile) from often localized areas such as the nasolabial area, but trigger zones can occur elsewhere in the orofacial region. It is usually triggered by such activities as washing the face, shaving, brushing the teeth, chewing, and speaking. It is usually not associated with any physical findings on routine neurological examination, although the evoked attack manifests summation features and a refractory period, and typically persists beyond the period of stimulation [24, 42].

15.4.2 Pathology

Most autopsy studies of PHN have shown severe loss of neurons, demyelination, and scarring involving the peripheral nerve, DRG (Fig. 15.3), nerve root, and the spinal cord dorsal horn (Fig. 15.4, [75]). The situation is less clear for TPHN. An autopsy study of TPHN [76] did not show significant damage in the trigeminal ganglion but showed evidence of bilateral changes in the peripheral nerves and nerve roots. These were surprising findings, and more cases need to be studied.

The situation with TN is different. Following the observations of Dandy and Gardiner (see [42]), Peter Jannetta popularized the very successful microvascular

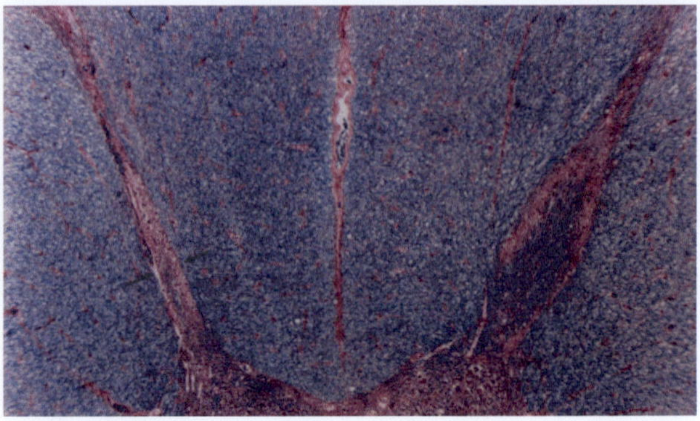

Fig. 15.4 Atrophy of the dorsal horn of the spinal cord in PHN seen in left side of image (MBPx2.5) [75]

decompression (MVD) operation [42]. This operation is based on the concept that vascular loops compressing the trigeminal nerve root account for many cases of this condition and that removing this compression can result in a cure. Maarbjerg et al. [51] concluded that neurovascular contact causing displacement or atrophy of the trigeminal nerve is highly associated with the symptomatic side in classical TN as opposed to neurovascular contact in general. They said their brain imaging findings demonstrated that severe neurovascular contact is involved in the etiology of classical TN and that TN is caused by contact by arteries located in the root entry zone. [51].

There have been pathological studies of this area of contact from 12 cases at MVD surgery which showed evidence of demyelination and axonal loss here which led to the proposition of the ignition theory (see below) postulating ectopic discharges at this site and in the trigeminal ganglion as a cause of TN. Nonetheliss, other causes of TN are probable such as those associated with multiple sclerosis, dental procedures, other more occult injuries to the nerve, and perhaps even causes related to viral persistence in the ganglion of both herpes simplex and HZ (Chap. 14). The above argues for a more minor injury in TN and perhaps a summation effect of two or more lesions in some cases (accounting for atypical, refractory cases) but with greater injuries to the central and peripheral nervous system in cases of TPHN. Although this is unproven pathologically with TPHN [76] , it does provide the substrate for the concept that it is the severity of injury that may account for these differences. The vascular loop type of minor injury may not occur below the neck (also unproven), and it is rare to see TN-like pain in those areas of the body below the head, and thus it makes TN appear to be a unique condition related to the trigeminal system.

15.5 Treatment

The medical and surgical treatments of TPHN and TN are very different, and not appreciating this by correct diagnosis will result in treatment failure with either condition.

15.5.1 Medical

As with other neuropathic pains, TPHN and other sites of PHN respond optimally to pharmacological treatment with TCAs, gabapentinoids, and opioids or their combination. The results of treatment at best are in the range of 50 % of patients having moderate (50 %) relief or better.

The pharmacological treatment of TN involves carbamazepine (which has been the most effective drug available for any neuropathic pain condition but only works well specifically for TN). A similar, more recently introduced drug, oxcarbazepine, is also quite effective but may have more side effects. These drugs, at least initially, may result in 70–80 % of patients with TN having good initial control if the drug is titrated carefully and with the correct formulation, timing, and dosing interval ([42]; Watson, personal observations).

15.5.2 Surgery

Surgical procedures for PHN, including TPHN, are usually desperate efforts in intractable patients, and the literature consists of case reports of variable quality and follow-up (Chap. 20). For TN there are several surgical options. These can be divided into minor and temporary but initially quite successful options such as the gamma knife, intracavernous glycerol, balloon compression, and radiofrequency lesions of the trigeminal ganglion. The major and potentially curative and frequently successful operation is MVD of vascular anomalies associated with the trigeminal sensory root [42]. This major operation is not always successful and carries a small morbidity and mortality. Patients need to know the risk of failure and especially of anesthesia dolorosa. Relief by operations on the trigeminal ganglion, as with Harvey Cushing's earliest surgical procedure of ganglionectomy, supports the possibility of ongoing ectopic discharges emanating from the ganglion and driving the TN pain. These pharmacological and surgical therapeutic differences argue for a difference in pathophysiological mechanisms between TN and TPHN, although perhaps the difference may simply reflect the severity of the lesion. The more minor injury of vascular compression occurring in the trigeminal system results in that kind of shocklike triggered pain which does not have its exact lesion counterpart (the vascular loop) below the neck where major nerve injury is more common, more severe, and more varied (in PHN, causalgia, phantom limb pain, peripheral neuropathies) and the pain different pathophysiologically.

Other management approaches that are often useful in some other neuropathic pain states, such as cognitive behavioral therapy, acupuncture, and deep brain stimulation, lack good evidence-based data to support their use for TN and PHN, but clinical experience also indicates they are not very useful.

15.6 Pathophysiology: TN and TPHN Mechanisms

Since TN and PHN (including TPHN) are neuropathic pain states, an overview of mechanisms leading to these states is first considered, followed by findings and concepts specifically proposed of the mechanisms underlying TN and those underlying TPHN. In addition, because much of the current knowledge of these mechanisms has been derived from animal studies of the spinal nociceptive system following injury to spinal nerves and since the trigeminal subnucleus caudalis and dorsal horn of the spinal cord are continuous and share many features, a brief outline of spinal mechanisms is initially presented.

15.6.1 Spinal Neuropathic Pain Models and Mechanisms

Although some have used experimental manipulation of central nerosus system (CNS) sites to model TN or PHN (or other less common neuropathic pain states, e.g., thalamic syndrome), an overriding feature of most concepts of the mechanistic basis of either PHN or TN is some change in the input into the CNS from primary afferents that result in structural and functional alterations in somatosensory processing in the CNS. Thus, it is usually conceived that the CNS changes are initiated by alterations in primary afferents but that the CNS changes, perhaps coupled with persistent alterations in the afferents, are the dominant factors leading to the chronic neuropathic pain state.

Several types of injury of spinal afferent fibers can lead to behavioral alterations reflecting spontaneous pain and hypersensitivity to mechanical and sometimes thermal stimulation of peripheral tissues. These behavioral effects are associated with changes in primary afferent fibers and CNS nociceptive neurons (for review, see [9], [14], [17]). The primary afferent alterations may include demyelination of myelinated afferents and even degeneration of unmyelinated (i.e., C-fiber afferents) as well as myelinated afferents. These peripheral changes include an increased excitability (so-called peripheral sensitization) of the nociceptive afferents and also may be associated with alterations in the neuronal cell bodies of the afferents in the DRG. Sprouting of afferent nerve fibers also can occur from the damaged site toward the peripheral tissues, leading to neuroma formation. Ephaptic transmission has been described in the neuroma; this process allows for transfer of nerve impulses (i.e., action potentials) between the damaged afferents (and perhaps to undamaged afferents) and can result in the generation of ectopic or abnormal impulses that are conducted through the ganglion into the CNS. The ganglion neurons themselves may also generate abnormal impulses, perhaps through ephaptic transmission or through processes that involve non-neural cells, especially the satellite glial cells that wrap around the neuronal cell bodies. As a result of the nerve injury, signals are transmitted to the cell bodies of the afferents and cause the altered expression and release of several mediators (e.g., substance P, ATP). These mediators activate the satellite glial cells that generate calcium waves among themselves to produce subsequent activation of not only the cell bodies of damaged afferents but also those of undamaged afferents, thus providing a peripheral process that may contribute to the spread of pain following nerve injury. There is also an abnormal expression of sodium ion channels in the afferents and their neuronal cell bodies, and this is thought to be one of the processes by which the abnormal impulses are generated. Another peripheral process that may occur with some forms of nerve injury is upregulation of alpha adrenoreceptors in the afferents as well as coupling of the afferents or their neuronal cell bodies in the ganglion with sympathetic efferents, providing a mechanism by which the sympathetic nervous system may modulate the activity of the afferents and their input into the CNS.

These various peripherally based changes in the properties of the primary afferents can lead to the following alterations in the CNS. 1) There can be central sprouting of the afferents such that large-diameter (e.g., mechanosensitive) afferents that normally synapse in the deeper laminae of the spinal dorsal horn make synaptic connections in more superficial dorsal horn laminae where the nociceptive neurons predominate. This is thought to provide a mechanism whereby tactile-evoked, large-fiber input can activate nociceptive neurons and evoke pain (e.g., mechanical allodynia). 2) The functional properties of the nociceptive neurons may also be altered such that they undergo neuroplastic changes and become hyperexcitable (so-called central sensitization) 3) Several chemical processes have been implicated in this process reflecting neuroplasticity in the dorsal horn, including glutamate (released from the central terminals of the afferents) activation of N-methyl-D-aspartate (NMDA) and alpha-amino-3-hydroxy-5-isoxazole-proprionic acid (AMPA) receptors on the neurons. There is also considerable evidence from studies in the past 15 years that neurotrophins and non-neural cells (astroglia, microglia, and immune cells) in the dorsal horn play key roles in the development and maintenance of the central sensitization induced by nerve injury. 4) Another associated key event are alterations in central inhibitory as well as facilitatory processes operating through segmental modulatory circuits or descending projections from higher brain centers; such alterations can also contribute to the neuronal hyperexcitability.

Some models of neuropathic pain include those that focus on an intrinsic central nervous system (CNS) etiology, such as in central pain and thalamic syndrome. Injury or inflammation of key structures in the CNS have been shown to lead to the development of an increased neuronal excitability akin to central sensitization (e.g., [1, 16, 17, 30, 77]).

15.6.2 Trigeminal Somatosensory System: Normal Features and Neuropathic Processes

In the case of the trigeminal somatosensory system, studies of the past 50 years have provided a substantial knowledge base of the normal structural characteristics and functional processes of this system (for review, see [12, 23, 64, 67]). The following provides a brief outline of these features.

15.6.2.1 Normal Features

Trigeminal Primary Afferent Mechanisms

Most of the sensory (i.e., primary afferent) nerve fibers innervating orofacial tissues occur in the trigeminal nerve. The larger diameter (e.g., A-beta) afferents end in the tissues as mechanoreceptors which are activated by tactile or proprioceptive stimuli. Some of the smaller afferents (A-delta and C-fiber afferents) terminate as receptors sensitive to warming or cooling stimuli, but the majority end as so-called free nerve endings. These endings function as nociceptors since they are activated by noxious stimulation of the tissues. In addition, the nociceptive afferent endings and their cell bodies (which are predominantly located in the trigeminal ganglion) may also develop a prolonged increase in excitability after injury or inflammation of orofacial tissues. This process reflects a peripheral sensitization and may contribute to the allodynia (pain resulting from a stimulus not normally evoking pain) and the hyperalgesia (increased sensitivity and/or excessive response to a stimulus that is normally painful) that occur in certain pain conditions.

Trigeminal Brainstem Mechanisms

The trigeminal primary afferents pass through the trigeminal ganglion into the brainstem where the vast majority terminate in the trigeminal brainstem sensory nuclear complex (VBSNC; see Fig. 15.5). The larger-diameter afferents that conduct tactile or proprioceptive information end throughout the VBSNC and activate low-threshold mechanosensitive neurons. Those afferents that are activated by warming or cooling orofacial stimuli, or by noxious orofacial stimuli, terminate in the most caudal component (subnucleus caudalis) of the VBSNC. Subnucleus caudalis merges caudally with the cervical dorsal horn of the spinal somatosensory system and is often referred to as the medullary dorsal horn because of its many morphological and physiological similarities with the spinal dorsal horn. Like their spinal afferent counterparts, the central endings of the trigeminal nociceptive afferents release excitatory neurochemicals (e.g., glutamate and substance P) that may excite the second-order nociceptive neurons with which the afferent endings synapse. Also, like the nociceptive neurons existing in the spinal dorsal horn, nociceptive neurons in caudalis and the upper cervical spinal cord (C1–C2) that receive and process trigeminal afferent inputs are of two main types: nociceptive specific (NS) which normally respond only to noxious stimuli and wide dynamic range (WDR) neurons which respond to non-noxious stimuli (e.g., tactile) as well as to noxious stimuli. It is noteworthy that some of these nociceptive neurons receive afferent inputs only from facial skin or oral tissues and have properties indicative of a role in encoding superficial pain. Others in contrast have convergent nociceptive afferent inputs from deep tissues (e.g., temporomandibular joint (TMJ), muscle) as well as from cutaneous or oral tissues, and it appears that they contribute to the CNS processes underlying deep pain, including the referral of pain that is typical of orofacial pain conditions involving deep tissues (e.g., temporomandibular disorders).

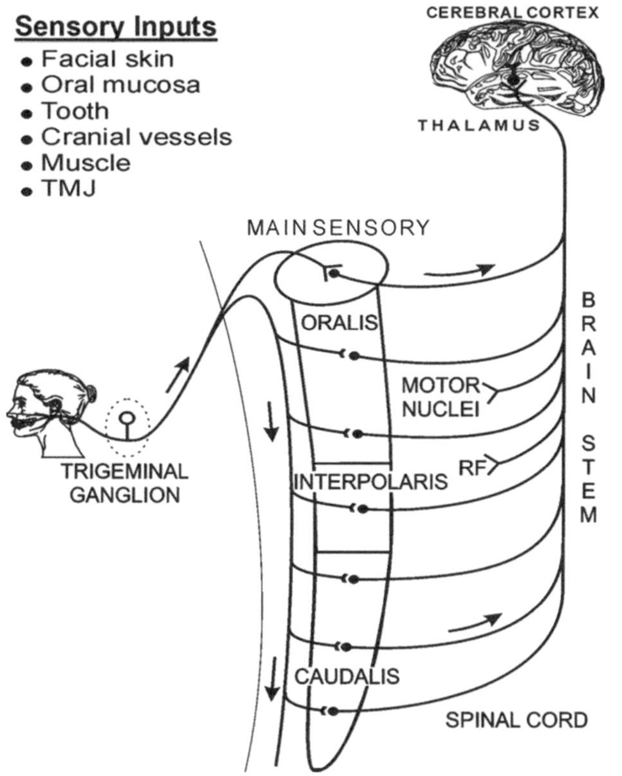

Fig. 15.5 Schema of major somatosensory pathway from the orofacial region. The cell bodies of most primary afferents in the trigeminal nerve are in the trigeminal ganglion and project to second-order neurons in the VBSNC which is made up of the trigeminal main sensory nucleus and the trigeminal spinal tract nucleus; the latter has three subnuclei: oralis, interpolaris, and caudalis. These neurons may project to neurons at higher levels of the brain (e.g., in the somatosensory thalamus) or to brainstem regions such as the reticular formation (RF) or the cranial nerve motor nuclei. Not shown are the projections of some afferents in cranial nerves VII, IX, X, and XII and in cervical nerves to the VBSNC and the projections of many V, VII, IX, and X afferents to the solitary tract nucleus [64]

The axons of some VBSNC neurons project to other brainstem areas including the cranial nerve motor nuclei and reticular formation and thereby participate in the CNS circuitry underlying skeletal muscle reflex responses and in autonomic nervous system-based changes in salivary, cardiovascular, and respiratory functions evoked by stimulation of orofacial tissues. The axons of many VBSNC neurons ramify within the VBSNC itself and influence the activity of other VBSNC neurons. Some projections pass to the reticular formation and adjacent regions (e.g., raphe nuclei) which are involved in pain modulation and in wakefulness and sleep (see below). Many VBSNC neurons may also contribute to CNS pathways ascending directly or indirectly to the ipsilateral and contralateral somatosensory thalamus.

Thalamocortical Mechanisms

Orofacial somatosensory information that is relayed from the brainstem to the thalamus mainly passes to the ventrobasal complex in animals (which is analogous to the ventroposterior nucleus in the human), the posterior nuclear group, and the medial thalamic nuclei. Many of the low-threshold mechanoreceptive and thermosensitive neurons occurring in these somatosensory thalamic nuclei project to analogous neurons in overlying regions of the somatosensory cerebral cortex where their relayed signals are processed to provide for the detection and localization of tactile and non-noxious orofacial thermal stimuli. Nociceptive neurons also occur in these thalamic nuclei, and most of them in the ventrobasal thalamus have functional properties and connections to neurons in the somatosensory cortical regions that indicate their role in coding the spatiotemporal features of orofacial noxious stimuli. In contrast, most of the nociceptive neurons in the posterior nuclear group and the medial thalamic nuclei have properties and connections to cortical areas such as the prefrontal cortex, insula, and anterior cingulate cortex that are involved in the cognitive, affective, or motivational dimensions of pain.

Modulation of CNS Nociceptive Processes

Some of the convergent nociceptive afferent inputs mentioned above may induce neuroplasticity in the VBSNC nociceptive neurons. These neuroplastic changes are reflected in an increase in excitability (central sensitization) of the neurons and are produced by neurochemicals released from the nociceptive afferent endings in the VBSNC. Nociceptive neurons at the higher levels of the CNS (e.g., somatosensory thalamus, somatosensory cortex) may also exhibit central sensitization. The central sensitization of the nociceptive neurons has features that include increases in neuronal spontaneous activity, receptive fields size, and responses to stimulation of peripheral tissues, plus a decrease in activation threshold, including phenotypic switching of NS neurons that become excited by tactile inputs and thus take on properties typical of WDR neurons. The neuroplasticity reflected in central sensitization emphasizes the point that the nociceptive circuits in the CNS are not "hard-wired"; instead they are "plastic" and modifiable by events in peripheral tissues related to injury or inflammation as well as by CNS circuits underlying a variety of functions, as noted below.

Recent studies have revealed that central sensitization involves non-neuronal (i.e., glial) as well as neural processes [11, 20, 66]. This is noteworthy since these glial processes may provide novel targets for developing new or improved analgesic approaches to control pain. Central sensitization occurs in acute as well as chronic pain states, including neuropathic pain conditions, and along with peripheral sensitization (see above), its features can explain the hyperalgesia, allodynia, and pain spread and referral that are features of many orofacial pain states.

Central sensitization is an example of how orofacial nociceptive transmission can be modified at brainstem and thalamocortical levels, in this case reflected in facilitation of transmission. The variety of inputs and interconnections in subnucleus caudalis and other components of the VBSNC provide the basis for interactions between the afferent inputs to the VBSNC from peripheral tissues and projections to the VBSNC from a variety of CNS areas including the reticular formation, locus coeruleus, periaqueductal gray, rostroventral medial medulla, and cerebral cortex. Several endogenous chemical mediators such as glycine, gamma amine butyric acid (GABA), noradrenaline, serotonin (5-HT), dopamine, and opioids (e.g., enkephalins and endorphins) provide a chemical substrate by which these inputs may modulate nociceptive transmission and central sensitization. The modulatory influences of behavioral factors, including state of alertness, attention, and distraction, are examples where higher brain areas involved in these states give rise to descending projections to the VBSNC and thereby contribute to the effects of these behavioral factors on pain. These effects include inhibitory influences on nociceptive neurons and represent intrinsic CNS mechanisms contributing to the analgesic effects of several therapeutic measures that control pain, such as deep brain stimulation, acupuncture, opiate-related drugs (e.g., morphine), and 5-HT agonist drugs (e.g., amitriptyline), as well as to the phenomenon of placebo analgesia. The thalamocortical transfer of sensory information from the orofacial region (and other parts of the body) can also be modulated or "gated" as a result of inhibitory and facilitatory processes involving local neuronal circuits or inputs to the thalamus and cerebral cortex from other CNS regions such as the reticular formation. Gating also occurs during changes in behavioral state and consciousness such as during sleep (see [48]).

15.6.2.2 Trigeminal Neuropathic Pain Models and Mechanisms

Several animal models of trigeminal neuropathic pain have been developed in the past 25 years, and some of these seek to represent specific models of TN or TPHN. In general, these trigeminal neuropathic pain models have revealed several comparable mechanisms to those described above for the spinal somatosensory system. And, like the neuropathic pain models involving alterations to spinal nerves and DRG, most trigeminal neuropathic pain models have involved damage to one or more branches of the trigeminal nerve or the trigeminal ganglion, typically in rats or mice (for review, see [20, 22, 40]). Common features of most of these models are facial hypersensitivity to mechanical and/or thermal stimulation of the face or mouth that is indicative of orofacial allodynia or hyperalgesia. Spontaneous pain and aversive behaviors and changes in grooming, facial expression, and exploratory activity may also occur. Also notable in many studies has been the documentation of bilateral changes (e.g., bilateral facial hypersensitivity) following unilateral nerve damage as seen in humans (54, 59, 69, 76]).

Several studies have utilized chronic constriction injury (CCI) of a trigeminal nerve branch, the infraorbital nerve CCI model being a prime example. A major advantage of this model over the CCI of a spinal nerve used in many spinal models is that the infraorbital nerve is a pure sensory nerve as compared to the mixed sensory and motor nature of spinal nerves used (e.g., sciatic). Transection or other forms of damage of one or more branches of the infraorbital nerve or inferior alveolar nerve or even cervical nerves have also been used in models of trigeminal neuropathic pain, as has compression of the trigeminal ganglion or sensory root and cancerous lesions (for review, see [10, 20, 40]). Lesions in parts of the trigeminal system in the CNS, as well as the introduction of viral vectors into the trigeminal system, have also been used, as outlined below. These models, especially those involving injury to the trigeminal nerve or ganglion, have revealed a number of cellular and molecular changes in the peripheral and central components of the trigeminal system that accompany the orofacial hypersensitivity and spontaneous pain and other behaviors manifested in these models.

Peripheral Trigeminal Neuropathic Pain Processes

In most of the models, an increase in nociceptive afferent excitability (peripheral sensitization) has been documented within just a few days following trigeminal nerve injury. Many afferents and their cell bodies in the trigeminal ganglion exhibit irregular spontaneous activity, and there may be a decrease in the number of A-beta and C-fiber afferents (large and small nerve fibers) and an increase in the number of A-delta fibers (small myelinated fibers), although the normal functional phenotype of the affected afferents (e.g., nociceptive, thermal, tactile) is unclear. Some of the afferent neurons may also become responsive to stimuli applied to trigeminal nerve territories outside the innervation territory of the injured nerve (see [20, 40, 73]), as illustrated in Fig. 15.6. This suggests a possible peripherally based mechanism contributing to extraterritorial spread of sensitivity that may occur in clinical cases (also see below). There are also changes in the expression of intracellular protein kinases and channel proteins. Changes especially in sodium and potassium channels can affect the excitability of the trigeminal afferent neurons. In addition, changes in neuropeptides (e.g., substance P, calcitonin gene-related peptide (CGRP)), receptors (e.g., P2X3), cytokines, and growth factors in trigeminal ganglion neurons are additional mechanisms following trigeminal nerve injury that could possibly cause an increased excitability of the neurons (see [20, 40, 49, 57]).

Also noteworthy are recent findings of the involvement of satellite glial cells in the trigeminal ganglion in the peripheral sensitization process of afferent neurons (Fig. 15.6). The number of ganglion neurons encircled by satellite glial cells is increased after trigeminal nerve injury (see [20, 22, 40]). Changes in intracellular signaling processes including increased nitric oxide synthesis and activation of $P2Y_{12}$ receptors in the satellite glial cells following trigeminal nerve injury also appear to be involved in the enhancement of trigeminal afferent neuron activity and the development and maintenance of nociceptive behavior.

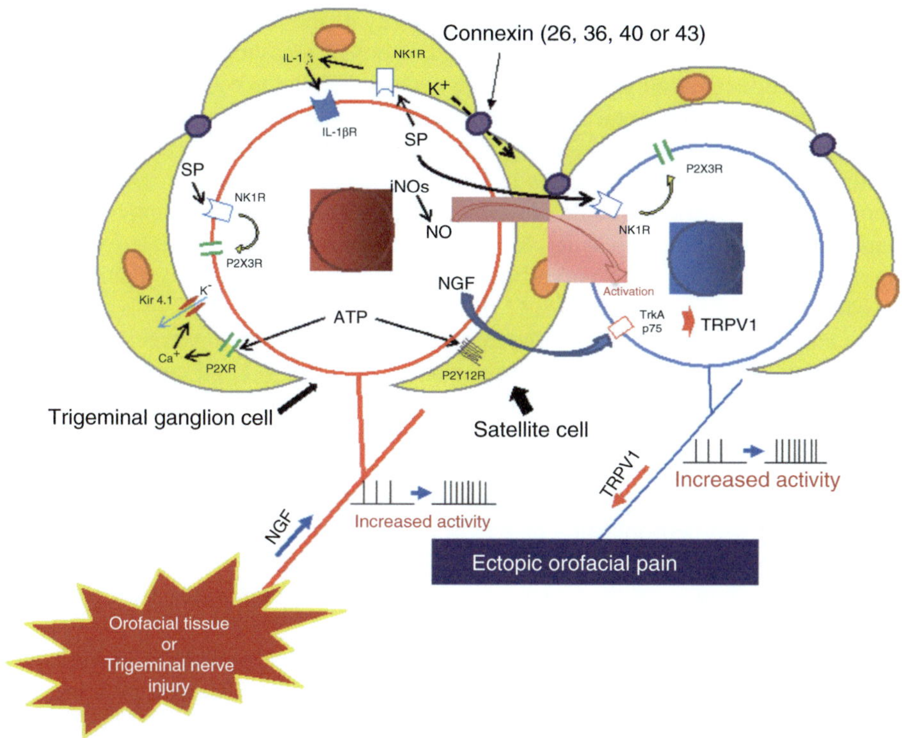

Fig. 15.6 Diagram showing neuron-neuron or neuron-glia interactions in the trigeminal ganglion. Following orofacial tissue injury or trigeminal nerve injury, primary afferent activity is considerably increased. Expression of various molecules such as SP, NO, or ATP also increases in the neurons or satellite glial cells, and the neurons or satellite glial cells are strongly activated, resulting in the peripheral sensitization of the neurons. NGF is also expressed in the injured tissues and then transported to the ganglion neurons where it is released and is involved in enhancement of TRPV1 expression in the neurons. (*ATP* adenosine triphosphate, *IL-1β* interleukin-1 beta, *IL-1βR* interleukin-1 beta receptor, *iNOs* inducible nitric oxide synthase, *Kir 4.1* inwardly rectifying potassium channel subfamily 4.1, *NGF* nerve growth factor, *NK1R* neurokinin-1 receptor, *NO* nitric oxide, *P2XR* P2X purinergic receptor, *P2X3R* purinergic P2X3 receptor, *P2Y12R* purinoceptor P2Y12, *p75* low affinity neurotrophin receptor, *SP* Substance P, *TrkA* TRK1-transforming tyrosine kinase protein, *TRPV1* transient receptor potential vanilloid) [20]

Central Trigeminal Neuropathic Pain Processes

Brainstem Processes

Injury to the trigeminal nerve, as well as to tissues that it innervates (e.g., tooth pulp), has been found to lead to morphological changes in the VBSNC including subnucleus caudalis (e.g., [32, 40, 44]). Several investigations in animals have also documented that both the NS and the WDR neurons that occur in trigeminal subnucleus caudalis (see above) exhibit a hyperexcitability (central sensitization) following trigeminal nerve injury (see [20, 22, 40]). This central sensitization is manifested as an increase in spontaneous activity of the NS and WDR neurons, as well as by increased receptive field size and responsiveness to orofacial stimuli and a decrease in activation threshold. These nerve injury-induced changes reflect the neuroplasticity of the nociceptive neuronal processes. It is noteworthy that it is accompanied by the animal developing nociceptive behavior indicative of clinical features in neuropathic pain states in humans such as hyperalgesia, allodynia, and pain spread or referral. It should however also be noted that the nociceptive behavior and the associated neuroplastic changes reflecting a central sensitization state in animals are not specific to nerve injury, but these features are also manifested in inflammatory pain states (see [18, 66]).

In addition, it is possible that neuroplastic changes following trigeminal nerve injury may also occur in other components of the VBSNC and contribute to trigeminal neuropathic pain states. For example, nociceptive neurons have been documented in subnuclei oralis and interpolaris of the VBSNC, and in orofacial inflammatory pain models, hyperexcitability can occur in oralis nociceptive neurons (see [18, 22, 62, 64]). The low-threshold mechanoreceptive neurons in oralis can also undergo neuroplastic changes following manipulations producing deafferentation of afferent inputs from the tooth pulp or other orofacial tissues; the possible contribution of these neurons to neuropathic pain states such as TN is considered further below.

Recent studies have revealed that non-neural cells including astroglia and microglia and immune cells contribute to the central sensitization in models of acute and chronic orofacial inflammatory as well as neuropathic pain states (for review, see [11, 20, 22, 40, 66]). It appears that glia, immune cells, and neurons form an integrated network that modulates the excitability of neurons in the nociceptive pathways. The central sensitization is indeed dependent on the functional integrity of glial cells, e.g., the central sensitization of the trigeminal nociceptive neurons and the associated nociceptive behavior can be overcome by the administration of inhibitors of astroglial or microglial function. It is also notable that these glial cells become activated over a wide area of subnucleus caudalis following a trigeminal (or cervical) nerve injury, e.g., injury of the inferior alveolar nerve (a branch of the V3 division of the trigeminal nerve) can lead to glial cell activation in the V1 and V3 regions of subnucleus caudalis. Such findings suggest that glia in subnucleus caudalis may play a role in mechanisms underlying the development of extraterritorial sensitivity in neuropathic pain states. Along with the spread of activation in the trigeminal ganglion beyond the trigeminal division (see above), these features following trigeminal nerve injury may at least in part explain the abnormal pain sensations that accompany peripheral neuropathies and are sometimes manifested in a distribution that does not coincide with the territories of nerves or sensory roots directly affected by the injury; these features could account for pain spread beyond the zone of injury, resulting in hypersensitivity in nearby or even more remote regions.

Another related CNS-based effect has been revealed in recent studies, namely, that the nociceptive behavior in animal models of trigeminal neuropathic pain is closely linked to a long-lasting activation in the CNS of so-called descending modulatory pathways (for review, see [18, 20, 21, 55]). This activation appears to contribute to the development of persistent pain after trigeminal nerve injury (and peripheral inflammation) and involves a 5-HT-dependent descending facilitatory projection from parts of the raphe system (e.g., rostroventral medial medulla) that appears to contribute to the neuropathic pain state. Astroglia and microglia in the rostroventral medial medulla have been shown to be involved in the initiation and maintenance of allodynia and hyperalgesia following trigeminal nerve injury, and cytokines released from the glia may mediate the coupling between glial activation and increased neuronal excitability in the rostroventral medial medulla following trigeminal nerve.

These various findings in animals in the trigeminal ganglion and VBSNC indicate that both peripheral alterations in primary afferents and their inputs into the brainstem, as well as changes in the CNS such as trigeminal central sensitization and associated alterations in CNS inhibitory and facilitatory circuits, may underlie the development and maintenance of chronic pain states following trigeminal nerve injury and the allodynia and hyperalgesia in humans that are clinical features of many of these states. Pain spread and referral likely also involves these mechanisms, since glial cell activation in the trigeminal ganglion and subnucleus caudalis can spread and so contribute to the development of persistent extraterritorial sensitivity associated with trigeminal nerve injury. Contributions may also be made by excitability changes in the higher levels of the trigeminal somatosensory and related systems, as noted below. In addition, the effectiveness of drugs used for the management of trigeminal neuropathic pain states (e.g., carbamazepine, gabapentinoids, TCAs, opioids) can be attributed, at least in part, to their actions on these cellular processes reflecting heightened excitability of the trigeminal nociceptive circuits and the intrinsic systems modulating them (e.g., [10, 11, 20, 26, 27, 68, 73]).

Thalamocortical Processes

Only limited investigation has been made of processes in the thalamus and cerebral cortex in trigeminal neuropathic pain states, although it is clear that increased excitability reflected in changes in spontaneous activity and response properties that have the characteristics of central sensitization do occur in the neurons of the somatosensory thalamus and cortical areas such as somatosensory cortex and anterior cingulate and insula in animal models of neuropathic pain (see [1, 4, 13, 17, 20, 41, 72, 81]) and in neural recordings from CNS sites (e.g., somatosensory thalamus) of humans (see [1, 16, 17]). These include sites giving rise to descending projections that influence nociceptive transmission at lower levels of the CNS such as the VBSNC and spinal dorsal horn. Given the strategic role of the somatosensory thalamus and somatosensory cortex in somatosensory processing, such changes likely contribute to the development of trigeminal neuropathic pain states and their clinical features. More investigation is warranted of these excitability changes in thalamic and cerebral cortical cells as well as their possible modulation by therapeutic approaches such as pharmacological agents used for the management of neuropathic pain.

Several of these studies of trigeminal neuropathic pain processes have utilized some of the abovementioned peripheral manipulations of afferent inputs into the CNS, or more direct manipulation of CNS processes themselves, as animal models of TN or TPHN. These will now be considered in turn.

15.6.2.3 TN Models and Mechanisms

In the case of TN, there has been a tendency from a historical perspective to view its etiology in terms of either having a peripheral basis or a CNS origin, although there has usually been general agreement that its pathogenesis involves CNS processes. The concept of a peripherally based mechanism goes back many decades to clinical observations of what appeared to be peripheral causes (e.g., tumors, infections, nerve compression by an abnormal vascular loop) and the pain-relieving effects of peripherally based therapeutic approaches in some TN cases. More recently, authors such as Calvin [9] and Devor and colleagues ([14, 15, 56]; see also Chap. 13) have also championed a peripheral process, in particular emphasizing the critical role that peripheral nerve injury and deafferentation may play in TN (and in TPHN). They drew attention to nerve injury-induced abnormal or ectopic sensory inputs into the CNS that could result from cross talk between injured and non-injured afferent fibers and evoke an explosive input. Such an "ignition" hypothesis as proposed by Devor and colleagues (Chap. 13) appears to have been largely based on studies of spinal nerve injury and neuroma formation in animals, yet a clinical neuropathic pain state replicating TN is not apparent in the spinal somatosensory system.

Nonetheless, there have been findings that do fit with the ignition hypothesis. There are reports that many TN patients do show evidence of compression, although as noted above, findings are mixed (e.g., [42, 51]). In addition, ultrastructural changes including demyelination, axonal alterations, and close apposition of some axons have been shown in trigeminal sensory root biopsies taken during decompression surgery in humans with TN, but the absence of comparable tissue from non-TN cases "muddies" the interpretation. In animal studies, Tal and Devor [70] reported that injury to the infraorbital nerve produced neuromas that were associated with some spontaneous activity and mechanosensitivity of both myelinated and unmyelinated infraorbital afferents. However, these abnormal properties were much less than those recorded from afferents ending in sciatic nerve neuromas. Other investigations in animals have also produced evidence that peripheral injury of trigeminal sensory nerves may induce abnormal spontaneous or mechanically evoked activity in the injured afferents and in trigeminal ganglion neurons (e.g., [57]; for review, see [40]).

Several more recent studies have produced evidence (as noted above) for several chemical mediators and interactions between trigeminal ganglion neurons and satellite glial cells in these effects of nerve injury and indeed that functional changes may even occur in trigeminal afferent neurons not directly affected by the nerve injury (for review, see [11, 20, 28, 40]). Interestingly, the animals showing these neuropathic changes in trigeminal afferents also manifest trigeminal central sensitization and facial nociceptive behavior as reflected in mechanical allodynia evoked by tactile stimulation of facial skin. These various findings suggest peripheral mechanisms that may contribute not only to neuropathic pain states themselves but also to extraterritorial pain spread following trigeminal nerve injury. The findings are also not inconsistent with the "ignition" hypothesis that an abnormal sensory input evoked, for example, by an orofacial tactile stimulus may elicit CNS changes reflected in abnormal pain behavior that has some features resembling those manifested in TN (and possibly TPHN). Another recent TN model involving chronic compression of the trigeminal sensory root also has produced findings suggestive of abnormal sensory inputs that are associated with facial hypersensitivity and alterations in trigeminal subnucleus caudalis [43, 50]; this model is built on the clinical reports (see above) that decompression of the trigeminal sensory root is often effective in relieving TN.

Another school of thought has directed attention away from peripheral processes toward a dysfunction of CNS mechanisms as the cause of TN as well as their involvement in the pathogenesis of TN. Such a concept in fact harkens back to earlier viewpoints such as that of Trousseau [71] who hypothesized that the pain attacks of TN result from paroxysms of neural activity in the VBSNC. Such a concept led serendipitously to the best current drug treatment for TN which is the anticonvulsant carbamazepine. Studies in more recent decades have lent some support to such a concept, buttressed also by some of the clinical features of TN noted above, e.g., the pain of TN is usually elicited by a tactile stimulus and can manifest summation features and a refractory period and persist beyond the period of stimulation; it may also radiate or refer to more distant sites (Watson personal observations). And in animal studies, Black and King and their colleagues (e.g., see [5, 45, 60]) reported that application of epileptogenic agents (e.g., alumina gel, strychnine) to the subnucleus caudalis produced hypersensitivity to facial tactile stimulation. Young and King [79] and Greenwood and Sessle [33] proposed that such effects in TN and the pain-relieving effects of trigeminal tractotomy could be explained by these manipulations affecting the normal modulatory effects that they (and subsequently others [e.g., see [64, 78]) showed are exerted by caudalis on more rostral brainstem neurons, e.g., in trigeminal subnucleus oralis.

More recently, a more harmonized viewpoint of the etiology and the pathogenesis of TN has gained a hold by unifying aspects of the peripheral and CNS processes in TN, namely, that the etiology of TN is peripherally initiated, but its pathogenesis involves peripherally induced neural changes in the CNS. The possibility that these CNS changes might sometimes be maintained by ongoing abnormal peripheral afferent drive could explain the success of procedures (e.g., MVD gangliolysis, rhizotomy, alcohol block) directed at peripheral generators in the ganglion or sensory root in some TN patients. On the other hand, independence of the CNS changes from peripheral afferent drive in other TN patients may explain the failure of a proportion of peripherally based approaches (see [27, 42]). As noted above, CNS changes have been shown to occur in the animal studies utilizing injury or deafferentation of trigeminal nerves. These changes include morphological changes as well as alterations in the physiological properties of trigeminal caudalis nociceptive neurons and oralis low-threshold mechanoreceptive neurons. Such neuroplastic changes appear critical to the expression of pain behavior, in this case induced originally by altered trigeminal sensory inputs produced by damage or other alterations to trigeminal afferent neurons.

The so-called hyperactivity in neurons in trigeminal subnucleus caudalis following injury of the trigeminal sensory root or following injury associated with deafferentation of tooth pulps was reported and invoked as TN mechanisms [2, 5], but subsequent detailed studies revealed only limited changes reflecting hyperexcitability in the receptive field and response properties of functionally identified WDR and NS neurons in subnucleus caudalis; curiously, it was the low-threshold mechanoreceptive neurons particularly in subnucleus oralis that showed evidence of such neuroplasticity changes induced by pulp deafferentation [37–39, 47].

A role for trigeminal subnucleus oralis in the pathogenesis of TN was also proposed by Fromm and his colleagues (see [26, 27]) on the basis of their findings of excitatory and inhibitory influences on oralis low-threshold mechanoreceptive neurons and the effects of drugs (carbamazepine, phenytoin) used in the management of TN. They suggested that injury or chronic irritation of trigeminal afferents produces an increased afferent input to the brainstem that leads to degeneration of brainstem inhibitory neurons and consequently failure of segmental presynaptic and postsynaptic inhibitory processes within subnucleus oralis that normally serve to control the level of excitatory activity in nociceptive (WDR) neurons in subnucleus caudalis. They viewed the abnormal ectopic afferent inputs, coupled with the impaired efficacy of segmental inhibition in subnucleus oralis, as resulting in paroxysmal discharges in oralis low-threshold mechanoreceptive neurons. They considered that such discharges would then affect caudalis WDR neurons, causing them to show hyperexcitability to a tactile stimulus to such an extent that it leads to a TN attack.

There is evidence that abnormal nociceptive afferent inputs can bring about loss of inhibitory neurons in the VBSNC (at least in subnucleus caudalis; see [19, 64]), and there are connections and influences between subnuclei oralis and caudalis neurons (e.g., [33, 79]; for review, see [64, 78]) which collectively are consistent with the findings of Fromm and colleagues and provide support for such a proposal. Also supportive are findings that oralis neurons do show neuroplastic changes and hyperexcitability following injury of orofacial tissues (e.g., [37, 38, [39]; for review, see [64, 78]). In the previous edition of this book, Sessle [65] also drew attention there to earlier proposals that WDR neurons in subnucleus caudalis as proposed by Fromm may indeed play a critical role given that they normally receive inputs both from large-diameter afferents carrying touch-related information as well as from small-fiber nociceptive inputs. Thus, hyperexcitability to a tactile stimulus could result in activity of WDR neurons to a level only normally achieved by noxious stimuli, and so excruciating pain may be experienced as a consequence.

However, Sessle [62, 63, 64, 65] also pointed out that a role for NS neurons cannot be discounted. As noted above, trigeminal caudalis NS neurons normally only receive orofacial nociceptive afferent inputs. However, following trigeminal nerve injury and even cervical nerve injury (e.g., [10, 40, 46]), they may develop sensitivity as well to tactile inputs from perioral and intraoral zones that are typical trigger sites and areas where TN pain is experienced Thus, he also raised the possibility that injury-induced neuroplastic changes in caudalis NS neurons may be a critical event in the pathogenesis of TN. In addition, since neurons in subnucleus caudalis, and also subnucleus oralis, are subject to descending modulation from higher brain centers (e.g., periaqueductal gray, rostroventral medial medulla, cerebral cortex) involved in pain control (see above), a role for disruption of these sources of inhibitory, or facilitatory, influences cannot be discounted.

Although it is clear that nerve injury (or inflammation) of orofacial tissues can produce neuroplastic changes in nociceptive (and non-nociceptive) neurons in sensory thalamus and cerebral cortex (see above), little emphasis has been placed in studies of TN or PHN mechanisms on the thalamocortical processes and the neuroplastic changes and hyperexcitability that can occur in these regions in nerve injury models in laboratory animals and in thalamic neural recordings from humans (for review, see [16, 17]). These processes include also those involved in sites giving rise to descending projections that modulate nociceptive transmission at lower levels of the CNS. Nerve injury can induce neuroplastic changes at these and other sources of descending influences. So it is conceivable that along with changes in segmental inhibition as proposed by Fromm, injury-induced changes in descending modulatory systems may also be involved in TN. Such changes have been implicated in chronic pain states and their maintenance, and so further study of these influences are warranted in models of TN (and TPHN; also see below).

It should be noted that the peripheral and CNS changes and behavioral alterations in animals that are suggestive of TN do not seem to be specific to any one type of nerve damage. They also are manifested in animal models as a result of any one of a whole variety of trigeminal nerve injuries and even in models of acute and chronic inflammatory orofacial pain states (for review, see 18, 20, 40, 66). This brings into question whether these nerve injuries described above are indeed credible models of TN per se or rather reflect behavioral and neural changes that are common to many non-neuropathic as well as neuropathic pain states. Thus uncertainty remains about the etiological and pathogenic factors that are specifically associated with TN.

15.6.2.4 TPHN Models and Mechanisms

In the case of TPHN, much attention has also focused on peripheral events in its etiology. In contrast to the uncertain nature of the etiological factors in TN, a peripheral etiology is more clear, the critical factor being the HZ virus that gains entry to the trigeminal nervous system. The subsequent features of the pain and pathology of PHN (see above) do however suggest that, like TN, CNS changes are involved in the pathogenesis of the disorder. As well as ganglionic scarring and cell loss [35, 75], dorsal horn atrophy extending several segments rostro-caudally has been seen in several autopsied cases [75]. These are sites of action of analgesic drugs, perhaps explaining the difficulty of treating PHN by pharmacological approaches such as TCAs, gabapentinoids, and opioids.

As noted above, TPHN results from reactivation of the HZ virus that earlier had gained entry into the trigeminal ganglion. There is evidence that through axoplasmic flow, the reactivated virus in ganglionic neurons is transported to skin sites innervated by those neurons (especially ganglionic neurons associated with the V1 division of the trigeminal nerve) and produces painful skin lesions at these sites. The pain of TPHN (and PHN elsewhere in the body, e.g., thoracic region) is reflected in neuropathic pain features such as spontaneous pain and mechanical allodynia. There have been several studies that have used animal models of TPHN or PHN in animals infected with a herpes virus (for review, see [3, 31, 53, 58]; Gilden et al.). It has been shown that such animals develop nociceptive behavior as well as the lodging of the virus in ganglion neurons in association with altered ganglionic expression of several chemical mediators and alterations in some specific sodium and calcium ion channels in ganglion neurons (e.g., [29]; Gilden et al. [31]); these changes may be reversed by administration of sodium channel blockers and gabapentin (which acts on calcium channels). However, it is notable that the virus-induced effects in the ganglion are not limited to the neurons; the satellite glial cells that encircle the neurons may also be involved. These cells were noted above to play an important role in peripheral mechanism of neuropathic pain. In the case of herpes infection, Warwick and Hanani [74], for example, have shown that short-term infection of mouse trigeminal ganglion cultures with herpes virus induces calcium waves and cell fusion among satellite glial cells and neurons and enhanced signaling in trigeminal ganglion neurons. In keeping with the effects of tissue injury or inflammation in the trigeminal ganglion (see above, and Fig. 15.6), this is likely to result in an augmented afferent input to nociceptive neurons in the CNS and the possibility of central sensitization of the neurons accompanied by mechanical allodynia and spontaneous pain. Indeed, it has been shown that animals infected with HZ virus develop mechanical allodynia that is associated with spinal central sensitization and upregulation of spinal astroglia; astroglial inhibitor administration can attenuate the allodynia and central sensitization [80]. The likelihood that CNS changes are integral to PHN is supported by the autopsy findings noted above of PHN cases showing structural changes in the spinal dorsal horn. The apparent involvement of satellite glial cells in the DRG and trigeminal ganglion and of astroglia in CNS nociceptive pathways in the pathogenesis raises the possibility that these non-neural elements may prove to be important novel targets for the development of new therapeutic approaches to control PHN, including TPHN.

The conceptualizations of TPHN share some similar features with several of the concepts of TN that were covered above and so will not be outlined in detail. Some concepts of the etiology and pathogenesis of TPHN and PHN in general have focused on the presence of peripheral sensitization or on reorganization and neuroplasticity in nociceptive pathways in the CNS because of central sprouting of damaged afferents (for review, see [3, 53, 58]; Oaklander [53]). Some concepts have emphasized other aspects such as peripheral nerve damage due to the HZ virus producing alterations in primary afferent neurons in the trigeminal ganglion that results in ectopic discharges and thus in an abnormal afferent input to the brainstem that then induces neuroplastic changes in the nociceptive pathways in the CNS (e.g., [3, 53, 58]). Such concepts are based on experimental studies that have shown evidence for such changes in trigeminal primary afferent neurons, as noted above, as well as in nociceptive processes in subnucleus caudalis. The CNS changes are considered to reflect a loss of the inhibitory processes in the CNS that normally modulate nociceptive neurons, thus producing a hyperexcitability (central sensitization) of the neurons and the clinical features of TPHN.

It is noteworthy that many of the peripheral, CNS, and behavioral alterations in TPHN animal models do not seem to be specific for TPHN but (as also noted for the TN models; see above) are also produced in animal models of many types of trigeminal nerve injuries and even in inflammatory orofacial pain models. Thus, it is uncertain what processes can account for the specific clinical neurological features of TPHN compared with TN. Genetic factors could play a key role in the mechanisms of TPHN as well as TN, but knowledge of their contribution is as yet limited (see [52, 61]).

15.7 Final Perspectives, Future Directions, and Concluding Remarks

The preceding text has pointed out many uncertainties that still remain about the etiology and pathogenesis of TN and TPHN and their underlying mechanisms. These uncertainties can be couched as questions at which future research studies should be directed. They are as follows:

1. While the peripheral and CNS changes result in a sensitization of nociceptive neurons following nerve injury however produced (mechanical trauma, HZ virus) that could explain the exquisite sensitivity to tactile stimuli in both TPHN and TN (see above), what are the peripheral and central changes that can account for the distinctive clinical features that distinguish TPHN from TN? This is still unclear. Fromm and colleguess [26, 27] proposed that the degree of injury to afferent fibers and the resulting putative differences in changes in inhibitory processes is the significant factor that determines whether TN or PHN develops. This viewpoint however does not explain what peripheral and CNS changes occur that are specific for each pain state. More experimental studies in animals of the neuronal and non-neural (e.g., glial) properties subsequent to different types of nerve injuries vis-à-vis HZ-related damage are needed to study this. Such studies need to be complemented by human brain imaging, autopsy material, and genotyping in TN and TPHN cases that address both the structural alterations that may occur and the chemical and genetic changes that may take place in peripheral and central elements of the trigeminal somatosensory system. The

advent of several different types of imaging and spectroscopic procedures, supplemented with immunohistochemical techniques and genetic profiling, make such research now feasible (see [52, 61]).

2. What functional types of afferents are affected in TN and TPHN? This is still not entirely clear. Future research should address whether there are differences in the functional types of afferents affected in each of these neuropathic pain states that could explain or be related to the different and distinctive clinical features of TN and TPHN and their pathogenesis. Likewise, a focus on nerve-injury-induced effects on the different components and the different types of neuronal and non-neural elements in trigeminal CNS circuits would greatly expand our knowledge base of these pain states. These approaches would also necessitate the use of new or improved animal models of both TN and TPHN.
3. Why is the ophthalmic division specifically affected in TPHN and the thoracic dermatomes specifically affected in PHN occurring outside the craniofacial region? Are these the sites that most efficiently transmit the virus to cause varicella? If so, why? And, why is the involvement usually unilateral? These are additional puzzling aspects of these conditions that require more research focus.
4. Why is carbamazepine so effective for TN but has little efficacy in other neuropathic pain states and other chronic pain conditions? Further investigation of the pharmacokinetics, cellular and molecular actions, and sites of action in the CNS of this drug vs other drugs (e.g., gabapentinoids) would likely provide important insights. Also, clinical trial and error approaches like those that unearthed phenytoin and carbamazepine being effective for TN may reveal other effective medications.
5. Why are different drugs effective for TN versus PHN? Is this a result of minor versus major nerve injury?
6. What role do non-neural elements play in TN and TPHN? HZ does clearly have a role in TPHN, but given the emerging evidence noted earlier of the involvement of non-neural elements (e.g., glial cells, immune cells) in peripheral or central processes in neuropathic pain, including animal models of TN and TPHN, further investigation of these elements promise to provide new insights into mechanisms bearing on the etiology and pathogenesis of these conditions and on new approaches to manage them.

In conclusion, this chapter has compared and contrasted TPHN and TN with regards to their clinical features, pathology, putative pathophysiological processes, and treatment. These two forms of neuropathic orofacial pain are very different and easy to distinguish clinically. These differences are critical to appreciate from a practical point of view so as to ensure an optimal management approach for each condition. However, their underlying mechanisms are still unclear, including the distinctive processes of each that lead to these distinguishable neuropathic pain conditions. Further research is called for to address their many puzzling aspects and answer the several questions that are raised about these conditions.

References

1. Albe-Fessard D, Berkley KJ, Kruger L et al (1985) Diencephalic mechanisms of pain sensation. Brain Res Rev 9:217–296
2. Anderson LS, Black RG, Abraham J, Ward AA Jr (1971) Neuronal hyperactivity in experimental trigeminal deafferentation. J Neurosurg 35(4):444–452
3. Bennett GJ (2004) Neuropathic pain in the orofacial region: clinical and research challenges. J Orofac Pain 18:281
4. Benoist J-M, Gautron M, Guilbaud G (1997) Alteration of neuronal activity in the cortical projection area of the pad vibrissae in rats with pain-related behaviors from the trigeminal area. In: Jensen TS, Turner JA, Wiesenfeld-Hallin Z (eds) Proceedings of the 8th world congress on pain. Progress in pain research and management. IASP Press, Seattle, pp. 561–573
5. Black RG (1974) A laboratory model of trigeminal neuralgia. Adv Neurol 4:651–659
6. Bergouignon M (1942) Cures hereuses de nevralgies faciales essentialles par le diphenylhydantoine de soude. Rev Laryngol Otol Rhinol 63:34–42
7. Blom S (1962) Trigeminal neuralgia; Its treatment with a new anticonvulsant drug (G-32883). Lancet 1:839–840
8. Burchiel KJ (1980) Abnormal impulse generation in focally demyelinated trigeminal roots. J Neurosurg 53(5):674–683
9. Calvin WH (1979) Some design features of axons and how neuralgias may defeat them. In: Bonica JJ, Liebeskind JC, Albe-Fessard D (eds) Advances in Pain Research and Therapy, vol 3. Raven Press, New York, pp. 297–309
10. Cao Y, Wang H, Chiang CY, Dostrovsky JO, Sessle BJ (2013) Pregabalin suppresses nociceptive behavior and central sensitization in a rat trigeminal neuropathic pain model. J Pain 14:193–204
11. Chiang CY, Dostrovsky JO, Iwata K, Sessle BJ (2011) Role of glia in orofacial pain. Neuroscientist 17:303–320
12. Darian-Smith I (1966) Neural mechanisms of facial sensation. Int Rev Neurobiol 9:301–395
13. Davis KD, Stohler CS (2014) Neuroimaging and orofacial pain. In: Sessle BJ (ed) Orofacial pain: recent advances in assessment, management and understanding of mechanisms. IASP Press, Washington, pp. 165–184
14. Devor M (1999) Unexplained peculiarities of the dorsal root ganglion. Pain Suppl 6:S27–S35
15. Devor M, Amir R, Rappaport ZH (2002) Pathophysiology of trigeminal neuralgia: the ignition hypothesis. Clin J Pain 18(1):4–1
16. Dostrovsky JO (2000) Role of thalamus in pain. Prog Brain Res 129:245–257
17. Dostrovsky JO, Craig AD (2013) Ascending projection systems. In: McMahon SB, Koltzenburg M, Tracey I, Turk DC (eds) Textbook of pain, 6th edn. Elsevier, Philadelphia, pp. 182–197
18. Dostrovsky JO, Sessle BJ, Lam K (2014) Inflammatory and cancer-related orofacial pain mechanisms: Insights from animal models. In: Sessle BJ (ed) Orofacial pain: recent advances in assessment, management and understanding of mechanisms. IASP Press, Washington DC, pp. 305–328
19. Dubner R, Bennett GJ (1983) Spinal and trigeminal mechanisms of nociception. Annu Rev Neurosci 6:381–418
20. Dubner R, Iwata K, Wei F (2014) Neuropathic orofacial pain mechanisms: Insights from animal models. In: BJ S (ed) Orofacial pain: recent advances in assessment, management and understanding of mechanisms. IASP Press, Washington, pp. 331–349
21. Dubner R, Ren KJ (2004) Brainstem mechanisms of persistent pain following injury. J Orofac Pain 18(4):299–305
22. Dubner R, Ren K, Sessle BJ (2013) Sensory mechanisms of orofacial pain. In: Greene C, Laskin D (eds) Treatment of TMDs: bridging the gap between advances in research and clinical patient management. Quintessence, Chicago, pp. 3–16
23. Dubner R, Sessle BJ, Storey AT (1978) The neural basis of oral and facial function. Plenum Press, New York, p. 483

24. Dubner R, Sharav Y, Gracely RH, Price DD (1987) Idiopathic trigeminal neuralgia: Sensory features and pain mechanisms. Pain 31:23–33
25. Fothergill J (1773) Of a painful affection of the face. Medical observations and inquiries by a society of physicians. London 5:129–142 [This publication was funded privately by Fothergill from 1771–1776. It is included in Complete Works of John Fothergill, a copy in library of RCP London.]
26. Fromm GH (1993) Facial pain with herpes zoster and postherpetic neuralgia. In: Watson CPN (ed) Herpes zoster and postherpetic neuralgia. Elsevier, Amsterdam
27. Fromm GH, Sessle BJ (eds) (1991) Trigeminal neuralgia: current concepts regarding pathogenesis and treatment. Butterworths, Stoneham
28. Garrett FG, Durham PL (2008) Differential expression of connexins in trigeminal ganglion neurons and satellite glial cells in response to chronic or acute joint inflammation. Neuron Glia Biol 4(4):295–306
29. Garry EM, Delaney A, Anderson HA, Sirinathsinghji EC, Clapp RH, Martin WJ, Kinchington PR, Krah DL, Abbadie C, Fleetwood-Walker SM (2005) Varicella zoster virus induces neuropathic changes in rat dorsal root ganglia and behavioral reflex sensitisation that is attenuated by gabapentin or sodium channel blocking drugs. Pain 118(1–2):97–111
30. Gerke MB, Duggan AW, Xu L, Siddall PJ (2003) Thalamic neuronal activity in rats with mechanical allodynia following contusive spinal cord injury. Neuroscience 117(3):715–722
31. Gilden D, Mahalingam R, Nagel MA, Pugazhenthi S, Cohrs RJ (2011) The neurobiology of varicella zoster virus infection. Neuropathol Appl Neurobiol 37(5):441–463
32. Gobel S (1984) An electron microscopic analysis of the trans-synaptic effects of peripheral nerve injury subsequent to tooth pulp extirpations on neurons in laminae I and II of the medullary dorsal horn. J Neurosci 4(9):2281–2290
33. Greenwood LF, Sessle BJ (1976) Inputs to trigeminal brain stem neurones from facial, oral, tooth pulp and pharyngolaryngeal tissues: II. Role of trigeminal nucleus caudalis in modulating responses to innocuous and noxious stimuli. Brain Res 117:227–238
34. Guedon JM, Yee MB, Zhang M, Harvey SA, Goins WF, Kinchington PR (2015) Neuronal changes induced by Varicella Zoster virus in a rat model of postherpetic neuralgia. Virol 482:167–180
35. Head H, Campbell AW (1900) The pathology of herpes zoster and its bearing on sensory localization. Brain 23:353–523
36. Hope-Simpson R.E. (1965) The nature of herpes zoster: a long term study and a new hypothesis (First Albert Wander Lecture) Proc. R Soc Med 58:9–20
37. Hu JW, Dostrovsky J, Lenz Y, Ball G, Sessle BJ (1986) Tooth pulp deafferentation is associated with functional alterations in the properties of neurons in the trigeminal spinal tract nucleus. J Neurophysiol 56:1650–1668
38. Hu JW, Sessle BJ (1989) Effects of tooth pulp deafferentation on nociceptive and nonnociceptive neurons of the feline trigeminal subnucleus caudalis (medullary dorsal horn). J Neurophysiol 61:1197–1206
39. Hu JW, Sharav Y, Sessle BJ (1990) Effects of one or two-staged deafferentation of mandibular and maxillary tooth pulps on the functional properties of trigeminal brainstem neurones. Brain Res 516:271–279
40. Iwata K, Imamura Y, Honda K, Shinoda M (2011) Physiological mechanisms of neuropathic pain: the orofacial region. Int Rev Neurobiol 97:227–250
41. Iwata K, Tsuboi Y, Tashiro A, Sakamoto M, Sumino R (1999) Integration of tooth-pulp pain at the level of cerebral cortex. In: Nakamura Y, Sessle BJ (eds) Neurobiology of Mastication – from Molecular to Systems Approach. Elsevier, Tokyo, pp. 471–481
42. Jannetta PJ (2011) Trigeminal neuralgia. Oxford University Press, Oxford
43. Jeon HJ, Han SR, Park MK, Yang KY, Bae YC, Ahn DK (2012) A novel trigeminal neuropathic pain model: compression of the trigeminal nerve root produces prolonged nociception in rats. Prog Neuropsychopharmacol Biol Psychiatry 38(2):149–158
44. Johnson LR, Westrum LE, Henry MA (1991) Anatomic organization of the trigeminal system and the effects of deafferentation. In: Fromm GH, Sessle BJ (eds) Trigeminal neuralgia: current concepts regarding pathogenesis and treatment. Butterworths, Stoneham, pp. 27–70

45. King RB (1967) Evidence for a central etiology of tic douloureux. J Neurosurg 26(1):Suppl:175–80.
46. Kobayashi I, Masamichi S, Sessle BJ, Honda K, Imamura Y, Hitomi S, Tsuboi Y, Okada-Ogawa A, Iwata K (2011) Mechanisms involved in extraterritorial facial pain following cervical spinal nerve injury. Mol Pain 7:12
47. Kwan CL, Hu JW, Sessle BJ (1993) Effects of tooth pulp deafferentation on brainstem neurons of the rat trigeminal subnucleus oralis. Somatosens Mot Res 10:115–131
48. Lavigne GJ, Sessle BJ (2016) The neurobiology of orofacial pain and sleep and their interactions. J. Dent. Res. In press.
49. Luiz AP, Kopach O, Santana-Varela S, Wood JN (2015) The role of Nav1.9 channel in the development of neuropathic orofacial pain associated with trigeminal neuralgia. Mol Pain 11(1):72
50. Luo DS, Zhang T, Zuo CX, Zuo ZF, Liu H, Wu SX, Wang W, Li YQ (2012) An animal model for trigeminal neuralgia by compression of the trigeminal nerve root. Pain Physician 15:187–196
51. Maarbjerg S, Wolfram F, Gozalov A, Oleson J, Bendtsen L (2015) Significance of neurovasscular contact in trigeminal neuralgia. Brain 138(2):311–319
52. Meloto CB, Smith S, Maixner W, Seltzer S, Diatchenko L (2014) Genetic risk factors for orofacial pain: Insights from human experimental studies. In: Sessle BJ (ed) Orofacial pain: recent advances in assessment, management and understanding of mechanisms. IASP Press, Washington, pp. 455–480
53. Oaklander AL (2008) Mechanisms of pain and itch caused by herpes zoster (shingles). J Pain 1(Suppl 1):S10–S18
54. Oaklander AL, Brown JM (2004) Unilateral nerve injury produces bilateral loss of distal innervation. Ann Neurol 55:639–644
55. Ossipov MH, Morimura K, Porreca F (2014) Descending pain modulation and chronification of pain. Curr Opin Support Palliat Care 2:143–151
56. Rappaport ZH, Devor M (1994) Trigeminal neuralgia: the role of self-sustaining discharge in the trigeminal ganglion. Pain 56(2):127–138
57. Robinson PP, Boissonade FM, Loescher AR, Smith KG, Yates JM, Elcock C, Bird EV, Davies SL, Smith PL, Vora AR (2004) Peripheral mechanisms for the initiation of pain following trigeminal nerve injury. J Orofac Pain 18(4):287–292
58. Rowbotham MC, Lottrup Petersen K, and Fields H L (1999) Is postherpetic neuralgia more than one disorder? IASP Newsletter, ,3–7.
59. Rowbotham MC, Yosipovitch G, Connolly MK, Finlay D, Forde G, Fields HL (1996) Cutaneous innervation density in the allodynic form of postherpetic neuralgia. Neurobiol Dis 3:205–214
60. Sakai Y, Nishijima Y, Mikuni N, Iwata N (1979) An experimental model of hyper-irritability in the trigeminal skin field of the rat. Pain 7(2):147–157
61. Seltzer Z, Mogil JS (2008) Pain and genetics. In: Sessle BJ, Lavigne GJ, Lund JP, Dubner R (eds) Orofacial pain: from basic science to clinical management, 2nd edn. Quintessence Publishing, Hanover Park, pp. 69–75
62. Sessle BJ (1987) The neurobiology of orofacial and dental pain. J Dent Res 66: 962–981
63. Sessle BJ (1991b) Physiology of the trigeminal system. In: Fromm GH, Sessle BJ (eds) Trigeminal neuralgia: current concepts regarding pathogenesis and treatment. Butterworths, Stoneham, pp. 71–104
64. Sessle BJ (2000) Acute and chronic craniofacial pain: brainstem mechanisms of nociceptive transmission and neuroplasticity, and their clinical correlates. Crit Rev Oral Biol Med 11:57–91
65. Sessle BJ (2001) Commentary on Gerhard Fromm's 1993 chapter. In: Watson CPN, Gershon AA (eds) Herpes zoster and postherpetic neuralgia, 2nd Revised and Enlarged Edition, Pain Research and Clinical Management, vol 11. Elsevier, Amsterdam, pp. 161–165
66. Sessle BJ (2011) Peripheral and central mechanisms of orofacial inflammatory pain. Int Rev Neurobiol 97:179–206

67. Sessle BJ (ed) (2014) Orofacial pain: recent advances in assessment, management and understanding of mechanisms. IASP Press, Washington, p. 509
68. Sessle BJ, Greenwood LF (1975) Effects of trigeminal tractotomy and of carbamazepine on single trigeminal sensory neurones in cats. J Dent Res 54:B201–B206
69. Sharav Y, Benoliel R (eds) (2008) Orofacial pain and headache. Mosby Elsevier, Toronto, p. 441
70. Tal M, Devor M (1992) Ectopic discharge in injured nerves: comparison of trigeminal and somatic afferents. Brain Res 579(1):148–151
71. Trousseau A (1853) De la nevralgie epileptiforme. Arch Gen Med 1:33–44
72. Tseng WT, Tsai ML, Iwata K (2012) Yen CT (2012) Long-term changes in trigeminal ganglionic and thalamic neuronal activities following inferior alveolar nerve transection in behaving rats. J Neurosci 32:16051–16063
73. Tsuzuki K, Fukuoka T, Sakagami M, Noguchi K (2003) Increase of preprotachykinin mRNA in the uninjured mandibular neurons after rat infraorbital nerve transection. Neurosci Lett 345:57–60
74. Warwick RA, Hanani M (2016) Involvement of aberrant calcium signalling in herpetic neuralgia. Exp Neurol 277:10–18
75. Watson CPN, Morshead C, Van der Kooy D, Deck JH, Evans RJ (1988) Postherpetic neuralgia: Post-mortem analysis of a case. Pain 34:129–138
76. Watson CPN, Midha R, Devor M et al (2000) Trigeminal postherpetic neuralgia postmortem : clinically unilateral pathologically bilateral. In: Devor M, MC R, Wiesenfeld – Hallin Z (eds) Proceedings of the 9th world congress on pain. Progress in pain research and management. IASP Press, Seattle, pp. 733–739
77. Weng HR, Lenz FA, Vierck C, Dougherty PM (2003) Physiological changes in primate somatosensory thalamus induced by deafferentation are dependent on the spinal funiculi that are sectioned and time following injury. Neurobiologia 116(4):1149–1160
78. Woda A (2003) Pain in the trigeminal system: from orofacial nociception to neural network modeling. J Dent Res 82(10):764–768
79. Young RF, King RB (1972) Excitability changes in trigeminal primary afferent fibers in response to noxious and nonnoxious stimuli. J Neurophysiol 35(1):87–95
80. Zhang GH, Lv MM, Wang S, Chen L, Qian NS, Tang Y, Zhang XD, Ren PC, Gao CJ, Sun XD, Xu LX (2011) Spinal astrocytic activation is involved in a virally-induced rat model of neuropathic pain. PLoS One 6:e23059
81. Zhuo M (2007) A synaptic model for pain: long-term potentiation in the anterior cingulate cortex. Mol Cells 23(3):259–271

Part IV
Postherpetic Neuralgia: Treatment

Chapter 16
Treatment of Postherpetic Neuralgia: Subtypes and a Mechanism-Based Treatment

Ralf Baron, Paul Möller, and Philipp Hüllemann

16.1 Introduction

Patients with pain after herpes zoster may suffer from numerous, different, spontaneous, and evoked pain-related symptoms. Despite this heterogeneity of sensory abnormalities, the patients' complaints are summarized in an umbrella diagnosis of "postherpetic neuralgia," and they are treated according to the overall spontaneous pain intensity.

Various authors in the last 20 years have claimed that this approach might not be successful and, as a new strategy, proposed a mechanism-based classification of neuropathic pain, i.e., grouping patients according to the mechanism of pain generation [13, 23]. This classification scheme should further help to establish an individualized pharmacological treatment of postherpetic neuralgia not only by identifying new therapeutic targets but also in describing which patients are likely to respond to a treatment [13, 23].

Animal research has shown that different pathophysiological mechanisms in the central or peripheral nervous system can independently or in combination cause various signs and symptoms of neuropathic pain. Clinical studies support the notion that similar mechanisms are involved in patients suffering from pain [20]. As examples it was shown that channels and receptors located on primary afferent nociceptors can be up-regulated after nerve injury, causing abnormal sensitivity and spontaneous activity. This is thought to be the cause of spontaneous pain, shooting pain sensations, and

R. Baron (✉)
Division of Neurological Pain Research and Therapy, Department of Neurology, University Hospital Schleswig-Holstein, Campus Kiel, 24105 Kiel, Germany
e-mail: r.baron@neurologie.uni-kiel.de

P. Möller • P. Hüllemann
Division of Neurological Pain Research and Therapy, Department of Neurology, University Hospital, Kiel, Germany

C.P.N. Watson et al. (eds.), *Herpes Zoster: Postherpetic Neuralgia and Other Complications,* DOI 10.1007/978-3-319-44348-5_16

heat hyperalgesia described by patients. Hypersensitivity to light touch or pinprick, i.e., dynamic mechanical or pinprick allodynia, occurs if input from mechanoreceptive A-fibers is felt as pain due to hyperexcitability of projection neurons in the spinal cord, a phenomenon called central sensitization [11, 19, 20, 22].

Since it is difficult to unravel mechanisms of pain in human patients, one has to rely on surrogate markers of mechanisms. One promising approach claims that the expression of sensory signs and symptoms, the so-called sensory profile, of a patient might be a reflection of the underlying mechanisms. These specific patterns of signs and symptoms could be compared with the knowledge derived from animal experiments where the association of signs and symptoms and underlying mechanisms has been elucidated. This concept has led to the development of a symptom-oriented diagnostic approach to neuropathic pain and postherpetic neuralgia that supplements the etiology-based classification scheme, which recognizes the fact that neuropathic pains are usually a composite of several pain symptoms. A symptom-oriented approach does not negate the fact that distinct neuropathies present differently clinically and that some neuropathic disease states may predispose to certain constellations of pain symptoms (e.g., touch-evoked pain in postherpetic neuralgia). The rationale of this approach recognizes several principles:

(1) Clinically distinct pain signs and symptoms such as ongoing stimulus-independent pain may be caused by similar, if not identical, neural mechanisms, even if the underlying neuropathies differ. (2) More than one pain mechanism is usually present in an individual patient. (3) Some signs and symptoms, such as mechanical hypersensitivity, can be explained by several distinct neural mechanisms that may even coexist in an individual patient.

As said, a symptom-based approach to neuropathic pain and postherpetic neuralgia can be useful for dissecting the underlying neural mechanisms, and this knowledge may eventually be harnessed for the development of novel analgesic drugs that differentially target these mechanisms.

Almost 20 years ago, the first attempt was made to differentiate patients with postherpetic neuralgia based on their thermal sensitivity of the skin, the cutaneous reaction to histamine, the results of skin biopsies, and the presence or absence of allodynia. Two main subtypes of patients were proposed: patients with irritable nociceptors and patients with deafferentation. The latter group was further divided into patients with and without dynamic mechanical allodynia [8] or in patients with decreased or increased pinprick sensation [4].

16.2 Where Do We Stand Today?

A mechanism-based classification and an individualized treatment approach of postherpetic neuralgia is still not part of a clinician's daily routine. Nevertheless, considerable progress has been made. In this article, we aim to outline the current state of research in patients with neuropathic pain with special emphasis on postherpetic neuralgia on the ambitious, yet promising, path toward an individualized pharmacological treatment.

Table 16.1 Frequency of different combinations of abnormal values in patients with PHN

Loss	Gain 0 (no)	1 (thermal)	2 (mechanical)	3 (both)	All
0	(0 %)	1 (1.4 %)	(0 %)	4 (5.6 %)	5 (6.9 %)
1	2 (2.8 %)	(0 %)	6 (8.3 %)	2 (2.8 %)	10 (13.9 %)
2	3 (4.2 %)	1 (1.4 %)	5 (6.9 %)	6 (8.3 %)	15 (20.8 %)
3	10 (13.9 %)	4 (5.6 %)	21 (29.2 %)	7 (9.7 %)	42 (58.3 %)
All	15 (20.8 %)	6 (8.3 %)	32 (44.4 %)	19 (26.4 %)	72 (100 %)

Maier et al. [12]
L0 no loss of detection, *L1* only thermal loss, *L2* only mechanical loss, *L3* mixed loss of detection, *G0* no gain (no hyperalgesia), *G1* with only thermal hyperalgesia, *G2* with only mechanical hyperalgesia, *G3* with both thermal and mechanical hyperalgesia

16.3 Classification of Patients Based on Sensory Abnormalities

The first step in establishing a mechanism-based classification scheme is the proof that patients can be subgrouped based on pain-related sensory abnormalities and based on their individual sensory profile.

In order to analyze the different sensory abnormalities and their combinations, several tools are available.

A standardized quantitative sensory testing (QST) protocol for clinical trials has been introduced by the German Research Network on Neuropathic Pain (DFNS) in 2006 as standardization is crucial in order to compare study results [18]. Sensory stimuli are applied to the skin or deep somatic structures to elicit a painful or non-painful sensation that can be quantified on a rating scale. QST uses a standardized battery of mechanical and thermal stimuli (graded von Frey hairs, several pinprick stimuli, pressure algometer, quantitative thermal testing, etc.) and assesses both minus signs (loss of function) and positive signs (gain of function) in the nociceptive and non-nociceptive afferent nervous systems.

With quantitative sensory testing (QST), relatively simple QST subgrouping schemes have been described focusing on the relative preservation or loss of small fiber functions (e.g., hypoesthesia to thermal stimuli [16]). One study introduced a novel classification and subgrouping method based on a large battery of QST results [12]. Similar to the tumor grading system, patients were classified according to loss or gain of function of their small and large afferent fibers. By this arbitrary organization, 12 different and relevant QST profiles for subgrouping of neuropathic pain patients were identified (Table 16.1). Although this sophisticated organizing principle allows a very precise characterization of the sensory function, it is difficult to use it in large clinical trials since the numbers of patients in a single subgroup tend to be small.

There are additional techniques to subgroup patients regarding their sensory QST profile. These studies have applied a hierarchical cluster analysis for statistical segmentation of patients. This approach identifies patterns or dimensions of sensory signs that occur most frequently without using a priori hypothesis or other

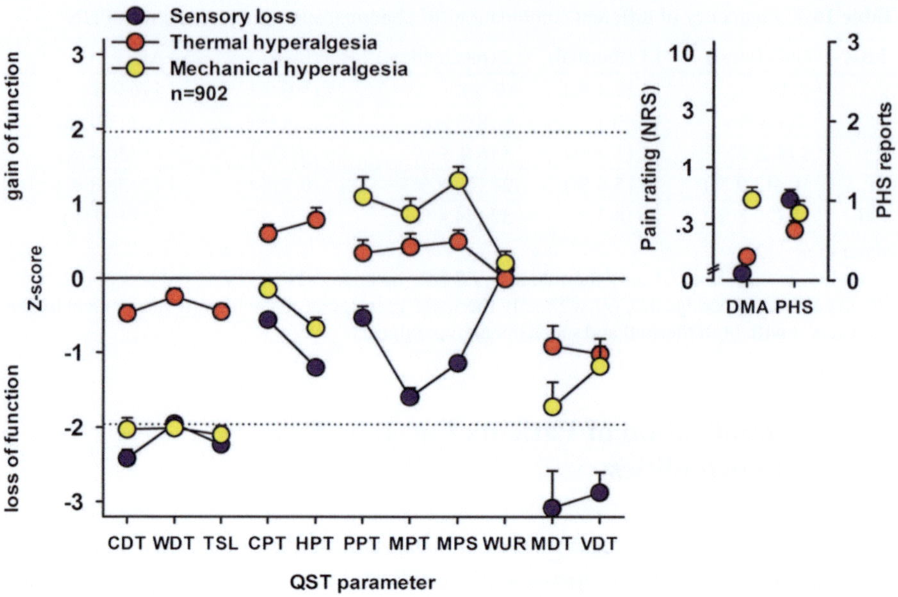

Fig. 16.1 Sensory profiles of patients with peripheral neuropathic pain. Sensory profiles of the three clusters presented as mean z-scores ± 95 % confidence interval (n = 902). Note that z-transformation eliminates differences due to test site, gender, and age. Positive z-scores indicate positive sensory signs (hyperalgesia); negative z-values indicate negative sensory signs (hypesthesia, hypalgesia). Dashed lines – 95 % confidence interval for healthy subjects ($-1{\cdot}96 < z < +1{\cdot}96$). Note that if the mean of a cluster is within the shaded area, this does not imply that it does not differ from a healthy cohort. Values are significantly different from those of healthy subjects, if their 95 % confidence interval does not cross the zero line. Inserts show numeric pain ratings for DMA: dynamic mechanical allodynia on a logarithmic scale (0–100) and frequency of PHS: paradoxical heat sensation (0–3). *Blue symbols*: subgroup "sensory loss." *Red symbols*: "thermal hyperalgesia." *Yellow symbols*: "mechanical hyperalgesia"

predefined assumptions. Using QST data from more than 900 patients with peripheral neuropathic pain, three distinct subgroups with characteristic sensory profiles were identified (Fig. 16.1). Subgroup 1 (thermal hyperalgesia) was characterized by preserved sensory functions in combination with thermal hyperalgesia and mild mechanical allodynia. Subgroup 2 (sensory loss) showed a loss of small and large fiber function. Subgroup 3 (mechanical hyperalgesia) demonstrated a loss of small fiber function in combination with pinprick hyperalgesia, marked dynamic mechanical allodynia, and deep somatic allodynia. About 31 % of the PHN patients fall into group 1, 22 % into group 2, and 47 % into group 3 (Baron et al. 2016).

Another possibility to asses symptoms in neuropathic pain patients are "patient-reported outcomes" (PROs, questionnaires such as the pain*DETECT* questionnaire (PDQ) [9] and the Neuropathic Pain Symptom Inventory (NPSI) [5]). Pain symptoms are evaluated directly by the patients and therefore can be used to phenotype patients on the basis of their perceived sensory abnormalities.

With this approach we found five distinct subgroups of sensory symptoms profiles assessed with pain *DETECT* in 498 patients with postherpetic neuralgia. In contrast to the QST, which evaluates stimulus-evoked sensations, questionnaires mainly assess spontaneous pain-related sensations such as burning pain or prickling [3].

16.4 Potential Mechanisms Operating in Different Subgroups

As an example the results of the QST-based dataset allow some speculation about potential underlying pathophysiological mechanisms operating in the different subgroups of patients.

16.4.1 Subgroup 1

The fact that in this subgroup the cutaneous innervation is relatively preserved despite documented nerve damage is somewhat surprising and clearly indicates that postherpetic neuralgia may be associated with hyperactive and irritable afferent neurons with only mild signs of degeneration. The pattern of symptoms of subgroup 1 is likely reflected by a pathological sensitization of the peripheral and central nervous system as demonstrated in animal research. Sensitized nociceptors are associated with pathological spontaneous discharges and develop a lowered activation threshold for thermal (heat and cold) and mechanical stimuli. Ongoing nociceptor hyperactivity produces secondary changes in the spinal cord dorsal horn, i.e., a prolonged enhancement of responses to afferent stimuli in multi-receptive spinal cord dorsal horn neurons, so that light tactile stimuli become capable of activating spinal cord pain signaling neurons via A-delta- and Aß-low-threshold mechanoreceptors. By this mechanism mechanical stimuli induce pain, i.e., pinprick and dynamic mechanical allodynia. It is interesting that these types of mechanical hyperalgesia are only present in about 20 % of the patients of subgroup 1. Obviously, peripheral nociceptor drive does not always induce central sensitization.

16.4.2 Subgroup 2

In this subgroup negative sensory signs of all sensory modalities dominate the clinical picture indicating severe deafferentation. These patients do not suffer from any type of evoked sensation except a mild dynamic mechanical allodynia in few patients. The spontaneous pain in this subgroup is likely generated in proximal sites, e.g., in the dorsal root ganglion or in deafferented central pain signaling neurons.

16.4.3 Subgroup 3

The third subgroup of patients has a profound sensory deficit for temperature stimuli (cold and heat) in combination with severe hyperalgesia or allodynia to mechanical pinprick stimuli and light touch. Furthermore, deep somatic hyperalgesia to mechanical pressure on muscles is more pronounced than in the other subgroups. This combination of thermal loss and mechanical gain occurs in the same innervation territory of the affected nerves. Whereas the physician may have difficulty recognizing this paradoxical finding, patients are even more confused by the complexity of their sensory experiences. This sensory profile indicates that cold- and heat-sensitive small fibers are impaired or degenerated within the skin regions with marked mechanical hyperalgesia and allodynia. It is most commonly present in patients with PHN (47 %) who characteristically suffer from severe allodynia.

The generation of the combination of thermal hyposensitivity and mechanical hypersensitivity was repeatedly explained by anatomical alterations in the synaptic structure of non-nociceptive mechanoreceptive central terminals after degeneration of thermoreceptive small fibers. In animal models of peripheral nerve injury, it was shown that preserved Aß-low-threshold mechanoreceptors that normally project into deeper layers of the spinal cord dorsal horn (lamina III–IV) sprout into more superficial nociceptive areas (lamina II) that are deprived of their normal C-fiber input and form abnormal connections with central pain transmission neurons. However, this interpretation was repeatedly questioned. Therefore, alternative explanations for pain and allodynia in the setting of impaired cutaneous small fiber function must be considered. Since cold and warm perceptions were only studied in the skin, ongoing activity that drives central sensitization and allodynia might originate in intact nociceptors of deep somatic tissues (e.g., muscle, ligaments, etc.). The finding that patients in this subgroup show signs of deep somatic hyperalgesia points to this possibility.

16.5 Treatment Studies Using Mechanism-Based Classification

The next important step in establishing an individualized, mechanism-based therapy is the proof that patients classified into different mechanistic groups respond differently to pharmacological pain treatments.

Retrospective analyses of signs and symptoms of neuropathic pain patients at baseline in several studies demonstrate that patients with distinct underlying mechanisms respond differently to a certain therapy.

In a double-blind, placebo-controlled study in 2004, Attal et al. found in 22 patients with pain due to traumatic nerve injury or postherpetic neuralgia that dynamic or static mechanical allodynia assessed by QST appeared to be predictive of the response to intravenously administered lidocaine [2]. In 2006, Edwards et al. observed in 64 patients suffering from postherpetic neuralgia that the heat pain threshold at baseline predicted the response to the treatment with opioids but not to tricyclics or placebo [7]. A study by Ranoux et al. examining the effect of intradermally injected botulinum A in 29 patients with postherpetic neuralgia or posttraumatic/postoperative neuropathy showed a correlation between the analgesic effect and preserved thermal sensibility at baseline, indicating intact cutaneous innervations [16]. In 2005, Wasner et al., examining the association between the treatment response to topical lidocaine and the function of thermosensitive and histamine-sensitive cutaneous afferents in patients with postherpetic neuralgia, even showed that there was a significant pain reduction in patients with impaired nociceptor function [21].

A study by Höper et al. investigating the relation between the response to topical treatment with capsaicin 8 % patches in patients with peripheral neuropathic pain and their symptoms as assessed by pain*DETECT* questionnaire at baseline could only show weak associations between pain reduction and high pain*DETECT* scores, burning pain, and pressure-evoked pain, respectively [10].

Post hoc stratification according to the sensory profile or sensory phenotype showed promising results for subgroups of patients. As a further step toward an individualized pharmacological treatment of neuropathic pain, studies **prospectively** stratifying patients according to their sensory symptoms and signs are needed [17].

Recently, Demant et al. examined in a randomized, double-blinded, placebo-controlled trial the pain-relieving effect of oxcarbazepine in 72 patients with postherpetic neuralgia, surgical or traumatic nerve injury, or polyneuropathy. They performed quantitative sensory testing in the beginning of the trial and stratified the patients according to their sensory profile into two following groups: (1) "irritable nociceptor" with predominantly a "gain of function" and a preserved small fiber nerve function and (2) "deafferentation type" dominated by sensory loss. This stratification is based on the assumption that ectopic activity from up-regulated sodium channels is mainly responsible for hyperalgesia ("irritable nociceptor"), and therefore oxcarbazepine as a sodium channel blocker should have an effect in these patients. Although oxcarbazepine is recommended as first-line therapy for trigeminal neuralgia, it plays a minor role for the treatment of other neuropathic pain syndromes due to controversial study results [1]. This study showed positive results and a treatment response depending on the sensory phenotype. For all patients the number needed to treat (NNT) for a 50 % pain relief was 6.9, and the NNT in the group with the "irritable nociceptor phenotype" was only 3.9, whereas for the "nonirritable nociceptor" phenotype the NNT was 13 [6].

16.6 Limitations of the Approach

As attractive a subtype classification based on sensory profiles might be, it should be emphasized that not all PHN patients fit exactly into one category or the other. It rather seems to be a continuum. Furthermore, in a large group of PHN patients, many heterogeneous patterns of sensory dysfunction were detected [14]. Accordingly, detailed testing of sensory function, chemical stimulation, and cutaneous innervation in one PHN patient clearly showed areas of relative preservation in close vicinity to impaired thermal sensation, both within the affected dermatome [15]. Furthermore, the sensory patterns showed a variation over the time course of PHN. However, by classification of PHN patients due to sensory abnormalities within the most painful skin area, it is possible to detect the predominant individual sensory profile and the most likely underlying pain-generating mechanism.

16.7 Future Implications for Therapy and Clinical Trials

The examples presented above highlight the importance of an individualized pain therapy to tailor the treatment to the right patients. The challenge to implement this individualized therapy concept is to identify appropriate subgroups of patients. At this point of time, the most promising approach to achieve this aim is to use the individual sensory profile as a surrogate of pain mechanisms. With QST and pain symptom questionnaires, very promising results have already been revealed. Other diagnostic techniques, like laser-evoked potentials, heat evoked potentials (CHEPS), functional imaging techniques, skin biopsies, and microneurography, are also potentially valuable tools and should be further explored.

Given the problem of negative outcomes in many recent trials in PHN and other neuropathic pain syndromes as well as the promising results of the symptom-based classification scheme so far, researchers and pharmaceutical companies are encouraged to implement this obvious and rational approach in future clinical trial design. A multiple step approach should be envisaged.

Step 1: Sensory profiling should be performed at baseline of clinical trials. Based on these results, a post hoc analysis of the responders will be calculated, and the results will be used to generate new hypotheses.

Step 2: Patients should be prospectively stratified according to the results of sensory profiling.

Step 3: The results have to be replicated in several independent large-scale trials to show that a study drug really performs superior in predefined sensory subgroups than in the entire cohort or in the alternative subgroups.

16.8 Conclusions

Sensory profiling of patients with postherpetic neuralgia is a promising technique to subgroup patients and to predict treatment response. This assessment should be implemented in future trial designs to ultimately proof or disproof the mechanism-based treatment concept. Furthermore, assessment tools have to be developed that simply and reliably identify subgroups of patients in the general practitioner setting. Only if these prerequisites are fulfilled, the method of targeting treatment to the appropriate patients will be implemented in the general practice.

References

1. Attal N, Cruccu G, Baron R, Haanpaa M, Hansson P, Jensen TS, Nurmikko T (2010) EFNS guidelines on the pharmacological treatment of neuropathic pain: 2010 revision. Eur J Neurol:Off J Eur Federation Neurolog Soc 17(9):1113–e1188. doi:10.1111/j.1468-1331.2010.02999.x
2. Attal N, Rouaud J, Brasseur L, Chauvin M, Bouhassira D (2004) Systemic lidocaine in pain due to peripheral nerve injury and predictors of response. Neurology 62(2):218–225
3. Baron R, Maier C, Attal N, Binder A, Bouhassira D, Cruccu G, Finnerup NB, Haanpää M, Hansson P, Hüllemann P, Jensen TS, Freynhagen R, Kennedy JD, Magerl W, Mainka T, Reimer M, Rice ASC, Segerdahl M, Serra J, Sindrup S, Sommer C, Tölle T, Vollert J, Treede RD. Peripheral neuropathic pain: a mechanism-related organizing principle based on sensory profiles. Pain, 2016, in press.
4. Baumgartner U, Magerl W, Klein T, Hopf HC, Treede RD (2002) Neurogenic hyperalgesia versus painful hypoalgesia: two distinct mechanisms of neuropathic pain. Pain 96(1–2): 141–151
5. Bouhassira D, Attal N, Fermanian J, Alchaar H, Gautron M, Masquelier E, Rostaing S, Lanteri-Minet M, Collin E, Grisart J, Boureau F (2004) Development and validation of the Neuropathic Pain Symptom Inventory. Pain 108(3):248–257. doi:10.1016/j.pain.2003.12.024
6. Demant DT, Lund K, Vollert J, Maier C, Segerdahl M, Finnerup NB, Jensen TS, Sindrup SH (2014) The effect of oxcarbazepine in peripheral neuropathic pain depends on pain phenotype: a randomised, double-blind, placebo-controlled phenotype-stratified study. Pain 2014 Nov;155(11):2263–73
7. Edwards RR, Haythornthwaite JA, Tella P, Max MB, Raja S (2006) Basal heat pain thresholds predict opioid analgesia in patients with postherpetic neuralgia. Anesthesiology 104(6): 1243–1248
8. Fields HL, Rowbotham M, Baron R (1998) Postherpetic neuralgia: irritable nociceptors and deafferentation. Neurobiol Dis 5(4):209–227. doi:10.1006/nbdi.1998.0204
9. Freynhagen R, Baron R, Gockel U, Tolle TR (2006) painDETECT: a new screening questionnaire to identify neuropathic components in patients with back pain. Curr Med Res Opin 22(10):1911–1920. doi:10.1185/030079906X132488
10. Hoper J, Helfert S, Heskamp ML, Maihofner CG, Baron R (2014) High concentration capsaicin for treatment of peripheral neuropathic pain: effect on somatosensory symptoms and identification of treatment responders. Curr Med Res Opin 30(4):565–574. doi:10.1185/03007995.2013.869491
11. Lai J, Hunter JC, Porreca F (2003) The role of voltage-gated sodium channels in neuropathic pain. Curr Opin Neurobiol 13(3):291–297
12. Maier C, Baron R, Tolle TR, Binder A, Birbaumer N, Birklein F, Gierthmuhlen J, Flor H, Geber C, Huge V, Krumova EK, Landwehrmeyer GB, Magerl W, Maihofner C, Richter H, Rolke R, Scherens A, Schwarz A, Sommer C, Tronnier V, Uceyler N, Valet M, Wasner G, Treede RD (2010) Quantitative sensory testing in the German Research Network on Neuropathic Pain (DFNS): somatosensory abnormalities in 1236 patients with different neuropathic pain syndromes. Pain 150(3):439–450. doi:10.1016/j.pain.2010.05.002
13. Max MB (1990) Towards physiologically based treatment of patients with neuropathic pain. Pain 42(2):131–137
14. Pappagallo M, Oaklander AL, Quatrano-Piacentini AL, Clark MR, Raja SN (2000) Heterogenous patterns of sensory dysfunction in postherpetic neuralgia suggest multiple pathophysiologic mechanisms. Anesthesiology 92(3):691–698
15. Petersen KL, Fields HL, Brennum J, Sandroni P, Rowbotham MC (2000) Capsaicin evoked pain and allodynia in post-herpetic neuralgia. Pain 88(2):125–133
16. Ranoux D, Attal N, Morain F, Bouhassira D (2008) Botulinum toxin type A induces direct analgesic effects in chronic neuropathic pain. Ann Neurol 64(3):274–283. doi:10.1002/ana.21427

17. Reimer M, Helfert SM, Baron R (2014) Phenotyping neuropathic pain patients: implications for individual therapy and clinical trials. Curr Opin Support Palliat Care 8(2):124–129. doi:10.1097/SPC.0000000000000045
18. Rolke R, Baron R, Maier C, Tolle TR, Treede RD, Beyer A, Binder A, Birbaumer N, Birklein F, Botefur IC, Braune S, Flor H, Huge V, Klug R, Landwehrmeyer GB, Magerl W, Maihofner C, Rolko C, Schaub C, Scherens A, Sprenger T, Valet M, Wasserka B (2006) Quantitative sensory testing in the German Research Network on Neuropathic Pain (DFNS): standardized protocol and reference values. Pain 123(3):231–243. doi:10.1016/j.pain.2006.01.041
19. Siqueira SR, Alves B, Malpartida HM, Teixeira MJ, Siqueira JT (2009) Abnormal expression of voltage-gated sodium channels Nav1.7, Nav1.3 and Nav1.8 in trigeminal neuralgia. Neuroscience 164(2):573–577. doi:10.1016/j.neuroscience.2009.08.037
20. von Hehn CA, Baron R, Woolf CJ (2012) Deconstructing the neuropathic pain phenotype to reveal neural mechanisms. Neuron 73(4):638–652. doi:10.1016/j.neuron.2012.02.008
21. Wasner G, Kleinert A, Binder A, Schattschneider J, Baron R (2005) Postherpetic neuralgia: topical lidocaine is effective in nociceptor-deprived skin. J Neurol 252(6):677–686. doi:10.1007/s00415-005-0717-z
22. Woolf CJ (2011) Central sensitization: implications for the diagnosis and treatment of pain. Pain 152(3 Suppl):S2–15. doi:10.1016/j.pain.2010.09.030
23. Woolf CJ, Bennett GJ, Doherty M, Dubner R, Kidd B, Koltzenburg M, Lipton R, Loeser JD, Payne R, Torebjork E (1998) Towards a mechanism-based classification of pain? Pain 77(3):227–229

Chapter 17
Interventional Approaches to Postherpetic Neuralgia

Shane E. Brogan and Perry G. Fine

17.1 Introduction

The role of interventional techniques including nerve blocks and neuromodulation in the management of pain due to acute herpes zoster (HZ) eruption and the prevention or control of postherpetic neuralgia (PHN) continues to be evolving. There has been very little done in the way of methodologically sound studies to clarify the role or value of interventional therapies, although many pain management physicians continue to prescribe nerve block therapies, spinal cord stimulation, and intrathecal therapy as preventative or therapeutic adjuncts.

With nerve blocks, the controversy over the timing (when and how often), type (what block with what drug(s)), and other variables is compounded by the potential risks and economic costs of such therapies. This latter but timely issue is not addressed in the medical literature dealing with the prevention and management of zoster-related pain, so there is little to draw upon in the way of references. Nevertheless, upon considering the literature that focuses on interventional therapies, it is important to keep this issue in mind. Balancing the potential risks versus benefits of such therapies in this light may help direct us toward reasonable therapeutic dispositions. For instance, due to the extremely debilitating nature of intractable postherpetic neuralgia, and the economic costs of chronic treatment, an intervention that is *possibly* effective but has some potential risk and is somewhat costly in the short run may be worth pursuing. This chapter explores these issues in depth and concludes with suggestions for dealing with the present controversies raised by this unresolved subject.

S.E. Brogan, MB, BCh (✉) • P.G. Fine, MD
Department of Anesthesiology, University of Utah Health Sciences Center, Salt Lake City, UT, USA
e-mail: shane.brogan@hsc.utah.edu; perry.fine@hsc.utah.edu

C.P.N. Watson et al. (eds.), *Herpes Zoster: Postherpetic Neuralgia and Other Complications*, DOI 10.1007/978-3-319-44348-5_17

17.2 General Considerations

In all branches of medicine, mythology and science are entwined. The lore of nerve blocks and other interventional techniques as potentially useful tools in the management, prevention, or treatment of pain due to herpes zoster is a richly laden case in point. Documentation of this form of therapy as a useful treatment modality dates back to the beginning of this century [21].

In view of the terribly demoralizing nature of this pain-producing disorder, it is easy to see, through the eyes of the sanguine, albeit, frustrated clinician, how hope and wishful thinking might promote inordinately positive claims and unproven conclusions. And, after all, who would be so cruel as to cast shadows with the skeptical lamp of empirical science, when there might be some chance, however remote, of benefit to the sufferer. It is within this quite understandable emotional context coupled with the difficulties inherent in carrying out prospective controlled clinical studies for this type of problem that, almost a century and many reports later, we have not moved beyond the limits of our impressions in this particular field.

Although unintended at its inception, in its evolution, this chapter may now appear to be a study in disillusionment for those who believe there is an empiric basis and well-defined role for the prescription of nerve block therapies in those with herpes zoster. Rigorous review and honest evaluation of the existing literature in its entirety challenges such notions. The art of medicine may be served by adhering to such beliefs and practices, but the science is not. And so, with sincere apologies to those firm believers, this chapter will serve more as a critique of inadequate study design and a plea for correcting such deficiencies in future studies, than as a clinical guide of what to do and how, when, and where to do it.

17.3 Nerve Blocks

17.3.1 Timing of Interventions

There are three time periods or phases during the presentation of the herpes zoster syndrome when useful treatment information regarding pain can be derived. These include treatment of pain during the acute infectious/inflammatory period, preemptive treatments during the acute period to prevent chronicity, and treatment of protracted neuralgic pain. Nerve block therapies have been applied with a view toward each of these.

17.3.2 Acute Pain Control

Rosenack [30] reported the results of treating 22 patients with symptoms of herpes zoster for 2 to 14 days. He performed somatic or sympathetic nerve blocks with procaine and described 90 % success in achieving pain relief and vesicular healing after 48 h. No long-term follow-up is provided. Street [32] and Findley and Patzer [12] report a total of seven cases who obtained marked pain relief and rapid healing from sympathetic block using procaine. Similarly, Ferris and Martin [11] used stellate ganglion and lumbar sympathetic blocks in 22 patients with acute zoster. Of the 20 patients seen within 5 days of symptoms, complete pain relief and crusting over of lesions after one or two blocks is reported. One of the two patients who were treated 4 weeks into their disease had similar results, but the other did not respond to nerve blocks. Again, no long-term follow-up is provided.

Using steroid infiltration around the painful lesions in lieu of local anesthetic, Lefkovits [20] described complete or marked relief in four patients with symptoms for less than 3 months. No follow-up information is provided.

Colding has submitted two reports in the English language literature investigating the efficacy of sympathetic blocks in controlling pain during the acute phase of zoster. In the first [5], 243 patients were treated with regional sympathetic blocks, and results were analyzed according to the duration of precedent symptoms. In 204 patients with zoster for less than 14 days, 15 % experienced no relief compared to 30 % in the 38 patients who had symptoms longer than 14 days. Incomplete relief in these two groups was 23 % and 26 %, respectively. In his subsequent report [6], he reviews the results of sympathetic blocks in 483 patients with acute zoster pain. He found that 10 % of patients with symptoms less than 2 weeks old fail to obtain pain relief compared with a 60 % nonresponse rate in those treated who had symptoms longer than 2 weeks. Similar results were reported by LaFlamme et al. [19]. These investigators used sympathetic blocks to treat acute zoster pain of the trigeminal distribution. Three of five patients studied had complete relief after 5–12 blocks, applied at 2-day intervals.

Riopelle et al. [29] studied the effects of different types of nerve blocks (epidural, stellate ganglion, peripheral nerve) in 72 patients with symptoms of herpes zoster for 30 days or less. Pain relief was described as prompt with all these techniques, and relief lasted for hours to days. Patients could return as often as desired for repeat blocks. Patients' clinical courses were followed for 6 months. Bonica and Buckley [4] gave a testimonial of complete relief of pain during the acute phase of zoster in 126 patients treated with several nerve blocks daily followed by intermittent blocks thereafter. Unfortunately, this is an unpublished account without extensive details.

Perkins and Hanlon [25] describe the use of serial epidural injections of local anesthetic with steroid in 12 patients with acute zoster pain. They report virtually complete relief of pain with this method.

Using a rather novel technique at the time, Marmer [22] provided a single case report of continuous epidural analgesia in a 70-year-old man for the control of acute zoster pain. This treatment lasted 11 days with reportedly excellent pain control, but no follow-up information was provided. Using a different continuous technique, Reiestad et al. [28] report the efficacy of local anesthetic-mediated analgesia using pleural catheters in 18 patients with acute zoster. Catheters remained in place for 2–3 weeks during the acute eruptive phase. No significant complications arose.

What can be concluded from these reports? Most evident is the fact that none of these represent controlled clinical trials. A variety of techniques are described with varying degrees of efficacy reported, but there are no controls against which one can attribute specificity to the technique in question. Similarly, the nerve blocking or infiltrating methods described are not compared to other conventional treatment modalities such as peripheral- or central-acting analgesics, either by efficacy, cost, or risk. Nevertheless, in severe cases where the risk of other treatments or their lack of efficacy exceeds that of nerve blocking, it may be quite reasonable to proceed with these more invasive means.

Contrary to the frequently encountered yet unsubstantiated statements which appear in the literature, there is no well-demonstrated "technique of choice" for this indication. Based upon the known and hypothetical mechanisms by which acute zoster produces pain, the location and severity of symptoms, the underlying health status of the patient, the experience of the treating physician, and the resources available, a rational plan can be determined. In these cases, the best measure of "success" will be the subjective response of the patient and objective functional changes that are pertinent and measurable. Placebo-controlled randomized clinical trials in clinically similar populations or in blinded "double-dummy" crossover trials will be the only means of discerning the relative merits of any technique.

17.3.3 Preemptive Nerve Blocks to Prevent Postherpetic Neuralgia

Two essential criteria which need to be met in order to assess the efficacy of nerve block therapy as a prophylaxis against persistent neuralgia following herpes zoster infection are (1) the performance of nerve block(s) during the acute phase of the disease and (2) follow-up for a period of time long enough to determine whether postherpetic neuralgia has developed. Unfortunately, there are few studies in the literature which meet even these minimal criteria.

Colding's reports (see [5, 6]) are the earliest which setout to answer this question. Combining results, in those patients who provided information 5–6 months after onset of zoster symptoms and treatment with sympathetic blocks, 10–20 % had continued pain. Follow-up information from less than one third of the initial study population was obtainable, and this sample was not selected at random. This introduces an insurmountable source of bias. In addition, stratification by age is not reported. Since it is well established that advancing age is a significant factor in persistent pain after zoster, lumping results without regard to age is not illuminating. Thus, these data cannot be critically compared to other studies evaluating the "natural history" of the disease process. Without concurrent controls, conclusions regarding treatment specificity, even under better sampling conditions, are not valid.

The nonrandomized, uncontrolled studies from Perkins and Hanlon [25] using epidural anesthetics with steroids; Riopelle et al. [29] using sympathetic, epidural, and peripheral nerve blocks; Milligan and Nash [23] using stellate ganglion blocks; Dan et al. [7] using various nerve block techniques; Reiestad et al. [28] using continuous pleural analgesia; and the unpublished accounts of Bonica (see [4]) suffer from the same limitations.

Yanagida et al. [35] retrospectively analyzed the incidence of postherpetic neuralgia in two groups of patients receiving stellate ganglion, epidural, or caudal blocks, depending upon the location of their acute zoster pain. One group received blocks prior to eruption of the typical zoster rash, and the other received treatment immediately after the rash appeared. Surveys were completed after 1 year by 91.8 % and 83.2 % of these two groups, respectively. There was no difference in the incidence of persistent pain between groups. The authors conclude that preemptive blocks early in the course of the disease do not prevent the syndrome of postherpetic neuralgia. Although this statement would appear to be an honest appraisal, the study is flawed in its design. Without concurrent controls, any difference in outcome compared with the natural history as it might unfold in these patients is impossible to assess.

This study was criticized by Bauman [3]. He takes issue with the conclusion that preemptive blocks do not contribute significantly to a reduction in the incidence of postherpetic neuralgia, citing his own experience to the contrary. His criticism is based mostly upon technical questions surrounding the performance of the nerve blocks. Although these points are debatable, this type of attack is misplaced, distracting from the real issue of drawing conclusions from noncontrolled, nonrandomized, retrospective data. Yanagida's conclusions may or may not be true. However, technical disqualifications and the refutation of a study do not serve as proof of equally unsubstantiated opinions.

Tenicela et al. [33] published a study which may shine some light on this question. In a randomized, double-blind, placebo-controlled, crossover trial, the investigators studied 20 patients over the age of 50 with herpes zoster for less than 6 weeks. Subjects were treated with sympathetic blocks or placebo blocks during 4 consecutive days. All patients with continued pain went on to have another series of blocks with known active drug (bupivacaine). There was no other therapy with the exception of the occasional use of low doses of oral analgesics. Follow-up was via questionnaire every 2 months for 1 year and then at 6-month intervals. The groups were demographically similar, and there was one subject from the placebo group who, although there was no substantial relief of pain, refused further blocks.

Ninety percent of subjects from the sympathetic block group had long-term pain relief, whereas only 20 % from the placebo group had initial pain relief. About 55 % of the remaining control group went on to have long-term resolution of pain after sympathetic blocks. This left 30 % from the initial control group who continued to have long-term pain. Again, the only difference in this group was the use of placebo blocks for 4 days prior to sympathetic blocks. This translates into a 25 % postherpetic neuralgia incidence overall. The limitation of this study is in the numbers of subjects, especially when stratified by age. Were the differences in numbers of subjects from each group who went on to have postherpetic neuralgia a function of the interventions, or a function of coincidental uneven distribution of predisposed patients in the two groups, in spite of randomization? When the differences boil down to one patient versus three patients, as was the case in this study, larger numbers are required in order to be conclusive.

17.3.4 Treatment of Postherpetic Neuralgia with Nerve Blocks

Published accounts of injection therapy for postherpetic pain date back to Russell et al. [31]. These authors describe the use of various techniques, including "injection therapy," in 100 patients with postherpetic neuralgia and present a few descriptive cases. There is no data given, and no formal study invoked. Lefkovits [20] used local infiltration of steroid in three patients with symptoms of 5 months, 2 years, and 5 years duration. The shorter-term patients are reported to have had either complete or partial relief, with indeterminate results in the patient with symptoms for 5 years. There is no information given regarding long-term results, and there were no controls.

Colding [5] studied the effect of sympathetic blocks in 34 patients with PHN for an average of 2 years. Thirteen patients were available for follow-up 5–6 months later, and none had complete pain relief, although three reported less pain than pre-block. Four years later, the same author reported on the results of treating 67 patients with postherpetic pain (2 months to 11 years duration) with up to 10 sympathetic blocks (see [6]). He reports that 50 % of patients experienced short-term pain relief. He presented follow-up information on 34 of these patients, 6–12 months after treatment. Half of this follow-up group experienced no pain relief with block therapy, but two patients from this group experienced spontaneous remission of pain. Of the group that had experienced some pain relief (17 patients), about 30 % had maintained complete relief of painful symptoms, while 50 % had gone back to their previous pain levels, and the remainder reported some improvement compared to pre-block levels. Putting this all together reveals ten patients out of the follow-up group of 34 with some perceived long-term pain reduction (two spontaneous, eight attributed to blocks), or a figure of 29 % improvement. It needs to be recalled that this figure is generated from a follow-up sample of a nonrandomized, noncontrolled treatment group.

Perkins [25] treated five patients with PHN with 1–3 epidural injections consisting of local anesthetic and steroid. One patient reported 50 % pain reduction several months later, and the others reported 0–25 % reduction in pain. The authors conclude that this form of therapy is ineffective in treating postherpetic neuralgia. Forrest [14] also tested epidural steroids for the control of PHN, but came to a different conclusion. Thirty-seven patients with pain for more than 6 months were enrolled after screening with psychometric tests, drug detoxification, and an appropriate response to a graduated epidural analgesia test. Subjects received weekly epidural injections of methylprednisolone acetate for 3 weeks. Follow-up occurred for 12 months. Eighty-nine percent were reported being pain-free at the end of the follow-up period. Although this is an intriguing finding, what is missing is the use of concurrent controls assigned with randomization. This is needed in order to attribute this result specifically to the therapeutic intervention with any degree of statistical certitude.

Other case reports using techniques such as gasserian ganglion block with steroid (see [34]), cryoanalgesia (see [2, 16]), and stellate ganglion block (see [19, 23]) have either failed to prove efficacy or have been inconclusive. Although they suggest potentially useful ideas, they universally fail to follow a randomized controlled methodology.

As well, anecdotal reports of techniques such as stellate ganglion phenol neurolysis (see [26]) or epidural phenol injections (see [27]) for this indication are suspect. Without data being provided in well-controlled studies, and without publication of such trials in peer-reviewed journals, it is not possible, no less responsible, to assign any scientific merit to such suggestions. Fair evaluation by clinicians and the patients they serve is simply not possible under these circumstances.

In one case of seemingly intractable postherpetic pain, the use of local anesthetic with the opioid, fentanyl, in a stellate ganglion block, repeatedly led to protracted relief of several week's duration (see [13]). Although only a single case study, the study design (double-blind, crossover, placebo controlled) adds credibility to the findings. The use of an opioid to supplement the effects of local anesthetic is a novel approach. This is predicated upon the finding of opioid receptors within ganglionic tissue and the notion that some degree of sympathetically maintained pain may be involved in the pathophysiology of pain in some patients with PHN. Arias et al. [1] have demonstrated a similar result, using sufentanil, in the treatment of refractory reflex sympathetic dystrophy. In the case of the individual referenced above, the use of opioid stellate ganglion blocks was not enduring. Until the time of his natural death, he was still burdened by the slow recurrence of pain for which he elected to undergo a block almost every month.

17.4 Neuromodulation

Neuromodulation refers to highly technical interventional pain management options, including intrathecal therapy and spinal cord stimulation (SCS). Because these techniques are costly and require extensive physician training and expertise, they are typically only provided in tertiary referral centers or highly specialized practices, and for patients who have failed all other management strategies. Consequently, given that neuromodulation is usually offered late, its role has been confined to the management of PHN, though acute zoster pain has also been treated with SCS.

As discussed elsewhere in this text, there is broad agreement that PHN results from a complex interaction of both ganglionic and more central abnormal neuronal activity. Therefore, it should not be too surprising that peripheral nerve blocks, as discussed above, have limited success in improving longer-term outcomes. Neuromodulation techniques, on the other hand, work more centrally and therefore offer a greater *theoretical* advantage over peripheral techniques. However, as will be discussed for each modality below, robust data to support this assumption are lacking, but studies to date are perhaps marginally better than for peripheral nerve block techniques.

17.4.1 Spinal Cord Stimulation

Spinal cord stimulation, also known as dorsal column stimulation, is an advanced pain management technique used to treat refractory neuropathic and visceral pain. Specialized electrodes are placed in the posterior epidural space and used to electrically stimulate the dorsal columns and other spinal cord elements. The electrodes can be introduced in two ways: first, using percutaneous needle access to the epidural space and, second, using a neurosurgical technique involving a laminotomy. The percutaneous technique typically allows for two SCS leads to be placed, each housing four to eight electrodes. The leads can be steered, using fluoroscopic guidance, to the area of the spinal cord that subserves the painful area. Each electrode can be designated a cathode or an anode, and electrical impulse parameters such as amplitude, pulse width, and frequency can be readily programed to provide a paresthesia in the painful region. This relatively pleasant paresthesia "masks" the nociceptive signaling and offers analgesia in the stimulated region. Electrophysiologically, SCS is thought to primarily act by stimulating large fiber sensory afferents in the dorsal columns of the spinal cord, inhibiting small fiber nociceptive activity, and possibly by enhancing the intrinsic descending inhibitory pathways in the spinal cord. [24] Typically, a percutaneous trial using an external pulse generator is used as a test of efficacy for a number of days. If the trial is deemed successful, a permanent generator (similar in size to a pacemaker) is placed subcutaneously for long-term use.

SCS has good evidence for the management of leg and back pain associated with failed back surgery syndrome, and in complex regional pain syndrome. In postherpetic neuralgia there are numerous positive case reports and one prospective trial.

Harke et al. [15] prospectively followed 28 patients with PHN treated with SCS over a median period of 29 months. All patients had been suffering from PHN for over 2 years and had failed standard neuropathic therapies for PHN. Interestingly, only patients with intact sensation in the painful dermatome were studied – patients with loss of sensation were deemed to have deafferentation pain and were not included. Long-term pain relief was reported in 23 (82 %) subjects.

An unnecessary methodological flaw in this study was that outcome data was only presented for the 23 responders, showing the median visual analogue scale dropping from 9 to 1 ($p < 0.001$). Functional gains were also noted in this group, as measured by a marked improvement in the pain disability index. Thirteen of the 23 long-term responders did not require any analgesics at follow-up. Notwithstanding the aforementioned reporting flaw, this study provided some promising data on the efficacy of SCS in PHN. However, in the absence of a randomized trial, with better controls and reporting parameters, it is difficult to argue a strong case for SCS in all cases. Nevertheless, when PHN has failed appropriate medical management, consideration should be given to an SCS trial.

The Harke paper also presented some limited data on four patients with *acute* herpes zoster (HZ) that was refractory to standard therapy – all patients reportedly had near complete pain relief, and the SCS was discontinued after several months.

An advantage of SCS therapy is that patients may be offered a "test-drive" percutaneous trial for 5–7 days, encouraging resumption of normal activities to fully assess the analgesic and functional benefits. If the trial is unsuccessful, the leads are removed in the office, and there is no commitment to proceed to the implantation of the definitive system.

17.4.2 Intrathecal Therapy

Intrathecal therapy (ITT) involves the delivery of drugs directly into the cerebrospinal fluid surrounding the spinal cord where they can act directly on central nervous system receptors with minimal systemic activity. A catheter is placed in the intrathecal space and tunneled under the skin, then connected to a subcutaneous electronic pump that is fully programmable and refillable. Opioids are typically the mainstay of ITT, but adjuncts such as bupivacaine and clonidine are also used, offering a theoretical advantage in the treatment of neuropathic pain conditions. The conotoxin ziconotide is a unique N-type calcium channel blocker that is approved for intrathecal use in refractory pain and may be considered as a first-line intrathecal agent in refractory neuropathic pain conditions such as PHN [8].

There are case reports and two small case series [10, 36] showing benefit, but unfortunately there are no prospective studies on ITT in PHN.

ITT for PHN may be considered when pain is suboptimally managed with more conventional approaches or when there are unacceptable side effects related to standard analgesic approaches or contraindications to SCS.

17.4.3 Radiofrequency Procedures

An emerging area in the field of interventional pain medicine is the use of radiofrequncy (RF) techniques to alter neural function. Before proceeding to RF, diagnostic test injections with local anesthetic are typically performed on the target nerve to assess for analgesic and functional efficacy. Using specialized needles capable of emitting radiofrequency energy from the tip, discrete lesions of neural targets may be achieved. A thermistor in the tip of the probe allows for precise monitoring and control of temperature. Modern RF units also allow for 50 and 2 Hz testing stimulation to stimulate sensory and motor nerve fibers, respectively, and allow for optimum needle tip positioning.

High-energy lesions induce a thermal lesion causing a controlled area of heat destruction and loss of axonal function. A less destructive technique is the use of pulsed radiofrequency (PRF) which emits energy in short bursts, allowing for cooling in between bursts, and limiting temperature rise to 42°C. While controversial, it is claimed that PRF is less anatomically destructive, yet still attenuates neural nociceptive function, but importantly, has minimal effect on motor fibers. [9].

RF has an established role in the treatment of spinal degenerative joint pain through the denervation of the medial branch nerves that innervate the facet joints. Other pain indications are less well established, but there has been some work in post herpetic neuralgia.

Kim et al. [17] prospectively studied 49 patients with PHN refractory to conservative therapy. Using fluoroscopic guidance, pulsed radiofrequency lesioning of the dorsal root ganglia of the corresponding painful level was performed, and the patients were reassessed at 4, 8, and 12 weeks. Although there was no control group, at 4 weeks there was a 55 % pain reduction (mean VAS reduced from 7.2 +/−1.8 to 3.4+/−1.5, $p < 0.05$), and this response was sustained until the 12-week visit.

Ke et al. [18] used a prospective, randomized, partially blinded design to compare PRF of the intercostal nerves to a sham injection (though the sham procedure was not described) in 96 patients with PHN. The affected PHN level, plus a level above and below, were lesioned, and the two groups were followed for 6 months. No complications were reported. A significant reduction in visual analog pain scores was observed with the treatment effect waning toward 6 months. The Short-Form Health Survey 36 (SF36) demonstrated significant functional improvement in several domains during the 6-month follow-up period. Medication usage was also slightly decreased in the treatment group.

17.5 Conclusions

Given that the randomized controlled trial is the gold standard for clinical studies, a certain form of analytical alchemy would be required to infer a well-defined role for interventional therapies in the management of pain due to zoster, based upon the available literature. This rather bleak summary notwithstanding certain positive conclusions are possible. Clearly, during the acute phase of zoster infection, nerve blocks can be used to provide acute pain relief. This is not a surprising result, given that much of the pain experienced during the acute phase of the virus is inflammatory, activating usual nociceptive pathways that can be interrupted by various nerve blocking techniques. Weighing risks and benefits, costs, availability of resources, and patient acceptance, nerve blocks may provide a viable alternative to other medical therapies. This places herpes zoster in the same domain as many other acutely painful conditions (infectious, traumatic, postoperative, and so forth) which might be palliated by this treatment modality. And, in a similar vein to acute pain management for other conditions, the circumstances that would point to a specific type of nerve block as the *best* choice have not been clearly defined in the literature. Until the time that this is the case, such decisions would appear to be a matter of personal choice between patient and physician.

Scientifically based data convincingly supporting the efficacy of nerve blocks for either prevention or long-term treatment of postherpetic neuralgia are not available. Regrettably, some conclusions drawn from the existing literature can be misleading. Statements such as "instant relief from pain and rapid healing of lesions" (see [12]), "epidural injection of bupivacaine with or without methylprednisolone acetate is the treatment of choice for the pain of cutaneous herpes zoster" (see [25]), or "the use of sympathetic block, alone or combined with somatic nerve blocks, if begun early after the onset of pain or the eruption of acute herpes zoster, results in a high incidence of prompt pain relief, appears to decrease the severity and duration of the eruption, and accelerates healing" (see [4]) are unsubstantiated. From an academic viewpoint, these must be regarded with healthy skepticism and held up to scientific scrutiny. These hopeful and tempting, but unproven, messages cannot yet serve as standards for routine clinical practice.

From a practical standpoint, is there nonetheless a justification for performing nerve blocks in an effort to prevent or treat the horrible aftereffects of zoster infection? Might this be especially reasonable in the susceptible elderly? Should the anecdotal experiences and impressions of many clinicians be discounted simply because of a non-compelling literature? These are important questions, and they can only be answered within the context of one's own ethical deliberations. Without a solid scientific literature to back clinical decisions, clinicians must rely on practicality, balancing of ethical principles (informed consent, nonmaleficence, beneficence), and judgment.

With the advent of more sophisticated techniques such as intrathecal therapy, spinal cord stimulation, and radiofrequency ablation, an interesting set of pathophysiologic questions is presented. It is appealing to think that these techniques that modulate neural function in the central nervous system, where the pathology of PHN putatively resides, may be more efficacious. As the literature stands, spinal cord stimulation and radiofrequency ablation techniques have some encouraging prospective data, but larger, controlled, and randomized studies are needed to confirm and strengthen initial results.

Patients with PHN are understandably desperate and ingenuous; compassionate physicians are compelled to do what they can to help. Within the limits of our present knowledge and capabilities, the irreducible processes of caring for patients and maintaining the integrity of medicine are best served by taking each patient's circumstances, goals, and values into account, providing fair and ample counsel, and then applying the most safe, least costly, and *potentially* effective therapies in successive order.

References

1. Arias LM, Schwartzman RJ, Bartkowski R, Tom CM, Grossman KL (1989) Sufentanil stellate ganglion injection in the treatment of refractory reflex sympathetic dystrophy. Reg Anesth 14:90–92
2. Barnard D, Lloyd J, Evans J (1981) Cryoanalgesia in the management of chronic facial pain. J Max Fac Surg 9:101–102
3. Bauman J (1987) Prevention of postherpetic neuralgia. Anesthesiology 67:441–442
4. Bonica JJ, Buckley FP (1990) Regional analgesia with local anesthetics. In: Bonica JJ (ed) The management of pain, vol 2, 2 edn. Lea and Febiger, Philadelphia/London, pp. 1938–1939

5. Colding A (1969) The effect of regional sympathetic blocks in the treatment of herpes zoster. Acta Anaesthesiol Scand 13:133–141
6. Colding A (1973) Treatment of pain: organization of a pain clinic: treatment of acute Herpes zoster. Proc R Soc Med 66:541–543
7. Dan K, Higa K, Noda B (1985) Nerve block for herpetic pain. In: Fields HL (ed) Advances in pain research and therapy, vol 9. Raven Press, New York, pp. 831–838
8. Deer TR et al (2012) Polyanalgesic consensus conference 2012: recommendations for the management of pain by intrathecal (intraspinal) drug delivery: report of an interdisciplinary expert panel. Neuromodulation 15(5):436–464
9. Erdine S, Yucel A, Cimen A (2005) Effect of pulsed versus conventional radiofrequency current on rabbit dorsal root ganglion morphology. Eur J Pain 9:251–256
10. Fabiano AJ, Doyle C, Plunkett RJ (2012) Intrathecal medications in post-herpetic neuralgia. Pain Med 13(8):1088–1090
11. Ferris LM, Martin GH (1950) The use of sympathetic nerve block in the ambulatory patient with special reference to its use in Herpes Zoster. Ann Intern Med 32:257–260
12. Findley T, Patzer R (1945) Treatment of herpes zoster by paravertebral procaine block. JAMA 128:1217–1219
13. Fine PG, Ashburn MA (1988) Effect of stellate ganglion block with fentanyl on postherpetic neuralgia with a sympathetic component. Anesth Analg 67:897–899
14. Forrest JB (1980) The response to epidural steroid injections in chronic dorsal root pain. Can Anaesth Soc J 27:40–46
15. Harke H, Gretenkort P, Ladleif HU, Koester P, Rahman S (2002) Spinal cord stimulation in postherpetic neuralgia and in acute herpes zoster pain. Anesth Analg 94:694–700
16. Jones MJT, Murrin KR (1987) Intercostal block with cryotherapy. Ann R Coll Surg Engl 69:261–262
17. Kim YH, Lee CJ, Lee SC, Huh J, Nahm FS, Kim HZ, Lee MK (2008) Effect of pulsed radiofrequency for postherpetic neuralgia. Acta Anaesthesiol Scand 52(8):1140–1143
18. Ke M1, Yinghui F, Yi J, Xeuhua H, Xiaoming L, Zhijun C, Chao H, Yingwei W (2013) Efficacy of pulsed radiofrequency in the treatment of thoracic postherpetic neuralgia from the angulus costae: a randomized, double-blinded, controlled trial. Pain Physician 16(1):15–25. Pain Med 14(12):1944–1953
19. LaFlamme MY, LaBreque B, Mignault G (1979) Zona ophtalmique: traitement de la nevralgie zonateuse par infiltrations stellaires repetees. Can J Ophthalmol 14:99–101
20. Lefkovits AM (1961) Postherpetic neuralgia. Neurology 11:170–171
21. Loeser JD (1986) Herpes zoster and postherpetic neuralgia. Pain 25:149–164
22. Marmer MJ (1965) Acute Herpes zoster: successful treatment by continuous epidural analgesia. Calif Med 103:277–279
23. Milligan NS, Nash TP (1985) Treatment of post-herpetic neuralgia. A review of 77 consecutive cases. Pain 23:381–386
24. Oakley JC, Prager JP (2002) Spinal cord stimulation mechanisms of action. Spine 27(22):2574–2583
25. Perkins HM, Hanlon PR (1978) Epidural injection of local anesthetic and steroids for relief of pain secondary to Herpes zoster. Arch Surg 113:253–254
26. Racz GB, Holubec JT (1989) Stellate ganglion phenol neurolysis. In: Racz GB (ed) Techniques of neurolysis. Kluwer Academic Publishers, Boston, p. 140
27. Racz GB, Heavner J, Haynworth R (1989) Repeat epidural phenol injections in chronic pain and spasticity. In: Racz GB (ed) Techniques of neurolysis. Kluwer Academic Publishers, Boston, pp. 205–206
28. Reiestad F, Kvalheim L, McIlvaine WB (1989) Pleural analgesia for the treatment of acute severe thoracic herpes zoster. Reg Anesth 14:244–246
29. Riopelle JM, Naraghi M, Grush KP (1984) Chronic neuralgia incidence following local anesthetic therapy for herpes zoster. Arch Dermatol 120:747–750
30. Rosenak S (1938) Procaine injection treatment of Herpes Zoster. Lancet 5:1056–1058

31. Russell WR, Espir MLE, Morganstern FS (1957) Treatment of post-herpetic neuralgia. Lancet 2:242–245
32. Street A (1943) Use of sympathetic nerve block in herpes zoster, Bell's palsy. Mississippi Doctor 20:480–481
33. Tenicela R, Lovasik D, Eaglstein W (1985) Treatment of Herpes zoster with sympathetic blocks. Clin J Pain 1:63–67
34. Yamashiro H, Hara K, Gotoh Y (1990) Relief of intractable postherpetic neuralgia with gasserian ganglion block using methyl prednisolone. Masui 39:1239–1244
35. Yanagida H, Suwa K, Corssen G (1987) No prophylactic effect of early sympathetic blockade on postherpetic neuralgia. Anesthesiology 66:73–76
36. Zacest A, Anderson VC, Burchiel KJ (2009) The glass half empty or half full-how effective are long-term intrathecal opioids in post-herpetic neuralgia? A case series and review of the literature. Neuromodulation 12(3):219–223

Chapter 18
Treatment of Postherpetic Neuralgia: The Role of Opioids

Jane C. Ballantyne

18.1 Introduction

In many clinical situations, there is no medication that controls pain better than an opioid. Its effects are immediate, based in key areas of the central nervous system that process pain, and produce a dose-dependent reduction in pain intensity whatever the source or type of pain. Thus, opioids have a time-honored place in the treatment of severe acute pain and in the treatment of pain at the end of life. In these situations, other medications can be used as adjuncts, to improve pain relief and reduce opioid doses, but there is no complete substitute for an opioid that will effectively and quickly reduce pain's intensity and render pain tolerable. Medicine has come to accept the supreme role for opioids for treating severe time-limited pain and the duty of clinicians to provide opioids in such a situation. What is much more difficult is what the role is for opioids in treating pain that is not time limited – chronic pain. It is always tempting to want to prolong opioid therapy on the basis that its early effects are dramatic; neither patients nor providers want to go back to pain remembered from before opioids were initiated. And it has been argued recently that centuries' old caution about opioids' addictive properties was not justified and should not interfere with attempts to relieve pain, even chronic pain.

Pain medications are often chosen on the basis of how they perform in randomized controlled trials (RCTs). We use RCTs to approve drugs for marketing, as the basis for clinical guidelines and as a basis for clinical practice. Neuropathic pain was traditionally thought to be nonresponsive to opioids. But more recently (in the mid-2000s), although there was some suggestions that higher doses may be needed than for nociceptive pain, RCTs showed good sensitivity of neuropathic pain to

J.C. Ballantyne
Department of Anesthesiology and Pain Medicine, University of Washington, Seattle, WA 19128, USA
e-mail: jcb12@uw.edu

C.P.N. Watson et al. (eds.), *Herpes Zoster: Postherpetic Neuralgia and Other Complications,* DOI 10.1007/978-3-319-44348-5_18

opioids [1–4]. Thus, opioids were included as recommended treatment for neuropathic pain in guidelines. They did not replace drugs that had previously been adopted as first-line treatments – antidepressants and anticonvulsants [5] – but they were recommended as "second-line treatment that can be considered for first-line use in select clinical circumstances" [6]. This recommendation has changed more recently, since newer methods for summarizing and reporting evidence (Grades of Recommendation Assessment, Development and Evaluation [GRADE] methodology) are more sensitive in balancing risk against benefit, and opioids for neuropathic pain were relegated to third-line treatment largely on the basis of safety concerns [7]. The newer guidelines also state "strong opioids should be strictly monitored, particularly for patients requiring high doses." Numbers needed to treat (NNTs) were 3.6 for tricyclic antidepressants, 4.3 for strong opioids, 4.7 for tramadol, 6.4 for SNRIs, 7.2 for gabapentin, and 7.6 for pregabalin, making strong opioids second only to tricyclic antidepressants in terms of efficacy, according to RCTs. Studies contributing to the newer systematic review assessed neuropathic pain broadly, which is pain caused by any lesion or disease of the somatosensory system (IASP definition) [8]. However, the most common neuropathic pain conditions studied were diabetic neuropathy and postherpetic neuralgia (PHN). Although a mechanism-based – rather than disease-based – approach might be more appropriate when selecting primary, non-opioid medications, the ability of opioids to produce a dose-dependent reduction in pain intensity regardless of the type of pain means that for opioids, findings in neuropathic pain in general are probably as applicable to postherpetic neuralgia as to any other type of neuropathic pain. For the purpose of this chapter, many of the principles described will apply to neuropathic pain in general as well as to PHN. The chapter will focus mostly on the issues surrounding the treatment of chronic pain with opioids. The role of opioids in multimodal pain therapy during the acute phase of herpes zoster is more straightforward.

Unfortunately, the idea that opioids are a reasonable third-line medication for the treatment of chronic neuropathic pain oversimplifies the complexity of opioid decision-making for chronic pain. What we see as the result of opioid therapy in trials is not always reflected in what we see in real life. Safety is hard to evaluate in trials because trial subjects are highly selected to predict both efficacy and safety: patients with known risk factors are excluded. Many adverse outcomes are only manifest after months or years of therapy and not during the conduct of trials, most of which last only weeks. Efficacy is not always maintained over time due to factors such as tolerance, dependence, hyperalgesia, and loss of placebo [9]. Overall, the results may not be as robust as those demonstrated in trials, and thus the principle of embarking on chronic opioid therapy simply because other treatments have failed or are inadequate may need further thought. The suitability of opioids for chronic pain depends on many factors other than whether or not other treatments have provided relief. It depends also on the degree to which pain interferes with an individual's ability to function, and enjoy life. It depends on an individual's risk of developing an opioid use disorder. It depends on the pain prognosis and likely progression. It depends on the personal goals set for opioid treatment and whether they are likely to be met. It depends on accepting a certain amount of risk and balancing risk against uncertain benefit. The remainder of this chapter will explore what we know

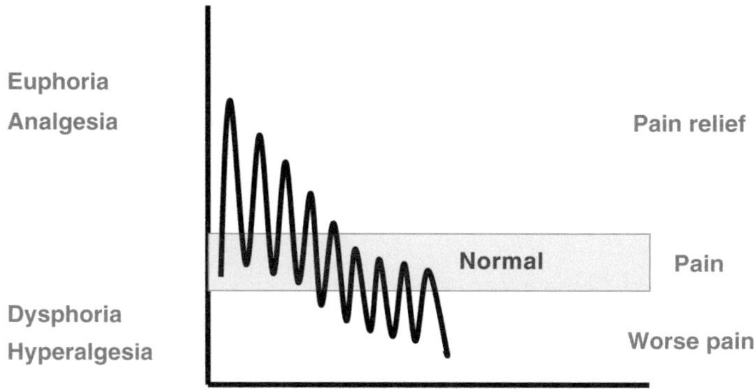

Fig. 18.1 Natural progression of opioid dependence. For illicit opioid users, as dependence develops, opioids no longer produce euphoria but are needed to avoid dysphoria. A similar phenomenon occurs with analgesia. Opioids produce good analgesia when first used, but in time, opioids are needed to avoid withdrawal and associated hyperalgesia

about chronic opioid actions and the adaptations that affect efficacy and harm, which patients are suitable candidates for chronic opioid therapy, whether there are limitations in terms of opioid selection or dose, how to protect against harm, and finally, what are the controversies that surround opioid pain treatment.

18.2 Adaptations to Chronic Opioid Use

It is an old observation that people who become "habituated" to opioids feel better if they continue to take opioid. To the point that their doses become very high and toxic, continued opioid therapy is helpful. What was not understood, until relatively recently (mid-1990s), is why opioids produce the state, whereby with continued use, opioids are needed in order to feel normal [10] (Fig. 18.1). What is now understood is that through their actions on an endogenous opioid system, opioid drugs produce cellular and molecular adaptations in the nervous system that are manifested as two phenomena that are crucial to the changes that are commonly seen with continued and continuous opioid use – loss of analgesic efficacy and worsening pain if opioids are withheld. These two phenomena are tolerance and dependence.

18.2.1 Tolerance

Tolerance arises to both the analgesic and euphoric properties of opioids, albeit by different mechanisms. Tolerance can be described as the need to get a higher dose to achieve the same effect. Importantly, tolerance is both a pharmacological phenomenon and a psychological phenomenon. A number of different mechanisms are proposed as the underlying basis for pharmacological or *nonassociative tolerance*, including receptor internalization, recycling, desensitization, or downregulation [11]. The psychological component, *associative tolerance*, is perhaps more important during the treatment of pain with opioids because it is amenable to treatment in a way that nonassociative tolerance is not. Associative tolerance can involve learning and is changed by environment or circumstance [12–14].

18.2.2 Physical Dependence

Dependence on opioids is manifested as the need to continue opioid in order to feel normal and prevent withdrawal. Withdrawal from opioids results in a classic withdrawal syndrome (agitation, nausea and diarrhea, abdominal pain, pupillary dilation, goose flesh, shivering, pain). For pain patients, the fact that pain worsens during withdrawal (a phenomenon known as *withdrawal hyperalgesia*), contributes to the desperation opioid-treated pain patients feel if they cannot obtain enough opioid to meet their needs [15, 16]. This can also be, and often is, misinterpreted as needing opioid to treat the underlying pain. Although the pain of withdrawal is a whole body muscular pain reminiscent of influenza or fibromyalgia, and therefore not necessarily the same as the underlying pain, opioid withdrawal can alter underlying pain, including neuropathic pain and hyperalgesia, by mechanisms that are not fully understood [17]. It should be noted here that there is also a psychological component to dependence, but unlike *associative tolerance* and *physical dependence,* the manifestations are entirely psychological. Dependence may be purely physical and relatively easy to reverse when opioid treatment has not been prolonged, but when the treatment is prolonged, dependence can become complex and persistent [18].

18.2.3 Tolerance and Dependence Cannot Be Separated

We have seen that the mechanisms underlying analgesic tolerance are complex. Importantly, tolerance is not simply produced because of continued use but can also be altered by a patient's mood or circumstance. Any increase in tolerance, even if imperceptible, could result in withdrawal symptoms, unless met with a dose adjustment. This means that tolerance can be manifested as symptoms of withdrawal and that it is impossible to separate tolerance from dependence (Fig. 18.2). Tolerance and dependence should be considered linked phenomena [18].

18.2.4 What Is Addiction?

Because tolerance and physical dependence are inevitable consequences of continuous opioid use, there has been a school of thought that suggests that for opioid-treated pain patients, tolerance and dependence should not be considered criteria for

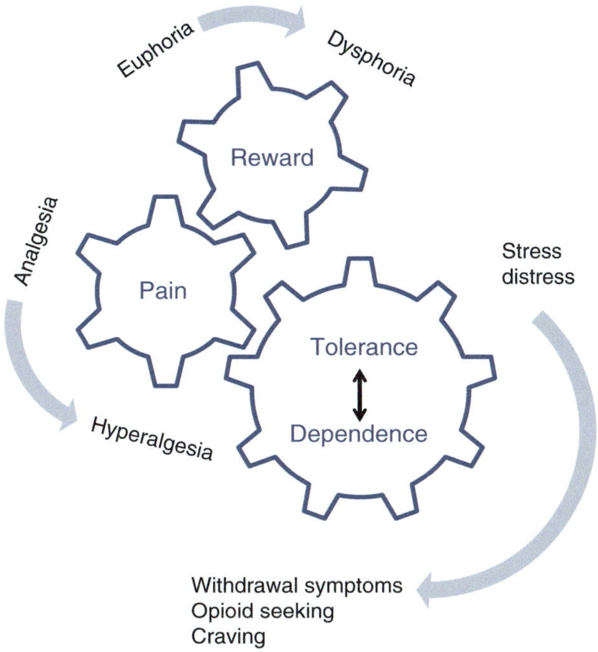

Fig. 18.2 Interdependence of mood, tolerance/dependence, and pain. Even in normal individuals, pain and mood are interdependent, in part through endogenous opioid mechanisms. Individuals who are taking exogenous opioids continuously over the long term adapt by developing tolerance and dependence. Psychological factors can alter tolerance and thereby induce withdrawal symptoms. For dependent individuals, the need for more opioids becomes the predominant reaction to stress. Although pain is seen as the primary reason to dose escalate, dependence may underlie the need for opioid (Figure was published Ballantyne et al. [18])

addiction. In other words, during opioid pain treatment, tolerance and dependence can occur without addiction therefore should not be included in addiction criteria. So the thinking runs [19]. This conflicts with established thinking about addiction wherein tolerance and physical dependence are fundamental components. Considering that tolerance and dependence are linked and that withdrawal symptoms are powerful drivers of opioid seeking [14], this separation between tolerance, physical dependence, and addiction in pain patients is somewhat contrived. Addiction arises when opioid-seeking behaviors are combined with opioid use, and such behaviors become established and irreversible (as in memory) [14, 20]. With long-term use, it is withdrawal that drives opioid seeking not the quest for euphoria, since loss of euphoric effect (and pain relief) is one of the consequences of continued use (Fig. 18.1). But if a patient is seeking opioids for the treatment of pain, even if the behaviors appear aberrant, is this addiction or simply a need for pain relief? This question presents us with significant difficulty when it comes to establishing rates of iatrogenic addiction [21, 22], the need for specific addiction treatment in any individual, the suitability of continuing opioids in the face of aberrant behaviors, and risk analysis. It is highly likely, in the view of this author, that underlying any aberrant behavior, whether focused on procuring illicit opioids or focused on

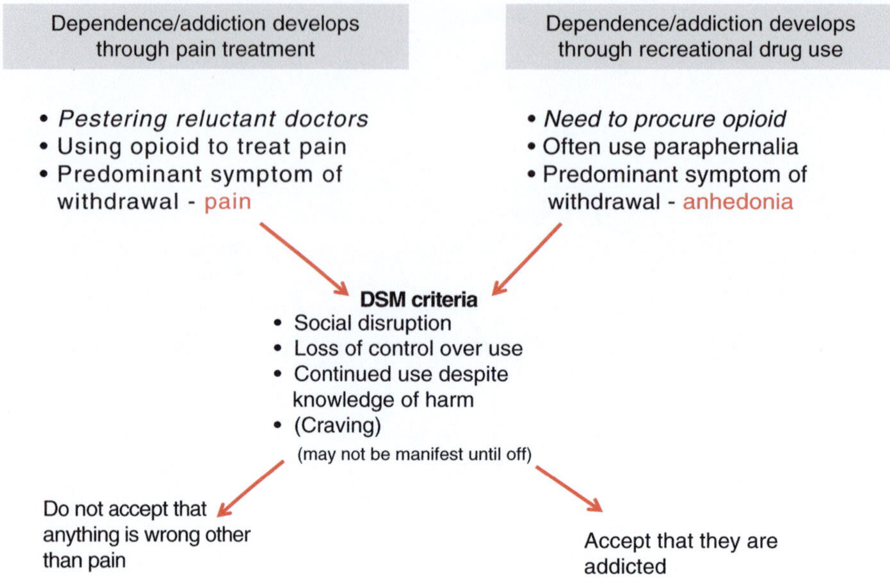

Fig. 18.3 Opioid-seeking behaviors. Although there are some behaviors shared between pain patients and recreational opioid users (*center of figure*), there are also behaviors that are markedly different (*top of figure*). Quite apart from wanting to avoid stigma, the fact that opioid-seeking behaviors could be accounted for by the existence of pain even when they meet criteria for addiction makes it difficult to make the diagnosis of addiction in pain patients. The predominant clinical challenge for those treating dependent or addicted pain patients is that patients often have difficulty accepting that anything is wrong other than pain. Those seeking addiction treatment, on the other hand, have already accepted that they need treatment for addiction (*bottom of figure*). *DSM* diagnostic and statistic manual of mental disorders of the American Psychiatric Association [65]

getting doctors to prescribe opioids, there is a similar neuroadaptive process [10]. This may be true whether the behaviors are typical procurement behaviors (meeting DSM or ICD ten criteria for addiction) or whether they consist of pestering reluctant doctors (Fig. 18.3). Thus, dependence on opioid analgesics may not look like addiction as defined by addiction criteria but is similar enough to be difficult to distinguish, to warrant similar treatment, and to interfere with the success of chronic opioid therapy (Fig. 18.4).

18.2.5 Implications

The important take-home message is that tolerance and dependence (dependence that may or may not reach criteria for addiction) are the root causes of chronic opioid treatment failure. The patient who is failing chronic opioid treatment needs frequent dose increases, often to the point of toxicity, does not get good pain relief, and yet has great difficulty with tapering. Withdrawal symptoms, including

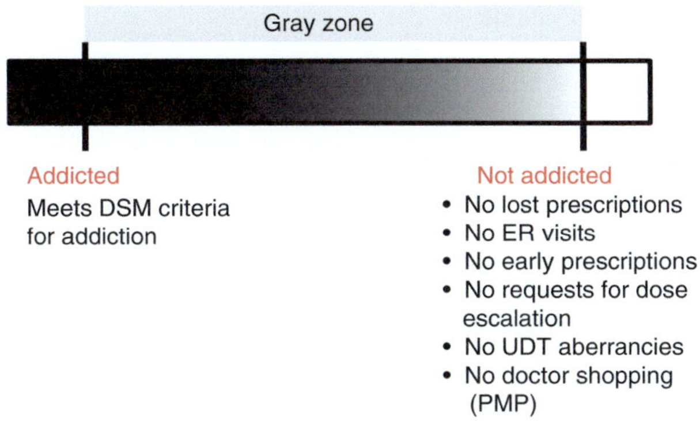

Fig. 18.4 The gray zone between clearly addicted and not addicted. The vast majority of patients seen in the clinic are neither clearly addicted nor clearly not addicted. It is difficult to determine addiction rates, risks, and when an individual warrants treatment and possible opioid discontinuation. *ER* emergency room, *UDT* urine drug testing, *PMP* prescription monitoring program

hyperalgesia, are interfering with the ability to taper. The individual cannot conceive of being better without opioids and is trapped by the expectation based on early efficacy that there is a dose that will improve the pain.

18.3 Opioid Harms

There are two types of opioid harms: side effects and the much more concerning long-term harms that tend to develop insidiously. Side effects of opioids are well known and need not be fully explored in this context. These are listed in Table 18.1 together with recommended treatment approaches. It should be mentioned that tolerance develops to most side effects (bowel effects being the exception), including that most concerning of all side effects – respiratory depression. This means that with chronic use, side effects generally subside leaving clinically useful analgesia without bothersome side effects. A laxative regime for all chronic opioid users is recommended, however. The elderly and frail tend to be highly sensitive to opioid side effects, and particularly to cognitive impairment, which may actually worsen rather than subside with chronic use. Since side effects are dose dependent, elderly and frail patients warrant cautious dosing. Respiratory depression is not generally a problem for chronic opioid users, but it is important to recognize that respiratory depression can cause hypoxic brain injury or even death and is a lurking problem should overdose occur.

Other harms are not strictly side effects but rather the consequences of long-term opioid use. They develop insidiously, sometimes imperceptibly, so that it may be difficult to identify the time of onset or the point at which to intervene. Dependence and addiction are the most important and may underlie all other long-term harms. The dependent or addicted patient is less likely to be able to stop opioid therapy even if it is causing harm and is much more likely to lose control of usage and possibly

Table 18.1 Management of side effects

Constipation	Prophylactic mild peristaltic stimulant (e.g., bidocodyl or senna)
	If no bowel movement for 48 h, increase dose of bowel stimulant
	If no bowel movement for 72 h, perform rectal exam
	If not impacted, provide additional therapy (suppository, enema, magnesium etc.)
Nausea or vomiting	Consider prophylactic antiemetic therapy
	Add or increase non-opioid analgesic
	Decrease opioid dose, pain permitting
Sedation	Eliminate causes other than opioid
	Eliminate other CNS depressants where possible, especially benzodiazepines
	Add or increase non-sedating, non-opioid analgesics
	Add stimulant such as caffeine
	Change opioid
Pruritus	Consider treatment with antihistamines Change opioid
Hallucinations, delirium, or dysphoria	Evaluate underlying cause
	Eliminate other CNS-acting medications where possible
	May need to discontinue opioid at least temporarily to assess baseline
	May need antipsychotic
Sexual dysfunction	Reduce dose
	Testosterone replacement therapy may be helpful for men
	Erection-enhancing medications (e.g., sildenafil)

suffer accidental overdose and death. Related harms include high rates of endocrinopathy and immune dysfunction, cognitive impairment, danger with operating machinery or driving, higher than normal rates of falls and fractures, and less likelihood of returning to function or work. Not surprisingly, all these adverse outcomes are directly correlated with prolongation of opioid therapy [23]. What was not confirmed until recently is that these adverse outcomes are also directly correlated with dose [24–28]. Multiple studies are now showing that adverse outcomes are concentrated at doses of over 100 mg morphine equivalent daily dose (Table 18.2).

The North American experience with increased opioid prescribing should probably be mentioned here. As has been widely publicized in the lay and medical press, prescription opioid sales (thus opioid prescribing) increased severalfold in the USA and Canada between 1990 and 2010 and produced parallel increases in related abuse treatment admissions and deaths [24, 29, 30] (Fig. 18.5). Both the number of patients treated and the doses used account for the increased sales. The expansion in patient numbers treated occurred largely because of new teaching recommending opioids for chronic pain, whereas before the 1990s, opioids were considered neither

Table 18.2 Longer duration and higher-dose associations

Longer duration and higher doses are associated with
Higher rates of dependence and addiction
Higher rates of overdose and death
Less likelihood of being able to wean if necessary
Higher rates of mental health and substance use disorder, less able to control usage
Higher rates of falls and fractures in the elderly
Less likelihood of returning to function or work
Higher rates of endocrinopathy affecting fertility, libido, and drive
Higher rates of immune dysfunction

effective nor safe for chronic pain. It is important to reiterate that many of the adverse outcomes, particularly dependence, addiction, and death, are tending to occur in high-dose users and less in the majority low- to moderate-dose users [26–28, 31, 32]. Almost all the evidence of long-term harm comes from population data examined retrospectively. Until prospective studies (which are likely to be observational given the difficulty of studying long-term outcomes in RCTs) can be completed [33, 34], it will remain hard to pinpoint exactly which factors account for the increases in abuse and death. What is already clear is that concomitant use of CNS sedatives, particularly benzodiazepines, adds significant risk.

18.4 Which Patients Are Suitable Candidates?

Herpes zoster (HZ) presents with prodromal, acute, subacute, and chronic phases [35] (Fig. 18.6). Considering that pain can be severe and disabling in each of the acute, subacute, and chronic phases, opioids may be needed during each of these phases. There is some evidence that pain relief can reduce the likelihood of progression to PHN, although it is not clear whether any single analgesic is protective [6, 7]. Postherpetic neuralgia occurs in 10–34 % of HZ cases [36, 37]. Any hesitation about using opioids to treat HZ pain is due to the safety considerations that were discussed in Sect. 18.3, knowing also that opioid efficacy may diminish over time, as discussed in Sect. 18.2. Before starting opioids, whether for acute or longer-term use, it is always a good idea to make some assessment of addiction risk, however rudimentary. A majority (at least 60 %) of the patients who are started on opioids will abandon the treatment, often due to lack of efficacy or unacceptable side effects [38]. Still more decline the treatment. Those that continue on opioids represent a demographic with increased risk warranting special caution. More details on precautions are detailed in Sect. 18.6.

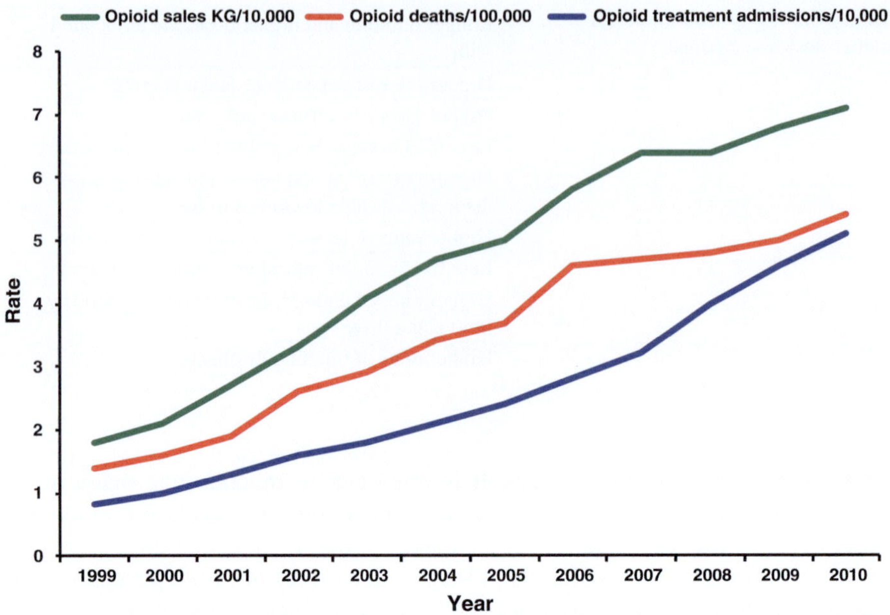

Fig. 18.5 Rates of opioid overdose deaths, sales, and treatment admission, USA, 1999–2010 (Note that growth in prescription opioid abuse treatment rates and prescription opioid related deaths increased in parallel with the growth in opioid sales. CDC. *MMWR* 2011. http://www.cdc.gov/mmwr/preview/mmwrhtml/mm60e1101a1.htm?s_cid=mm60e1101a1_w. Updated with 2009 mortality and 2010 treatment admission data [61])

	Rash onset	30 days	3 months	6 months
Varicella zoster virus reactivation	Rash healing			Pain can persist for several years
Prodromal symptoms 7–10 days	Acute herpetic neuralgia	Subacute herpetic neuralgia	PHN	Well established PHN

Fig. 18.6 Natural progression of herpes zoster and postherpetic neuralgia (PHN) (Adapted from Jeon [35])

18.4.1 Prodromal Phase

Approximately 75 % of herpes zoster presents with pain and other sensory changes before the appearance of the rash. Early intervention with an antiviral agent has been shown to reduce acute HZ pain and the occurrence of PHN. In the rare cases where the prodrome is correctly diagnosed, an antiviral agent may be all that is needed to control pain.

18.4.2 Acute Phase

Again, an antiviral agent is the first-line therapy for acute HZ pain. Combining this with oral or intravenous corticosteroids significantly improves the pain relief but does not reduce the likelihood of developing PHN. Neural blockade, including sympathetic block, may also be useful. Recently, the value of epidural local anesthetic plus corticosteroid in treating pain and preventing PHN has been recognized [39–41]. Oral analgesics may be needed if these primary approaches do not provide adequate relief. As with all acute pain, opioids have an important role for treating refractory pain. Even so, because it is unknown whether or not acute HZ pain will become chronic, it is advisable to reserve strong opioids as a last resort and be cognizant of opioid risks especially in patients with risk factors. It is reasonable to base the choice of oral analgesic for acute pain on the WHO stepladder approach, starting with acetaminophen or nonsteroidal anti-inflammatory drugs (NSAIDs), adding adjuvants such as antidepressants or an anticonvulsant if necessary and then progressing first to “mild” opioids (e.g., opioid-acetaminophen or NSAID combination, or tramadol) and finally to a strong opioid such as morphine or oxycodone [42]. Since many patients with HZ are elderly or immunosuppressed, this may limit the use of NSAIDs.

18.4.3 Subacute Phase

Continued efforts must be made to try and prevent PHN. Whatever nerve blocks have been helpful during the acute phase can be repeated to the limits of acceptable dosing, especially for corticosteroids. If opioids were not needed during the acute phase, it would rarely be necessary or advisable to start them during the subacute phase. Per the recommendations for the treatment of PHN, antidepressants and anticonvulsants should be optimized. This may also be the time to consider discontinuing NSAIDs and systemic corticosteroids, which are not good long-term treatments because of their side effects. If opioids were needed during the acute phase, it is reasonable to start a conversation about whether they should be tapered or continued, based on whether or not the pain is improving. Since the considerations for chronic use of opioids are very different than for acute opioid therapy, this is the time that the conversation about opioid limitations and risks should take place with the patient and family, and the time that safety precautions should be set in motion as described in Sect. 18.6 if the decision is to continue the opioid.

18.4.4 Postherpetic Neuralgia (PHN)

A lot of chronic opioid therapy starts by default – acute opioid therapy continues on to become chronic opioid therapy – and that is especially true for PHN because it develops from an acute pain syndrome. Occasionally, however, the need for opioid treatment arises after the onset of PHN as the pain becomes more difficult to tolerate and begins to interfere with function and enjoyment of life. Opioid therapy helps in a unique way. Not only does it relieve pain it also has a feel-good effect that blunts distress. As we know, opioids are supremely helpful in people who are dying, for the same reasons. But if opioids are used for prolonged periods, especially if they are taken continuously (i.e., round the clock), the adaptations described in Sect. 18.2 begin to interfere. Pain relief is not as good, higher doses are needed, and dependence makes it hard to stop treatment even in the face of diminishing benefit. It becomes increasingly clear that the people who do best with long-term opioid are those that do not develop problems with controlling usage and use of opioids at low to moderate doses, as needed and not round the clock [43, 44].

Because of the aforementioned concerns that long-term opioid therapy has reduced efficacy and increased risk, opioids are considered third-line treatment for PHN. Whether opioids are already being used or are being considered for first usage once the diagnosis of PHN has been made, it is helpful to do a thorough review of risks and benefits for the individual and to have an in-depth conversation with the patient and family about the limitations of chronic opioid therapy before it becomes established. Opioid decisions should be shared. Low-risk patients can be reassured that provided they take their opioids as prescribed and keep them in a safe place, there is little risk to them or others. The threshold for treating higher-risk patients may be higher, and more active safety measures may be needed (see Sect. 18.6), but high risk does not preclude receiving chronic opioid therapy when warranted. In fact, HZ patients tend to be older and sicker than normal, and this is a population with generally low risk of dependence, addiction, and treatment deterioration. It is also the population that tends to progress to PHN [36].

18.5 Which Opioid, Which Dose?

There is no intention to describe individual opioids in detail here. Such information can be found in any standard pharmacology texts, package inserts, pharmacopeias, or reliable web sites. The aim of this section is to first describe and delineate categories of opioids (Table 18.3) and then make recommendations regarding the dose.

18.5.1 "Mild" Opioids

In many cases, the so-called "mild" opioids are actually strong opioids combined with mild analgesics such as acetaminophen or an NSAID, which produces a ceiling dose and places them in less-restrictive categories in drug schedules. Tramadol

Table 18.3 Broad categories of opioid choices

"Mild analgesics"	As a class, these opioids are considered "weak" either because side effects limit dose or because they are not full mu opioid agonists. May be useful as the first opioid. Easier to prescribe than pure mu opioid agonists because they are not as restricted in drug regulations
Codeine	Dose generally limited by constipation. Available in combination drugs. Metabolized to morphine except in non-metabolizers in whom it is ineffective
Combination opioid plus acetaminophen or NSAID	There are dose limitations for acetaminophen and NSAID that produce a ceiling dose for these combination drugs
Tramadol	Mild mu agonist with additional norepinephrine and serotonin reuptake inhibition. Specific benefit for neuropathic pain and grouped a second-line choice. Available as a long-acting preparation
Tapentadol	Novel mu opioid agonist with additional norepinephrine reuptake inhibition. Expensive. RCTs in diabetic neuropathy but not PHN. Conceptually useful for PHN but evidence inconclusive and therefore not recommended in guidelines
Pure mu opioid agonists	Strong opioids, most common are morphine, hydrocodeine, oxycodone, dilaudid, methadone, and fentanyl
Short acting	Generally preferred, taken as needed and not round the clock
Long acting	May be preferred for patients with control issues, including risk factors for abuse or confusion (e.g., elderly). Taken round the clock
Short with long acting	In the view of this author, there are no indications for combing long- and short-acting opioids except at the end of life
Mixed agonist antagonists (e.g., pentazocine, butorphanol, nalbuphine)	Apart from buprenorphine, these have limited clinical utility and can produce complications when switching opioids
Buprenorphine	Approved for office-based treatment of addiction because of its safety. Respiratory depression is rare. It is underused for the treatment of dependence in patients dependent on opioid analgesics. It has good analgesic efficacy for chronic pain and can be considered a good choice when treating pain with opioid dependence (buprenorphine SL) or chronic pain without opioid dependence (buprenorphine TD)

is a mild opioid because it has only partial mu opioid activity and is less addictive than strong opioids. It is categorized as a second-line treatment for neuropathic pain and has particular benefit in neuropathic pain because of its noradrenergic and serotonergic actions and lower abuse potential [7, 45]. It is available in both short- and long-acting formulations. Codeine can be addictive but is generally used in lower doses or in combinations drugs, and in some countries, it is available over the counter. These drugs are useful first-line opioids. It is important to remember the toxicity of drugs included in combination preparations which can become problematic if a patient is self-medicating at higher than recommended doses or if a patient is using either prescribed or over-the-counter acetaminophen or NSAIDs unwittingly.

18.5.2 Tapentadol

Tapentadol is a novel and expensive analgesic. It needs its own category because it is neither a mild opioid nor a pure mu opioid agonist. It has combined mu opioid agonism plus norepinephrine reuptake inhibition. It is a schedule II opioid considered as addictive as the pure mu opioid agonists. It may have advantages in neuropathic pain because of its norepinephrine effects but has not been studied in PHN and is not recommended in guidelines for neuropathic pain because of inadequate evidence [7, 46]. It is available in both short- and long-acting formulations.

18.5.3 Pure Mu Opioid Agonists

As described in Sect. 18.2, there are inevitable neuroadaptations to chronic and continuous opioid use. The most important of these are tolerance and dependence, which can ultimately be the basis of treatment failure. Conceptually, if opioid is given continuously, as in regimes consisting of long-acting opioid given round the clock, there is no time for recovery, thus tolerance and dependence are likely to become established and increase over time. This is borne out by clinical and population studies that show greater satisfaction and better pain relief for patients using low to moderate doses of opioid as needed, compared to high-dose users who tend to be on long-acting opioids and whose usage tends to be continuous. For this reason, short-acting opioids used as needed are recommended for all but the patients who may have difficulty controlling their usage, particularly those at risk of addiction or those who may be confused, as may many elderly and ill. Combining long- and short-acting opioids using the concept of "breakthrough pain" is not a good idea when pain is chronic.

The idea of breakthrough pain came into being while treating cancer pain at the end of life where pain might "break through" whatever underlying regime (usually long-acting opioid round the clock) was providing acceptable pain relief most of the time. Thus, a short-acting opioid to be taken as needed was often added to the long-acting opioid in dying patients where the goal was to keep them as comfortable as possible. The approach tends to lead to dose escalation, but for patients who are being palliated and have a short life expectancy, dose escalation is less of a problem. For chronic pain patients whose pain could be open-ended, dose escalation leads to eventual treatment failure and high risk (see Sect. 18.3). The breakthrough pain concept is best avoided when pain is chronic. Whatever pain occurs can be treated by the prescribed short-acting opioid within the limits of the prescribed dose, by non-opioid medications, or by using techniques such as distraction, relaxation, or exercise.

18.5.4 Mixed Agonist Antagonists

Most of these have limited clinical utility, except buprenorphine. Although buprenorphine may not be the most obvious choice of opioid for the treatment of PHN, it is developing an important role in chronic pain treatment, especially where this is accompanied by opioid dependence. Transdermal buprenorphine is a popular first-line opioid for chronic pain treatment in some countries (e.g., Norway), while in others (e.g., USA), its high price precludes common use. Buprenorphine can be complicated because it is a partial mu agonist with high affinity for the mu receptor, making it difficult to use other stronger mu agonists should they be needed (e.g., for acute or chronic pain). Buprenorphine use for the treatment of addiction requires special training. In most countries, sublingual buprenorphine (usually combined with naloxone) is not approved as an analgesic, even though it has good analgesic properties.

18.5.5 Dose

There is now overwhelming evidence that in the case of chronic pain, high-opioid dosing is neither safe nor effective [25–28]. Of course, individuals are all different, and some may benefit where others fail. Most of the evidence we base our recommendations upon comes from populations and not from individuals. One of the most difficult conflicts we face is that despite the emergence and visibility of increasing numbers of individuals who become dependent or addicted to opioids, many of whom still have severe pain despite high doses, and some of whom progress to death, there are still many people championing access to opioids for the treatment of pain without dose limitation. There is an enormous amount of fear that dose restrictions will leave patients in pain. Indeed, for those already taking high doses, it may, such is the difficulty of weaning opioids described in Sect. 18.2. Nevertheless, it is becoming clear, both from population data and from 30 years clinical experience of prescribing with open-ended dosing, that the policy of open-ended dosing has resulted in much harm to society and to pain patients themselves, without any clear benefit in terms of reducing the incidence of uncontrolled chronic pain. It is hard to argue with the statistics that show clearly that most long-term harms of opioid therapy are occurring for high-dose users. Thus, several bodies, including several US states and the US Centers for Disease Control are developing guidelines that suggest dose limitations. The State of Washington introduced legislation in 2011 that limits high-dose prescribing by requiring documented functional improvement at high doses or requiring that high-dose prescribers have received pain training, have consulted pain specialists, or are themselves pain specialists [47]. The Washington State stated dose limitation is 120 mg morphine equivalent per day (MED). Others have suggested lower daily doses, down to 50 mg [29, 48]. Epidemiological evidence suggests that 100 mg MED may be an inflection point where adverse effects increase exponentially with dose [27, 49, 50]. Other evidence suggests that the safest and most effective doses are below 50 mg MED [29, 51]. It is unlikely that consensus will be reached as to what is a reasonable upper limit for chronic opioid dosing, but these figures give some idea about what is meant by a high dose and suggest a dose that begins to be unsafe. It is important to recognize that while this order of dosing is recommended for new chronic opioid therapy patients, established patients who have difficulty tapering may need higher doses.

18.5.6 *Titrating to Effect: A Bad Idea When Pain Is Chronic*

When pain is acute, or occurs at the end of life, in both cases short lived, it tends to progress in a predictable and linear fashion. In not so chronic pain, the intensity of which is unpredictable and variable, often varying according to circumstance. Titrating opioids to the stated pain intensity level work well for acute and end-of-life pain, which apart from being predictable, also tends to respond well and predictably to opioids. Applying the titrate-to-effect principle to chronic pain can have undesired consequences if doses are increased solely according to the report of pain as in the titrate-to-effect principle [52–54]. Chronic pain, even if it has a known pathological basis, as does PHN, is influenced less by nociception than by emotional and psychosocial factors [55]. When pain is chronic, it may be hard to find any correlation between the severity of the causative disease process and the reported pain intensity level. A report of high pain intensity is often a cry for help, but opioids may not be the help needed. Those reporting high pain scores are often those with cofactors that increase the risk of developing dependence and addiction leading to "adverse selection," the tendency for self-selection to high-dose opioid treatment by the patients at highest risk for problematic use [32, 49, 56, 57]. Opioids, and in fact all other analgesics, are only capable of reducing pain by a few points on a pain scale [38, 58, 59], and repeated dose escalation without sustained pain improvement leads to toxicity. It seems intuitive to treat pain according to its reported intensity level. It is simple, but it oversimplifies something that is complex. When a patient reports a high pain intensity level despite being on a reasonable dose of opioid, ask what it is that is producing the high level. There may be many treatments that are preferable to repeated opioid dose escalation, including behavioral and physical treatments.

18.6 Sensible Precautions

As described in Sect. 18.4, HZ can produce acute, subacute, and chronic pain. Even though the complications of opioid therapy may differ at each stage, whenever opioids are started, the risks pertaining to the individual patient, both immediately and in the future, should be elicited and discussed. On the one hand, we should not frighten patients to the extent that they would rather stay in pain than accept opioids. On the other, patients and family members should be included in a decision that could have significant implications. The clinician's role is to help patients understand the potential benefits as set against the risks of the treatment.

18.6.1 *At the Start of Opioid Therapy/Acute Pain*

When using opioids for acute pain, it is not necessary to be overly concerned about long-term harms, but the potential for long-term harm should not be ignored since every time opioid treatment is started, it could extend into a chronic treatment.

Table 18.4 High level of caution needed

High level of caution needed
Caution regarding side effects
Elderly and frail
Dementia
Unsteady gait
Alcohol and drug use
Concomitant sedative use
Caution regarding control issues
Young
Personal or family history of substance abuse
History of being abused/PTSD
Current substance use or abuse
Smoker
Psychiatric comorbidity, especially anxiety and depression
Personality disorder

There may be particular concerns regarding short-term harms and side effects, especially in the elderly and frail, and there may be particular concerns about long-term harms and losing control of opioid use, especially in younger patients and those with substance use disorders and other psychiatric comorbidities (Table 18.4). The need for caution regarding side effects will be obtained from the normal clinical assessment, but the need for caution regarding control issues may need a few extra steps. The first step can be a simple conversation, asking questions such as:

- Have you ever had a drug or alcohol problem?
- Has anyone in your family had a drug or alcohol problem?
- Do you have any concerns about addiction?
- Have you ever suffered extreme anxiety?

If no concerns are identified with these simple questions that would suggest a low-level concern about long-term harm. However, if concerns are raised, then it would be prudent to consider further steps such as using a screening tool for substance abuse risk; using simple measurement tools for depression, anxiety, and post-traumatic stress disorder (PTSD); asking for a baseline urine toxicology screen; initiating a written patient agreement; and checking prescription monitoring data if available.

Most importantly, once the assessment of potential harms has been completed, consider whether or not to use opioids, and if using, consider choice of drug and dose in the light of potential risk.

18.6.2 When Opioid Is Selected as Chronic Pain Treatment

Even though opioid treatment may have been initiated to treat acute zoster pain and evolved into chronic treatment, it is advisable to set into motion additional precautions aimed at preventing long-term harm once it is clear that it is chronic pain

Table 18.5 Precautions during long-term use

Precautions during long-term opioid use
Work with a care agreement that includes stated goals
Monitor goal achievement
Test urine at baseline, and thereafter as needed or annually
Check prescription monitoring data at baseline, if available, then as needed, or at every prescription if practical
Require in person prescription pick up and do not give early prescriptions
Do not dose escalate
Assess cognitive function in the elderly
Do not use concomitant sedatives, especially benzodiazepines

(PHN) that is being treated, or when treating PHN with opioids for the first time. Recommended precautions are listed in Table 18.5.

18.6.2.1 The Care Agreement

The utility of the care agreement cannot be overstressed. It is not simply a means of educating patients about risk but should be used to develop and document an understanding about realistic goals for opioid therapy and establish which goals, particularly functional goals, are important to the individual patient. A goal of simply reducing pain intensity is likely to be disappointed [38, 52–54, 58, 59]. When it comes to evaluating therapy as it progresses, the achievement or nonachievement of the individual's goals is the single most important factor in determining whether or not to continue therapy. This is especially important if the therapy is not going well.

18.6.2.2 Urine Toxicology

Testing of urine for the presence of prescribed and nonprescribed dangerous drugs has become a recommended standard during chronic opioid therapy. It is beyond the scope of this chapter to describe or discuss urine testing in detail, but a there are a few important general points to be made. Many clinicians feel that urine testing affects the relationship of trust they have with their patients. This is a legitimate fear. In some circumstances, and for some patients (particularly long-established patients with very low risk), urine testing may be such low yield as to be considered unnecessary. However, even low risk patients can surprise one, and studies suggest that urine screening identifies a substantial number of aberrancies that are not detected by interview [60]. Urine testing policies will vary according to the clinical setting, and in general, it is wise to follow whatever policy is in place locally. One policy that works well is to test all patients at the start of opioid therapy, testing thereafter either for cause, or annually. The *all patient* policy removes stigma and loss of trust to a large extent, and *testing before opioid therapy begins* may reveal red flags.

Urine dipsticks are notorious for their false negatives and false positives. Any positive result from a dipstick test must be followed up with a laboratory test. Laboratory results can also be confusing, especially since opioid metabolites confuse the results. Any concerning result should be discussed with the laboratory before discussing with the patient.

18.6.2.3 Prescription Monitoring

Prescription monitoring is in its infancy and not available everywhere. In the USA, prescription monitoring is rapidly becoming available across all states, and guidelines are beginning to recommend mandatory checking of prescription monitoring data at least at the first prescription. Prescription data monitoring programs (PDMPs) record every opioid prescription, drug, dose, prescriber, and patient. Data are entered by the pharmacist when the medication is picked up. There are currently several glitches in the systems that mean that PDMP data is not always fully accurate. Another problem at present is that not all states have programs that have advanced to the stage of being practical to use in busy practice settings, especially if wanting to use the system for every prescription. Ideally, these programs will eventually have the capacity to provide prescribers with the necessary data when writing electronic prescriptions, but neither electronic prescribing nor PDMPs have advanced to that stage yet.

18.7 Conclusions and Controversies

The pendulum has swung between underuse and overuse of opioids throughout recorded history, but the extremes have never been as great as those we have witnessed over the past few decades. Periods of overuse culminate in fears that the drugs are causing unacceptably high rates of addiction and need to be restricted. Periods of underuse culminate in fears that people in pain are suffering needlessly because restrictions are reducing access. The current state of affairs, at least in the developed world, is that we are coming out of a three-decade period of responding to perceived underuse with enormously increased use. In Northern America, the use has increased severalfold in the last few decades, and other developed countries are somewhere along the same trajectory. Increased use was largely accounted for by increased prescribing for chronic pain and was based on a change in teaching that stressed that fears of addiction were unfounded: when used as pain therapy, opioids were unlikely to produce addiction, and on balance, they were more helpful than harmful. Partly because of this teaching, a whole culture changed to one that perceives it a right for patients to receive opioids and duty for clinicians to prescribe. This is the culture within which most of us were trained and have practiced, within which all recent guidelines and textbooks have been written, and the culture that forms our beliefs. When faced with someone in pain and distress, especially when that pain has a clear generator as it does in the case of HZ, and when other medical treatments have failed, we would not hesitate to prescribe opioids.

But now the pendulum is beginning to swing back, and, yet again, it is because we have become alarmed by the scale of harm caused by liberal prescribing of opioids, especially in the USA where increases in prescribing were greatest and where prescription opioids have caused an "epidemic" of abuse and deaths [61]. We are at a difficult crossroads. We cannot eradicate our beliefs, yet new evidence – new scientific knowledge about opioid adaptations that explain why opioids do not work well for pain in the long term and new epidemiological evidence on harms – suggests that long-term opioid therapy is not as helpful or safe as we once thought.

18.7.1 A Palliative Care Construct

When opioid treatment was first popularized in the late 1980s, it was actually palliative care specialists who during prior decades had succeeded in liberalizing opioids for cancer pain treatment, who became strong advocates for using opioids to treat chronic pain. Their arguments were largely ethical and based on the idea that chronic pain caused suffering equal to end-of-life pain. Yet their experience had been with cancer patients, often at the end life. Not only was this generally short-lived pain, it was also a situation where certain risks are acceptable because comfort and not function is the goal of treatment, and even inadvertent death is acceptable in the principle of "double effect" [62]. Perhaps they envisaged patients whose lives were simply intolerable without the comfort and numbing effects of opioids, and who did not have any alternatives, as being rightful recipients of opioid therapy. And perhaps this is still true. Unfortunately, their message was taken up and spread to the treatment not only of patients who did not have any alternatives but to all types of chronic pain and for all types of situations, probably 90 % of which do have viable and often more effective alternatives which begin to be less viable once opioid dependence has set in. A useful construct, which may also be useful when considering the role of opioids in the treatment of zoster pain, is that chronic opioid therapy should be thought of as a palliative treatment that provides comfort when other treatments are not viable and when certain risks are acceptable because life is otherwise intolerable [63, 64]. Using this construct, comfort would be a reasonable goal for opioid treatment for which goal achievement is of paramount importance (Table 18.6).

18.7.2 What Is Different About Herpes Zoster Pain

One aspect of HZ we should not forget is that HZ, particularly HZ that progresses to PHN, rarely occurs in isolation. It is a pain-producing disease that tends to occur in the elderly and immunocompromised. There is a high likelihood that there are other disease processes in play and that those disease processes are also associated with pain – pain that would be of a different etiology possibly with different treatment priorities. There is also a high likelihood that the patients are debilitated and frail.

Table 18.6 Summary recommendations

Summary recommendations
Do not embark on opioid therapy without an assessment and discussion of risks and limitations
For acute HZ, provide analgesics according to the WHO stepladder (opioids last) if primary treatments have not controlled the pain
For PHN, use opioids as a third-line therapy but with clearly stated functional goals
Prefer short acting as needed
Use long-acting round the clock if there are control issues or confusion
Do not use short and long acting together
Use lowest effective dose
Avoid dose escalation
Continuously review pain and opioid usage for a stop point
Be prepared to wean and discontinue if goals are not met
Early discontinuation can prevent dependence
Complex-persistent dependence is akin to addiction and destroys lives

Where even small doses of opioid can increase confusion and the likelihood of falls and fractures in the elderly and debilitated, they are also the patients where risks of abuse seem unimportant set against the need to provide comfort. Yet some PHN patients, particularly the less debilitated, will go on to develop dependence and will find it hard to wean from opioid therapy despite poor pain relief and side effects such as confusion. Fortunately, the less debilitated are less likely to progress to PHN and protracted neuropathic pain [36]. Another important difference for PHN is that unlike many other chronic pain states, PHN can have a limited time course (Fig. 18.6). This makes the continuous reviewing of pain and opioid usage for a stop point particularly important to avoid needing opioids to treat dependence and not pain (Table 18.6).

This chapter has emphasized safety and efficacy considerations advisedly. There can be no simple paradigm to direct opioid decision-making, especially in complex situations. But an understanding of the processes that underlie the clinical effects of opioids can help clinicians navigate through a decision process that will always be challenging and will always be unique to each patient and each situation.

18.8 Commentary

C. Peter Watson

This chapter should be required reading for those prescribing opioids for postherpetic neuralgia and other neuropathic pain as should the recent CDC opioid guideline [66]. It is this editor's opinion that opioids do remain an option for extremely severe postherpetic neuralgia with significant disability and often suicidal rumination because of the severity and poor quality of life which can result from this pain. This is based on four decades of clinical experience of treating carefully selected, otherwise medically intractable patients with severe, longstanding, disabling PHN. This should be considered after other medical approaches (gabapentinoids, antidepressants, etc.; see Chap. 19) have failed and must be done with extreme caution, being careful to exclude known drug addicts and significant psychiatric illness but also with great caution, if at all, with those with a family and personal history suggesting an elevated risk of addiction, such as the questions suggested in the Opioid Risk Tool [71]. If this is done, it is, in this editor's opinion, a viable option if the patients are followed, monitored, and recordings made with attention to pain severity, pain relief, function, quality of life, mood, adverse effects, tolerance, escalating doses, and a copy of the prescription for future reference which indicates the drug, the dose, and the amount prescribed [66–70]. It is possible there may be an elevated risk with long-term use of physical and psychological complications but I believe this is not large if patients are selected carefully, but it should be explained to the patient that the long-term effects over months and years are probably not known. For some patients, this may best be done in a pain clinic setting. An opioid contract and urine testing for a variety of drugs of abuse can be useful adjuncts.

References

1. Hempenstall K, Nurmikko TJ, Johnson RW, A'Hern RP, Rice AS (2005) Analgesic therapy in postherpetic neuralgia: a quantitative systematic review. PLoS Med 2:e164
2. Watson CP, Moulin D, Watt-Watson J, Gordon A, Eisenhoffer J (2003) Controlled-release oxycodone relieves neuropathic pain: a randomized controlled trial in painful diabetic neuropathy. Pain 105:71–78
3. Raja SN, Haythornthwaite JA, Pappagallo M et al (2002) Opioids versus antidepressants in postherpetic neuralgia: a randomized, placebo-controlled trial. Neurology 59:1015–1021
4. Rowbotham MC, Twilling L, Davies PS, Reisner L, Taylor K, Mohr D (2003) Oral opioid therapy for chronic peripheral and central neuropathic pain. N Engl J Med 348:1223–1232
5. McQuay HJ, Tramer M, Nye BA, Carroll D, Wiffen PJ, Moore RA (1996) A systematic review of antidepressants in neuropathic pain. Pain 68:217–227
6. Dworkin RH, O'Connor AB, Backonja M et al (2007) Pharmacologic management of neuropathic pain: evidence-based recommendations. Pain 132:237–251
7. Finnerup NB, Attal N, Haroutounian S et al (2015) Pharmacotherapy for neuropathic pain in adults: a systematic review and meta-analysis. Lancet Neurol 14:162–173
8. IASP (1994) Taxonomy classification of chronic pain. In: Merskey H, Bogduk N (eds) IASP task force on taxonomy, 2nd edn. IASP Press, Seattle. http://www.iasp-painorg/Taxonomy
9. Ballantyne JC, Shin NS (2008) Efficacy of opioids for chronic pain: a review of the evidence. Clin J Pain 24:469–478
10. Ballantyne JC, LaForge SL (2007) Opioid dependence and addiction in opioid treated pain patients. Pain 129:235–255
11. South SM, Smith MT (2001) Analgesic tolerance to opioids. Pain Clin Updates IASP Press 9:1–4
12. Koob GF, Le Moal M (2001) Drug addiction, dysregulation of reward, and allostasis. Neuropharm 24:97–129
13. Nestler EJ, Aghajanian GK (1997) Molecular and cellular basis of addiction. Science 278:58–63

14. Hyman SE, Malenka RC, Nestler EJ (2006) Neural mechanisms of addiction: the role of reward-related learning and memory. Annu Rev Neurosci 29:565–598
15. Koob GF, Le Moal M (1997) Drug abuse: hedonic homeostatic dysregulation. Science 278:52–58
16. Cami J, Farre M (2003) Drug addiction. N Engl J Med 349:975–986
17. Compton P, Athanasos P, Elashoff D (2003) Withdrawal hyperalgesia after acute opioid physical dependence in nonaddicted humans: a preliminary study. J Pain 4:511–519
18. Ballantyne JC, Sullivan MD, Kolodny A (2012) Opioid dependence vs addiction: a distinction without a difference? Arch Intern Med 172:1342–1343
19. Heit HA (2003) Addiction, physical dependence, and tolerance: precise definitions to help clinicians evaluate and treat chronic pain patients. [Review] [30 refs]. J Pain Palliative Care Pharmacother 17:15–29
20. Nestler EJ (2004) Molecular mechanisms of drug addiction. Neuropharmacology 47(Suppl 1):24–32
21. Palmer RE, Carrell DS, Cronkite D et al (2015) The prevalence of problem opioid use in patients receiving chronic opioid therapy: computer-assisted review of electronic health record clinical notes. Pain 156:1208–1214
22. Ballantyne JC (2015) Assessing the prevalence of opioid misuse, abuse, and addiction in chronic pain. Pain 156:567–568
23. Martin BC, Fan MY, Edlund MJ, Devries A, Braden JB, Sullivan MD (2011) Long-term chronic opioid therapy discontinuation rates from the TROUP study. J Gen Intern Med 26:1450–1457
24. Franklin GM, Mai J, Wickizer T, Turner JA, Fulton-Kehoe D, Grant L (2005) Opioid dosing trends and mortality in Washington State workers' compensation, 1996–2002. Am J Ind Med 48:91–99
25. Paulozzi LJ (2012) CDC grand rounds: prescription drug overdose – a U.S. epidemic Morbidity and Mortality Weekly Report (MMWR). www.cdcgov/mmwr/preview/mmwrhtml/mm6101a3htm
26. Gomes T, Mamdani MM, Dhalla IA, Paterson JM, Juurlink DN (2011) Opioid dose and drug-related mortality in patients with nonmalignant pain. Arch Intern Med 171:686–691
27. Dunn KM, Saunders KW, Rutter CM et al (2010) Opioid prescriptions for chronic pain and overdose: a cohort study. Ann Intern Med 152:85–92
28. Bohnert AS, Valenstein M, Bair MJ et al (2011) Association between opioid prescribing patterns and opioid overdose-related deaths. JAMA 305:1315–1321
29. Draft CDC (2016) Guideline for prescribing opioids for chronic pain – United States. http://www.regulationsgov/ - !documentDetail;D=CDC-2015-0112-0002
30. Paulozzi LJ, Kilbourne EM, Shah NG et al (2012) A history of being prescribed controlled substances and risk of drug overdose death. Pain Med 13:87–95
31. Edlund MJ, Martin BC, Devries A, Fan MY, Braden JB, Sullivan MD (2010) Trends in use of opioids for chronic noncancer pain among individuals with mental health and substance use disorders: the TROUP study. Clin J Pain 26:1–8
32. Seal KH, Shi Y, Cohen G, Maguen S, Krebs EE, Neylan TC (2012) Association of mental health disorders with prescription opioids and high-risk opioid use in US veterans of Iraq and Afghanistan. JAMA 307:940–947
33. Ballantyne JC (2015) What can the POINT study tell us? Pain 156:201–202
34. Campbell G, Nielsen S, Bruno R et al (2015) The pain and opioids IN treatment study: characteristics of a cohort using opioids to manage chronic non-cancer pain. Pain 156:231–242
35. Jeon YH (2015) Herpes Zoster and postherpetic neuralgia: practical consideration for prevention and treatment. Korean J Pain 28:177–184
36. Opstelten W, Van Essen GA, Schellevis F, Verheij TJ, Moons KG (2006) Gender as an independent risk factor for herpes zoster: a population-based prospective study. Ann Epidemiol 16:692–695
37. Watson CP, Watt VR, Chipman M, Birkett N, Evans RJ (1991) The prognosis with postherpetic neuralgia. Pain 46:195–199

38. Kalso E, Edwards JE, Moore RA, McQuay HJ (2004) Opioids in chronic non-cancer pain: systematic review of efficacy and safety. Pain 112:372–380
39. van Wijck AJ, Opstelten W, Moons KG et al (2006) The PINE study of epidural steroids and local anaesthetics to prevent postherpetic neuralgia: a randomised controlled trial. Lancet 367:219–224
40. Pasqualucci A, Pasqualucci V, Galla F et al (2000) Prevention of post-herpetic neuralgia: acyclovir and prednisolone versus epidural local anesthetic and methylprednisolone. Acta Anaesthesiol Scand 44:910–918
41. Kumar V, Krone K, Mathieu A (2004) Neuraxial and sympathetic blocks in herpes zoster and postherpetic neuralgia: an appraisal of current evidence. Reg Anesth Pain Med 29:454–461
42. Ventafridda V, Saita L, Ripamonti C, De Conno F (1985) WHO guidelines for the use of analgesics in cancer pain. Int J Tissue React 7:93–96
43. Ballantyne JC (2011) Opioids around the clock? Pain 152:1221–1222
44. Von Korff M, Merrill JO, Rutter CM, Sullivan M, Campbell CI, Weisner C (2011) Time-scheduled vs. pain-contingent opioid dosing in chronic opioid therapy. Pain 152:1256–1262
45. Boureau F, Legallicier P, Kabir-Ahmadi M (2003) Tramadol in post-herpetic neuralgia: a randomized, double-blind, placebo-controlled trial. Pain 104:323–331
46. Elling C, Galic M, Steigerwald I (2015) Tapentadol prolonged release in the treatment of neuropathic pain related to diabetic polyneuropathy. Lancet Neurol 14:684–685
47. Washington State Department of Health, Health systems quality assurance, pain management adopted rules. 2012. http://www.dohwagov/hsqa/professions/painmanagement/meetings.htm
48. Chou R, Turner JA, Devine EB et al (2015) The effectiveness and risks of long-term opioid therapy for chronic pain: a systematic review for a National Institutes of Health Pathways to Prevention Workshop. Ann Intern Med 162:276–286
49. Edlund MJ, Martin BC, Fan MY, Devries A, Braden JB, Sullivan MD (2010) Risks for opioid abuse and dependence among recipients of chronic opioid therapy: results from the TROUP study. Drug Alcohol Depend 112:90–98
50. Fulton-Kehoe D, Garg RK, Turner JA et al (2013) Opioid poisonings and opioid adverse effects in workers in Washington State. Am J Ind Med 56:1452–1462
51. Naliboff BD, Wu SM, Schieffer B et al (2011) A randomized trial of 2 prescription strategies for opioid treatment of chronic nonmalignant pain. J Pain 12:288–296
52. Sullivan MD, Ballantyne JC (2016) Must we reduce pain intensity to treat chronic pain? Pain 157:65–69
53. Ballantyne JC, Kalso E, Stannard C (2016) WHO analgesic ladder: a good concept gone astray. BMJ 352:i20
54. Ballantyne JC, Sullivan MD (2015) Intensity of chronic pain – the wrong metric? N Engl J Med 373:2098–2099
55. Hashmi JA, Baliki MN, Huang L et al (2013) Shape shifting pain: chronification of back pain shifts brain representation from nociceptive to emotional circuits. Brain 136:2751–2768
56. Phifer J, Skelton K, Weiss T et al (2011) Pain symptomatology and pain medication use in civilian PTSD. Pain 152:2233–2240
57. Weisner CM, Campbell CI, Ray GT et al (2009) Trends in prescribed opioid therapy for non-cancer pain for individuals with prior substance use disorders. Pain 145:287–293
58. Lunn MP, Hughes RA, Wiffen PJ (2014) Duloxetine for treating painful neuropathy, chronic pain or fibromyalgia. Cochrane Database Syst Rev 1:Cd007115
59. Zhang SS, Wu Z, Zhang LC et al (2015) Efficacy and safety of pregabalin for treating painful diabetic peripheral neuropathy: a meta-analysis. Acta Anaesthesiol Scand 59:147–159
60. Katz NP, Sherburne S, Beach M et al (2003) Behavioral monitoring and urine toxicology testing in patients receiving long-term opioid therapy. Anesth Analg 97:1097–1102
61. CDC (2012) CDC grand rounds: prescription drug overdoses – a US epidemic. MMWR Morb Mortal Wkly Rep 61(01):10–13. http://www.cdc.gov/mmwr/preview/mmwrhtml/mm6101a3.htm
62. Fohr SA (1998) The double effect of pain medication: separating myth from reality. J Palliat Med 1:315–328

63. Ballantyne JC, Sullivan MD (2012) Is chronic opioid therapy comfort care? In: Tracey I (ed) Pain 2012: refresher courses, 14th world congress on pain. IASP Press, Seattle
64. Sullivan MD, Ballantyne JC (2012) What are we treating with chronic opioid therapy? Arch Intern Med 172:433–434
65. American Psychiatric Association DSM-5 development, substance use disorder (2012) http://www.dsm5org/ProposedRevision/Pages/proposedrevisionaspx?rid=431. Last accessed 3/31/2012
66. Frieden TC, Houry D (2016) Reducing the risks of relief- the CDC opioid- prescribing guideline. N Eng J Med 374:1501–1504
67. Watson CPN, Babul N (1998) Oxycodone relieves neuropathic pain: A randomized trial in postherpetic neuralgia. Neurology 50:1837–1841
68. Watson CPN, Watt-Watson JH, Chipman MA (2010) The long-term efficacy and safety of opioids: a survey of 84 selected patients with intractable chronic non-cancer pain. Pain Res Manage 15(4):213–218
69. Watson CPN (2012) Opioids in chronic non-cancer pain: Some blind men and an elephant. Scand J Pain 3:5–13
70. Watson CPN (2012) Opioids in chronic non-cancer pain: more faces from the crowd. Pain Res Manage 17(4):213–217
71. Webster LR (2005) Predicting aberrant behaviors in opioid-treated patients: Preliminary validation of the opioid risk tool. Pain Medicine 6(6):432–442

Chapter 19
Postherpetic Neuralgia: Difficult to Treat, Easier to Prevent

C. Peter N. Watson

For it happens…in the beginning the malady is easy to cure but difficult to detect but in time, not having been detected or treated, it becomes easy to detect but difficult to cure.
Nicola Machiavelli, The Prince

A kindly understanding doctor who will keep in touch with his patients with intractable postherpetic neuralgia even for a very long time is extremely valuable therapeutically.
Edgar Hope-Simpson [25]

The efficient physician is he who amuses his patients while nature effects a cure.

The philosophical dictionary, Voltaire (1694–1778)

19.1 Introduction

The first quotation describes the situation as the virus lies dormant and asymptomatic and difficult to detect but likely easier to prevent by vaccination, whereas later with the eruption of the characteristic painful rash of herpes zoster (HZ), it becomes easy to diagnose but difficult to treat.

The second quote by one of the pioneers in varicella-zoster virus (VZV) research [24, 25] describes what, in some patients, may be all we have to offer in some instances for treating severe postherpetic neuralgia (PHN).

Some good news and a cautionary note are reflected by Voltaire as, by virtue of the natural course of the disease, many patients with PHN, especially the young, recover which is a good thing for the afflicted but something which has plagued uncontrolled studies or inadequately screened and powered randomized controlled trials. This improvement with time even occurs with long duration PHN but much more slowly than in the first months after acute HZ onset [51].

C.P.N. Watson
Department of Medicine, University of Toronto,
99 Avenue Rd, Toronto, ON M5R 2G5, Canada
e-mail: peter.watson@utoronto.ca

C.P.N. Watson et al. (eds.), *Herpes Zoster: Postherpetic Neuralgia and Other Complications*, DOI 10.1007/978-3-319-44348-5_19

Fig. 19.1 Hope-Simpson's graph of the increasing incidence of herpes zoster and postherpetic neuralgia with age (Hope-Simpson 1965) [24, 58]. The increase in herpes zoster and PHN after age 50 (*arrow*) is the rationale for the use of the zoster vaccine at this time

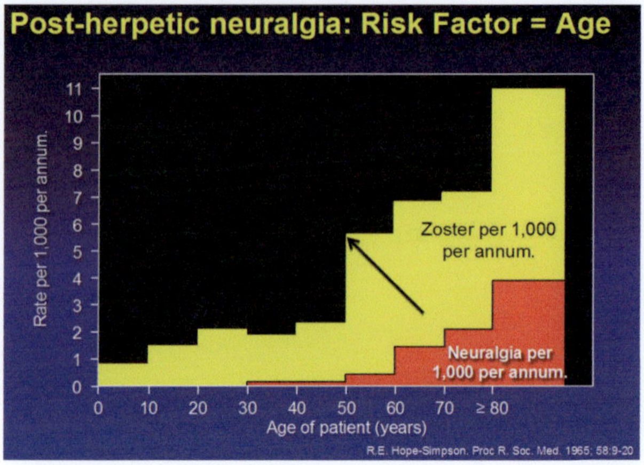

PHN is a neuropathic (nerve injury) pain and is the most common and feared complication of HZ. PHN can be defined arbitrarily in different ways and for different purposes. It is in general terms pain that persists after rash healing. This may be tallied at 1 month or, for clinical trials with PHN, at 3 or 6 months since many patients improve in the weeks following the initial eruption and therefore a definition of a longer duration means greater stability for all RCTs but especially for those of a crossover design.

HZ, the precursor of PHN, is due to reactivation of the varicella-zoster virus (VZV) usually in one of the spinal or cranial sensory ganglia often a half century or more following a primary infection with varicella (chickenpox) usually occurring during childhood. HZ is characterized by a unilateral, painful, vesicular rash typically in a single dermatome (most commonly in midthoracic dermatomes or in the trigeminal ophthalmic division) often resulting in PHN which is the commonest neurological disease [30].

Most children (95 %) have been exposed to varicella and are vulnerable to developing herpes zoster in later life. In Canada (population nearly 35,000,000) there are 130,000 cases of HZ and 17,000 of PHN per year [4]. In the USA, the figure is 1,000,000 cases of zoster per annum [26]. The incidence is directly related to age [24] (Fig. 19.1) and due to decreased cell-mediated immunity. The increase in HZ and PHN that begins at ages 50 to 60 years provides the rationale for vaccination commencing at this time. Other risk factors are greater severity of the rash and severe pain at onset [12, 13]. Overall about 10 % of HZ cases will have pain at 1 month after the rash and this may rise to as much as 50 % at age 60 [10, 11]. About 20 % with HZ of all ages have pain at 3 months from onset with 15 % suffering significantly after a year [48]. If one looks at PHN at 3 months from rash onset at age 50–54 years of age, 8 % have PHN and rising to 20 % at age 80–84 [21]. Of patients having PHN at a year or more about 50 % will slowly improve [51]. Because PHN fails to resolve within a

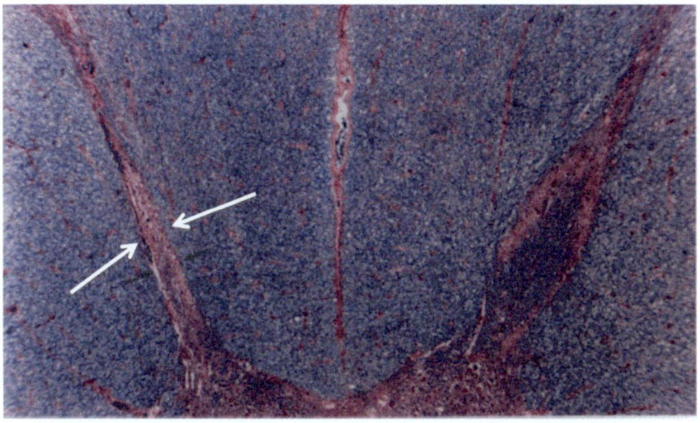

Fig. 19.2 Atrophy of the dorsal horn of the spinal cord in postherpetic neuralgia (*arrows*) (Watson et al. 1988) [49]

year in a proportion, the prevalence of PHN is cumulative and higher. Because the population is aging and with the increase in immunosuppressed groups afflicted with cancer and HIV, HZ will likely increase. Recurrence of herpes zoster is uncommon at about 5 % often occurring in the same dermatome ([24, 38].

PHN may be difficult and even impossible to treat even with opioids. Pathological evidence suggests that VZV causes permanent damage to the central and peripheral nervous system probably destroying sites of intrinsic pain-inhibitory mechanisms where analgesics act especially in the dorsal horn of the spinal cord [49, 50] (Fig. 19.2). Postherpetic neuralgia has a major effect on the quality of life of sufferers [12, 13].

This article will outline the best current therapy of PHN, the possible prevention at the stage of acute zoster (Chap. 23), and the exciting promise of zoster prevention vaccines (Chap. 24, [31, 37]). The reader should be advised that the practical guidelines for the treatment of PHN outlined here differ somewhat from previous excellent reviews [28] in that, besides reviewing the science, it includes recommendations (some perhaps controversial) based on the author's experience with treating a large number of PHN sufferers over 4 decades.

There are two approaches to preventing the irreparable end-stage state of PHN. One is with early and aggressive treatment of HZ with oral antivirals (valaciclovir, famciclovir) and intravenous acyclovir (thought best used within 72 h of rash onset) and analgesics such as gabapentinoids (gabapentin, pregabalin), analgesic antidepressants [tricyclic antidepressants (TCAs), serotonin norepinephrine reuptake inhibitors (SNRIs)], and opioids. The difficulties with this approach are (1) delays with treatment within 72 h when pain precedes the rash by days and when the rash does not appear at all (zoster sine herpete) and (2) with the limited or lack of efficacy of antiviral agents ([6], Chap. 23) and of the largely unproven, early use of analgesics (even opioids if severe) in preventing severe PHN (Chap. 23). Although this is good practice for the acute often severe pain of HZ, the most promising preventative approach appears to be the use of a zoster prevention vaccine (Chap. 24).

19.2 Clinical Features (Fig. 19.7)

19.2.1 Pain Assessment

The initial and follow-up evaluations of the neuropathic pain of PHN are critical. PHN may be assessed in different ways and for different purposes, for example, at initial and at follow-up assessment, for research purposes, and if a patient has to be placed on chronic opioid therapy.

At initial and follow-up assessments in daily practice, it may be useful to capture the temporal profile of the pain severity throughout the day and its effect on such activities and sleep since timed intake of analgesics may be important to limit or minimize doses, to avoid daytime drowsiness and improve sleep. (An elderly patient may suffer mostly nocturnally and only require analgesia in a low dose.) A 0–10 scale rating is easily done by most patients, does not require a paper scale, and can rate three major components of this pain separately, i.e., steady burning pain, electric shocks, and skin sensitivity or pain from touch (allodynia). This can be rated before and after an analgesic or other measures especially for follow-up evaluations.

For opioid use regulatory guidelines suggest rating change in function (activities of daily living), quality of life, mood, adverse effects (usual side effects and altered behavior suggesting psychological dependency, tolerance, or physical withdrawal), and sleep.

For research purposes other scales may be employed such as the visual analogue scale (VAS) for the three main pain components, the McGill Pain Questionnaire, and the Zoster Brief Pain Inventory. Other useful assessments for different facets of the effect of the pain are for mood (Hospital Anxiety and Depression Scale, Beck Depression Inventory), function (Pain Disability Index, Brief Pain Inventory Interference Scale), and quality of life (SF12v2).

19.2.2 Physical Examination

When the acute rash has healed, the affected skin often exhibits a reddish, purple, or brownish hue (Fig. 19.6). As this subsides, pale scarring often remains (Fig. 19.7). Occasionally severe pain with no residual scar may occur or the scars in very long-duration cases are barely perceptible. A steady burning or aching may occur and also a paroxysmal, lancinating pain. Both may occur spontaneously and are often aggravated by any contact with the involved skin such as friction from even the lightest clothing or even wind blowing on the skin (allodynia). Firm pressure on the skin may curiously be soothing. Some patients describe unbearable itch, formication (like ants crawling on the skin), or other forms of dysesthesia. As well as light touch, these symptoms may be exacerbated by physical activity, temperature change, and emotional upset.

The scarred areas are usually at least hypesthetic and often anesthetic to punctate touch, and yet paradoxically the skin often exhibits marked pain on moving tactile stimulation (dynamic mechanical allodynia), increased pain to the noxious stimulation of a pinprick (hyperalgesia), a disagreeable feeling (dysesthesia), or an increased sensitivity(hyperesthesia) to moving touch stimuli (Figs. 19.6 and 19.7). The affected, scarred skin often reveals a loss of sensation to pinprick, temperature, and touch over a wider area than the scars and even wider area of sensitive or painful skin (Fig. 19.7). This sensitive skin may paradoxically include the area anesthetic to punctate touch where it is elicited by light stroking or skin traction between thumb and forefinger, an effect which may be caused by summation on hypersensitive, deafferented spinal dorsal horn neurons with expanded receptive fields. The Zoster Brief Pain Inventory is a valid tool for the assessment of this pain for clinical trials [9]. There is no need for an exhaustive search for underlying malignancy in the usual patient with herpes zoster [39].

19.2.3 Putative Pain Mechanisms Based on Pathology, Clinical Features and Pharmacodynamics, and Implications

There is considerable information about the pathology and possible pathogenesis of PHN. It has been known for more than 100 years that pathologically there is an acute hemorrhagic inflammation in one dorsal root ganglion at the stage of the eruption of HZ (Fig. 19.3 top image) [23]. Inflammation then extends proximally and distally. Proximally it extends via the nerve root into the dorsal horn of the spinal cord. Distally it travels into the skin of the affected segment. After months, there is significant scarring and loss of neurons in the dorsal root ganglion (Fig. 19.3 bottom image and Fig. 19.4) and atrophy and scarring of the dorsal horn of the spinal cord (Fig. 19.2), [23, 49, 50]. Some of these cases have persistent inflammatory cells [50]. An assessment of the nerve fiber population in the peripheral nerve after the eruption of HZ shows a predominance of small (some pain-conducting) fibers and a deficiency in large myelinated (pain-inhibitory fibers) [35, 49]. This predominance of small fibers is probably due in part to regenerating sprouts from a variety of sensory neurons transmitting pressure and vibration as well as pain and temperature. Further, although shingles and PHN are associated with unilateral clinical findings with the rash, distribution, and residual scarring associated with only one ganglion, contralateral findings in the same nerve, nerve root, and skin segment have been shown pathologically [36, 57].

The history of a patient with PHN often demonstrates three main features to the pain. There is a constant, steady burning pain and electric shock-like pains reminiscent of trigeminal neuralgia, and the skin is often very sensitive or painful to summating touch stimuli such as skin stroking (allodynia or pain from moving tactile stimuli such as clothing). The problem is that this knowledge has not led to useful changes in medical or surgical therapy.

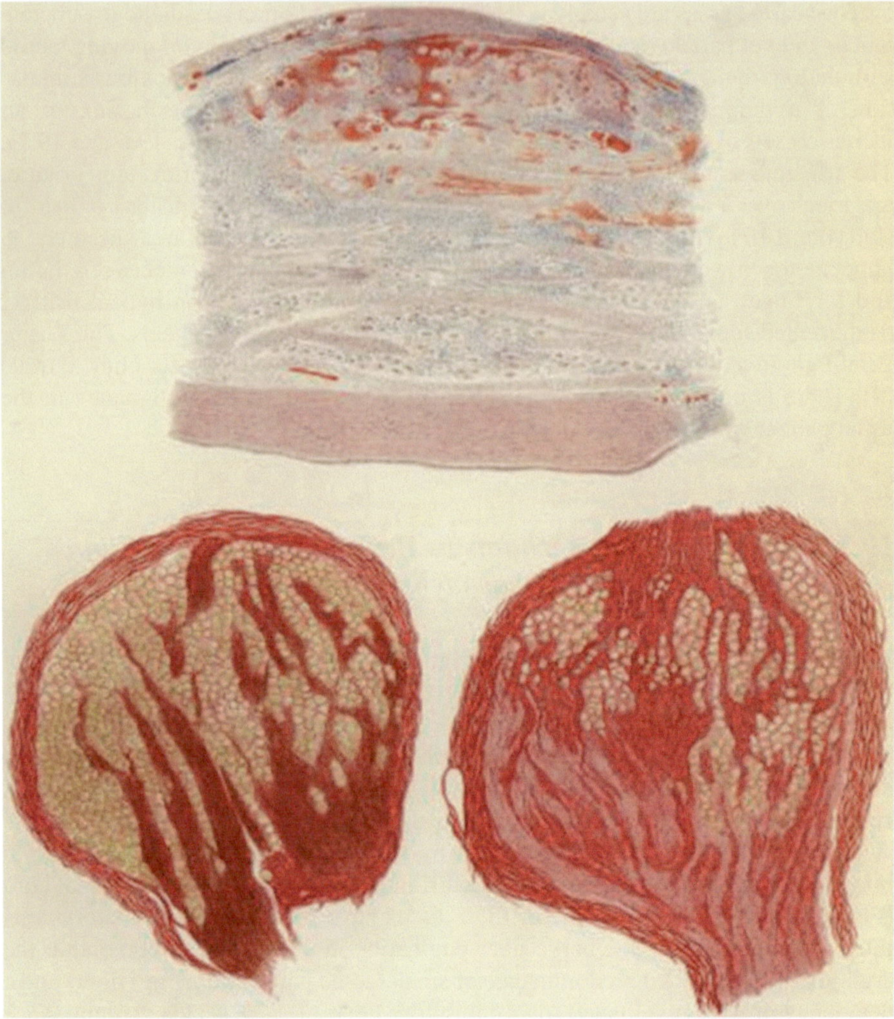

Fig. 19.3 Hemorrhagic inflammation of a dorsal root ganglion (*upper* image sagittal view). Scarring of a dorsal root ganglion (sagittal view *right lower* image) and normal ganglion (sagittal view: left lower image) (Head and Campbell 1900)

For most patients surgery is generally unsuccessful at solving this problem (Chap. 20). Surgery may relieve the sensitivity of the skin but usually does not solve the problem of the steady and shock-like pain. This is likely because of the damage to the spinal cord, nerve root, and ganglion so that most surgical procedures cannot access the area at or proximal to the injury. Surgical treatment can even worsen the situation producing anesthesia dolorosa (pain in a numb area) or provide temporary relief at best.

Differing pharmacodynamics of the various drugs used to treat PHN and the limitations of monotherapy provide a rationale for the use of combinations of drugs,

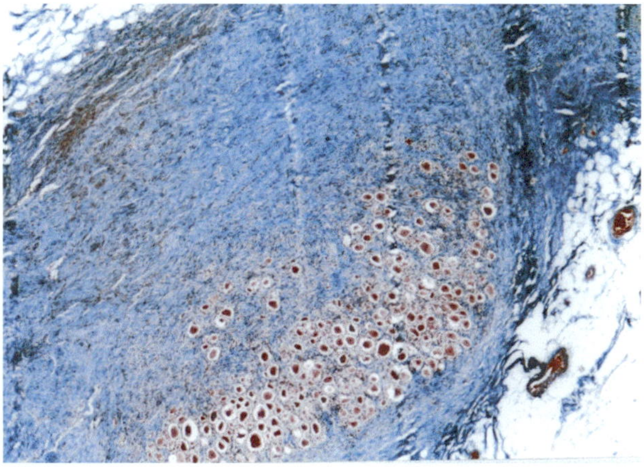

Fig. 19.4 Scarring in a dorsal root ganglion associated with postherpetic neuralgia. Surviving ganglion cells in red (Watson et al. 1988)

which may also limit adverse effects because of lower doses. TCAs and SNRIs potentiate the inhibitory neurotransmitters noradrenaline and serotonin in pain-inhibitory pathways descending from the brainstem to the spinal cord. Gabapentinoids are alpha-2-delta calcium channel modulators, and opioids act on opioid receptors. Despite this specific knowledge regarding pharmacodynamics, a good mechanism-based monotherapy continues to elude us. Although the shock-like pain component resembles trigeminal neuralgia (TN), the sodium channel blocker carbamazepine (the closest we have to a mechanism-based treatment for TN and so successful in TN) is usually a failure in PHN. Drugs such as TCAs, gabapentinoids, and opioids affect indiscriminately all features of the pain, that is, the steady-burning, shock-like pain, and sensitivity of the skin (allodynia) [54, 55]. We can achieve moderate or better improvement in only half to two-thirds of patient with established PHN in RCTs and few have complete relief. Results are likely not as good when generalized to clinical practice (external validity) where patients are older, have other diseases, and are on other drugs [52]. Perhaps one reason for the intractability is the severe damage to the dorsal horn of the spinal cord (Fig. 19.2) so that receptors where pain-inhibitory drugs such as opioids, TCAs, and gabapentinoids drugs might act have been destroyed or damaged. This scenario argues very strongly for early, aggressive treatment of HZ with antivirals and analgesics but especially by zoster prevention vaccines.

19.3 Management Options

There are three possible approaches to managing PHN: (1) the treatment of established PHN, (2) the prevention of PHN by early and aggressive treatment of HZ, and (3) the prevention of HZ and PHN by vaccination.

The treatment of PHN remains difficult and follows consistent guidelines for neuropathic pain (NP) from Canada (Fig. 19.5), Europe, and the USA [1, 15, 17, 18,

TCA ←→ gabapentin or pregabalin (add additional agents sequentially if partial but inadequate relief ++)

↓

SNRI and Topical lidocaine*

↓

Tramadol or CR opioid analgesic

↓

Fourth line agents+

Fig. 19.5 Stepwise pharmacological management of neuropathic pain [34]. *5 % gel or cream – useful for focal neuropathy such as postherpetic neuralgia (the lidocaine patch is not available in Canada), + cannabinoids, methadone, lamotrigine, topiramate, and valproic acid; ++ do not add serotonin noradrenaline reuptake inhibitors (SNRIs) to tricyclic antidepressants (TCAs). CR controlled release

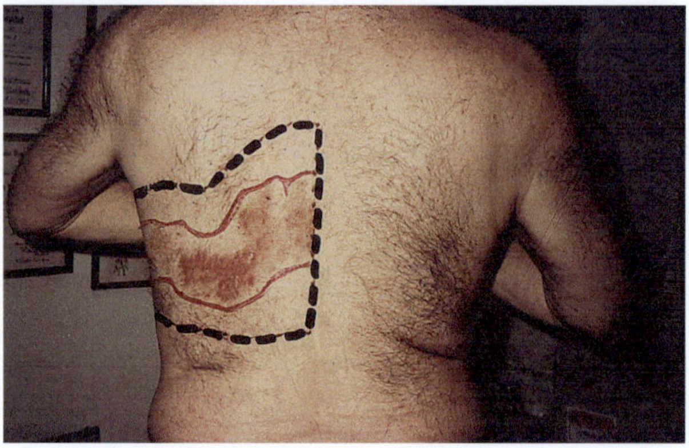

Fig. 19.6 Postherpetic neuralgia 3 months after the rash: skin lesions soon after rash healing surrounded by an area of anesthesia to punctate touch [*solid line*] and pin with wider area of pain on moving touch of cotton or tissue [interrupted line]. Moving the hair on this hirsute individual is exquisitely painful. Firm pressure is soothing

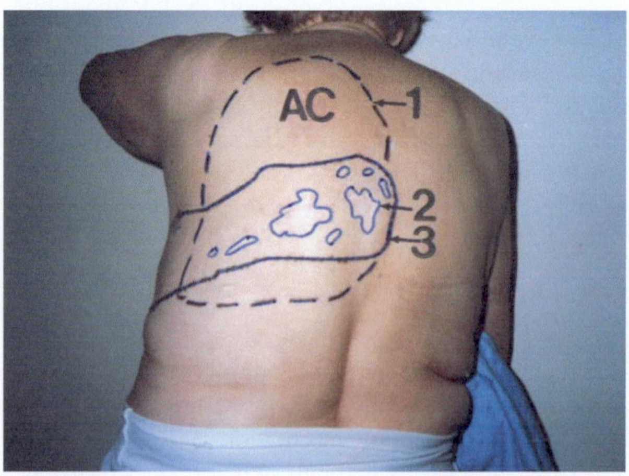

Fig. 19.7 Long duration postherpetic neuralgia 12 months after the rash. (*1*) Margin of allodynia (pain from stroking with cotton), (*2*) scarring, and (*3*) area of sensory loss

19, 22, 34]. These view gabapentinoids, TCAs, SNRIs, and topical agents as first choices and other drugs including opioids for refractory cases. Surgical procedures have generally no role in PHN (Chap. 20). Because of the limitations of the medical and surgical treatment of PHN, it is important to consider preventative measures.

Prevention by treating acute zoster (Chap. 23) is problematic because this approach presumably works better if given in a timely fashion. Often the pain occurs days before the rash onset, making the diagnosis difficult, or one might have pain without a rash (zoster sine herpete). The oral choices are famciclovir or valaciclovir. Valaciclovir is a prodrug for acyclovir but is better absorbed orally. These are safe drugs. For severely affected patients or immunocompromised patients, acyclovir can be given intravenously. The problem is that the data indicate that antivirals are not effective (acyclovir) or unproven (famciclovir, valaciclovir) at preventing severe PHN ([6], Chap. 23). Additionally one can also and concurrently treat acute zoster aggressively by giving a TCA, such as nortriptyline or amitriptyline [3], or a gabapentinoid or both as soon as HZ occurs. It is good medicine to relieve severe, acute HZ pain with strong medications including opioids, and this may also have a preventative effect, but this is largely unproven (Chap. 23). With acute severe, burning, and shock-like unilateral radicular pain without rash (zoster sine herpete) in typical locations for zoster such as the forehead or midthoracic area, it has been the author's approach to begin a course of antiviral medication even without the rash as famciclovir and valaciclovir are safe drugs.

Vaccination by a shingles prevention vaccine [31, 33, 37] is the first truly preventative measure for any neuropathic pain problem, specifically PHN. The first approved vaccine reduces the incidence of HZ by about 50 % and the occurrence of PHN by two-thirds; thus many vaccinated individuals, if they get HZ, may experience reduced and/or shortened symptoms. This live, attenuated vaccine that is 14 times the potency of varicella vaccine has few adverse effects [primarily injection site reactions] and is approved in Canada for immunocompetent adults aged 50 years and older. There are some logistical problems with the vaccine in Canada [it is frozen], but in other countries it is refrigerator stable. The vaccine may not be covered by government health plans or by private insurance, but in many countries and in Canada, it is about $200.00. The frozen vaccine has to be reconstituted in the physician's office and must be given within 30 min or it loses efficacy. Frequently asked questions about the vaccine, such as duration of protection, efficacy, effective age, previous HZ, concomitant administration with other vaccines, used in immunocompromised patients, and others, can be obtained from the reference [44].

It is important to note that the natural history of HZ and PHN is one of improvement with time, so that overall for patients of all ages with HZ, only about 10 % will have significant pain at a month; however, if one is 60 years old, the figure may be 50 % [10]. If one looks at patients who have had PHN for a year or more, about 1 year later, about half of the pain will be mild or gone, but the rest will continue to suffer [51]. PHN affecting the forehead has the additional problem of the threat to vision and the cosmetic disturbance of facial scarring. Thus there is even more impetus here for preventing the problem in this common site for viral eruption.

19.4 How Effective Are Pharmacological Agents for PHN in Clinical Practice?

An important question for clinicians regarding the favorable results of randomized controlled trials is how satisfactory these drugs are for PHN patients in ordinary practice (the external validity or generalizability of these data) where patients are older with comorbidities and on other drugs regarding pain relief, function, side effects, and long-term benefit [52]. Many RCTs answer questions focused on regulatory requirements rather than clinical use. Clinical trials, which demonstrate a statistically significant difference in rating scales, may not clearly convey effectiveness. The best clinical trial is unlikely to duplicate clinical practice because of factors such as subject selection even if the analysis reports data on all enrolled subjects (intent to treat analysis).

As well, studies comparing a new drug with a standard drug are few in number since they pose obvious problems, if industry sponsored, of numbers of patients required and also the risk that although the new agent may be better than placebo, it may not be as effective and safe as a standard therapy [53]. In the absence of comparative head-to-head trials, the number needed to treat (NNT) has been suggested to convey the clinical meaningfulness of a trial and to provide a comparison of different drugs [8]. This evaluation is a description of an arbitrary therapeutic effect for a desired outcome such as 50 % improvement or more. It describes the difference between interventions, such as a drug and a control treatment. It is expressed as the number of patients required to be treated for a favorable response (Table 19.1). Tricyclic antidepressants and oxycodone may have an advantage in relieving pain in PHN according to NNT data (although opioids must be a final choice (Chap. 18)). Numbers needed to harm (NNH) figures do not clearly indicate that newer drugs such as gabapentin, pregabalin, or duloxetine are safer than tricyclic antidepressants or opioids, although they have been marketed that way without comparative studies. NNT and NNH figures should be treated with some caution because of differing study designs, numbers, and data analyses.

19.5 Practical Guidelines for the Prevention and Treatment of Postherpetic Neuralgia

19.5.1 The Prevention of PHN by Treatment of HZ (Also Chap. 23)

Most of the putative preventive approaches by treating HZ at onset can be regarded as not conclusively established by more than one controlled trial with adequate numbers of patients. Pending final proof, it is still reasonable to treat patients aggressively to relieve the acute, severe pain of HZ (even if suspected before the rash appears) and to try to prevent severe PHN if the therapy is safe and well tolerated. It is important to

Table 19.1 Number needed to treat (NNT) data for at least 50 % relief in postherpetic neuralgia and some other neuropathic pain conditions

Drug	PHN	PDN	PN,NP	Central pain	FM	Comments
TCAs						
Collins 2000 [7]	2.1	3.5				Review
Sindrup, Jensen 1999 [45]			2.6 PN			Review
Saarto, Wiffen 2010 [43]	2.7	1.3	3.6 NP			Review
Finnerup 2010 [19]	2.8		2.1 PN	2.7		Review
Imipramine						
Sindrup 2003 [47]			2.7 PN			RCT
Saarto, Wiffen 2010 [43]			2.2 NP			Review
SSRIs						
Sindrup, Jensen 2000 [46]			6.7 PN			Review
Finnerup 2010 [19]			6.8 PN			Review
SNRIs (venlafaxine, duloxetine)						
Finnerup 2010 [19]			5.0			Review
Venlafaxine						
Sindrup 2003 [47]			5.2 PN			RCT
Rowbotham 2004 [41]		4.5				RCT
Saarto,Wiffen2010 [43]			3.1 NP			Review
Duloxetine						
Kajdasz 2007 [29]						
60 mg/day		5.3				Review
120 mg/day		5.7				
Lunn 2009 [32]						Review
60 mg/day		6			8	
Gabapentin						
SindrupJensen 2000 [46]			4.1 PN			Review
Finnerup 2010 [19]	4.3		6.4 NP			Review
Pregabalin						
Dworkin 2003 [14]	3.4					RCT
Finnerup 2010 [19]	4.2		4.5 PN	5.6		Review
Oxycodone						
Watson 1998 [54]	2.5					RCT
Tramadol						
Sindrup, Jensen 2000 [46]			3.4 PN			Review
Finnerup 2010 [19]	4.8		4.9 NP			Review

Caution should be used in interpreting these figures as they involve studies of differing experimental designs, numbers of patients, and data analyses

PHN postherpetic neuralgia, *FM* fibromyalgia, *PDN* painful diabetic neuropathy, *NP* neuropathic pain, *PN* painful neuropathy, *RCT* randomized controlled trial, *TCA* tricyclic antidepressants

recognize that the population at highest risk for PHN is the age group 50 to 60 years and over, who may have a risk of up to 50 % or more of developing this complication. Valaciclovir and famciclovir are best given within the first 72 h for 7 days but may not prevent severe PHN [6]. Amitriptyline, if used early, may also help to prevent ongoing pain [3] as may a gabapentinoid (gabapentin, pregabalin), although this is unproven (Chap. 23). Although no well-conducted RCT has been done of nerve blocks to treat HZ pain or prevent PHN, they are reasonable and safe in experienced hands, may help to relieve acute severe pain, and may be repeated, if effective, as symptoms dictate. The use of nonsteroidal anti-inflammatory drugs, acetaminophen and even opioids (not codeine), is also justified to relieve severe pain with the acute illness. With opioids patients need to be carefully screened for a family and personal history of addiction and psychiatric illness, given the guidelines, a contract considered and need to be carefully followed (see Chap. 18). There is no good evidence supporting the use of corticosteroids for HZ to prevent PHN [60]. Whether better control of acute pain will reduce the occurrence of severe PHN needs to be further evaluated.

19.5.2 Prevention by Vaccination

The future will likely include the routine vaccination of those over age 50 to prevent HZ and hence PHN [37] (reduces HZ by 50 % and PHN by 2/3) and may be the best way of dealing with these frequently intractable disorders. A vaccine currently undergoing trials appears more effective and applicable to the significantly immunosuppressed who cannot have the current live vaccine [31, 33].

19.5.3 A Practical Approach to the Treatment of Postherpetic Neuralgia

Although there is an increasing evidence base for treating PHN, there are limits to the generalizability of these RCT data, and the clinical management of severe PHN especially in patients with concomitant diseases and on other drugs remains very much an art. The pharmacodynamics and effectiveness (NNT, NNH) of the drugs used to treat PHN have been dealt with in previous sections of this chapter. Because of the variability of responsiveness and of adverse events in different individuals, it is important to consider some general rules in this author's opinion as it is in other conditions. There is no "cookie cutter" (one size fits all) approach. Precision medicine calls for the customization of treatment tailored to the individual. This in ordinary practice currently is a trial-and-error process.

1. *Consider the patient and his or her history with a view to age, gender, body weight, concomitant disorders, and current drug therapy and choose the first drug accordingly.*
2. *Start low and go slow.*

3. *Inform the patient of the common side effects: constipation (TCAs, opioids), drowsiness (most drugs), weight gain (TCAs, gabapentinoids), dry mouth (TCAs), rash (most drugs), cardiac risk (TCAs), limb swelling (gabapentiniods), and urinary retention in elderly males with prostatism (TCAs).*
4. *Consider preemptive use of a laxative, mouth spray, daily weight records, and diet.*
5. *Consider drugs in combination with different actions (gabapentinoids, TCAs, opioids) and topical agents (lidocaine, capsaicin).*
6. *Consider opioids screen with the Opioid Risk Tool.*
7. *Arrange initial follow-up in about 2 weeks and regularly thereafter record data, such as pain severity, drugs, dose and frequency, dose adjustment, adverse effects, function, and quality of life, with opioids any signs of misuse and a copy of the prescription.*

For established PHN a first-line drug may be monotherapy with a TCA, SNRI, gabapentinoid (gabapentin, pregabalin), or a topical agent (lidocaine patch, capsaicin, an NSAID). A TCA may not be first choice with heart disease (re arrhythmia), glaucoma, and a male with prostatism or in any elderly patient because of other side effects. Both antidepressants and gabapentinoids cause weight gain in some and patients need to be informed of this. Several RCTs indicate that pain may be taken from moderate or severe to mild in about one-half of two-thirds of patients by a TCA (amitriptyline, nortriptyline) in RCTs if not contraindicated. In suitable patients one can commence with nortriptyline (less side effects than amitriptyline) [55] in a dose of 10 mg before bed in those over 65 years and with 25 mg in those 65 or under. Amitriptyline may be preferable if insomnia is an issue as it is more sedating. The dose of either may be increased by similar increments in a single bedtime dose every 7–10 days until relief is obtained or side effects supervene. If this fails, one can try a more noradrenergic agent, such as desipramine or maprotiline. An SNRI antidepressant (duloxetine, venlafaxine [41]) is another choice supported by RCTs but with a higher NNT [59]. Antidepressants proven useful in other NP states by RCT may be worth trying. In patients with contraindications (older, cardiac risk, prostatism) to the analgesic antidepressants, an alternative approach is to use the anticonvulsant gabapentin titrated from as little as 100 mg/day as much as 3600 mg/day in divided doses or pregabalin titrated slowly up to 150 mg twice daily [14, 42]. An extended release form of gabapentin has come on the market in the USA based on placebo-controlled RCTs [27]. Combining drugs with differing actions (a TCA, gabapentinoid, topical agent) is reasonable. Occasional patients failing these may benefit from a more serotonergic antidepressant such as trazodone, clomipramine, fluoxetine, or other SSRIs, but no good controlled trial of these drugs has been done in PHN, and the author has not been impressed with these agents for most individuals. A trial-and-error approach in refractory patients may also include the anticonvulsants, carbamazepine, oxcarbazepine, topiramate, lamotrigine, phenytoin, clonazepam, valproate, and other new anticonvulsants all not proven by RCT [61]. Patients must be cautioned about possible adverse effects as listed above.

In the author's opinion, with severe and intractable PHN, analgesics including strong opioids (short acting on an as needed and/or long acting and round the clock basis) may be necessary. This is a major decision perhaps best made by a pain specialist or by a pain clinic (See Chap. 18) [20]. These patients must be screened with such as the Opioid Risk Tool (personal and family addiction history, psychiatric illness) and carefully monitored at follow-up re benefit, side effects, and misuse. Codeine (such as acetaminophen/codeine/caffeine) is a poor choice because its action in most patients is due to its unpredictable and poor metabolic conversion (PM) to usually small amounts or no morphine at all unless the patient is from northeast Africa [Somalia, Kenya] where ultrametabolizers (UM) may produce large amounts of morphine. Both PM and UM are because of pharmacogenetics [56]. The moderately effective tramadol (short or long acting) is reasonable as it appears to have a lower tendency for abuse, but NNT figures indicate it as less effective than stronger opioids (Table 19.1). A variety of short and long-acting opioids are available (morphine, oxycodone, hydromorphone). With opioids the start-low/go-slow approach is vital. I usually start with short acting used as needed to gain an impression with very severe PHN pain quickly as to whether opioids generally or one in particular may be more useful in a long-acting form. The dose can be very gradually titrated to satisfactory relief or unacceptable side effects. Concomitant dietary advice and the use of a stool softener or other laxatives are important (always) and sometimes an antinauseant such as metoclopramide initially.

The use of topical agents is attractive as it is simple and free of systemic effects. These can be used as monotherapy or as adjuvant agents. These include lidocaine, capsaicin, and nonsteroidal anti-inflammatory drugs (acetylsalicylic acid, diclofenac). A lidocaine patch has been shown to be useful by RCTs [40]. Topical capsaicin is commonly available as .025 and .075 % preparations given three or four times daily for 2 or 3 weeks as a trial [5]. An intolerable but harmless burning sensation is a limiting side effect. A high-potency 8 % patch of capsaicin is available in the USA [2], but requires care and experience and is best administered in a pain clinic setting.

A number of treatments can be regarded as of occasional use but scientifically unproven; however absence of evidence is not always evidence of absence. Responses of individuals or subgroups may be hard to capture by RCT, and a placebo effect is not to be denigrated if the treatment is safe and economical. Transcutaneous electrical nerve stimulation (TENS) may be worth trying. Electrode placement, frequency, intensity, and duration of stimulation are a matter of trial and error. Acupuncture is unproven but rarely of use. Some patients with severe PHN may benefit from nerve blocks, which, if efficacious, may be repeated at appropriate intervals; however, scientifically based data regarding the efficacy of nerve blocks and other interventional approaches for either prevention or long-term treatment of PHN are not available (Chap. 17). A new suggestion that PHN may be caused by a persistent VZV ganglionitis may result in the use of antiviral drugs at this stage (see Chap. 14).

At least 30 % of our patients remain totally refractory or unsatisfactorily relieved, and our approach with those is to see them regularly and try any new or older approach that seems reasonable and safe, hoping that with kind attention and time, as Voltaire said "nature will effect a cure," and keeping the initial quote of Edgar Hope-Simpson in mind.

19.6 Summary

Established postherpetic neuralgia remains a challenging problem. Although we have made moderate advances in drug treatment, a proportion of patients are inadequately or not relieved of this neuropathic pain. The principles of drug treatment of chronic postherpetic neuralgia follow a number of guidelines for neuropathic pain in general. The prevention of herpes zoster and postherpetic neuralgia appears to be key at this point in time. Attempted prevention at the stage of herpes zoster by an antiviral agent and analgesics is important and good practice for the relief of severe acute pain by a number of drugs but of uncertain value in the prevention of severe postherpetic neuralgia. Zoster prevention vaccines appear most important at present in the immunocompetent patient 50 years of age and older, and the current vaccine is safe and moderately effective at preventing postherpetic neuralgia [37]. A more effective vaccine applicable to the immunosuppressed is undergoing clinical trials at present [31].

References

1. Attal N, Cruccu G, Baron R et al (2010) EFNS guidelines on the pharmacological treatment of neuropathic pain: revision. Eur J Neurol 17:1113–1123
2. Backonja MD, Wallace MS, Bionski ER et al (2008) NGX-4010, a high concentration capsaicin patch for the treatment of postherpetic neuralgia: a randomized , double-blind study. Lancet Neurol 7(12):1106–1112
3. Bowsher D (1996) Postherpetic neuralgia and its treatment: a retrospective survey of 191 patients. J Pain Symptom Manage 12:327–331
4. Brisson M, Pellisier JM, Camden S, Quach C, DeWals S (2008) The potential cost-effectiveness of vaccination against herpes zoster and postherpetic neuralgia. Hum Vaccin 4:238–245
5. Capsaicin patch for postherpetic neuralgia (2011) The Medical Letter 53(Issue 1365):42–43
6. Chen N, Qifu L, Yang J (2014) Antiviral treatment for preventing postherpetic neuralgia, The Cochrane Collaboration.
7. Collins SL, Moore RA, McQuay HJ (2000) Antidepressants and anticonvulsants for diabetic neuropathy and postherpetic neuralgia: a quantitative systematic review. J Pain Symptom Manage 20:449–458
8. Cook RJ, Sackett DL (1995) The number needed to treat: a clinically useful measure of treatment effect. Br Med J ;310:452–454 (review). The Cochrane Collaboration
9. Coplan PM, Schmader K, Nikas A (2004) Development of a measure of burden of pain due to herpes zoster and postherpetic neuralgia for prevention trials; adaptation of the brief pain inventory. J Pain 5:344–356

10. Demoragas JM, Kierland RR (1957) The outcome of patients with herpes zoster. Arch Dermatol 75:193–119
11. Donahue JG, Choo PW, Manson JG et al (1995) The incidence of herpes zoster. Arch Intern Med 155:1605–1609
12. Drolet M, Brisson M et al (2010) A prospective study of the herpes zoster severity of illness. Clin J Pain 26:656–666
13. Drolet M, Brisson M, Schmader K et al (2010) Predictors of postherpetic neuralgia among patients with herpes zoster :a prospective study. J Pain 11:1211–1221
14. Dworkin RH, Corbin AE, Young JP et al (2003) Pregabalin for the treatment of postherpetic neuralgia; a randomized placebo-controlled trial. Neurology 60(8):1274–1283
15. Dworkin RH, O'Connor AB, Backonja M et al (2007) Pharmacologic management of neuropathic pain:evidence-based recommendations. Pain 132:237–251
16. Dworkin RH, O'Connor AB, Audette J et al (2010) Recommendations for the pharmacological management of neuropathic pain: an overview and literature update. Mayo Clin Proc 85(3 suppl):S3–S14
17. Dworkin RH, O'Connor AB, Backonja M et al (2007) Pharmacologic management of neuropathic pain:evidence-based recommendations. Pain 132:237–251
18. Dworkin RH, O'Connor AB, Audette J et al (2010) Recommendations for the pharmacological management of neuropathic pain: an overview and literature update. Mayo Clin Proc 85(3 suppl):S3–S14
19. Finnerup NNB, Sindrup SH, Jensen TS (2010) The evidence for pharmacological treatment of neuropathic pain. Pain 150:573–581
20. Frieden TR, Houry D (2016) Reducing the risks of relief- the CDC opioid- prescribing guideline. N Engl J Med 374:1501–1504
21. Gauthier A, Breuer J, Carrington D (2009) Epidemiology and cost of herpes zoster and postherpetic neuralgia in the United Kingdom. Epidemiol Infect 137:38–47
22. Gilron I, Watson CPN, Cahill CM, Moulin DE (2006) Neuropathic pain: a practical guide for the clinician. CMAJ 175(3):265–275
23. Head H, Campbell AW (1900) The pathology of herpes zoster and its bearing on sensory localization. Brain 23:353–523
24. Hope-Simpson RE (1965) The nature of herpes zoster: a long term study and a new hypothesis. Proc R Soc Med 58:9–20
25. Hope-Simpson RE (2001) Some early investigations into the nature of herpes zoster and postherpetic neuralgia. In: Watson CPN, Gershon AA (eds) Herpes Zoster and postherpetic neuralgia, 2nd revised and enlarged edition. Elsevier, pp 1–12
26. Insinga DRP, Itzler RF, Pellisier JM et al (2005) The incidence of herpes zoster in a US administrative database. J Intern Med 20(8):748–783
27. Irving G, Jensen M, Cramer M (2009) Efficacy and tolerability of gastric-retentive gabapentin for the treatment of postherpetic neuralgia. Clin J Pain 25:185–192
28. Johnson RW, Rice ASC (2014) Postherpetic neuralgia. N Engl J Med 371(16):1526–1532
29. Kajdasz DK, Iyengar S, Desaiah D et al (2007) Duloxetine for the management of diabetic painful neuropathic pain:evidence-based findings from post-hoc analysis of three multicenter, randomized, double-blind, placebo-controlled, parallel group studies. Clin Ther 29:2536–2546
30. Kurzke JR (1984) Neuroepidemiology. Ann Neurol 16:265–277
31. Lal H et al (2015) Efficacy of an adjuvanted herpes zoster subunit vaccine in older adults. N Engl J Med 372:2087–2096
32. Lunn MPT, Hughes RAC, Wiffen PJ (2009) Duloxetine for treating painful neuropathy or chronic pain (Review) The Cochrane Collaboration Issue (4):CD007115
33. Morrison V, Johnson GR, Schmader KE (2015) Long-term persistence of zoster vaccine efficacy. CID 60:900–909
34. Moulin DE, Clark AJ, Gilron I et al (2007) Pharmacological management of chronic neuropathic pain-Consensus statement and guidelines from the Canadian Pain Society. Pain Res Manag 12:13–21

35. Noordenbos W (1959) PAIN: problems pertaining to the transmission of nerve impulses which give rise to pain. Elsevier
36. Oaklander AL, Romans K, Horasdek S (1998) Unilateral postherpetic neuralgia is associated with bilateral sensory change. Ann Neurol 44:789–795
37. Oxman MN, Levin MD, Johnson JR (2005) A vaccine to prevent postherpetic neuralgia in older adults. N Engl J Med 352:2271–2284
38. Ragozzino MW, Melton LJ, Kurland LT (1982) Population-based study of herpes zoster and its sequelae. Medicine 61:310–316
39. Ragozzino MW, Melton RJ, Kurland LT (1982) Risk of cancer after herpes zoster: a population-based study. N Eng J Med 307:393–397
40. Rowbotham MC, Davies PS, Verkempinck C et al (1996) Lidocaine patch: a double-blind comtrolled study of a new treatment for postherpetic neuralgia. Pain 65:39
41. Rowbotham MC, Goli V, Kunz NR et al (2004) Venlafaxine extended release in the treatment of painful diabetic neuropathy: a double blind, placebo-controlled study. Pain 110:697–706
42. Rowbotham MC, Harden N, Stacey B et al (1998) Gabapentin for the treatment of postherpetic neuralgia: a randomized controlled trial. JAMA 280(21):1837–1842
43. Saarto T, Wiffen PJ 2010 Antidepressants for neuropathic pain. The Cochrane Collaboration Issue (1):CD005454
44. Shapiro M, Kvern B, Watson CPN et al (2011) Update on herpes zoster vaccination: A family practioner's guide. Can Fam Phys 57:1127–1132
45. Sindrup SH, Jensen TS (1999) Efficacy of pharmacological treatment of neuropathic pain: an update and effect related to mechanism of drug action. Pain 83:389–400
46. Sindrup SH, Jensen T (2000) Pharmacological treatment of pain in polyneuropathy. Neurology 55:915–920
47. Sindrup SH, Bach FW, Madsen C et al (2003) Venlafaxine versus imipramine in painful polyneuropathy: a randomized controlled trial. Neurology 60:1284–1289
48. Van Wijck AJ, Opstelten W, Moore KG et al (2006) The PINE study of epidural steroids and local anesthetics to prevent postherpetic neuralgia: a randomized controlled trial. Lancet 367:219–224
49. Watson CPN, Morshead C, Van Der Kooy D et al (1988) Postherpetic neuralgia: postmortem analysis of a case. Pain 34:129–138
50. Watson CPN, Deck JH, Morshead C, Van der Kooy D, Evans RJ (1991) Postherpetic neuralgia: further postmortem studies of cases with and without pain. Pain 44:105–117
51. Watson CPN, Watt VR, Chipman M, Birkett N, Evans RJ (1991) The prognosis with postherpetic neuralgia. Pain 46:195–199
52. Watson CPN (2007) External validity of pharmaceutical trials in neuropathic pain. In: Rothwell PM (ed) The Lancet: treating individuals from randomized trials to personalized medicine. Elsevier, pp 121–130
53. Watson CPN, Gilron I, Sawynok J (2010) A qualitative, systematic review of head-to-head randomized, controlled trials of oral analgesics in neuropathic pain. J Pain Res Manage 15(3):147–157
54. Watson CPN, Babul N (1998) Oxycodone relieves neuropathic pain: a randomized trial in postherpetic neuralgia. Neurology 50:1837–1841
55. Watson CPN, Vernich L, Chipman M, Reed K (1998) Amitriptyline versus nortriptyline in postherpetic neuralgia. Neurology 51:1166–1171
56. Watson CPN (2011) A death knell for codeine after craniotomy. Can Jnl Neur Sci 38(3):390–391
57. Watson CPN, Midha R, Devor M, et al (2000) Trigeminal postherpetic neuralgia postmortem: clinically unilateral, pathologically bilateral. In: Devor M, Rowbotham MC, Wiesenfeld–Hallin Z (eds) Proceedings of the 9th world congress on pain
58. Watson CPN (2001) A tribute to R. Edgar Hope-Simpson, O.B.E., F.R.C.G.P. In: Watson CPN, Gershon A (eds) Herpes zoster and postherpetic neuralgia, Elsevier Science, B.V. Amsterdam, p xxiii

59. Watson CPN, Gilron I, Sawynok J, Lynch M (2011) Antidepressant analgesics and pain: are serotonin norepinephrine reuptake inhibitors any better? Pain 152:2206–2210
60. Whitley RJ, Weiss H, Gnann JW et al (1996) Acyclovir with and without prednisone for the treatment of herpes zoster. Ann Intern Med 125:376–383
61. Wiffen P, Collins S, McQuay H et al (2001) Anticonvulsant drugs for acute and chronic pain. (Cochrane review). The Cochrane library Issue 3 Oxford update software

Chapter 20
Postherpetic Neuralgia: Are There Neurosurgical Options?

C. Peter N. Watson

> ***1993:*** *There are no controlled or prospective studies of surgical procedures for the management of the pain of postherpetic neuralgia. It is, unfortunately, exceedingly unlikely that any will be undertaken. Indeed, only a very small fraction of the patients who have undergone surgery of any type have been reported in the literature; hence, the true utility of any operation remains unknown. John D. Loeser [27]*
>
> ***2002****: Surgical procedures do not appear to be useful for most patients with postherpetic neuralgia (PHN)…The results of the few surgical procedures reported have been discouraging, and for most patients are not an option because of age, risk, and limited benefits. C. Peter N. Watson [44]*
>
> ***2015****: However, there will still be individuals, who develop medically intractable postherpetic neuralgia, and it is incumbent on neurosurgeons who treat this condition to report their outcomes honestly and completely….I continue to believe that there is a role for surgery in PHN. We just have to prove it. Kim J. Burchiel [6]*
>
> *The efficient physician is he who amuses his patients while nature effects a cure. The Philosophical Dictionary, Voltaire (1694–1778)*

20.1 Introduction

The first three quotations above span 23 years and outline the views of different chapter authors at particular times in regard to the surgical treatment of postherpetic neuralgia. The first quotation is by a senior and eminent pain neurosurgeon [27]. The second is my opinion based on a literature review for a book chapter in 2002 [44].

C.P.N. Watson
Department of Medicine, University of Toronto,
99 Avenue Rd, Toronto, ON M5R 2G5, Canada
e-mail: peter.watson@utoronto.ca

C.P.N. Watson et al. (eds.), *Herpes Zoster: Postherpetic Neuralgia and Other Complications,* DOI 10.1007/978-3-319-44348-5_20

The third quotation does inject a note of optimism by another very experienced and highly respected pain neurosurgeon [6]. This chapter will update the neurosurgical literature on this subject.

A search was conducted for articles in English for "postherpetic neuralgia and surgery," using PubMed, Medline, Ovid, and Embase. Regional anesthesia approaches will not be discussed here but are covered in Chap. 17. This literature consists of case reports and case series (Class III level of evidence). References chosen are of the most recent case reports and case series, previous reviews, and from original seminal books and articles.

The questions to be addressed by this chapter are whether things have changed and how these data may best be evaluated. Although individual case reports and case series occupy a lower tier of medical evidence, this may be, for practical reasons, the only way to approach the issue of the value of uncommonly performed procedures. A carefully chosen, medically intractable case (or cases), without a control, with well-documented outcome measures and good long-term follow-up can be compelling. A nonsurgical extreme example of a conclusive single case report in point is that it only takes one case of uniformly fatal meningococcal meningitis to be cured by penicillin to indicate efficacy.

The discussion here regarding surgery will focus on the specific challenge in surgically treating PHN based on the neuropathological changes that are associated with this condition and the pitfalls in selecting intractable patients for surgery and the assessment (using valid and reliable outcome measures) of published case reports and case series.

The last quotation by Voltaire highlights an important aspect of the interpretation of any uncontrolled study such as a neurosurgical procedure in PHN: that many patients with persistent pain after herpes zoster (HZ), some early and quickly (especially in younger age groups) and some more slowly, improve by virtue of the natural history of time-related improvement of PHN but which also has a strongly age-related increase [19]. An example is that any uncontrolled study of nerve blocks for acute zoster in patients which includes all ages will have good results based on the natural history of resolution which occurs as " nature effects a cure." Selecting patients with intractable PHN of at least 6 months duration in those over 60 years of age selects out a mostly intractable group, who are unlikely to improve with time. The natural history and other factors plague and confound the interpretation of many uncontrolled studies in many of these case reports of neurosurgical procedures. Reports of these operations and the results will be organized anatomically in a logical progression from the periphery to the cerebral cortex with these concerns in mind.

The surgeon reader particularly may find all of this somewhat discouraging, but this book will, as well, review here and refer to important advances in medical management covered more completely in book section IV, Chapters 18 and 19. Because of the increased incidence of HZ and PHN with age, as one approaches the fourth decade of life, there should be a personal interest in this common and disabling disease with its intractable pain and other complications.

Evidence-based medical advances involve (1) the treatment of PHN (Sect. IV), (the best medical treatment still leaves some patients who are virtually medically untreatable for whom surgical treatment needs to be considered almost as a "hail

Mary" option, but keeping in mind the principle of "primum non nocere"), (2) the aggressive, timely treatment of acute HZ (Chap. 23), and (3) excitement about successful prevention with zoster prevention vaccines (Chap. 24).

20.1.1 Definitions

PHN is neuropathic (nerve injury) pain (NP), which is the most common and feared complication following herpes zoster (HZ). PHN may be defined arbitrarily in different ways and for different purposes. It is pain that persists after rash healing. This may be tallied at 1 month or, for clinical trials, at 3 or 6 months since many patients improve in the weeks following the initial eruption, and therefore a definition of a longer duration means greater pain stability especially for randomized controlled trials (RCTs), particularly those of a crossover design but also for surgical procedures.

20.1.2 Epidemiology and Natural History *(Chap. 4)*

HZ, the precursor of PHN, is due to reactivation of the varicella zoster virus (VZV) in the spinal and cranial sensory ganglia often a half century following a primary infection with varicella (chicken pox) usually during childhood. HZ is characterized by a unilateral, cutaneous, painful, vesicular rash typically in a single dermatome (often midthoracic or trigeminal ophthalmic division) often resulting in PHN which has been found to be the commonest neurological disease [25].

In Canada (population nearly 35,000,000) there are 130,000 cases of HZ and 17,000 of PHN per year [3]. In the USA the annual incidence of herpes zoster is one million (cdc.gov, [20]). The risk of zoster over a lifetime is one in three (cdc.gov). The incidence is directly related to age (Fig. 20.1) [19] and due to decreased cell-mediated immunity. Overall about 10 % of HZ cases will have pain at 1 month after the rash, and this may rise to as much as 50 % at age 60 [3]. The increase in HZ and PHN that begins at ages 50–60 provides the rationale for vaccination commencing at this time (Fig. 20.1, [19]). Because PHN may fail to resolve within a year in a proportion, the prevalence of PHN is cumulative and higher. Because the population is aging and with the increase in immune-suppressed groups afflicted with cancer and HIV, HZ and PHN will likely increase.

20.1.3 Pathology and Putative Pain Mechanisms

In understanding a possible role for surgical procedures in PHN, it is critical to appreciate what is known about the pathology of this condition. There is considerable information about the pathology and possible pathogenesis of PHN. It has been

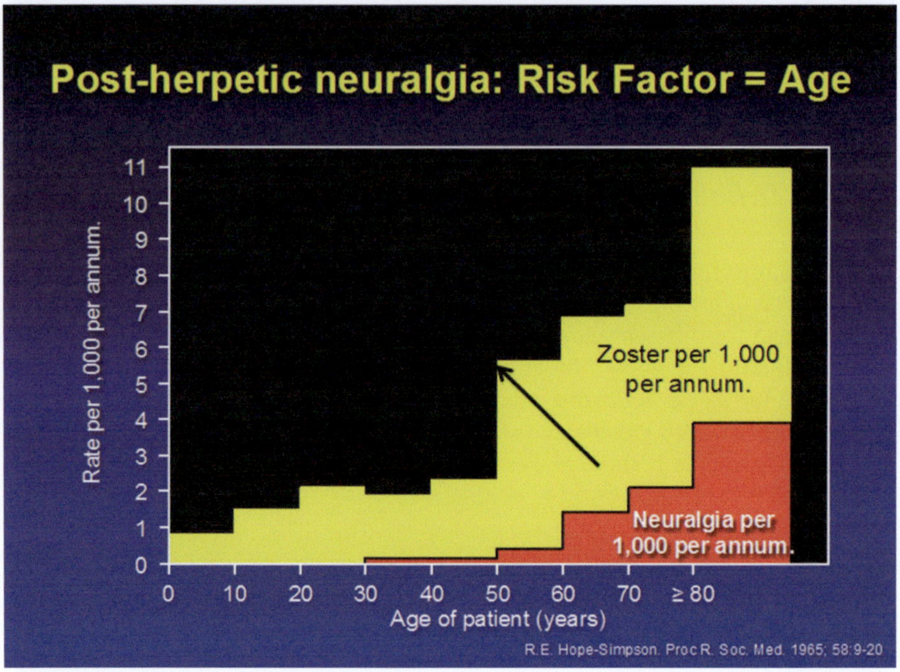

Fig. 20.1 Hope-Simpson's graph of the increasing incidence of herpes zoster and postherpetic neuralgia with age 2. The increase in herpes zoster and PHN after age 50 (*arrow*) is the rational for the use of the zoster vaccine at this time age [19]

known for more than 100 years that pathologically there is an acute hemorrhagic inflammation in one dorsal root ganglion at the stage of the eruption of HZ [17]. Inflammation then extends proximally and distally. Proximally this pathology extends into the spinal cord [42, 43]. After months, there is significant scarring and loss of neurons in the affected dorsal root ganglion [42] and atrophy and scarring of the dorsal horn of the spinal cord (Figs. 20.2 and 20.3) [42, 43]. Some of these long-standing cases have persistent inflammatory cells [43]. An assessment of the nerve fiber population in the peripheral nerve after the eruption of HZ shows a predominance of small (some probably pain-conducting) fibers and a deficiency in large myelinated (pain-inhibitory fibers) [29, 42] although many of these small fibers are likely regenerating neurites of all types.

Despite various and fairly consistent guidelines (Fig. 20.4) [12, 28] for NP for several drugs, PHN may be difficult and even impossible to treat even with opioids. Pathological evidence suggests that VZV causes permanent damage to the central and peripheral nervous system probably destroying sites of intrinsic pain inhibitory mechanisms where analgesics act especially the dorsal horn of the spinal cord (Fig. 20.2). This pathology if generalizable to other intractable cases may bear on the type of surgical procedure to be considered as it is difficult to access areas proximal to this damage.

Fig. 20.2 Atrophy of the dorsal horn of the spinal cord in postherpetic neuralgia [42]

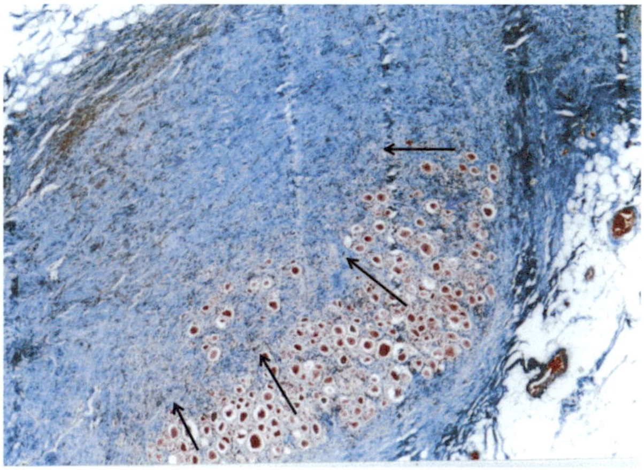

Fig. 20.3 Scarring in the dorsal root ganglion with postherpetic neuralgia (*arrows*) [42]

1) **First choice**: tricyclic antidepressant (amitriptyline/ nortriptyline) or gabapentin or pregabalin (add additional agents sequentially if partial but inadequate relief)

2) **Second choice**: serotonin/norepinephrine re-uptake inhibitor (SNRI) =(duloxetine) ++ and topical lidocaine*

3) **Third choice**: tramadol or opioid (morphine, oxycodone, hydromorphone, transdermal fentanyl)

4) **Fourth line agents+**

Fig. 20.4 Stepwise pharmacological management of neuropathic pain [28]. *Five percent gel or cream or lidocaine patch – useful for focal neuropathy such as postherpetic neuralgia (the lidocaine patch is not available in Canada), + cannabinoids, methadone, lamotrigine, topiramate, and valproic acid; ++ do not add serotonin-noradrenaline reuptake inhibitors (SNRIs) to tricylic antidepressants (TCAs) [28] (* for references see Chap. 19)

20.1.4 *Clinical Features (Figs 20.5 and 20.6)*

A good appreciation of the clinical aspects of the pain of postherpetic neuralgia is crucial to assessing patients for surgery and to evaluating the results of a surgical procedure for publication. An example is knowing that there may be three main components to this pain (e.g., steady burning, shocks, and pain on touch) which can be separately rated and which may not all be equally relieved by an operation. For the publication of cases, valid and reliable outcome measures are critical (Chaps. 19 and 21).

When the acute rash has healed, the affected skin often exhibits a reddish, purple, or brownish hue (Fig. 20.5). As this subsides, pale scarring often remains (Fig. 20.6). Occasionally, severe pain with no residual scar may occur, or the scars in very long duration cases are barely perceptible. A steady burning or aching may occur and also a paroxysmal, lancinating pain. Both may occur spontaneously and are often aggravated by any contact with the involved skin such as friction from even the lightest clothing. Firm pressure on the skin may curiously be soothing. Some patients describe unbearable itch, formication (ants crawling on the skin), or other forms of dysesthesia. As well as clothing contact, these symptoms may be exacerbated by physical activity, temperature change, and emotional upset.

The scarred areas are usually at least hypoesthetic and often anesthetic to *localized* touch, cold, and pain, and yet paradoxically this affected skin often exhibits marked pain on *moving* tactile stimulation (dynamic mechanical allodynia) or cold and/or increased *pain* to the noxious stimulation of a pinprick (hyperalgesia) or an increased *sensitivity* to moving touch stimuli (hyperesthesia) (Figs. 20.5 and 20.6). This affected, scarred skin often reveals a loss of sensation to pinprick, temperature, and touch over a wider area than the scars and even wider area of sensitive or painful skin (Figs. 20.5 and 20.6). This sensitive skin may paradoxically include the area anesthetic to punctate touch where it is elicited by light stroking or by skin

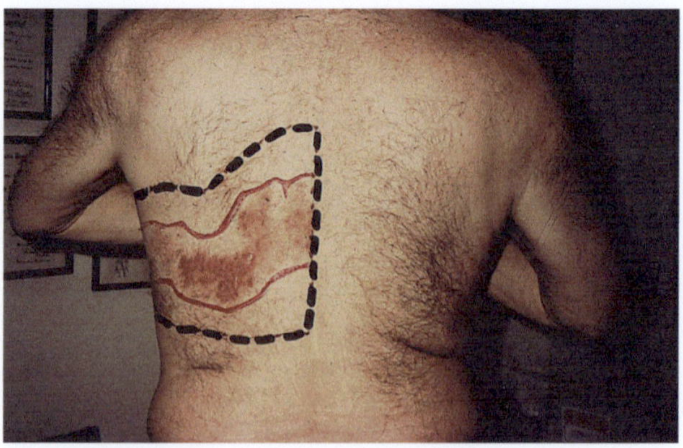

Fig. 20.5 Postherpetic neuralgia 3 months after the rash: skin lesions soon after rash healing surrounded by an area of anesthesia to punctate touch [solid line] and pin with wider area of pain on moving touch of cotton or tissue [interrupted line]. Moving the hair on this hirsute individual is exquisitely painful. Firm pressure is soothing

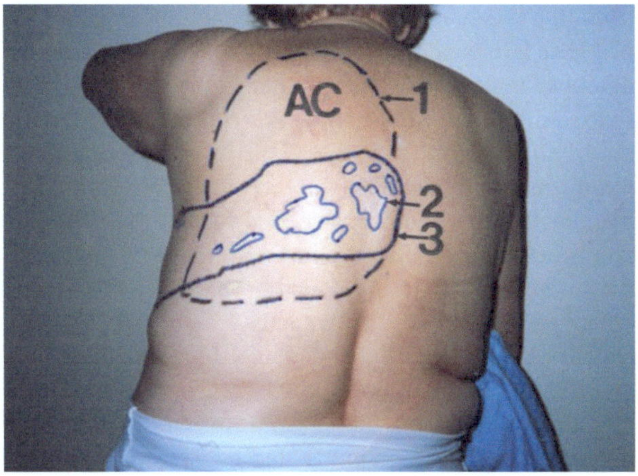

Fig. 20.6 Long-duration postherpetic neuralgia 12 months after the rash: (1) margin of allodynia (pain from stroking with cotton), (2) scarring, and (3) area of sensory loss

traction between the thumb and forefinger, an effect which may be caused by summation on hypersensitive, deafferented spinal dorsal horn neurons with expanded receptive fields.

20.1.5 *Outcomes: Management Options (Book Sects. IV and V)*

There are three possible approaches to managing PHN: (1) the treatment of established PHN, (2) the prevention of PHN by early and aggressive treatment of HZ, and (3) the prevention of HZ and PHN by vaccination.

The treatment of PHN remains difficult and follows reasonably consistent guidelines from Canada, Europe, and the USA (Fig. 20.4) [1, 9, 12, 28]. These view gabapentinoids, tricyclic antidepressants (TCAs), serotonin-norepinephrine reuptake inhibitors (SNRIs) [45], and topical agents as initial choices with the careful use of opioids (Chap. 18, [13]) and trial and error drugs for refractory cases. Differing pharmacodynamics of the various drugs used to treat PHN and the limitations of monotherapy provide a rationale for the use of combinations of these drugs, which may also limit adverse effects because of lower doses. TCAs and SNRIs potentiate the inhibitory neurotransmitters noradrenaline and serotonin in pain-inhibitory pathways descending from the brainstem to the spinal cord, gabapentinoids are spinal cord alpha-2-delta calcium channel modulators, and opioids act on spinal and brainstem opioid receptors. Despite this specific knowledge regarding pharmacodynamics, a good mechanism-based treatment continues to elude us (Chap. 19). Although the shock-like pain component of postherpetic neuralgia resembles trigeminal neuralgia (TN), the sodium channel blocker carbamazepine (the closest we have to a mechanism-based treatment and so successful in TN) is usually a failure in PHN. Drugs such as TCAs, gabapentinoids, and opioids ameliorate indiscriminately all features of the pain, that is, the steady burning, shock-like pain, and sensitivity of the skin (allodynia). We can achieve moderate or better improvement in only half to two-thirds of patient with established PHN, and few have complete relief. There are very few good comparative drug trials, and comparative clinically meaningful data [numbers needed to treat (NNT) for 50 % or more improvement] are presented in Table 19.1 in Chap. 19. Perhaps one reason for the intractability is the severe damage to the dorsal horn of the spinal cord (Fig. 20.2) so that receptors where pain-inhibitory drugs such as opioids, TCAs, and gabapentinoids drugs might act have been destroyed or damaged. This scenario all argues very strongly for prevention by vaccination and early, aggressive treatment of HZ in an attempt to prevent this situation.

Prevention is by the early and aggressive treatment of HZ and by vaccination. The former is problematic because this approach presumably works better if given in a timely fashion. Often the pain occurs days before the rash onset, making the diagnosis difficult, or there might be pain without a rash (zoster sine herpete). If unilateral, dermatomal, burning, and jabbing pain occurs suddenly without a rash and involves the forehead or midthoracic area [as these are common sites for zoster], it is reasonable to commence treatment with an antiviral agent as these are safe drugs, and early prevention of viral replication is probably important. Even with timely administration, the effect of this appears limited in preventing severe PHN (Chap. 23). The choices are the oral antivirals famciclovir or valaciclovir. Valaciclovir is a prodrug for acyclovir but is better absorbed orally. For severely affected patients or immune-compromised patients, acyclovir can be given intravenously. The problem is that the data indicate that these are not very effective or not useful at all at preventing severe PHN [7]. Additionally, one can also and concurrently treat acute zoster aggressively by giving a TCA, such as nortriptyline or amitriptyline, [2] or a gabapentinoid or both as soon as HZ occurs. It is good medicine to relieve severe, acute HZ pain with strong medications including short-term strong opioids, and this may also have a preventative effect but this is largely unproven.

The shingles prevention vaccine [30] is the first truly preventative measure for a NP problem, specifically PHN. It reduces the incidence of HZ by about 50 % and the occurrence of PHN by two-thirds; thus, many vaccinated individuals, if they get HZ, experience attenuated or shortened symptoms. This live, attenuated vaccine that is 14× the potency of varicella vaccine has few adverse effects (primarily injection-site reactions) and is approved by the FDA in the USA for immune-competent adults aged 50 years and older. There are some logistical problems with the vaccine in Canada (it is frozen), but in other countries, it is refrigerator stable. The vaccine may not be covered by government health plans or by private insurance in many countries, and in Canada, it is about $200.00. The frozen vaccine has to be reconstituted in the physician's office and must be given within 30 min or it loses efficacy. Frequently asked questions about the vaccine, such as duration of protection, efficacy, effective age, previous HZ, concomitant administration with other vaccines, use in immune-compromised patients, and others, can be obtained from the reference [36]. A more effective vaccine which is applicable to the immunosuppressed patient is on the horizon [26].

20.1.6 Neurosurgical Approaches

Early reports [4, 38, 46] documented poor results with cordotomy, rhizotomy, and sympathectomy. A review of surgical procedures for this disease in 1951 [38] concluded that almost every operation was said to work occasionally for this disease but none consistently. White and Sweet [46] came to similar conclusions. These procedures included retrogasserian rhizotomy, avulsion of the supraorbital nerve or gasserian ganglion, greater superficial petrosal neurectomy, trigeminal tractotomy, cordotomy, stereotactic thalamotomy and mesencephalotomy, sympathectomy, and sensory corticectomy. Resection of the underlying skin in the involved areas also rarely seemed to provide long-term pain relief, despite initial reports of good results [4, 39]. Stereotactic trigeminal tractotomy was reported successful in three patients but with less than a year follow-up [18]. Dorsal root entry zone (DREZ) lesions were reported useful in 10/17 cases [14, 15]. Stimulation of the nucleus ventroposteromedialis was suggested [37] with one in three a good result.

With respect to these and the updated reports to follow, one issue, as can be seen from the diffuse peripheral and central nervous system pathological changes (Figs. 20.2 and 20.3) [42, 43], is that there is no clearly localized, surgically accessible lesion (as often occurs with trigeminal neuralgia and vascular loop compression of the V nerve root). The pathological anatomy is messy with widespread inflammation and scarring involving the nerve, ganglion, nerve root, and the dorsal horn in spinal cases. We know little about dysfunction, but concepts of the pathophysiology (Chaps. 13, 14, and 15) can involve various mechanisms (ectopic discharges and increased excitation both peripherally and centrally and loss of central inhibitory mechanisms).

The literature reviewed since the previous chapter has revealed case reports and case series of various surgical options which include (a) peripheral procedures: skin excision, peripheral nerve stimulation, gamma knife radiosurgery (also central), and ganglionectomy (radiofrequency, surgical) and (b) central nervous system interventions: dorsal root entry zone (DREZ) lesions, spinal cord stimulation, trigeminal tractotomy, deep brain stimulation, and motor cortex stimulation.

There are a number of important issues for the surgeon to consider in both evaluating this literature and also in selecting suitable surgical candidates and in publishing the results. For credibility the interpretation of these case reports needs to bear in mind some important factors. Many case reports lack at least some of this information.

I shall begin by discussing these reports moving from the peripheral procedures centrally. A summary of these and possible deficiencies in the following articles are found in Table 20.1.

Table 20.1 A summary of the articles and comments regarding deficiencies in quality (see final column bold/italics)

Author, date, surgery	*PHN* age duration location number	Medical treatment	Outcome measures	Results	Follow-up	Conclusion and *comments*
Petersen et al. [31, 32] Skin resection	Age 70 8 years right T6 *N*=1	Gabapentin, nortriptyline, methadone, lidocaine patch	VAS daily pain 0–10 allodynia	Free of allodynia reduced meds, "50 % better at 1 year," pain worse at 5.5 years	5.5 years	Skin resection not advised by authors *a well-written case report*
Johnson and Burchiel [21] Peripheral nerve stimulation (PNS)	Ages: 44, 61,83,86, mean duration in all ten was 47.5 months, V1, V2 N=4/10	"Medical failure" "anticonvulsant, tricyclic antidepressant, gabapentin, topical anesthetic, neurectomy, gangliolysis, MVD"	50 % relief or better	2/4 had 50 % decrease in pain meds and were satisfied, 30 % adverse effects, 80 % had 50 % relief at 2 years	27 months	PNS of V1 or V2 effective, prospective trial needed 50 % (2/4) benefited
Kun et al. [24] Radiofrequency ganglionectomy	47–86 years of age, mean duration 30 months, spinal PHN, *N*=49	Not stated	VAS pain	Excellent in 3, in the rest pain improved "reduced meds" VAS from severe to mild or moderate	12 weeks	"Further research needed" *unclear details of PHN, medical treatment not clear, short follow-up*

Urgosik et al. [40] Leksell gamma knife	64–86 years, no duration stated, postherpetic V nerve neuralgia, *N*=16	Not stated	% ain (0 % = pain free, 100 % = no change)excellent = 0–20 %, very good = 21–40 %, good=41–60 %)	Median 44 % at least good (60 % pain relief or more)	Median 33 months (range 8–34 months)	"Relatively successful and safe for V PHN" *problems here with diagnosis, duration of PHN, outcome measures*
Keep et al. [23] Gamma knife	Ages 56, 61,83, duration 18 and 21 months, and unknown in one, V nerve PHN, *N*=3	"Intensive medical treatment"	0-10 scale before and after surgery	2/3 had "good result"	4.5 years, 6 months	"Effect promising, a larger study required"
Rath et al. [33, 34] DREZ lesions	Age mean = 73 (65–82) years, duration = 6 months to 13 years, spinal PHN, *N*=10	"Extensive medical treatment"	"Pain as percentage of pre-operative levels"	2/10 "good", major side effects in 6/10	47 and 54 months	DREZ surgery abandoned for PHN
Samreen and Friedman [35] Nucleus caudalis DREZ V1 PHN	Age 79 years 7-week duration V1 PHN 7 weeks *N*=1	"Large doses of narcotics" on "Tegretol, Lyrica, hydromorphone, hydrocodone"	0–10 scale	No pain at 1 year	1 Year	*Short duration of PHN means it could have improved by natural history but 60 % chance it would not*
Kanpolat et al. [22] V nerve tract and nucleus lesions	Age and duration unknown, V nerve PHN, *N*=3/65	Unknown	VAS	No or mild pain at follow-up	Specific follow-up duration not known for these three PHN cases	*Details missing as noted*

(continued)

Table 20.1 (continued)

Author, date, surgery	*PHN* age duration location number	Medical treatment	Outcome measures	Results	Follow-up	Conclusion and *comments*
Green et al. [16] Deep brain stimulation (DBS)	Age of onset 20, duration 10 years, V1 PHN, steady and jabbing pain	"Maximal medical therapy" (amitriptyline, carbamazepine, lamotrigine, phenytoin)	VAS	No pain	6 months	*An unusual case clinically for PHN (young at onset, no detail rerash, scarring to be convincing, intractability unusual at her age) longer follow-up would be optimal*
Brown and Pilitsis [5] Motor cortex stimulation (MCS)	Ages 51,59, duration of all 10 mean 6 years (1–12), N= 2/10 with V nerve PHN	"Prior medical therapy"	VAS. McGill Pain Questionnaire 50 % pain reduction at 10 months	No or mild pain at "most recent follow-up"	Mean follow-up entire group 10 months (range 3 months to 2 years)	*Clinical details incomplete, follow-up details for two cases lumped with entire group*
Esfahani and Pisansky [11] Motor cortex stimulation (MCS)	Age 41 10-year duration, "two HZ rashes and Ramsay Hunt" *N*=1	"Refractory to muscle relaxants, antiepileptics" "baclofen, carbamazepine, clonazepam, duloxetine, gabapentin, tramadol" "responded to occipital nerve blocks"	VAS	VAS 10/10 to 0/10 and off all meds at "most recent follow-up"	Unknown precisely	*An unusual case because of young age, recurrent zoster, response to blocks of C2, follow-up duration unknown*
Ebel et al. [10] Motor cortex stimulation (MCS)	Age 81, 12-month duration, allodynia, *N*=1	Carbamazepine 600 mgm	"Excellent pain control" no rating scales before	No rating scales after	At 6 months pain noted to become resistant	*Wrong drug for PHN, inadequate medical therapy, no rating scales used*

VAS visual analogue scale

Petersen et al. [31] have reported a case of the relief of PHN by surgical removal of the painful skin and reviewed the literature back to 1900. This is an exemplary case report since it deals with a 65-year-old man with severe and medically intractable PHN of 8-year duration, uses established pain rating scales, and provides follow-up of 1 year postoperatively at which time the patient said he was at least 50 % better, had no allodynia, and had reduced medication. Unfortunately, in a subsequent publication [32], they reported that at 5.5 years post-op, the pain exceeded presurgery ratings, and they concluded that this form of surgery was not recommended.

A retrospective pilot study of peripheral nerve stimulation for ten patients with medically refractory trigeminal neuropathic pain included four patients with PHN in V1 or V2 [21]. Of the four, two reported 50 % or better relief, a decrease in pain medication, and satisfaction at median 27 months follow-up.

Pulsed radiofrequency ganglionectomy was carried out in 49 subjects with PHN and assessed at 12 weeks post-op [24]. They reported "55 % pain reduction" and that "three had excellent pain reduction and eight had partial relief but needed more medication" and that the remainder "experienced pain improvement and maintained or reduced their medication" and had an "improved quality of life."

Gamma knife radiosurgery for trigeminal PHN [40] directed at the root of the trigeminal nerve alone in 16 patients reported good relief in eight at 6 months (onset of relief median of 1 month) and failure in eight. At 1 year 6/7, at 2 years 4/6, and at 3 years 3/5 persisted with significant relief. Another article [23] combined gamma knife surgery targets of the trigeminal nerve and centromedian nucleus in three patients with 2/3 having good results for 4.5 years and 6 months.

DREZ lesions were reported for ten cases of spinal PHN in two articles describing the same patients [33, 34] with only two having good results at 47 and 54 months but with major complications in 6/10. DREZ treatment was "abandoned" by this group because of these results. A case report describes nucleus caudalis DREZ in a 79-year-old female with V1 PHN for 7 weeks on large doses of narcotics with total relief at 1 year [35].

A review of the literature and experience with 65 patients undergoing computed tomography-guided percutaneous trigeminal tractotomy-nucleotomy for a variety of facial pain syndromes over 20 years has been reported [22]. Included were three patients with PHN who were described as having good relief at follow-up. Mean follow-up in the entire group was 5.3 years (6 months–6 years), but specific follow-up duration of the PHN cases was not available. There was considerable discussion appended to this article by eminent neurosurgeons KM Burchiel, NM Boulis, O Sagher, JM Henderson, and DM Long. Caution was registered about the need for experience and risk, and the need for continued and detailed follow-up and was generally complimentary.

A case report of deep brain stimulation [16] in a 30-year-old female with presumed V1 PHN reported no pain at 6 months follow-up. In a study of motor cortex stimulation, 2/10 neuropathic pain patients had trigeminal PHN [11]. Ten months after surgery, all ten patients had at least 50 % pain reduction. Moderate and severe ratings in PHN became 0 and mild in the two PHN sufferers "at most recent follow-up" (entire group mean was 10 months, range 3 months–2 years). A case report of motor cortex stimulation [5] of intractable V1, V2 PHN of 10-year duration reported no pain at "most recent follow-up." One 81-year-old female with V1, V2 PHN for 12 months was treated with motor cortex stimulation for severe allodynia with excellent pain control lasting 6 months but with recurrence at that time [10].

A wise quotation from Kim Burchiel regarding surgery for PHN is, in my view, a good way to conclude this chapter:

"Highly invasive major surgery is probably never going to be an easy option for their care. Areas for consideration of more organized prospective protocols for the surgical treatment of PHN include stimulation of the peripheral nerve, motor cortex and sensory thalamus, as well as image-guided tractotomy-nucleotomy. The former have the two advantages of being testable and reversible: the latter is selective and minimally invasive.... I continue to believe that there is a role for surgery in PHN. We just have to prove it" [6].

20.2 Summary

The response to the question in the chapter title is that neurosurgical procedures may help some patients with long-standing, medically intractable PHN. It is not possible to recommend any one procedure for all neurosurgical centers, but it is reasonable to begin with the safest and simplest and that which has some evidence of success.

A review of recent articles has revealed some potential deficiencies in these accounts (Table 20.1). The neurosurgical literature and the selection of refractory cases of PHN for surgical procedures must be evaluated very carefully, and suggestions are outlined (Table 20.2). Patients should be referred to a center with the technical facilities necessary, the surgery carried out by experienced neurosurgeons, and optimally where significant numbers of such cases of the specific neurosurgical procedure have been accumulated, the results published in good-quality journals and with full disclosure of the limited benefits and risks. Continued publication of well-worked up and good-quality case reports and case series is important.

Established PHN remains a challenging problem both medically and surgically. Although we have made moderate advances in drug treatment, a proportion of patients are inadequately or not relieved of this neuropathic pain. The principles of drug treatment of chronic PHN follow guidelines for neuropathic pain in general [1, 9, 12] (Chap. 19). The prevention of PHN appears to be key at this point in time. Attempted prevention at the stage of the rash and/or acute pain onset of HZ is important and good practice for the relief of severe acute pain but of uncertain value in the prevention of severe PHN [7]. The current zoster prevention vaccine appears

Table 20.2 Suggested optimal criteria for assessing a published case report or selecting patients for surgery for postherpetic neuralgia

1. Severe daily pain (7–10/10 for at least half the day)
2. A correct diagnosis:
(a) Usually segmental neuropathic pain in the same dermatome as the herpetic rash (vesicles on an erythematous base), usually in V1 or midthoracic dermatomes, the most common sites (other sites possible)
(b) Residual pale or pigmented dermatomal scarring in rash dermatome (not always present)
(c) Steady burning pain, +/− electric shocks, and +/− skin sensitivity with pain on moving touch (dynamic mechanical allodynia) in the affected and adjacent dermatomes (due to expanded receptive fields)
3. Age of 60 or more (these patients are less likely to improve with time)
4. Pain of more than 6 months duration (pain unlikely to improve with time)
5. Failure of appropriate medical therapy [gabapentinoids (gabapentin , pregabalin), tricyclic antidepressants (amitriptyline, nortriptyline), opioids, or combinations of these]
6. Rating scales (see points 7–9 below) before and after surgery
7. Pain rating scales such as 0–10, category (mild, moderate, severe), visual analogue scale (VAS)
8. Depression and anxiety scales: the Hospital Anxiety and Depression Scale (HADS) [47]
9. Function rating scales: Brief Pain Inventory (BPI) [8]
10. Quality of life rating: Short-Form Health Survey version 2(SF12v2) [41]
11. Follow-up status at least at 1 year
12. If a heterogeneous group of cases of neuropathic pain cases follow up data on specific disorders such as PHN, it should be reported

important at present in the immune-competent patient 50 years of age and older and is approved by the FDA in the USA as safe and moderately effective at preventing PHN [30]. A new vaccine is on the horizon with greater efficacy and also applicability to immune-compromised patients [26].

> Whoever saves a life saves a world.
> The Talmud.
> Whoso gives life to a soul it shall be as if he had given life to mankind altogether.
> The Quran.

References

1. Attal N, Cruccu G, Baron R et al (2010) 2010 EFNS guidelines on the pharmacological treatment of neuropathic pain: 2010 revision. Eur J Neurol 17:1113–1123
2. Bowsher D (1996) Postherpetic neuralgia and its treatment: a retrospective survey of 191 patients. J Pain Symptom Manag 12:327–333
3. Brisson M, Pellisier JM, Camden S et al (2008) The potential cost-effectiveness of vaccination against herpes zoster and postherpetic neuralgia. Hum Vaccin 4:238–245
4. Browder J, de Veer JA (1949) Herpes zoster: a surgical procedure for the treatment of postherpetic neuralgia. Am Surg 130:622–636

5. Brown JA, Pilitsis JG (2005) Motor cortex stimulation for central and neuropathic facial pain: A prospective study of 10 patients and observations of enhanced sensory and motor function during stimulation. Neurosurgery 56:290–297
6. Burchiel Kim J (2015) Postherpetic Neuralgia; are there neurosurgical options? In: Burchiel Kim J (ed) Surgical management of pain. Thieme Medical Publishers, New York, p 269. Editor's comments re Watson CPN
7. Chen N, Li Q, Yang T et al (2014) Antiviral therapy for the treatment of postherpetic neuralgia. Cochrane Database Syst Rev (2):CD006866
8. Daut IE, Cleeland CS, Flanery RC (1983) Development of theWisconsin Brief Pain Questionnaire to assess pain in cancer and other diseases. Pain 17:197–210
9. Dworkin RH, O'Connor AB, Audette J et al (2010) 2010 recommendations for the pharmacological management of neuropathic pain: an overview and literature update. Mayo Clin Proc 85(3 suppl):S3–S14
10. Ebel H, Rust D, Tronnier V et al (1996) Chronic precentral stimulation in trigeminal neuropathic pain. Acta Neurochir 138:1300–1306
11. Esfahani DR, Pisansky MT (2011) Motor cortex stimulation: functional magnetic resonance imaging-localized treatment for three sources of intractable facial pain. J Neurosurg 114:189–195
12. Finnerup NB, Sindrup NH, Jensen TS (2010) The evidence for the pharmacological treatment of neuropathic pain. Pain 150:573–581
13. Frieden TC, Houry D (2016) Reducing the risk of relief- the CDC opioid-prescribing guideline. N Engl J Med 374:1501–1502
14. Friedman AH, Nashold BS Jr (1984) Dorsal root entry zone lesions for the treatment of postherpetic neuralgia. Neurosurgery 15:969–970
15. Friedman AH, Nashold BSJ, Ovelmen-Levitt J (1984) Dorsal root entry zone lesions for the treatment of postherpetic neuralgia. J Neurosurg 60:1258–1262
16. Green AL, Nandi D, Armstrong G et al (2003) Post-herpetic trigeminal neuralgia treated with deep brain stimulation. J Clin Neurosci 10(4):512–514
17. Head H, Campbell AW (1900) The pathology of herpes zoster and its bearing on sensory localization. Brain 23:353–523
18. Hitchcock ER, Schvarz JR (1972) Stereotactic trigeminal tractotomy for postherpetic facial pain. J Neurosurg 37:412–417
19. Hope-Simpson RE (1965) The nature of herpes zoster; a long-term study and new hypotheses. Proc R Soc Med 58:9–20
20. Insinga DRP, Itzler RF, Pellissier JM et al (2005) 2005 The incidence of herpes zoster in a United States administrative base. J Gen Intern Med 20:748–753
21. Johnson MD, Burchiel KJ (2004) Peripheral stimulation for treatment of postherpetic neuralgia and trigeminal posttraumatic neuropathic pain: a pilot study. Neurosurgery 55(1):135–142
22. Kanpolat Y, Kahilogullari G, Ugur HC (2008) Computed tomography guided percutaneous trigeminal tractotomy-nucleotomy. Neurosurgery 63:147–155
23. Keep MF, DeMare PA, Ashby LS (2005) Gamma knife surgery for refractory trigeminal postherpetic neuralgia: targeting in one session the retrogasserian trigeminal nerve and the centromedian nucleus. J Neuroimmunol Suppl 102:276–282
24. Kim YH, Lee CJ, Lee SC et al (2008) Effect of pulsed radiofrequency for postherpetic neuralgia. Acta sthesiol Scand 52:1140–1143
25. Kurzke JR (1984) Neuroepidemiology. Ann Neurol 16:265–277
26. Lal H, Cunningham AL, Godeaux et al (2015) Efficacy of an adjuvanted herpes zoster subunit vaccine in older adults. N Engl J Med 372:2087–2096
27. Loeser JD (1993) Surgery for postherpetic neuralgia. In: Watson CPN (ed) Herpes zoster and postherpetic neuralgia. Elsevier Science, Amsterdam, pp. 255–264
28. Moulin DE, Clark AJ, Gilron I et al (2007) Pharmacological management of chronic neuropathic pain-consensus statement and guidelines from the Canadian pain Society. Pain Res Manag 12:13–21

29. Noordenbos W 1959 Pain: problems related to the transmission of nerve impulses which give rise to pain. Amsterdsam, Elsevier. ch 1, p 4–10: ch 10, p 68–80.
30. Oxman MN, Levin MD, Johnson JR et al (2005) A vaccine to prevent postherpetic neuralgia in older adults. N Engl J Med 352:2271–2284
31. Petersen K, Rice FL, Suess F et al (2002) Relief of post-herpetic neuralgia by surgical removal of painful skin. Pain 98:119–126
32. Petersen KL, Rowbotham MC (2007) Relief of post-herpetic neuralgia by surgical removal of painful skin.: 5 years later. Pain 131:214–218
33. Rath SA, Seitz K, Soliman N et al (1997) DREZ coagulations for deafferentation pain related to spinal and peripheral nerve lesions: indications and results of 79 consecutive procedures. Stereotact Funct Neurosurg 68:161–167
34. Rath SA, Braun V, Soliman N et al (1996) Results of DREZ coagulations for pain related to plexus lesions. spinal cord injuries and postherpetic neuralgia. Acta Neurochir (Wien) 138:364–369
35. Samreen N, Friedman WA (2009) Nucleus caudalis dorsal root entry zone lesions: a clinical-radiographic report. Stereotact Funct Neurosurg 87:314–321
36. Shapiro M, Kvern B, Watson CPN et al (2011) Update on herpes zoster vaccination: a family practioner's guide. Can Fam Physician 57:1127–1132
37. Siegfried J (1982) Monopolar electrical stimulation of nucleus ventroposteromedialis for postherpetic facial pain. Appl Neuropsychol 45:179–184
38. Sugar O, Bucy PC (1951) Postherpetic trigeminal neuralgia. Arch Neurol Psychiatr 65:1313–1135
39. Tindall GT, Odom GL, Vierth RG (1962) Surgical treatment of postherpetic neuralgia. Results of skin undermining in 14 patients. Arch Neurol 7:423–426
40. Urgosik D, Vymazal J, Vladyka V et al (2000) Treatment of postherpetic neuralgia with the Leksell gamma knife. J Neurosurg 93(Suppl3):165–168
41. Ware JE, Kosinski M, Keller SD (1996) A 12- item short-form health survey: construction of scales and preliminary tests of reliability and validity. Med Care 34(3):220–233
42. Watson CPN, Morshead C, Van Der Kooy D et al (1988) Postherpetic neuralgia: postmortem analysis of a case. Pain 34:129–138
43. Watson CPN, Deck JH, Morshead C, Van der Kooy D, Evans RJ (1991) Postherpetic neuralgia: further postmortem studies of cases with and without pain. Pain 44:105–117
44. Watson CPN (2002) Postherpetic neuralgia. Chapter 30. In: Burchiel KJ (ed) Surgical management of pain. Thieme Medical Publishers, New York, pp. 393–400
45. Watson CPN, Gilron I, Sawychok J, Lynch ME (2011) Nontricyclic antidepressants and pain: are the serotonin norepinephrine re-uptake inhibitors any better? Pain 21(152):2206–2221
46. White JC, Sweet WH (1969) Pain and the neurosurgeon. Charles C. Thomas, Springfield, pp. 1382–1384
47. Zigmond AS, Snaith RP (1983) The hospital anxiety and depression scale. Acta Psychiatr Scand 67:361–370

Chapter 21
Designing Randomized Controlled Trials of Oral Analgesics for Chronic Postherpetic Neuralgia

Ian Gilron and C. Peter N. Watson

21.1 Introduction

Postherpetic neuralgia (PHN) – pain persisting for more than 3 months after the resolution and healing of a varicellazoster eruption – is an important cause of chronic neuropathic pain and likely one of the most dreaded complications of the varicella zoster virus [31]. Recent estimates suggest a prevalence of PHN ranging from 3.9 to 42.0/100,000 person-years [61]. The prevalence of PHN may be expected to diminish with the development of a zoster vaccine [43] although widespread implementation of vaccination programs and long-term efficacy of this vaccine may be less than previously expected [41]. In addition to relative ease of documenting the diagnosis (e.g., acute onset, photograph, and follow-up of characteristic dermatomal zoster eruption), the unilateral nature and characteristic neuropathic features of burning or shock-like pain, allodynia, and sensory loss [35] make PHN one of the most frequently studied conditions for the development of new treatments for neuropathic pain [17]. Following earlier clinical observations and open-label studies, Watson and colleagues, in [63], conducted the first reported randomized controlled trial (RCT) of an oral agent (the antidepressant amitriptyline) to treat chronic pain associated with postherpetic neuralgia [63]. Since then, many dozens of RCTs have been reported involving many thousands of patients with PHN. Given the continuing prevalence of PHN and the potential for PHN trial results to be extrapolated to other neuropathic pain conditions [9], improvements in RCT design may facilitate

I. Gilron MD, MSc, FRCPC (✉)
Department of Anesthesiology and Perioperative Medicine,
Queen's University, Kingston, ON, Canada
e-mail: gilroni@queensu.ca

C.P.N. Watson
Department of Medicine, University of Toronto, Toronto, ON, Canada
e-mail: peter.watson@utoronto.ca

C.P.N. Watson et al. (eds.), *Herpes Zoster: Postherpetic Neuralgia and Other Complications*, DOI 10.1007/978-3-319-44348-5_21

the development of new and improved treatments for chronic PHN in particular and neuropathic pain in general.Ensuring *internal validity* of an RCT, evaluating the effects of a study treatment for PHN – in comparison with placebo or other treatment comparators – requires minimization of various sources of *bias* (systematic deviation from the true result), which can be largely accomplished through *random allocation* of each RCT participant to the different study treatments and effectively *blinding* trial participants and research personnel to treatment allocation [30]. Other potential risks of bias (Fig. 21.1) may persist despite randomization and blinding [25], and some of these may be mitigated by other trial design features (e.g., efforts to minimize trial dropouts and missing data). Another important feature of internal validity of a trial is *assay sensitivity* – defined as "the ability of an RCT to distinguish an effective treatment from a less effective or ineffective treatment" [13]. Several trial features that maximize internal validity (e.g., selection of motivated, compliant participants with a clear-cut diagnosis of PHN and no other comorbidities) can often conflict with the *external validity* or *generalizability* of the trial, and failure to balance internal vs. external validity runs the risk that the trial results have limited relevance in "real-world" practice [47]. This chapter will review the fundamental elements of clinical trials that should be considered in efforts to meet the trial goals and obtain an optimal balance between internal and external trial validity. Current challenges and future directions for trial designs of RCTs of treatments for chronic PHN will be discussed.

21.2 Elements of Clinical Trials

21.2.1 Selectionof Trial Participants

PHN may often be one of the most discrete and easily recognizable neuropathic pain conditions to diagnose. Nevertheless, several considerations should be given when selectingresearch patients with PHN in light of the intended purpose and generalizability of the analgesic clinical trial. Pain etiology is most often determined by history (with our without photographic documentation) of an acutely painful, unilateral, vesicular rash occurring in a dermatomal distribution [8]. Given the possibility that chronic pain due to some other non-neuropathic etiology could occur in the same bodily region as that of a previous zoster eruption, screening of prospective trial patients for a PHN trial could be further facilitated by completion of one of several previously validated neuropathic pain questionnaires/inventories including, but not limited to, the DN4, LANSS, and PainDETECT [3] to confirm the likelihood of neuropathic pain, e.g., by identifying characteristic features such as burning pain, allodynia, and/or electrical shock-like pains. In addition to confirming that chronic pain etiology is due to the preceding zoster rash, selecting trial patients according to duration and intensity of chronic pain has long been thought to be an important aspect of clinical trial design. Although PHN has been defined as

Fig. 21.1 Assessment of risk of bias in randomized controlled trials (To be reprinted – permission from publisher required – from: Moore et al. [39])

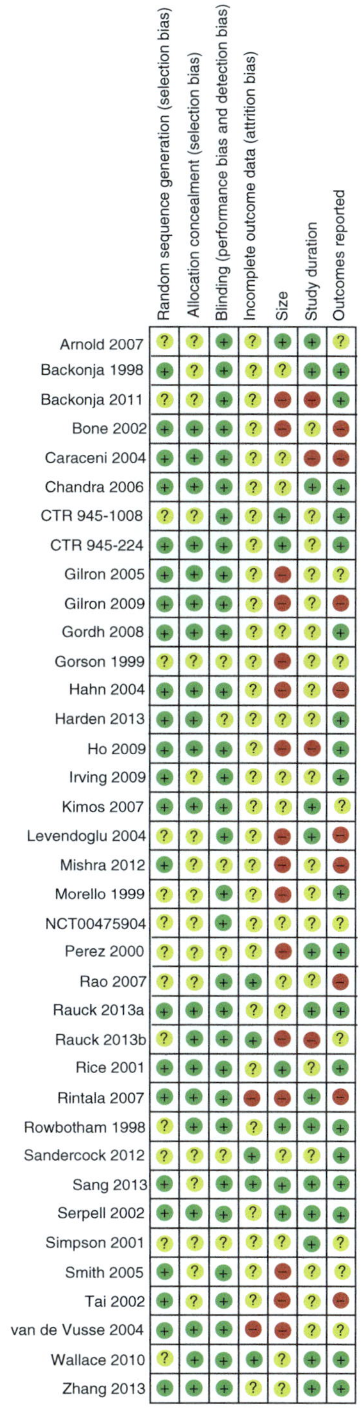

zoster-related pain that persists 4 months or more after onset of the zoster rash, it is suspected that a considerable number of individuals may experience pain resolution beyond this time point such that recruiting participants 6 months after rash onset may reduce the number of trial participants experiencing spontaneous pain resolution that would otherwise be falsely attributed to the study treatments [8]. Confirming also that chronic pain is experienced for more than half the day over that time period further reduces the likelihood of spontaneous pain remission during the trial. Regarding pain intensity, review of literature from trials of various chronic pain conditions suggests that selecting participants with at least moderate pain (i.e., ≥4/10) may reduce placebo response rates and identify larger treatment effects [12]. This, in part, could be due to avoidance of a "floor effect" whereby participants with only mild pain may be too close to "zero pain" to show a substantial treatment response vs. placebo. Exclusion of patients with other major medical comorbidities is a common practice since such participants may be unevenly distributed across treatment arms and major adverse events may be falsely attributed to study treatments.Similarly, careful consideration must be given as to what concomitant treatments participants should be allowed to receive during the trial. Exclusions may include any treatments known to adversely interact with the study treatments but also analgesic treatments that affect the outcome measures of the trial particularly if these treatments are not evenly distributed across treatment groups (i.e., for parallel group design trials) or taken consistently throughout the trial (i.e., for crossover trials). Other practical aspects of participant selection include adequate language and cognitive skills necessary to complete patient-reported assessments so as to provide meaningful and valid data for the trial.

21.2.2 Study Treatment, Placebo, and Comparator(s)

Selecting the treatment to study in an RCT should be directly related to the purpose of the trial. Until now, PHN trials have varied with respect to their intended purpose – from early-phase regulatory RCTs designed to provide the first-ever evidence of analgesic efficacy of a new molecular entity (e.g. [62],); to pragmatic RCTs designed to evaluate effectiveness of a known treatment in a broader, "real-world" patient population [48]; to "head-to-head" RCTs designed to compare the effectiveness of two or more known treatments for PHN (e.g. [45, 64],); and to testing the concept of mechanism-based pain treatment [7].In addition to the treatment of interest, other trial controlsto be evaluated in an RCT may include placebo, an alternative active comparator (e.g., a current standard therapy), and/or a lower dose of the study drug [12, 27, 28]. Recent evidence suggests increasing use of analgesic drug combinations in routine clinical practicegiven the recognized limitations of drug monotherapy [22]. Efforts to expand the evidence base to rigorously evaluate the safety and efficacy of various analgesic combinations have led to a growing number of analgesic combination trials in neuropathic pain [4] showing that some combination trials demonstrate superiority to monotherapy [19, 20, 23], some are equivocal with respect to the overall balance between efficacy and tolerability [26, 55], and, importantly, others suggest no added benefit or even inferior tolerability with the studied combination [14, 24, 34]. In addition to evaluating the potential for improved clinical outcomes with combination therapy, it is important to note that the additional value of a complete factorial combination trial design (i.e., placebo vs. drug A vs. drug B vs. A+B combination) is a built-in "head-to-head" comparison of each single agent [4], very few of which are available in neuropathic pain research [64].

Current regulatory thinking reflects that, in conditions where a delay in analgesic treatment is not expected to result in any long-term complications, use of a placebo comparator – an inactive intervention otherwise identical to the study intervention – is an optimal approach to evaluate a novel analgesic therapy [12].Investigational use of a placebo in chronic pain trials is considered ethical provided that study participants understand that they may withdraw from the trial at any time to pursue other pain treatments and/or that certain rescue analgesic treatment will be provided during the trial. Evidence indicates that pain improvement with placebo treatment can sometimes represent a clinically relevant treatment response, and considerable variability of placebo responses has been reported across trials [32, 60]. Positive responses to placebo may be related tononspecific factors including natural history of the pain condition, biopsychosocial effects of treatment expectation, and factors associated with clinical regression to the mean (e.g.,patients with fluctuating pain levels may more likely enroll in a trial when pain levels are highest) [58]. Since it is reasonable to expect that the same nonspecific"placebo responses" may also occur with active treatments, the use of a placebo comparator is essential for the quantification of treatment effect that can be specifically attributed to the actions of the investigational treatment. It should also be noted that patients treated with placebo also might report "nocebo" effects, i.e., adverse symptoms or responses that may be attributable to negative treatment expectations but possibly also underlying medical conditions or concomitant treatments [15]. Blinding of participants and study personnel to treatment allocation (e.g., to study treatment vs. placebo) is crucial in order to minimize bias in analgesic trials. Such blinding is facilitated by providing treatment and placebo interventions in an identical fashion, e.g., medication capsules identical in appearance, weight, smell, etc., as well as effective concealment of the treatment allocation schedule from study personnel [25, 30]. Since the majority of pain treatments often produce side effects (e.g., sedation, nausea, etc.) that can be readily recognized by study patients, the use of an inert placebo comparator – that does not directly produce any symptoms – can lead to partial unblinding. Therefore, administering a questionnaire to participants and study personnel that assesses blinding quality may serve to quantify the degree of unblinding [59]. Attempts to minimize the degree of unblinding related to recognizable side effects of the study treatment have included the use of an "active placebo," i.e., an agent with evidence of no therapeutic effect but with side effects similar to those of the study treatment so as to more closely mimic the study treatment. Examples of active placebos that have been used in previous trials of neuropathic pain include the benzodiazepine, lorazepam [20, 50]; the anticholinergic agent, benztropine [42, 65] and the antihistamine, diphenhydramine [67]. Although the critical importance of a placebo comparator to the internal validity of analgesic trials is well recognized, it should be noted that several phase 3 drug development programs have failed because – despite clinically relevant treatment improvement with the investigational treatment – comparable improvements in pain were reported during placebo treatment [11]. This observation has led to a hypothesis that modifying various clinical trial characteristics (e.g., standardizing research staff–patient interactions) might reduce placebo responses and increase the likelihood of identifying valid placebo-treatment differences [13]. However, currently there is no sufficient evidence to support this hypothesis, and any focused efforts on making such trial method modifications need to be weighed against possible adverse impact on external validity (i.e., generalizability).

Comparing a novel investigational treatment to an active treatment comparator (e.g., current standard therapy) can fulfill two important goals: (1) confirm "assay sensitivity" in the case of an otherwise negative trial and (2) compare therapeutic benefits of the study treatment with currently available therapy. Regarding the first goal, demonstrating a statistically significant difference between the active comparator and placebo may be crucial to confirm assay sensitivity of the trial [36] in situations where the study treatment fails to separate from placebo (i.e., fails to demonstrate efficacy). In terms of the second goal, providing a "head-to-head" comparison of the current standard with the study treatment provides important clinical relevance for the trial. For example, in situations where both active treatments are superior to placebo and, further, the study treatment is significantly superior to the standard treatment, such a result would serve to demonstrate new added benefit of the new treatment [64]. It should be noted that retrospective review of clinical trials of treatments for pain as well as depression suggests that trials with a larger number of treatment groups are, for various possible reasons, less likely to result in significant treatment vs. placebo differences [12]. Taken together, these considerations may lead to the recommendation of at least including one active treatment comparator to a placebo-controlled trial of a novel investigational treatment [12].

21.2.3 Trial Designs

The selection of trial design for the evaluation of a new analgesic clearly has a critical impact on its conduct, feasibility, and, possibly also, results. The most commonly used designs for analgesic clinical trials in postherpetic neuralgia include parallel groups, crossover, and enriched enrollment designs. A *parallel group design* (Fig. 21.2) – in which participants are randomly allocated (using a reliable method such as a computer-generated random sequence) to one of the two or more treatment groups in the trial – is the most common trial design and also the most accepted design for phase 3 regulatory trials (FDA [28]; EMA [27]).Within each treatment group, the study treatment is administered, and the outcome follow-up conducted, for a defined time period (e.g.,12-week treatment and outcome measurement

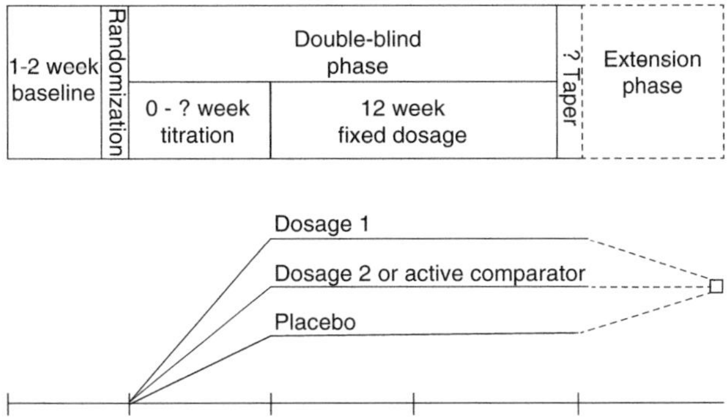

Fig. 21.2 Parallel group research design for confirmatory chronic pain clinical trials (To be reprinted – permission from publisher required – from: Dworkin et al. [12])

period). Since each participant in a parallel trial is randomized to only one of the treatment groups, a relatively larger trial size (e.g., compared to a crossover design) is needed to account for greater variability between treatment groups. It should be noted that not all trial participants are allowed the opportunity to try the study treatment, which presents potential ethical issues in placebo-controlled trials as well as limitations on trial recruitment and retention [12].

Alternatively, a *crossover trial design* involves random allocation of participants to a unique *sequence* of treatment periods occurring in consecutive order. For example, the most simple balanced crossover design comparing "treatment A" to placebo would randomize participants to receive either "treatment A" in the first treatment period followed by placebo in the second period or, in the other cohort, placebo first followed by "treatment A" second. Important threats to the validity of crossover designs include: (1) "period effect," where the natural history of the condition changes over time (e.g., postsurgical pain that diminishes over the days after surgery) such that outcomes in later treatment periods may falsely suggest greater treatment efficacy than those in earlier periods, and (2) "carryover effect," where effects of the treatment given in an earlier period persist into the subsequent period such that efficacy of the earlier treatment is falsely attributed to the subsequent treatment [18]. Period effects are likely minimal if the condition being studied with a crossover trial is known to be associated with stable pain levels over time and/or if return to baseline pain levels is documented in between each treatment period. Carryover effects are minimal if the study treatment has no long-lasting effects after discontinuation, if there is a washout period of suitable duration between each treatment period, and if the sequences of study treatments are balanced within the trial (e.g., A–B, B–A).

Relatively newer trial methodology involves an *enriched enrollment randomized withdrawal* (EERW) *design* whereby patients are randomized to *continue* the study treatment or switch to placebo (treatment withdrawal) only *after* having demonstrated a defined level of treatment response and/or tolerability to open-label (or single-blind) treatment with the investigational agent [33]. Potential benefits of the

EERW design include: (1) all trial participants are given the opportunity to try the study treatment, (2) the initial open-label (or single-blind) phase of the trial can provide estimates of treatment response and tolerability in "all comers" prior to enrichment,and (3) confirmation of efficacy in the double-blind phase is more efficient by restricting the trial cohort only to patients who tolerate and/or initially respond to the treatment. Concern has been raised that exposure to the study treatment during the open-label phase may partially unblind trial participants during the double-blind phase particularly if the study treatment produces distinct and easily recognizable side effects [53]. However, recent reviews suggest that partial or complete enrichment does not seem to bias in favor of active treatment [54] andavailable EERW analgesic trialssuggest "no gross differences" in efficacy results compared to those involving classical trial designs – although the amount of evidence for such comparisons is admittedly limited [40].Other trial designs used less commonly in chronic pain research include dose–response designs [49], n-of-1 trials [38], and add-on designs [68].

21.2.4 Outcome Measures

Based on decades of collective research, experience, and review of past clinical trials, the Initiative on Methods, Measurement, and Pain Assessment in Clinical Trials (IMMPACT) recommended six outcome domains that should be assessed in analgesic clinical trials: (1) pain, (2) physical functioning, (3) emotional functioning, (4) participant ratings of improvement and satisfaction with treatment, (5) symptoms and adverse events, and (6) participant disposition (e.g., adherence to the treatment regimen and reasons for premature withdrawal from the trial) [56].Subsequently, a wide range of different measurement tools were evaluated with respect to their validity and reliability leading to IMMPACT recommendations of core outcome measures (see Table 21.1 reprinted from [10]) for chronic pain clinical trials [10].In clinical conditions where pain is the most prominent symptom and in trials of interventions that are intended to reduce nociceptive transmission and/or perception, one or more measures of pain intensity or relief are generally accepted as the most relevant primary outcomes of an analgesic clinical trial [21]. However, currently available treatments provide clinically relevant reductions in pain in only 40–60 % of patients, and other outcomes such as physical, social, and occupational function are sometimes considered to have comparable importance. Within this context, a patient survey was conducted by IMMPACT to identify outcomes considered important to people suffering from pain, and the most prominent of those includedenjoyment of life, emotional well-being, fatigue, weakness, and sleep-related problems [57]. Given the multiple dimensions relevant to pain conditions, it is quite relevant to highlight a 2004 study by Coplan et al. [6] that adapted the Brief Pain Inventory [5] to more closely assess the burden of pain related to herpes zoster and postherpetic neuralgia [6]. Pain interference items used were similar to those of the original Brief Pain Inventory and included general activity, mood, walking, work, relations with

Table 21.1 Recommended core outcome measures for clinical trials of chronic pain treatment efficacy and effectiveness (To be reprinted – *permission required* – from: Dworkin et al. [10])

Pain
11-point (0–10) numerical rating scale of pain intensity
Usage of rescue analgesics
Categorical rating of pain intensity (none, mild, moderate, severe) in circumstances in which numerical ratings may be problematic
Physical functioning (either one of two measures)
Multidimensional Pain Inventory Interference Scale
Brief Pain Inventory interference items
Emotional functioning (at least one of two measures)
Beck Depression Inventory
Profile of Mood States
Participant ratings of global improvement and satisfaction with treatment
Patient Global Impression of Change
Symptoms and adverse events
Passive capture of spontaneously reported adverse events and symptoms and use of open-ended prompts
Participant disposition
Detailed information regarding participant recruitment and progress through the trial, including all information specified in the CONSORT guidelines

other people, sleep, and enjoyment of life [6], and this inventory has since been used in various trials of interventions for the prevention and treatment of PHN. In addition to evaluating outcomes of treatment efficacy, the assessment and reporting of adverse safety outcomes are critical for clinical trials of new interventions for chronic pain in general and PHN in particular. In 2004, an extension of the Consolidated Standards of Reporting Trials (CONSORT) statement was published describing several important elements of safety assessment and reporting (Table 21.2). Of great concern to the collective knowledge of pain treatment interventions, recent evidence suggests that, despite this 2004 extension, substantial deficiencies in safety assessment and reporting exist across studies and even in more recent trials of chronic pain treatments [52]. Continued improvements in this critical aspect of clinical trials require intensified awareness and implementation among clinical investigators and possibly also mandatory reporting requirements by journal editors.

21.2.5 Trial Duration and Sample Size

Selecting an optimal duration of treatment with study interventions requires a balance of feasibility and cost (that would favor a shorter trial duration) on one hand and maximizing validity and evaluating long-term exposure (that would favor a longer trial duration). Review of previous parallel-design neuropathic pain clinical trials suggests that pain reduction during placebo treatment often continues past

Table 21.2 Checklist of items to include when reporting harms in randomized, controlled trials[a] (To be reprinted – *permission required* – from: Ioannidis et al. [29])

Standard CONSORT checklist: paper section and topic	Standard CONSORT checklist: item number	Descriptor	Reported on page number
Title and abstract	1	If the study collected data on harms and benefits, the title or abstract should so state	
Introduction			
Background	2	If the trial addresses both harms and benefits, the introduction should so state	
Methods			
Participants	3		
Interventions	4		
Objectives	5		
Outcomes	6	List addressed adverse events with definitions for each (with attention, when relevant, to grading, expected vs. unexpected events, reference to standardized and validated definitions, and description of new definitions) Clarify how harms-related information was collected (mode of data collection, timing, attribution methods, intensity of ascertainment, and harms-related monitoring and stopping rules, if pertinent)	
Sample size	7		
Randomization			
Sequence generation	8		
Allocation concealment	9		
Implementation	10		
Blinding (masking)	11		
Statistical methods	12	Describe plans for presenting and analyzing information on harms (including coding, handling of recurrent events, specification of timing issues, handling of continuous measures, and any statistical analyses)	

(continued)

Table 21.2 (continued)

Results			
Participant flow	13	Describe for each arm the participant withdrawals that are due to harms and their experiences with the allocated treatment	
Recruitment	14		
Baseline data	15		
Numbers analyzed	16	Provide the denominators for analyses on harms Present the absolute risk per arm and per adverse event type, grade, and seriousness, and present appropriate metrics for recurrent events, continuous variables, and scale variables, whenever pertinent[b] Describe any subgroup analyses and exploratory analyses for harms[b]	
Outcomes and estimation	17		
Ancillary analyses	18		
Adverse events	19		
Discussion			
Interpretation	20	Provide a balanced discussion of benefits and harms with emphasis on study limitations, generalizability, and other sources of information on harms[c]	
Generalizability	21		
Overall evidence	22		

[a]This proposed extension for harms includes 10 recommendations that correspond to the original CONSORT checklist
[b]Descriptors refer to items 17, 18, and 19
[c]Descriptor refers to items 20, 21, and 22

4–5 weeks such that placebo-controlled, chronic pain treatment trials shorter than this could overestimate efficacy of the study intervention [44]. Based on these observations and the interest in maximizing the treatment observation period, current recommendations,at least for phase 3 confirmatory trials, suggest a 12-week treatment period [12]. Determining the necessary sample size is one of the first and most important tasks for designing a clinical trial. The fundamentals of sample size estimation [37] are first related to statistical hypothesis testing (e.g., to refute the null hypothesis that pain reduction with drug X is not different from pain reduction with placebo) and require also determinations of variance (e.g.,from previous studies involving similar populations) and statistical power (generally set between 80 and 90 %) and the treatment group difference to be detected (generally based upon previous evidence). However, beyond these considerations of statistical significance for a single study, collective review of multiple trials has underlined the importance of large trials (e.g., $n = 50–200$ or even greater) and/or meta-analysis of similar smaller trials in order to provide more robust estimates of treatment efficacy and also safety [1].

21.3 Current Challenges and Future Directions

Despite an explosion of knowledge surrounding the mechanisms of pain and potential analgesic targets over the past 50 years, the translation of preclinical science to new treatments for chronic pain has been disappointing [66]. Concerted efforts to address these challenges are being focused on enhancing the quality and predictive validity of both preclinical [2, 46] and clinical research [9]. A major challenge in clinical analgesic development is the evaluation of a new molecular entity that appears to be efficacious in earlier proof-of-concept trials (e.g.,"phase 2" regulatory trials) but then fails to demonstrate efficacy in larger phase 3 registration trials. It is conceivable that some negative phase 3 trials are actually "false negatives" such that an observed placebo-treatment difference (which might have been true) is not statistically significant, possibly due to excessive variability in outcome measurements, large pain reductions in the placebo group, and/or other causes of decreased assay sensitivity [9]. Such trial failures – if they are indeed false-negative results – prevent widespread access to a potentially beneficial new treatment and also represent a loss of millions of dollars invested in the development program for this failed treatment. Several strategies have been proposed [13], aimed at reducing the occurrence of false-negative trial results. Just some of these strategies, for which supportive research is ongoing, include training trial patients more carefully on how to more reliably rate their pain [51], limiting the number of clinical trial sites in order to reduce the magnitude of placebo response [13], excluding prospective trial candidates with highly variable baseline pain levels [16] and restricting the use of concomitant analgesic treatments during clinical trials [13]. It should be noted that some of these and other strategies might involve excluding certain prospective trial participants who suffer from the target condition and would otherwise require treatment in real-world practice. Therefore, implementation of such trial strategies for the purpose of increasing assay sensitivity of a clinical trial must be done cautiously and with careful attention to possible limitations in generalizability of trial results to broader populations of interest.

21.4 Conclusion

Randomized controlled clinical trials are essential tools for the evaluation of interventions for the treatment of zoster-related pain in general and PHN in particular. Such trials have been instrumental in providing new effective treatment for PHN and for providing valuable knowledge to help guide treatment choices for healthcare providers. Someaspects of current trial methodologies may lead to the occurrence of falsely negative trial results, while other factors may lead to positive results that do not necessarily generalize widely enough to the populations of interest. It is expected that continued research into refining trial methods would reliably improve both the validity and generalizability of clinical trials for zoster-related pain.

References

1. Moore RA, Eccleston C, Derry S, Wiffen P, Bell RF, Straube S, McQuay H, ACTINPAIN Writing Group of the IASP Special Interest Group on Systematic Reviews in Pain Relief; Cochrane Pain, Palliative and Supportive Care Systematic Review Group Editors (2010) "Evidence" in chronic pain – establishing best practice in the reporting of systematic reviews. Pain 150(3):386–389
2. Andrews NA, Latrémolière A, Basbaum AI, Mogil JS, Porreca F, Rice AS, Woolf CJ, Currie GL, Dworkin RH, Eisenach JC, Evans S, Gewandter JS, Gover TD, Handwerker H, Huang W, Iyengar S, Jensen MP, Kennedy JD, Lee N, Levine J, Lidster K, Machin I, McDermott MP, McMahon SB, Price TJ, Ross SE, Scherrer G, Seal RP, Sena ES, Silva E, Stone L, Svensson CI, Turk DC, Whiteside G (2016). Ensuring transparency and minimization of methodologic bias in preclinical pain research: PPRECISE considerations. Pain 157(4):901–909. Anesthesiology 2002;96(5):1053–1061
3. Bennett MI, Attal N, Backonja MM, Baron R, Bouhassira D, Freynhagen R, Scholz J, Tölle TR, Wittchen HU, Jensen TS (2007) Using screening tools to identify neuropathic pain. Pain 127(3):199–203
4. Chaparro LE, Wiffen PJ, Moore RA, Gilron I (2012). Combination pharmacotherapy for the treatment of neuropathic pain in adults. Cochrane Database Syst Rev (7):CD008943
5. Cleeland CS, Ryan KM (1994) Pain assessment: global use of the Brief Pain Inventory. Ann Acad Med Singapore 23(2):129–138
6. Coplan PM, Schmader K, Nikas A, Chan IS, Choo P, Levin MJ, Johnson G, Bauer M, Williams HM, Kaplan KM, Guess HA, Oxman MN (2004) Development of a measure of the burden of pain due to herpes zoster and postherpetic neuralgia for prevention trials: adaptation of the brief pain inventory. J Pain 5(6):344–356
7. Demant DT, Lund K, Vollert J, Maier C, Segerdahl M, Finnerup NB, Jensen TS, Sindrup SH (2014) The effect of oxcarbazepine in peripheral neuropathic pain depends on pain phenotype: a randomised, double-blind, placebo-controlled phenotype-stratified study. Pain 155(11):2263–2273
8. Dworkin RH, Gnann JW Jr, Oaklander AL, Raja SN, Schmader KE, Whitley RJ (2008) Diagnosis and assessment of pain associated with herpes zoster and postherpetic neuralgia. J Pain 9(1 Suppl 1):S37–S44
9. Dworkin RH, Turk DC, Basch E, Berger A, Cleeland C, Farrar JT, Haythornthwaite JA, Jensen MP, Kerns RD, Markman J, Porter L, Raja SN, Ross E, Todd K, Wallace M, Woolf CJ (2011) Considerations for extrapolating evidence of acute and chronic pain analgesic efficacy. Pain 152(8):1705–1708
10. Dworkin RH, Turk DC, Farrar JT, Haythornthwaite JA, Jensen MP, Katz NP, Kerns RD, Stucki G, Allen RR, Bellamy N, Carr DB, Chandler J, Cowan P, Dionne R, Galer BS, Hertz S, Jadad AR, Kramer LD, Manning DC, Martin S, McCormick CG, McDermott MP, McGrath P, Quessy S, Rappaport BA, Robbins W, Robinson JP, Rothman M, Royal MA, Simon L, Stauffer JW, Stein W, Tollett J, Wernicke J, Witter J, IMMPACT (2005) Core outcome measures for chronic pain clinical trials: IMMPACT recommendations. Pain. 113(1–2):9–19
11. Dworkin RH, Turk DC, Katz NP, Rowbotham MC, Peirce-Sandner S, Cerny I, Clingman CS, Eloff BC, Farrar JT, Kamp C, McDermott MP, Rappaport BA, Sanhai WR (2011) Evidence-based clinical trial design for chronic pain pharmacotherapy: a blueprint for ACTION. Pain 152(3 Suppl):S107–S115
12. Dworkin RH, Turk DC, Peirce-Sandner S, Baron R, Bellamy N, Burke LB, Chappell A, Chartier K, Cleeland CS, Costello A, Cowan P, Dimitrova R, Ellenberg S, Farrar JT, French JA, Gilron I, Hertz S, Jadad AR, Jay GW, Kalliomäki J, Katz NP, Kerns RD, Manning DC, McDermott MP, McGrath PJ, Narayana A, Porter L, Quessy S, Rappaport BA, Rauschkolb C, Reeve BB, Rhodes T, Sampaio C, Simpson DM, Stauffer JW, Stucki G, Tobias J, White RE, Witter J (2010) Research design considerations for confirmatory chronic pain clinical trials: IMMPACT recommendations. Pain. 149(2):177–193

13. Dworkin RH, Turk DC, Peirce-Sandner S, Burke LB, Farrar JT, Gilron I, Jensen MP, Katz NP, Raja SN, Rappaport BA, Rowbotham MC, Backonja MM, Baron R, Bellamy N, Bhagwagar Z, Costello A, Cowan P, Fang WC, Hertz S, Jay GW, Junor R, Kerns RD, Kerwin R, Kopecky EA, Lissin D, Malamut R, Markman JD, McDermott MP, Munera C, Porter L, Rauschkolb C, Rice AS, Sampaio C, Skljarevski V, Sommerville K, Stacey BR, Steigerwald I, Tobias J, Trentacosti AM, Wasan AD, Wells GA, Williams J, Witter J, Ziegler D (2012) Considerations for improving assay sensitivity in chronic pain clinical trials: IMMPACT recommendations. Pain. 153(6):1148–1158
14. Eichenberger U, Neff F, Sveticic G, Björgo S, Petersen-Felix S, Arendt-Nielsen L, Curatolo M (2008) Chronic phantom limb pain: the effects of calcitonin, ketamine, and their combination on pain and sensory thresholds. Anesth Analg 106(4):1265–1273
15. Enck P, Benedetti F, Schedlowski M (2008) New insights into the placebo and nocebo responses. Neuron 59(2):195–206
16. Farrar JT, Troxel AB, Haynes K, Gilron I, Kerns RD, Katz NP, Rappaport BA, Rowbotham MC, Tierney AM, Turk DC, Dworkin RH (2014) Effect of variability in the 7-day baseline pain diary on the assay sensitivity of neuropathic pain randomized clinical trials: an ACTTION study. Pain 155(8):1622–1631
17. Finnerup NB, Attal N, Haroutounian S, McNicol E, Baron R, Dworkin RH, Gilron I, Haanpää M, Hansson P, Jensen TS, Kamerman PR, Lund K, Moore A, Raja SN, Rice AS, Rowbotham M, Sena E, Siddall P, Smith BH, Wallace M (2015) Pharmacotherapy for neuropathic pain in adults: a systematic review and meta-analysis. Lancet Neurol 14:162
18. Gewandter JS, McDermott MP, McKeown A, Hoang K, Iwan K, Kralovic S, Rothstein D, Gilron I, Katz NP, Raja SN, Senn S, Smith SM, Turk DC, Dworkin RH (2016) Reporting of cross-over clinical trials of analgesic treatments for chronic pain: Analgesic, Anesthetic, and Addiction Clinical Trial Translations, Innovations, Opportunities, and Networks systematic review and recommendations. Pain 157(11):2544–2551
19. Gilron I, Bailey JM, Tu D, Holden RR, Jackson AC, Houlden RL (2009) Nortriptyline and gabapentin, alone and in combination for neuropathic pain: a double-blind, randomised controlled crossover trial. Lancet 374(9697):1252–1261
20. Gilron I, Bailey JM, Tu D, Holden RR, Weaver DF, Houlden RL (2005) Morphine, gabapentin, or their combination for neuropathic pain. N Engl J Med 352(13):1324–1334
21. Gilron I, Jensen MP (2011) Clinical trial methodology of pain treatment studies: selection and measurement of self-report primary outcomes for efficacy. Reg Anesth Pain Med 36(4):374–381
22. Gilron I, Jensen TS, Dickenson AH (2013) Combination pharmacotherapy for management of chronic pain: from bench to bedside. Lancet Neurol 12(11):1084–1095
23. Gilron I, Tu D, Holden RR, Jackson AC, DuMerton-Shore D (2015) Combination of morphine with nortriptyline for neuropathic pain. Pain 156(8):1440–1448
24. Graff-Radford SB, Shaw LR, Naliboff BN (2000) Amitriptyline and fluphenazine in the treatment of postherpetic neuralgia. Clin J Pain 16(3):188–192
25. Higgins JP, Altman DG, Gøtzsche PC, Jüni P, Moher D, Oxman AD, Savovic J, Schulz KF, Weeks L, JA S, Cochrane Bias Methods Group; Cochrane Statistical Methods Group (2011) The Cochrane Collaboration's tool for assessing risk of bias in randomised trials. BMJ 343:d5928
26. Holbech JV, Bach FW, Finnerup NB, Brøsen K, Jensen TS, Sindrup SH (2015) Imipramine and pregabalin combination for painful polyneuropathy: a randomized controlled trial. Pain 156(5):958–966
27. http://www.ema.europa.eu/docs/en_GB/document_library/Scientific_guideline/2015/12/WC500199242.pdf. Accessed 29Feb 2016
28. http://www.fda.gov/downloads/Drugs/GuidanceComplianceRegulatoryInformation/Guidances/UCM384691.pdf. Accessed 29Feb 2016.
29. Ioannidis JP, Evans SJ, Gøtzsche PC, O'Neill RT, Altman DG, Schulz K, Moher D, CONSORT Group (2004) Better reporting of harms in randomized trials: an extension of the CONSORT statement. Ann Intern Med 141(10):781–788

30. Jadad AR, Moore RA, Carroll D, Jenkinson C, Reynolds DJ, Gavaghan DJ, McQuay HJ (1996) Assessing the quality of reports of randomized clinical trials: is blinding necessary? Control Clin Trials 17(1):1–12
31. Johnson RW, Rice AS (2014) Clinical practice. Postherpetic neuralgia. N Engl J Med 371:1526.153(6):1148–1158
32. Katz J, Finnerup NB, Dworkin RH (2008) Clinical trial outcome in neuropathic pain: relationship to study characteristics. Neurology 70(4):263–272
33. Katz N (2009) Enriched enrollment randomized withdrawal trial designs of analgesics: focus on methodology. Clin J Pain 25(9):797–807
34. Khoromi S, Cui L, Nackers L, Max MB (2007) Morphine, nortriptyline and their combination vs. placebo in patients with chronic lumbar root pain. Pain 130(1–2):66–75
35. Maier C, Baron R, Tölle TR, Binder A, Birbaumer N, Birklein F, Gierthmühlen J, Flor H, Geber C, Huge V, Krumova EK, Landwehrmeyer GB, Magerl W, Maihöfner C, Richter H, Rolke R, Scherens A, Schwarz A, Sommer C, Tronnier V, Uçeyler N, Valet M, Wasner G, Treede RD (2010) Quantitative sensory testing in the German Research Network on Neuropathic Pain (DFNS): somatosensory abnormalities in 1236 patients with different neuropathic pain syndromes. Pain 150:439
36. Max MB, PortenoyRK, LaskaEM (eds)(1991) Advances inpain research and therapy: the design of analgesic clinical trials, vol. 18, Raven Press, Ltd., New York
37. McKeown A, Gewandter JS, McDermott MP, Pawlowski JR, Poli JJ, Rothstein D, Farrar JT, Gilron I, Katz NP, Lin AH, Rappaport BA, RowbothamMC, Turk DC, Dworkin RH, Smith SM (2015) Reporting of sample size calculations in analgesic clinical trials: ACTTION systematic review. J Pain 16(3):199–206.e1–7
38. McQuay HJ, Carroll D, Jadad AR, Glynn CJ, Jack T, Moore RA, Wiffen PJ (1994) Dextromethorphan for the treatment of neuropathic pain: a double-blind randomized controlled crossover trial with integral n-of-1 design. Pain 59(1):127–133
39. Moore RA, Wiffen PJ, Derry S, Toelle T, Rice AS 2014 Gabapentin for chronic neuropathic pain and fibromyalgia in adults. Cochrane Database Syst Rev (4):CD007938
40. Moore RA, Wiffen PJ, Eccleston C, Derry S, Baron R, Bell RF, Furlan AD, Gilron I, Haroutounian S, Katz NP, Lipman AG, Morley S, Peloso PM, Quessy SN, Seers K, Strassels SA, Straube S (2015) Systematic review of enriched enrolment, randomized withdrawal trial designs in chronic pain: a new framework for design and reporting. Pain 156(8):1382–1395
41. Morrison VA, Johnson GR, Schmader KE, Levin MJ, Zhang JH, Looney DJ, Betts R, Gelb L, Guatelli JC, Harbecke R, Pachucki C, Keay S, Menzies B, Griffin MR, Kauffman CA, Marques A, Toney J, Boardman K, Su SC, Li X, Chan IS, Parrino J, Annunziato P, MN O, Shingles Prevention Study Group (2015) Long-term persistence of zoster vaccine efficacy. Clin Infect Dis 60:900
42. Moulin DE, Iezzi A, Amireh R, Sharpe WK, Boyd D, Merskey H (1996) Randomised trial of oral morphine for chronic non-cancer pain. Lancet 347(8995):143–147
43. Oxman MN, Levin MJ, Johnson GR, Schmader KE, Straus SE, Gelb LD, Arbeit RD, Simberkoff MS, Gershon AA, Davis LE, Weinberg A, Boardman KD, Williams HM, Zhang JH, Peduzzi PN, Beisel CE, Morrison VA, Guatelli JC, Brooks PA, Kauffman CA, Pachucki CT, Neuzil KM, Betts RF, Wright PF, Griffin MR, Brunell P, Soto NE, Marques AR, Keay SK, Goodman RP, Cotton DJ, Gnann JW Jr, Loutit J, Holodniy M, Keitel WA, Crawford GE, Yeh SS, Lobo Z, Toney JF, Greenberg RN, Keller PM, Harbecke R, Hayward AR, Irwin MR, Kyriakides TC, Chan CY, Chan IS, Wang WW, Annunziato PW, Silber JL (2005) A vaccine to prevent herpes zoster and postherpetic neuralgia in older adults. N Engl J Med. 352(22):2271–2284
44. Quessy SN, Rowbotham MC (2008) Placebo response in neuropathic pain trials. Pain 138(3):479–483
45. Raja SN, Haythornthwaite JA, Pappagallo M, Clark MR, Travison TG, Sabeen S, Royall RM, Max MB (2002) Opioids versus antidepressants in postherpetic neuralgia: a randomized, placebo-controlled trial. Neurology 59(7):1015–1021
46. Rice AS, Cimino-Brown D, Eisenach JC, Kontinen VK, Lacroix-Fralish ML, Machin I, Preclinical Pain Consortium, JS M, Stöhr T (2008) Animal models and the prediction of effi-

cacy in clinical trials of analgesic drugs: a critical appraisal and call for uniform reporting standards. Pain 139(2):243–247

47. Rothwell PM (2005) External validity of randomised controlled trials: "to whom do the results of this trial apply?". Lancet 365(9453):82–93
48. Rowbotham MC, Gilron I, Glazer C, Rice AS, Smith BH, Stewart WF, Wasan AD (2013) Can pragmatic trials help us better understand chronic pain and improve treatment? Pain 154(5):643–646
49. Rowbotham MC, Twilling L, Davies PS, Reisner L, Taylor K, Mohr D (2003) Oral opioidtherapy for chronic peripheral and central neuropathic pain. N Engl J Med 348(13):1223–1232
50. Sang CN, Booher S, Gilron I, Parada S, Max MB (2002) Dextromethorphan and memantine in painful diabetic neuropathy and postherpetic neuralgia: efficacy and dose-response trials. Anesthesiol 96(5):1053–1061
51. Smith SM, Amtmann D, Askew RL, Gewandter JS, Hunsinger M, Jensen MP, McDermott MP, Patel KV, Williams M, Bacci ED, Burke LB, Chambers CT, Cooper SA, Cowan P, Desjardins P, Etropolski M, Farrar JT, Gilron I, Huang IZ, Katz M, Kerns RD, Kopecky EA, Rappaport BA, Resnick M, Strand V, Vanhove GF, Veasley C, Versavel M, Wasan AD, Turk DC, Dworkin RH (2016) Pain intensity rating training: results from an exploratory study of the ACTTION PROTECCT system. Pain 157(5):1056–1064
52. Smith SM, Wang AT, Katz NP, McDermott MP, Burke LB, Coplan P, Gilron I, Hertz SH, Lin AH, Rappaport BA, Rowbotham MC, Sampaio C, Sweeney M, Turk DC, Dworkin RH (2013) Adverse event assessment, analysis, and reporting in recent published analgesic clinical trials: ACTTION systematic review and recommendations. Pain 154(7):997–1008
53. Staud R, Price DD (2008) Long-term trials of pregabalin and duloxetine for fibromyalgia symptoms: how study designs can affect placebo factors. Pain 136(3):232–234
54. Straube S, Derry S, McQuay HJ, Moore RA (2008) Enriched enrollment: definition and effects of enrichment and dose in trials of pregabalin and gabapentin in neuropathic pain. A systematic review. Br J Clin Pharmacol 66(2):266–275
55. Tesfaye S, Wilhelm S, Lledo A, Schacht A, Tölle T, Bouhassira D, Cruccu G, Skljarevski V, Freynhagen R (2013) Duloxetine and pregabalin: high-dose monotherapy or their combination? The "COMBO-DN study" – a multinational, randomized, double-blind, parallel-group study in patients with diabetic peripheral neuropathic pain. Pain 154(12):2616–2625
56. Turk DC, Dworkin RH, Allen RR, Bellamy N, Brandenburg N, Carr DB, Cleeland C, Dionne R, Farrar JT, Galer BS, Hewitt DJ, Jadad AR, Katz NP, Kramer LD, Manning DC, McCormick CG, McDermott MP, McGrath P, Quessy S, Rappaport BA, Robinson JP, Royal MA, Simon L, Stauffer JW, Stein W, Tollett J, Witter J (2003) Core outcome domains for chronic pain clinical trials: IMMPACT recommendations. Pain 106(3):337–345
57. Turk DC, Dworkin RH, Revicki D, Harding G, Burke LB, Cella D, Cleeland CS, Cowan P, Farrar JT, Hertz S, Max MB, Rappaport BA (2008) Identifying important outcome domains for chronic pain clinical trials: an IMMPACT survey of people with pain. Pain 137(2):276–285
58. Turner JA, Deyo RA, Loeser JD, Von Korff M, Fordyce WE (1994) The importance of placebo effects in pain treatment and research. JAMA 271(20):1609–1614
59. Turner JA, Jensen MP, Warms CA, Cardenas DD (2002) Blinding effectiveness and association of pretreatment expectations with pain improvement in a double-blind randomized controlled trial. Pain 99(1–2):91–99
60. Tuttle AH, Tohyama S, Ramsay T, Kimmelman J, Schweinhardt P, Bennett GJ, Mogil JS (2015) Increasing placebo responses over time in U.S. clinical trials of neuropathic pain. Pain 156(12):2616–2626
61. van Hecke O, Austin SK, Khan RA, Smith BH, Torrance N (2014) Neuropathic pain in the general population: a systematic review of epidemiological studies. Pain 155:654
62. Wallace MS, Rowbotham MC, Katz NP, Dworkin RH, Dotson RM, Galer BS, Rauck RL, Backonja MM, Quessy SN, Meisner PD (2002) A randomized, double-blind, placebo-controlled trial of a glycine antagonist in neuropathic pain. Neurology 59(11):1694–1700

63. Watson CP, Evans RJ, Reed K, Merskey H, Goldsmith L, Warsh J (1982) Amitriptyline versus placebo in postherpetic neuralgia. Neurology 32(6):671–673
64. Watson CP, Gilron I, Sawynok J (2010) A qualitative systematic review of head-to-head randomized controlled trials of oral analgesics in neuropathic pain. Pain Res Manag 15(3):147–157
65. Watson CP, Moulin D, Watt-Watson J, Gordon A, Eisenhoffer J (2003) Controlled-release oxycodone relieves neuropathic pain: a randomized controlled trial in painful diabetic neuropathy. Pain 105(1–2):71–78
66. Woolf CJ (2010) Overcoming obstacles to developing new analgesics. Nat Med 16(11):1241–1247
67. Wu CL, Tella P, Staats PS, Vaslav R, Kazim DA, Wesselmann U, Raja SN (2002) Analgesic effects of intravenous lidocaine and morphine on postamputation pain: a randomized double-blind, active placebo-controlled, crossover trial. Anesthesiology 96(4):841–848
68. Zakrzewska JM, Chaudhry Z, Nurmikko TJ, Patton DW, Mullens EL (1997) Lamotrigine (lamictal) in refractory trigeminal neuralgia: results from a double-blind placebo controlled crossover trial. Pain 73:223–230

Part V
The Prevention of Herpes Zoster and Postherpetic Neuralgia

Chapter 22
The Importance of Zoster Prevention Vaccines

C. Peter N. Watson

"An ounce of prevention is worth a pound of cure." Poor Richard's Almanack, Ben Franklin 1732

Herpes zoster is the commonest neurological disease [11] and is likely to increase as the population ages [15]. There are many complications of herpes zoster of which severe postherpetic neuralgia (PHN) is the most common and much feared. Other related consequences are blindness and facial scarring with ophthalmic zoster, facial paralysis (Ramsay Hunt syndrome), meningitis, encephalitis, myelitis, and the potential for an increased vascular risk of granulomatous angiitis, stroke, and myocardial infarction. Postherpetic neuralgia specifically may be compared with other end-stage disorders in which damage to an organ occurs and that structure is unable or not fully able to repair itself and respond to treatment. Several categories of evidence support this view and argue for prevention of herpes zoster by vaccination. These data are clinical, pathological, therapeutic, preventative, and epidemiological.

22.1 Clinical

Clinical findings indicate that this disease alters the nervous system such that there are often areas of sensory loss and widespread areas of sensitive skin, which respond to the lightest touch with severe pain [20] (Fig. 22.1). There are three main types of pain associated with PHN: steady, often burning pain, lancinating shock-like pain, and pain on non-painful stimulation of the skin (allodynia [20]). These are often all present in a single individual. The physical findings of extensive areas of sensory loss and pain on moving touch such as skin stroking (dynamic mechanical

C.P.N. Watson
Department of Medicine, University of Toronto,
Toronto, ON, Canada
e-mail: peter.watson@utoronto.ca

C.P.N. Watson et al. (eds.), *Herpes Zoster: Postherpetic Neuralgia and Other Complications,* DOI 10.1007/978-3-319-44348-5_22

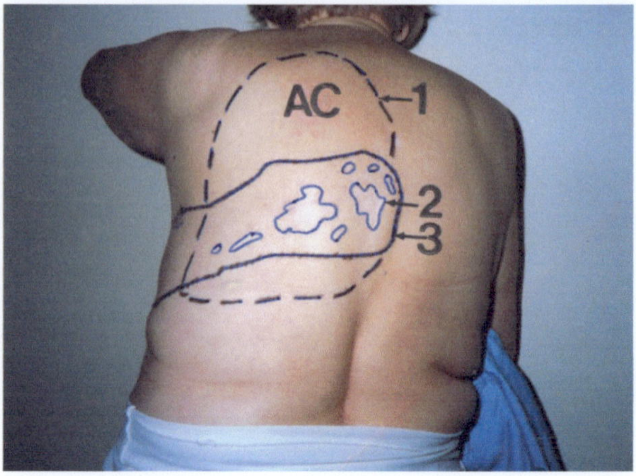

Fig. 22.1 Postherpetic neuralgia with areas of allodynia (*1*), scarring (*2*), and sensory loss (*3*)

allodynia) [20] suggest that central neurons have expanded their receptive fields and have lowered thresholds to sensory stimulation (Fig. 22.1). There is no evidence to date that these clinical changes are reversible.

22.2 Pathological

The few pathologic studies of PHN cases indicate that there is extensive damage to the peripheral and central nervous system [19, 21]. The peripheral nerve, ganglion, and sensory root are extensively scarred, leading to the loss of neurons and nerve fibers of all types. Surviving nerve fibers appear to shift to the smaller fiber population, which may be predominately excitatory and may fire spontaneously or in response to the lightest of tactile stimulation. Further, there is evidence of central nervous system damage in that the dorsal horn of the spinal cord is infiltrated by inflammatory cells and later becomes atrophic because of the destruction of nerve fibers and neurons. There is no evidence that any of these pathologic changes are reversible. There is some evidence of chronic persistent inflammation preceding and following the acute onset of the rash of herpes zoster [21].

22.3 Therapeutic

Table 22.1 summarizes controlled trials of antidepressant and opioid therapies in PHN. These data indicate that 30–50 % of the patients responded poorly or not at all to some antidepressants [10, 13, 18, 23, 24, 26] and to opioids [25]. Even with these

Table 22.1 Number of postherpetic neuralgia patients responding poorly or not at all to treatment with antidepressants or opioids in controlled trials

Study/year	Agent	% of subjects not responding to treatment with	
		Opioid or antidepressant %	Placebo
Watson et al. 1982 [18]	Amitriptyline	8/24, 33	100 %
Max et al. 1988 [13]	Amitriptyline	21/41, 53	84 %
Kishore-Kumar et al. 1990 [10]	Desipramine	14/26, 54	89 %
Watson et al. 1992 [23]	Amitriptyline and/or maprotiline	17/32, 53	No placebo
Watson et al. 1998 [26]	Amitriptyline and/or nortriptyline	10/31, 32	No placebo
Watson and Babul 1998 [25]	Oxycodone	17/38, 42	82 %

drugs, responses are usually incomplete, with total relief unusual. Adverse effects occur in nearly all treated patients. Randomized controlled trials of gabapentinoids (gabapentin and pregabalin) result in about two-thirds of patients having less than moderate relief in this type of selected population and over the short term of less than 12 weeks [7, 16]. Results will almost certainly not be as good when generalized to ordinary practice where patients are older, have concomitant diseases, and are on other drugs.

22.4 Preventative

A current preventative avenue is to treat herpes zoster early and aggressively with multimodal therapy, which would include antivirals such as acyclovir, famciclovir, and valacyclovir [1, 17], tricyclic antidepressants [2], gabapentinoids, opioids, and regional anesthesia. However, there are considerable difficulties with this approach in that it is suggested for optimal efficacy to use the antiviral drugs within 72 h of rash onset. This can be very difficult. Often disease activity starts with pain before the rash appears and then there is the issue of being seen by a physician in a timely fashion in order to begin the antiviral agent. A complete discussion of the limited evidence and difficulties with this approach is covered in Chap. 23; therefore, there is no consistent or good evidence for supporting antivirals [5] for any of these other drugs or for interventions such as nerve blocks (Chap. 17) to prevent severe postherpetic neuralgia. However good practice involves treating aggressively if necessary in this multimodal fashion to deal with the relief of the acute pain, in the hope that there will also be a reduction in severe postherpetic neuralgia.

22.5 Epidemiological

There are 130,000 cases of herpes zoster in Canada [3] and 1,000,000 annually in the United States [9]. Postherpetic neuralgia occurs 1 month after herpes zoster infection in about 10 % (overall incidence) of those with the infection [4, 6, 8, 15]. The incidence of postherpetic neuralgia, however, increases with age, affecting 50 % of herpes zoster patients at 1 month by age 60, and the incidence rises steadily with increasing age [6, 15]. The number of those affected will likely increase in the future with the shift of demographics to an increase in the aged population. The prognosis for patients with established PHN is poor, since at least half continue to suffer for many years – some even until death [22]. Of the 50 % who seem to be reasonably well, many require continual pharmacotherapy, often with unpleasant side effects.

22.6 Conclusion

Herpes zoster is difficult to manage, which is emphasized here and elsewhere in this book; therefore, there is a compelling argument for preventing the disease in the first place with either the current live zoster prevention vaccine [14] or new vaccines which appear to be much more effective and applicable to the immunosuppressed population [12]. Poor vaccine uptake in some countries may occur in part because of the expense if payment is out of pocket. In the province of Ontario, Canada, the government health plan now covers this cost of about 200 dollars for people 65–70 years old.

References

1. Beutner KR, Friedman DJ, Porszpaniak D, Anderson PL, Wood MJ (1995) Valacyclovir compared acyclovir for improved therapy for herpes zoster in immunocompetent adults. Antimicrob Agents Chemother 39:1546–1533
2. Bowsher D (1996) Postherpetic neuralgia and its treatment: a retrospective survey of 191 patients. J Pain Symp Manage 12:327–331
3. Brisson M, Pellissier JM, Camden S et al (2008) The potential cost effectiveness of the vaccine against herpes zoster and post herpetic neuralgia. Hum Vaccin 493(3):238–245
4. Burgoon CP, Burgoon JS, Haldridge GD (1957) The natural history of herpes zoster. JAMA 164:265–269
5. Chen N, Li Q, Yang T et al (2014) Antiviral therapy for the treatment of postherpetic neuralgia. Cochrane Database Syst Rev (2):CD006866
6. De Moragas JM, Kierland RR (1957) The outcome of patients with herpes zoster. Arch Dermatol 75:193–196
7. Dworkin RH, Corbin AE, Young JP et al (2003) Pregabalin for postherpetic neuralgia: a randomized placebo-controlled study. Neurology 60(8):1174–1283
8. Hope-Simpson RE (1975) Postherpetic neuralgia. J R Coll Gen Pract 25:571–575

9. Insinga DRP, Itzler RF, Pellissier JM et al (2005) The incidence of herpes zoster in a United States administrative database. J Gen Intern Med 20(8):748–653
10. Kishore-Kumar R, Max MB, Schafer SC et al (1990) Desipramine relieves postherpetic neuralgia. Clin Pharmacol Ther 47:305–312
11. Kurtzke JF (1984) Neuroepidemiology. Ann Neurol 16:265–277
12. Lal H, Cunningham AL, Godeaux O et al (2015) Efficacy of an adjuvanted herpes zoster subunit vaccine in older adults. N Engl J Med 372(22):2087–2296
13. Max MB, Schafer SC, Culnane M, Smoller B, Dubner R, Gracely R (1988) Amitriptyline but not lorazepam relieves postherpetic neuralgia. Neurology 38:1427–1432
14. Oxman MN, Levin M, Johnson GR et al (2005) A vaccine to prevent herpes zosterand postherpetic neuralgia in older adults. N Eng J Med 352:2271–2284. 13
15. Ragozzino MW, Melton IJ, Kurland IT et al (1982) Population based study of herpes zoster and its sequelae. Medicine 21:310–316
16. Rowbotham MC, Harden N, Stacey B et al (1998) Gabapentin for the treatment of postherpetic neuralgia: a randomized controlled trial. JAMA 280(21):1837–1842
17. Tyring S, Barbarash RA, Nahlek JE et al (1995) Famciclovir for the treatment of acute herpes zoster: Effects on acute disease and postherpetic neuralgia. Ann Intern Med 123:89–96
18. Watson CPN, Evans RJ, Reed K, Merskey H, Goldsmith I, Warsh J (1982) Amitriptyline versus placebo in postherpetic neuralgia. Neurology 32:671–673
19. Watson CPN, Morshead C, Van der Kooy D, Deck J, Evans RJ (1988a) Postherpetic neuralgia: post-mortem analysis of a case. Pain 34:129–138
20. Watson CPN, Evans RJ, Watt VR, Birkett N (1988b) Postherpetic neuralgia: 208 cases. Pain 35:289–298
21. Watson CPN, Deck JH, Morshead C, Van der Kooy D, Evans RJ (1991a) Postherpetic neuralgia: further post-mortem studies of cases with and without pain. Pain 44:105–117
22. Watson CPN, Watt RR, Chipman M, Birkett N, Evans RJ (1991b) The prognosis with postherpetic neuralgia. Pain 45:195–199
23. Watson CPN, Chipman M, Reed K, Evans RJ, Birkett N (1992) Amitriptyline versus maprotiline in postherpetic neuralgia; a randomized double-blind, crossover trial. Pain 48:29–36
24. Watson CPN (1995) The treatment of postherpetic neuralgia. Neurology 45:58–59
25. Watson CPN, Babul N (1998) Oxycodone relives neuropathic pain: a randomized trial in postherpetic neuralgia. Neurology 50:1837–1841
26. Watson CPN, Chipman M, Reed K (1998) Amitriptyline versus nortriptyline in postherpetic neuralgia. Neurology 51:1166–1171

Chapter 23
Aggressive Noninvasive Treatment of Acute Herpes Zoster for the Prevention of Postherpetic Neuralgia

Sigrun Alba Johannesdottir Schmidt and Michael C. Rowbotham

23.1 Introduction

Postherpetic neuralgia (PHN) is the most frequent complication of herpes zoster [15]. The debilitating pain of PHN persists for over 1 year in 30 % of patients [52] and may have negative impact on patients' quality of life [47]. It may result in depression, sleep disturbances, social isolation, and decreased functional capacity [47]. Options for pain relief in PHN include oral treatment with tricyclic antidepressants, anticonvulsants (gabapentin, pregabalin), opioid analgesics, topical treatment (lidocaine patches), and nerve blocks [3]. Unfortunately, at least 30 % of patients with PHN are partially or completely refractory to currently available analgesic treatments [96], and treatment satisfaction is only approximately 15 % [47].

According to observational data, the absolute risk of PHN after an episode of herpes zoster varies between 3 and 30 % [52]. Even higher risks have been reported among placebo recipients in trials on treatment of acute herpes zoster. This discrepancy between studies is largely explained by differences in study design and definitions of PHN. Differences in age distribution between trials are another important source of variation, as the risk of PHN is highly dependent on age, increasing by a factor 1.22–3.11 for each consecutive decade of age [31].

With the increasing longevity of the world population [108], the overall prevalence of PHN is expected to grow, increasing the burden to public health. Because of the difficulties in managing PHN, efforts have been directed at identifying

S.A.J. Schmidt (✉)
Department of Clinical Epidemiology, Aarhus University Hospital, Aarhus, Denmark
e-mail: saj@clin.au.dk

M.C. Rowbotham
California Pacific Medical Center Research Institute, San Francisco, CA, USA
e-mail: mcrowbotham@gmail.com

C.P.N. Watson et al. (eds.), *Herpes Zoster: Postherpetic Neuralgia and Other Complications*, DOI 10.1007/978-3-319-44348-5_23

preventive strategies. Although a vaccine to prevent herpes zoster (and thus PHN) has been available for years, vaccine utilization has been limited [53]. Furthermore, some vaccinated individuals will go on to experience herpes zoster PHN despite vaccination [71, 74]. A closer look at effective measures to prevent development of PHN in persons with herpes zoster eruption is therefore warranted.

It has been suggested that rigorous intervention at the time of presentation with acute herpes zoster may prevent the development of PHN, but the topic is controversial. Several treatment modalities have been proposed, including antiviral drugs, glucocorticoids, antidepressants, anticonvulsants, opioid analgesics, and various forms of local and regional anesthesia. The rationale behind the preemptive effects of these approaches lies within the current knowledge about important pathways in the pathophysiology of PHN, as outlined in Fig. 23.1. This chapter provides a summary of evidence pertaining to the effect of aggressive noninvasive treatment of acute herpes zoster on the risk and severity of PHN.

23.2 Systemic Antiviral Therapy

First-line therapy for acute herpes zoster includes a 7-day course with one of the antiviral nucleoside analogues acyclovir (800 mg five times daily), valacyclovir (1000 mg three times daily), famciclovir (250 or 500 mg three times daily depending on country), or brivudine (125 mg once daily, licensed in selected countries only). These antiviral drugs have been proven efficacious in reducing the time to cessation of viral shedding, new lesion formation, and acute pain [23] (reference to the chapter on treatment of acute zoster by Anne Gershon). Because the extent of dermatomal disease is a predictor of persistence of pain, such that a high number of lesions are associated with slower rate of acute and long-term pain resolution [31, 100], antiviral drugs could also carry a potential to relieve or prevent subsequent PHN. However, data on this topic are sparse and conflicting.

Most trials on the efficacy of antiviral drugs have been designed with the primary aim of investigating the effects on acute herpes zoster and have thus been underpowered to study persistence of pain beyond the first month. Table 23.1 provides characteristics and detailed results for previous antiviral trials reporting any data pertaining to prolonged pain. For completeness, the table includes trials reporting on zoster-associated pain or presence of pain at any specified time point beyond 1 month after rash onset. Randomized controlled trials were selected because this design is generally considered the gold standard and because the spontaneous improvement of PHN with time requires an appropriate control group, which is typically difficult to identify in observational studies. Only placebo-controlled trials and trials comparing currently licensed antiviral drugs are included in the table. All studies were double-blinded, except one trial which was open-labeled and in addition did not explicitly state that it was randomized [30].

Today, the convention is to define PHN as any pain >0 on a 0–100 scale persisting for 90 days or more after rash onset [2, 34]. This definition is based on studies

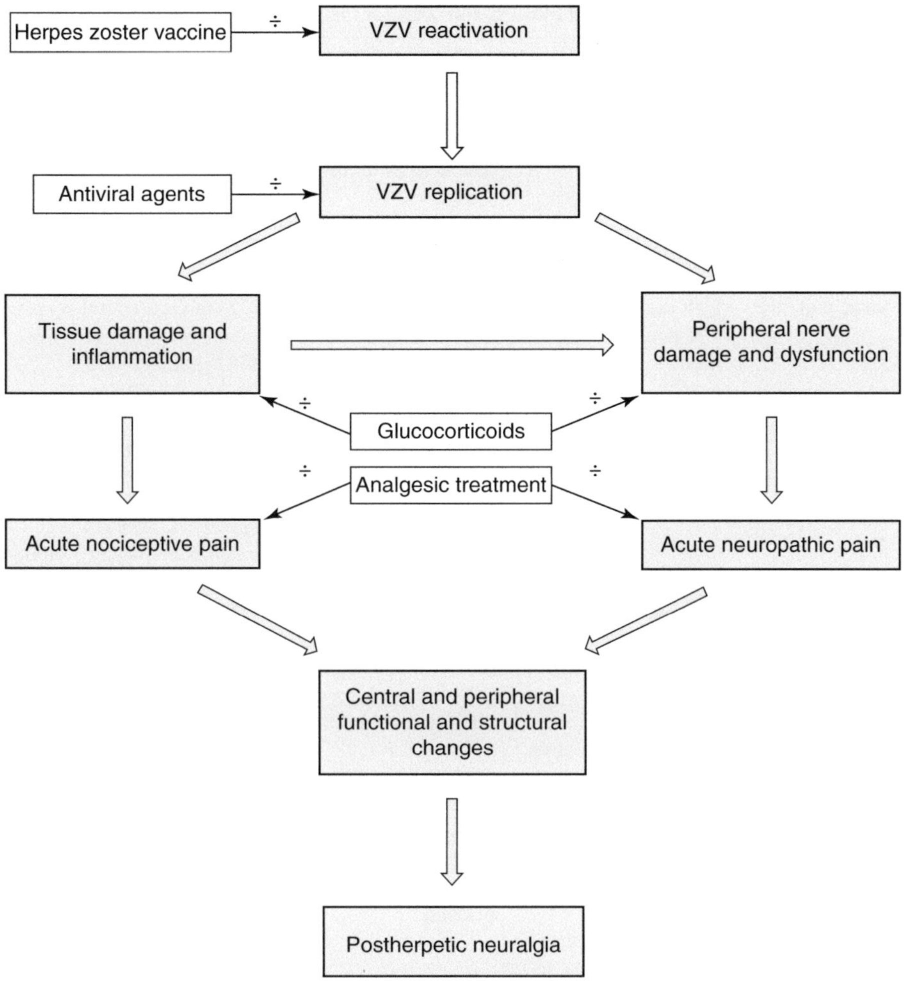

Fig. 23.1 Proposed mechanisms for reducing the risk and severity of postherpetic neuralgia (Modified from Dworkin et al (2000) *Clin J Pain* 16(2 Suppl):S90–S100)

of data on the neuralgia of herpes zoster, which suggest that it may advantageously be categorized in three phases: acute (0–30 days), subacute (31–90 days), or chronic neuralgia (>90 days). However, definitions employed by previous studies have varied largely, including pain after rash healing, presence of pain at 30 days or later after rash onset, and duration of zoster-associated pain, which is measured as a pain continuum from the time of acute herpes zoster until pain resolution. Some studies have further differentiated between any pain >0 on a 0–100 scale and clinically significant pain defined as pain of at least 30 on a 0–100 scale [17, 88].

Comparison of results from the trials is hampered by methodological dissimilarities, including differences in types of assigned treatments, follow-up methods, and

Table 23.1 Randomized controlled trials on antivirals for the prevention of PHN

Reference	Study population	Treatment groups	Results (investigational drug vs. comparator)[a]
Placebo-controlled acyclovir trials			
Balfour et al. [4]	Immunocompromised; Any age; <72 h of rash onset/ ongoing vesicle formation; $n = 94$	500 mg/m^2 i.v. acyclovir t.i.d. for 7 days vs. placebo	Localized zoster ($n = 52$): 1/15 (6.7 %) vs. 4/13 (31 %) pain at 28 days and 1/13 (7.7 %) vs. 4/14 (29 %) at 49 days
Bean et al. [5]	≥18 years <72 hours of rash onset $n = 29$	500 mg/m^2 i.v. acyclovir t.i.d. for 5 days vs. placebo	7/19 (37 %) vs. 6/10 (60 %) pain at 1 month and 6/19 (32 %) vs. 5/10 (50 %) at 2 months
Cobo et al. [14]	HZO Adults <7 days of rash onset $n = 71$	600 mg acyclovir 5× daily for 10 days vs. placebo	17/36 (47 %) vs. 20/35 (57 %) any pain after rash healing, among whom 41 vs. 35 % still in pain at 3 months 7/36 (19 %) vs. 5/35 (14 %) moderate/severe pain (≥3 on 0–4 scale) after rash healing
Esmann et al. [27], Peterslund et al. [79]	>30 years <96 hours of rash onset $n = 56$	5 mg/kg i.v. acyclovir t.i.d. for 5 days vs. placebo	14/27 (52 %) vs. 14/29 (48 %) at 1 month after admission and 10/27 (37 %) vs. 4/29 (14 %) at 3 months[b]
Finn and Smith [30]	31–89 years <7 days of rash onset $n = 27$	200 mg acyclovir 5× daily for 10 days vs. placebo	No patients reported pain at 6 weeks and 3 months
Harding and Porter [38][b]	HZO without intraocular complication ≥20 years <72 hours of rash onset $n = 46$	800 mg acyclovir 5× daily for 5 days vs. placebo	2/16 (13 %) vs. 7/15 (47 %) pain at 2 months after enrollment, 1/15 (7 %) vs. 6/16 (38 %) at 3 months, and 1/21 (5 %) vs. 5/19 (26 %) at 6 months Reduced mean severity (0–100 scale) during all 8 months
Huff et al. [41], [42][b], and [43]	Localized ≥18 years <72 hours of rash onset $n = 252$	400 mg acyclovir 5× daily for 10 days vs. placebo 800 mg acyclovir 5× daily for 10 days vs. placebo	400 mg acyclovir: no differences (data N.R.) 800 mg acyclovir: 3/72 (4.2 %) vs. 13/78 (16.7 %) persistent pain at 1–3 months and 2/51 (3.9 %) vs. 3/57 (5.3 %) at 4–6 months 13/27 (48.1 %) vs. 17/40 (42.5 %) analgesic use at 1–3 months and 5/11 (45.4 %) vs. 6/17 (35.2 %) at 4–6 months No difference (data N.R.) in severity (0–3 scale) Median pain duration in those with pain at enrollment 20 vs. 62 days

(continued)

Table 23.1 (continued)

Reference	Study population	Treatment groups	Results (investigational drug vs. comparator)[a]
Juel-Jensen et al. [48]	>18 years <72 hours of rash onset $n = 40$	10 mg/kg i.v. acyclovir t.i.d. for 5 days vs. placebo	8/20 (20 %) vs. 10/20 (40 %) pain at follow-up (mean 5 months after discharge)
McGill et al. [62]	≥18 years <96 hours of rash onset $n = 37$	5 mg/kg i.v. acyclovir t.i.d. for 5 days vs. placebo	2/16 (13 %) vs. 5/19 (26 %) pain at 3 months after enrollment
McKendrick et al. [64]	>50 years <72 hours of rash onset $n = 41$	400 mg acyclovir 5× daily for 5 days vs. placebo	3/18 (17 %) vs. 5/23 (22 %) pain at 6 months after enrollment
Morton and Thomson [72][b]	≥16 years <72 hours of rash onset $n = 83$	800 mg acyclovir 5× daily for 7 days vs. placebo	33 vs. 60 % pain at 1 month after enrollment, 10 vs. 40 % at 3 months, 13 vs. 25 % at 4 months, 10 vs. 20 % at 5 months, and 6 vs. 19 % at 6 months Similar trend for analgesic use
Wassilew et al. [94]	>18 years <72 hours of rash onset $n = 60$	400 mg acyclovir 5× daily for 5 days vs. placebo	8/29 (28 %) vs. 6/31 (19 %) pain at 1 month after enrollment, 7/29 (3.4 %) vs. 7/31 (23 %) at 2–3 months, 1/29 (3.4 %) vs. 0/31 (0 %) at 6 months, 1/29 (3.4 %) vs. 0/31 (0 %) at 9 months, and 0/29 (0 %) vs. 0/31(0 %) at 1 year Median duration of continuous pain 12 vs. 13 days No difference in mean severity (0–3 scale) during treatment
Whitley et al. [99][b]	Localized >50 years <72 hours of rash onset $n = 100$	800 mg acyclovir 5× daily for 21 days vs. placebo[c]	ITT: HR 1.39 (0.84–2.32) for pain cessation during 6 months since enrollment
Wood et al. [105] and [106][b], McKendrick et al. [65–67], McGill and White [63]	>60 years <72 hours of rash onset $n = 376$	800 mg acyclovir 5× daily for 7 days vs. placebo	39/161 (24 %) vs. 38/156 (24 %) pain at 3 months after treatment, 26/157 (17 %) vs. 26/149 (17 %) at 4 months, and 22/162 (14 %) vs. 20/155 (13 %) at 6 months No difference in severity (scale 0–4) during 6 months Conflicting results from participating centers

(continued)

Table 23.1 (continued)

Reference	Study population	Treatment groups	Results (investigational drug vs. comparator)[a]
Placebo-controlled famciclovir trials			
Tyring et al. [89][b], Dworkin et al. [22]	Uncomplicated ≥18 years <72 hours of rash onset $n = 419$	500 or 750 mg famciclovir t.i.d. for 7 days vs. placebo	ITT: RRs of PHN for 500 mg dose 0.93 (0.76–1.15) at 1 month after enrollment (41 vs. 44 %), 0.85 (0.73–1.00) at 3 months (28 vs. 33 %), 0.83 (0.72–0.96) at 4 months (25 vs. 30 %), and 0.63 (0.57–0.70) at 6 months (15 vs. 24 %) ITT: similar results for 750 mg dose (RR 0.83 at 3 months and 0.56 at 6 months) ITT: if pain after rash healing, median pain duration 63 days for 500 mg (HR 2.0 [1.3–3.3]), 61 days for 750 mg (HR 2.0 [1.2–3.1]), and 119 days for placebo (ref.)
Comparative trials of antivirals			
Beutner et al. [8]	>50 years <72 hours of rash onset $n = 1141$	1000 mg valacyclovir t.i.d. for 7 or 14 days vs. 800 mg acyclovir 5× daily for 7 days	ITT: 19.9 % in 7-day valacyclovir group, 18.6 % in 14-day valacyclovir group, and 25.7 % in acyclovir group pain at 6 months after enrollment ITT: median pain duration 38 days (HR 1.34 [1.12–1.60]) for 7-day valacyclovir vs. acyclovir, 44 days (HR 1.22 [1.03–1.46]) for 14-day valacyclovir vs. acyclovir, 51 days (HR 1.10 [0.92–1.30]) for 7-day vs. 14-day valacyclovir ITT: mean pain score (Gracely scale) 10.3, 10.2, and 12.3 at 8 weeks
Colin et al. [16]	HZO ≥18 years <72 hours of rash onset $n = 110$	1000 mg valacyclovir t.i.d. vs. 800 mg acyclovir 5× daily for 7 days	ITT: 14/56 (25 %) vs. 17/54 (31 %) pain at 1 month after enrollment, 8/56 (14 %) vs. 10/54 (19 %) at 2 months, 3/56 (5 %) vs. 6/54 (11 %) at 4 months, and 3/56 (5 %) vs. 3/54 (6 %) at 6 months ITT: no difference in ocular pain severity (0–3 and 0–100 scales) or analgesic use

(continued)

Table 23.1 (continued)

Reference	Study population	Treatment groups	Results (investigational drug vs. comparator)[a]
Degreef and Famciclovir Herpes Zoster Clinical Study Group [19]	Uncomplicated >40 years <72 hours of rash onset $n = 545$	250 mg, 500 mg, or 750 mg famciclovir t.i.d. vs. 800 mg acyclovir 5× daily for 7 days	ITT: HRs for pain resolution 1.4 for 250 mg famciclovir, 1.8 for 500 mg famciclovir, and 1.4 for 750 mg famciclovir vs. acyclovir
Tyring et al. [90]	Excluding HZO >50 years <72 hours of rash onset $n = 597$	1000 mg valacyclovir t.i.d. vs. 500 mg famciclovir t.i.d. for 7 days	ITT: HR for pain resolution 1.01 (0.84–1.23) for valacyclovir vs. famciclovir ITT: 64 vs. 62 % pain at 1 month after enrollment, 32 vs. 34 % at 3 months, and 19 vs. 19 % at 6 months
Wassilew and Collaborative Brivudine PHN Study Group [93]	Uncomplicated, painful >50 years <72 hours of rash onset and new vesicles within last 24 hours $n = 2025$	125 mg brivudine q.d. vs. 250 mg famciclovir t.i.d. for 7 days	ITT: 109/980 (11.1 %) brivudine vs. 90/974 (9.2 %) famciclovir moderate pain (≥3/10 or need for analgesics) at 3 months after enrollment ITT: median pain duration 47 vs. 54 days ITT: severity of pain comparable

Abbreviations: *5×* five times, *HR* hazard ratio, *HZO* herpes zoster ophthalmicus, *ITT* intention to treat, *i.v.* intravenous, *PHN* postherpetic neuralgia, *q.d.* quaque die (once a day), *RR* risk ratio, *t.i.d.* ter in diē (three times a day)

[a]If study explicitly stated that an ITT method was used, this is specified. When available, 95 % confidence intervals for HRs and RRs are provided in parenthesis

[b]Included in Cochrane review by Chen et al. [12]

[c]Also examined is the effect of prednisone, as presented in Table 23.3

outcome definitions. For example, some of the listed studies examined dosing schedules other than that licensed for herpes zoster today. As an illustration of the complexity of the topic, several meta-analyses on acyclovir for the prevention of PHN have been published providing contradictory conclusions [12, 18, 45, 57, 82, 104]. The results of the meta-analyses are summarized in Table 23.2, showing that although the magnitude of effect estimates vary, they overall support a preventive effect on PHN. Most recently, the Cochrane Collaboration, which regularly produces systematic reviews that are widely considered to be the gold standards of health evidence, published a review on systemic antiviral therapy initiated for herpes zoster within 72 h of rash onset for the prevention of PHN [12]. The review included six trials, and the primary outcome in the review was the risk of PHN at 6 months after rash onset. Five trials considered oral acyclovir and followed participants for at least 6 months from enrollment. In an intention-to-treat meta-analysis comparing acyclovir and placebo, the risk ratio of PHN was 0.83 (95 % confidence interval [CI]: 0.71–0.96, four trials) at 4 weeks, 0.75 (95 % CI: 0.51–1.11, three trials) at 4 months, and 1.05 (95 % CI: 0.87–1.27, two trials) at 6 months after rash

Table 23.2 Meta-analyses on systemic acyclovir for the prevention of PHN

Studies included	Sample size	Outcome	Results (acyclovir vs. placebo)[a]
Chen et al. [12]			
Studies [38], [42], [99] and [106]	692	Pain at 1 month	ITT: RR 0.83 (0.71–0.96)
Studies [38], [42], [99] and [106]	609	Pain at 4 months	ITT: RR 0.75 (0.51–1.11)
Studies [99] and [106]	476	Pain at 6 months	ITT: RR 1.05 (0.87–1.27)
Crooks et al. [18]			
Studies [38], [42], [106], and [72]	689	Pain at 6 months	11 vs. 19 %: RR 0.58 (confidence interval N.R.)
Studies [38], [42], [106], and [72]	689	Time to first cessation of pain	Mean 50 vs. 60 days
Studies [38], [42], and [72]	313	Time to complete cessation of pain	Mean 49 vs. 86 days
Jackson et al. [45]			
Studies [38], [42], [99], [106], and [72]	792	Pain at 6 months	OR 0.54 (0.36–0.81)
Lancaster et al. [57]			
Unclear[b]	N.R.	Pain at 1 month	Any dose: RR 0.85 (0.61–1.19) 800 mg dose: RR 0.83 (0.58–12.21)
Unclear[b]	N.R.	Pain at 4 months	Any dose: RR 0.65 (0.46–0.93) 800 mg dose: RR 0.62 (0.43–0.90)
Unclear[b]	N.R.	Pain at 6 months	Any dose: RR 0.70 (0.47–1.06) 800 mg dose: RR 0.68 (0.47–1.07)
Schmader and Studenski [82]			
Studies [42], [106], [5], [27], [62], [48], and [64]	N.R.	Pain at 1 month or later	OR 0.81 (0.56–1.11)
Wood et al. [104]			
Studies [38], [42], [106], and [72]	691	Pain at 4 months	Any pain: 21 vs. 43 %; RD 22 % (11–33 %) Moderate/severe pain: 6 vs. 13 %; RD 7 % (0–14 %)
Studies [38], [42], [106], and [72]	691	Pain at 6 months	Any pain: 12 vs. 24 %; RD 13 % (4–22 %) Moderate/severe pain: 2 vs. 9 %; RD 7 % (1–12 %)
Studies [38], [42], and [72]	316	Time to complete cessation of pain persisting after day 30	ITT: HR 1.81 (1.35–2.43)
Studies [38], [42], and [72]	316	Time to complete cessation of moderate/severe pain	ITT: HR 1.46 (1.11–1.93)
Studies [38], [42], [106], and [72]	691	Time to first cessation of pain	ITT: HR 1.31 (1.08–1.60)

Abbreviations: *HR* hazard ratio, *N.R.* not reported, *OR* odds ratio, *PHN* postherpetic neuralgia, *RD* risk difference, *RR* risk ratio

[a]If study explicitly stated that an ITT method was used, this is specified. When available, 95 % confidence intervals for estimates are provided in parenthesis

[b]Unclear, but studies [38], [42], [106], [72] [64], [14], and [94] were referenced in the review

onset. Based on these results, authors concluded that there is high quality evidence that oral acyclovir does not reduce the incidence of PHN significantly. The antiviral drugs were well tolerated in all trials, with no serious adverse effects observed during or within 2 weeks of treatment. Nonserious side effects were also equally frequent in treatment and placebo groups. The most common adverse effects in the acyclovir trials included nausea, vomiting, diarrhea, and headache.

The Cochrane review provides a good summary of evidence on antiviral drugs and risk of PHN. However, several aspects should be kept in mind when interpreting the results. The meta-analysis suggested that the risk of PHN was reduced by one-fourth at 4 months after rash onset. This estimate was based on only 88 events in total, which may have been insufficient for achieving statistical significance. Notwithstanding, the effect may be clinically relevant and should not be discarded by reducing the interpretation to dichotomy at the 0.05 level [81]. In contrast, the statistically significant albeit lower effect estimate observed at 1 month after rash onset received more attention in the review. Inspection of the primary data also calls into question the interpretation of the effect on the presence of PHN at 6 months after rash onset. The meta-analysis at this time point was based on the results from two trials. In a study by Wood et al. [106], there was no difference in the prevalence of PHN at 6 months (risk ratio 1.03; 95 % CI: 0.84–1.27), whereas Whitley et al. [99] reported a 6-month hazard ratio of 1.39 (95 % CI: 0.84–2.32) for acyclovir vs. placebo recipients, indicating that acyclovir accelerated pain resolution. However, the Cochrane review reported a risk ratio of 1.15 (95 % CI: 0.70–1.91) for the study by Whitley et al. The details of the analysis behind this estimate were not specified [12].

Although the effect of antiviral drugs on incidence of PHN is important, other patient-reported outcome measures of efficacy, such as pain intensity and quality of life, need also to be considered. As pointed out in the Cochrane review, previous antiviral trials have rarely considered these aspects. On a 0–100 scale, Harding and Porter [38] reported that, compared with placebo recipients, participants randomized to acyclovir had a lower mean pain score at both 3 months (0.6 vs. 9.7) and 6 months (1.0 vs. 9.3). However, the differences are small in absolute terms. Huff et al. [42] found no difference in pain severity, but data were not reported. Wood et al. [106] reported a greater mean reduction in pain during treatment compared with the placebo group, but did not perform analyses of mean reductions at later time points. Similarly, Whitley et al. [99] considered the effect only during the first month, reporting increased, albeit statistically imprecise, hazard ratios for time to cessation of acute neuritis (1.47; 95 % CI: 0.67–3.21), to uninterrupted sleep (1.18; 95 % CI: 0.68–2.05), to return to 100 % usual activity (1.63; 95 % CI: 0.96–2.76), and to no use of analgesics (1.27; 95 % CI: 0.66–2.49). The effects of acyclovir on pain severity and quality of life measures at 3 months or later thus remain uncertain.

The Cochrane review included a thorough risk of bias assessment, which unfortunately was hampered by insufficient information about methods in the primary studies [12]. Five out of six trials had unclear risks for one or more of the parameters assessed, which included random sequence generation, allocation concealment, blinding of participants and personnel, blinding of outcome assessment, incomplete outcome data, selective reporting, and other bias. Only the negative study by Wood et al. [106] provided data necessary for critical evaluation and was considered at low risk of bias on all parameters. However, the study had other unnoted shortcomings, such as loss to follow-up after first cessation of pain. Indeed, a small follow-up study estimated that recurrence of pain had occurred in 12 % of participants in the trial [67]. Thus, each study included in the review, as well as remaining studies listed in Table 23.1, had potential risks of bias, which complicate their interpretation and comparability.

Acyclovir has been studied more extensively than the newer antiviral drugs valacyclovir and famciclovir. In the Cochrane review, no eligible valacyclovir trials and only one famciclovir trial were identified [12]. In the famciclovir trial, participants were randomized to 500 mg famciclovir, 750 mg famciclovir, or placebo for 7 days and followed up for 5 months after rash healing [89]. According to the Cochrane review, the risk of prolonged pain was only reported as pain after rash healing with a risk ratio of 1.15 (95 % CI: 0.87–1.52) for 500 mg and 1.31 (95 % CI: 1.01–1.71) for 750 mg famciclovir. However, a critical review of the literature reveals different and more detailed results. In a reanalysis of the data, Dworkin et al. [22] showed that famciclovir recipients had a decreased risk ratio of pain throughout all 6 months of follow-up compared with placebo (Table 23.1). The effect was most pronounced among those aged 50 years or older. These data were not considered in the Cochrane review.

The lack of valacyclovir and famciclovir trials in the Cochrane review is mainly explained by a restriction to placebo-controlled studies. With the establishment of acyclovir as mainstay therapy for acute herpes zoster, placebo-controlled trials of the newer antiviral drugs have quickly become unethical. For this reason, the efficacy of newer antiviral drugs has mainly been investigated using acyclovir as active comparator. The results from such studies consistently demonstrate superiority of valacyclovir and famciclovir (Table 23.1). Furthermore, similar efficacy has been reported for valacyclovir, famciclovir, and brivudine. Because valacyclovir is a prodrug of acyclovir, it can be deduced that its superiority stems from more favorable pharmacokinetics (increased bioavailability) rather than differences in biochemical and physiological effects. Indeed, it is possible to attain four times higher plasma levels for valacyclovir compared with acyclovir, and dosing schedules are more convenient [8]. Despite potential superiority of valacyclovir, large price differences have favored the use of acyclovir for many years. However, with the increasing availability of generic drugs on the market, the price of valacyclovir is approaching that of acyclovir. Hopefully, the same can be expected for famciclovir in the coming years.

In summary, existing systematic reviews and individual trials on the effect of antiviral therapy of acute herpes zoster and the subsequent risk of PHN can singly be challenged. Importantly, there is a profound paucity of data on the effect on severity of pain and quality of life. Although this insufficiency in evidence precludes any firm conclusions regarding the preemptive effects of antiviral drugs, the majority of studies have been positive. In particular, the newer antiviral drugs have demonstrated promising effects. Regardless of the long-term outcomes, antiviral drugs reduce viral shedding, accelerate rash healing, and reduce the acute pain of herpes zoster, which in combination with few adverse events and a low rate of antiviral resistance justifies their use in the treatment of herpes zoster.

23.3 Oral Glucocorticoids

Inflammation definitely accompanies acute zoster, as pathologic studies have demonstrated the presence of inflammation in peripheral nervous system structures, viral ganglionitis, and destruction of neuronal structures in the dorsal horn of the spinal cord with infiltration of inflammatory cells [39, 95]. The theory that inhibition of these inflammatory processes by the use of anti-inflammatory agents decreases the risk of PHN was the primary motivation behind the extensive use of glucocorticoids for the treatment of acute herpes zoster in the past century. Today, the use of glucocorticoids is more controversial.

Table 23.3 describes trials of glucocorticoids for the prevention of PHN. Evidence on this association has been collated by the Cochrane Collaboration [36]. In brief, the Cochrane review included all trials on intravenous, intramuscular, or oral glucocorticoids given within 7 days after rash onset compared with no treatment or placebo. Five trials fulfilled the inclusion criteria, but only two trials comprising 114 participants in total were included for quantitative assessment in a meta-analysis [25, 26], which showed that the risk ratio of PHN at 6 months after onset was 0.95 (95 % CI: 0.46–1.99) for glucocorticoids compared with placebo. The remaining studies lacked detailed data on PHN at more than 1 month after rash onset and were thus deemed unsuitable for the pooled analysis. In all five trials (755 participants in total), there was a potentially increased risk ratio of serious (1.65; 95 % CI: 0.51–5.29) and nonserious events (1.30; 95 % CI: 0.90–1.87), although no cases of disseminated infection were found among glucocorticoid recipients. Based on these data, the Cochrane review thus concluded that there was moderate quality evidence that treatment of acute herpes zoster with glucocorticoids does not reduce the risk of PHN. The studies by Benoldi et al. [6] and Keczkes and Basheer [54], which are shown in Table 23.3, were not included in the Cochrane review, because they compared glucocorticoids to carbamazepine. In addition, authors did not describe whether blinding of patients or investigators was used.

It is important to note that the results from the Cochrane review included also studies evaluating the effect of glucocorticoids compared with placebo. With the wide acceptance of antiviral drugs as first-line therapy for acute herpes zoster, data on this comparison have become virtually obsolete. It is reasonable to believe that combination therapy with antiviral drugs and glucocorticoids is not comparable to monotherapy with glucocorticoids. For example, the proposed favorable anti-inflammatory effect of glucocorticoids on PHN may be offset by an increased risk of viral dissemination in the absence of antiviral therapy. The appropriateness of pooling the two trials in the Cochrane meta-analysis may thus be questioned, as one was a placebo-controlled trial of triamcinolone [25] and the other considered prednisolone in combination with acyclovir compared with acyclovir alone [26]. In fact, risk ratios of PHN at 6 months were 1.33 (95 % CI: 0.21–8.41) and 0.88 (95 % CI: 0.39–1.98), respectively, pointing in each direction. Because of small sample sizes, firm conclusions regarding the difference in estimates are limited.

Table 23.3 Randomized controlled trials on glucocorticoids for the prevention of PHN

Reference	Study population	Treatment groups	Results (investigational drug vs. comparator)[a]
Glucocorticoids alone			
Benoldi et al. [6]	Severe pain >50 years <72 hours of rash onset $n = 18$	Prednisone tapered from 35 mg/day over 31 days vs. 100 mg carbamazepine q.i.d. for 28 days	3/9 (33 %) vs. 2/9 (22 %) pain at 2 months after rash healing and 1/9 (11 %) vs. 1/9 (11 %) at 6 months
Clemmensen and Anderson [13][b]	≥16 years <7 days of rash onset $n = 40$	Prednisone tapered from 45 mg/day over 21 days vs. placebo	4/19 (21 %) vs. 1/19 (5.3 %) persistent pain at 6 weeks after treatment start Mean pain score (scale 0–4) similar in both groups
Eaglstein et al. [25][b]	Severe, painful ≥21 years mean 5 days since rash onset $n = 35$	Triamcinolone, tapered from 48 mg/day over 21 days vs. placebo	8/15 (53 %) vs. 14/20 (70 %) pain at 1 month after rash onset, 2/15 (13 %) vs. 6/20 (30 %) at 4 months, and 2/15 (13 %) vs. 2/20 (10 %) at 6 months
Keczkes and Basheer [54]	Severe, painful >50 years mean 5 days since rash onset $n = 40$	Prednisolone tapered from 40 mg/day over 31 days vs. 100 mg carbamazepine q.i.d. for 28 days	3/20 (15 %) vs. 13/20 (65 %) pain at 2 months after pain onset and 0/20 (0 %) vs. 2/20 (10 %) at 1 year
Glucocorticoids in combination with acyclovir			
Esmann et al. [26][b]	≥60 years <96 h of onset $n = 79$[c]	Prednisolone tapered from 40 mg/day over 21 days vs. placebo (plus 800 mg acyclovir 5× daily for 7 days for all)	35 % for prednisolone + acyclovir vs. 50 % for placebo + acyclovir pain at 2 weeks after enrollment, 35 vs. 58 % at 3 months, and 23 vs. 24 % at 26 weeks
Whitley et al. [99][b]	Localized >50 years <72 hours of rash onset $n = 201$[d]	800 mg acyclovir 5× daily for 21 days and/or 60 mg prednisone daily tapered over 21 days vs. double placebo	ITT: HRs for time to pain cessation during 6 months since enrollment 1.56 (0.92–2.66) for acyclovir + prednisone vs. double placebo, 1.26 (0.72–2.21) for prednisone vs. double placebo, and 1.26 (0.91–1.75) for prednisone (±acyclovir) vs. no prednisone (double placebo or acyclovir)

(continued)

Table 23.3 (continued)

Reference	Study population	Treatment groups	Results (investigational drug vs. comparator)[a]
Wood et al. [103][b]	At least moderate pain >18 years <72 hours after rash onset *n* = 198	800 mg acyclovir 5× daily for 7 or 21 days + 40 mg prednisone tapered over 21 days or placebo	Median duration to complete pain cessation 120 days for 21-day acyclovir + prednisone, 120 days for 21-day acyclovir + placebo, 146 days for 7-day acyclovir + prednisone, and 147 days for 7-day acyclovir + placebo; HR 1.04 (0.81–1.34) for prednisone (±acyclovir) vs. no prednisone (double placebo/ acyclovir) Median duration 147 vs. 120 days (HR 1.09 [0.84–1.40]) for 7 vs. 21 days of monotherapy with acyclovir (*n* = 202)

Abbreviations: *5×* five times, *HR* hazard ratio, *ITT* intention to treat, *mos* months, *PHN* postherpetic neuralgia, *RR* risk ratio

[a]If study explicitly stated that an ITT method was used, this is specified. When available, 95 % confidence intervals for HRs, RRs, and RDs are provided in parenthesis

[b]Included in Cochrane review by Han et al. [36]

[c]Eighty-four patients were enrolled, but five patients were excluded because inclusion criteria were not fulfilled or because of lack of compliance in the first 1–2 weeks. It was not possible to restore the data for these patients for inclusion in the ITT analysis by Cochrane

[d]One hundred three for the comparison of prednisone + acyclovir vs. double placebo, 102 for prednisone vs. double placebo, and 201 for the main effect of prednisone

Only three trials have examined the effect of adding glucocorticoids to antiviral therapy [26, 99, 103]. Although all three studies found a potential effect on the acute neuralgia of herpes zoster, results are conflicting for persistent pain. Wood et al. [103] found that there was an effect on acute pain and rash healing, but not on time to first or complete pain resolution. A small study by Esmann et al. [26] reported a potentially decreased risk of PHN at both 3 and 6 months, but estimates were very imprecise. Whitley et al. [99] used a 2 × 2 factorial design, randomizing patients to receive acyclovir and prednisone, acyclovir and prednisone-placebo, prednisone and acyclovir-placebo, or double placebo. In the analysis, the effects of prednisone and acyclovir either alone or in combination were compared with double placebo. The hazard ratios were above unity for all these comparisons in the 6-month evaluation of time to cessation of pain, as well as the 1-month evaluation of quality of life, which comprised time to cessation of acute neuritis, time to uninterrupted sleep, time to return to 100 % usual activity, and time to no use of analgesics. Interestingly, the fastest improvement was found for acyclovir plus prednisone vs. double placebo (Tables 23.1 and 23.3). However, the small sample size and the lack of a direct comparison of combination therapy and acyclovir monotherapy preclude any firm conclusions regarding the effect of adding prednisone to acyclovir. No trials have examined the effect of combining glucocorticoids with valacyclovir, famciclovir, or brivudine.

In summary, only three studies have investigated the effect of glucocorticoids as a supplement to antiviral therapy in acute herpes zoster. Although they suggest a beneficial effect on acute herpetic neuralgia and quality of life, evidence is conflicting with regard to preventing long-standing pain.

23.4 Opioid and Non-opioid Analgesic Agents

Severe pain at the time of presentation with acute herpes zoster is a risk factor of PHN [31, 77, 78, 100]. Afferent bombardment from peripheral nerves due to tissue damage during acute herpes zoster may result in long-term enhancement of central excitability and loss of sensory function, as observed in established PHN [29, 77, 78, 80]. Theoretically, targeting of these mechanisms with aggressive analgesic treatment in the acute stage may prevent the development and maintenance of PHN. Several analgesic agents used for treatment of established PHN have been suggested in this context, including tricyclic antidepressants (amitriptyline, desipramine, nortriptyline), anticonvulsants (carbamazepine, gabapentin, pregabalin), and opioids (including tramadol).

Table 23.4 summarizes trials on oral analgesics for prevention of PHN. Bowsher [10] randomized 80 persons presenting with herpes zoster to their primary care practitioner within 48 h of rash onset to receive either 25 mg amitriptyline or placebo at bedtime for 90 days. Pain was present in 26 vs. 38 % at 3 months and 16 vs. 35 % at 6 months after rash onset, thus suggesting a substantial preventive effect of amitriptyline. However, the study has not been repeated, and interpretation of the results is complicated for several reasons. First, authors performed only a per-protocol analysis, which resulted in the exclusion of eight patients, among whom three had been randomized to amitriptyline, three to placebo, and two were untraceable. Second, blinding may have been inadequate, as patients were warned about dry mouth as a potential side effect. Third, supplemental treatments were not controlled, leading to large differences in the proportion of patients treated with acyclovir (24 % of amitriptyline and 50 % of placebo recipients). Although authors provided robust results in analyses stratified by acyclovir treatment, baseline differences cannot be ruled out. Finally, the study did not report on potential adverse effects of treatment and contraindications for treatment. This limitation is important, as many adverse effects have been established for amitriptyline, including orthostatic hypotension, arrhythmias, and electrocardiogram abnormalities, particularly among elderly and patients with cardiac comorbidity. These adverse effects warrant close monitoring with electrocardiograms during treatment initiation and dose increments.

The anticonvulsant carbamazepine at a dose of 100 mg four times a day for 28 days was compared with acyclovir in a small randomized trial including 18 patients aged >50 years presenting with herpes zoster within 72 h of rash onset and

Table 23.4 Randomized trials on opioid and non-opioid analgesics for the prevention of PHN

Reference	Study population	Treatment groups	Results (investigational drug vs. comparator)[a]
Tricyclic antidepressants			
Bowsher [10]	>60 years <48 hours of rash onset $n = 80$	25 mg amitriptyline q.d. for 90 days vs. placebo	10/38 (26 %) vs. 13/34 (38 %) pain at 3 months after rash onset and 6/38 (16 %) vs. 12/34 (35 %) at 6 months
Anticonvulsants			
Benoldi et al. [6]	Severe pain >50 years <72 hours of rash onset $n = 18$	100 mg carbamazepine q.i.d. for 28 days vs. 800 mg acyclovir 5× daily for 7 days[b]	2/9 (22 %) vs. 2/9 (22 %) pain at 2 months after rash healing and 1/9 (11 %) vs. 0/9 (0 %) at 6 months
Keczkes and Basheer [54]	Severe pain >50 years mean 5 days since rash onset $n = 40$	100 mg carbamazepine q.i.d. for 28 days vs. prednisolone tapered from 40 mg/day over 31 days	13/20 (65 %) vs. 3/20 (15 %) pain at 2 months after pain onset and 2/20 (10 %) vs. 0/20 (0 %) at 1 year
Krčevski Škvarč and Kamenik [56]	Pain intensity ≥4 (0–10 scale) on naproxen 30–80 years <7–14 days of rash onset $n = 29$	75–150 mg pregabalin b.i.d. vs. placebo for 3 weeks, plus oxycodone depending on pain	6/14 (43 %) vs. 7/15 (47 %) pain at 1–3 months after acute phase and 2/14 (14 %) vs. 3/15 (20 %) at 6 months
Lee et al. [59]	Pain intensity ≥4 (0–10 scale) ≥50 years <4 days of rash onset $n = 120$	300 mg gabapentin t.i.d. vs. no control for max. 24 weeks depending on pain, (plus valacyclovir and acetaminophen for all)	2/52 (3.8 %) vs. 3/49 (6.1 %) moderate-severe pain (≥4 on 0–10 scale) at 3 months Mean pain score 0.54 vs. 0.31 at 3 months (0–10 scale) Mean quality-of-life score (Dermatology Life Quality Index) 0.90 vs. 0.76 at 3 months

Abbreviations: *5×* five times, *PHN* postherpetic neuralgia; *q.d.* quaque die (once a day), *q.i.d.* quarter in die (four times a day)

[a]If study explicitly stated that an ITT method was used, this is specified. When available, 95 % confidence intervals for HRs, RRs, and RDs are provided in parenthesis

[b]Also examined is the effect of prednisone, as presented in Table 23.3

who had severe pain [6]. Twenty-two percent of patients in each group had pain at 2 months after rash healing. After 6 months, no patients in the acyclovir group and 1/9 (11 %) of patients in the carbamazepine group still had pain. Authors did not explicitly state if the study was blinded. The study also compared treatments with prednisone, as shown in Table 23.3. A similar trial of 40 patients randomized to carbamazepine or prednisolone reported that the prevalence of pain was 65 vs. 15 %

at 2 months and 10 vs. 0 % at 1 year after pain onset, respectively [54]. Unfortunately, the poor methodology and sample size limit the interpretation of both these studies.

Krčevski Škvarč and Kamenik [56] examined the efficacy of pregabalin in a randomized double-blind placebo-controlled trial. Twenty-nine patients aged 30–80 years who presented with moderate to severe pain within 7 to 14 days since rash onset were randomized to 75 mg pregabalin twice daily or placebo for 3 weeks. Dose was increased to 150 mg pregabalin or placebo, and oxycodone was given depending on pain and as side effects permitted. Pain was present in 6/14 (43 %) pregabalin-treated vs. 7/15 (47 %) of placebo-treated patients at 1 to 3 months after acute herpes zoster and in 2/14 (14 %) vs. 3/15 (20 %) patients at 6 months. The results indicate a positive effect, but the small sample size precludes firm conclusions. Furthermore, the interpretation is hampered by methodological issues, such as potential baseline differences, indicating unsuccessful randomization. For example, all pregabalin recipients had been treated with antivirals compared with only 80 % of placebo recipients. In addition, because use of other analgesics was allowed, patients in each arm achieved similar decrease in pain severity during the 3-week period of treatment. By not controlling use of other analgesics, the chance of observing a favorable effect of pregabalin may have been diminished.

In a randomized trial, Lee et al. [59] examined the preemptive effect of low-dose (900 mg/day) gabapentin initiated within 4 days of rash onset. One hundred and twenty patients with moderate or severe pain at presentation were enrolled in the study. Authors did not state whether the study was placebo-controlled and blinded. At 3 months, 3.8 % of gabapentin-treated and 6.1 % of untreated had PHN, which was defined as a pain score of 4 or more on a 10-point Likert scale. However, prevalence of such pain through week one to eight was higher in the gabapentin-treated group. Mean pain and quality-of-life scores did not differ between groups. These results indicate no effect of low-dose gabapentin on acute and subacute neuralgia and at most a slight effect on preventing PHN. However, it is possible that the gabapentin dose was too low to achieve sufficient analgesia in all patients. Furthermore, gabapentin was administered for up to 24 weeks depending on the presence of pain. Thus, treatment extended into the chronic phase, which makes it difficult to separate prevention and treatment of established PHN. More promising results were found in an uncontrolled open-label study, which found that the addition of gabapentin to antiviral therapy prevents clinically significant PHN when compared with findings of previously reported trials [58]. In addition, several randomized double-blind controlled trials have demonstrated that controlled-release oxycodone [21], pregabalin [46, 51], and gabapentin [7, 50] provide at least transient analgesia in acute herpetic neuralgia, although the effect of gabapentin was uncertain in one of the trials [21]. Assuming that reduction in acute pain prevents the development of PHN, these trials may provide some proof of concept that these agents carry a preemptive effect. Acetaminophen and nonsteroidal anti-inflammatory drugs are commonly used to treat mild pain with or without opioid agents. However, neither the effect on acute nor chronic pain of herpes zoster has been established for these agents.

In summary, oxycodone, gabapentin, and pregabalin may reduce acute herpetic neuralgia, but their effect on PHN as a supplement to antiviral therapy has not been examined. Preemptive effects of amitriptyline, carbamazepine, and pregabalin have been reported, but a high risk of bias and the small sample sizes in most of the studies raise the question of whether observed effects are causal.

23.5 Other Agents

Randomized controlled trials have been published on the efficacy of a wide range of other treatments for reducing PHN, including idoxuridine [1, 28, 49, 101, 102], topical acyclovir [61, 73], human leukocyte interferon-alpha [20, 68, 69], amantadine hydrochloride [32, 33], levodopa and benserazide [55], hyperbaric oxygen therapy [76], adenosine monophosphate [84], vidarabine [97, 98], netivudine [85], neuramide [92], isoprinosine [75], cimetidine [11, 60, 70], dehydroemetine [40], zoster hyperimmune globulin [44, 87], sorivudine [9, 35], radiotherapy [6], and adrenocorticotrophic hormone [13]. These agents were considered outside the scope of this chapter, because most have been associated with severe adverse effects and/or are considered obsolete. Interested readers are, however, encouraged to consult the studies referenced for more information. New antiviral drugs are currently being tested [91].

23.6 Summary and Recommendations

It has been suggested that aggressive treatment of acute herpes zoster with antiviral drugs, glucocorticoids, and analgesics may prevent PHN by reducing viral replication, inflammation, and central sensitization, respectively. However, this critical review of the literature reveals a paucity of high-quality data to either support or refute any preemptive effects of these agents. For acyclovir, which has been studied most extensively, several studies and systematic reviews support an effect, but firm conclusions are precluded by inadequate methods and reporting according to current standards. The newer antiviral drugs valacyclovir and famciclovir have primarily been compared against acyclovir, demonstrating their superiority with regard to both acute and chronic outcomes. Because antiviral drugs are considered first-line therapy for herpes zoster, further placebo-controlled trials on the preemptive effects are not to be expected. However, a well-designed observational study or a trial randomizing patients with late presentation to antiviral therapy or placebo could possibly contribute valuable and original information. When all comes to it, previous trials demonstrate that up to 20 % of patients who are treated with valacyclovir or famciclovir will experience PHN at 6 months, thus warranting the identification of other potentially preemptive treatments.

The effects of glucocorticoids and oral analgesic treatment have been examined in a few trials, which taken together are inconclusive. Furthermore, most of the trials are outdated, because treatment was provided alone rather than as a supplement to antiviral therapy. Despite the uncertainty of the long-term effects, glucocorticoids, oxycodone, gabapentin, and pregabalin appear to alleviate acute neuralgia at least temporarily, which may justify their use in the acute phase.

Pending scientific proof of whether aggressive treatment of herpes zoster has an effect on risk and severity of PHN, treatment in the acute phase is indicated and any resulting preemptive effect on prolonged pain would merely constitute an additional benefit. Consensus recommendations for the management of herpes zoster have previously been published, including detailed information on dosages and potential adverse effects [23]. Antiviral therapy should be offered to patients aged 50 years or older, younger patients with moderate or severe pain, patients with moderate or severe rash, immunosuppressed patients, and those with non-truncal herpes zoster. Valacyclovir or famciclovir is preferred because of their potential superiority to acyclovir. Brivudine provides an alternative option in some countries, but its use in immunosuppressed individuals is limited by potential interactions with 5-fluorouracil and other 5-fluoropyrimidines. Intravenous treatment with acyclovir is recommended for patients with severe ophthalmic herpes zoster, patients with severe immunosuppression, or when oral administration is precluded.

Despite limited evidence, antiviral therapy may also be considered for younger patients, as the adverse effects are limited. Similarly, patients presenting late may benefit from treatment, especially if vesicle formation is still ongoing and/or in the presence of complications. The usual rule of thumb is that treatment should be initiated within 72 h after rash onset (approximately 1 week for patients with ophthalmic herpes zoster or immune deficiency). However, this rule is arbitrarily based on the inclusion criterion applied in the early acyclovir trials. There are no biological data to support that viral replication has ceased completely after 72 h. In contrast to dermatological manifestations, data from clinical trials suggest that the effect on pain does not depend on whether treatment is initiated early or late within the 72 h time window [107]. This finding indicates that neuronal damage may continue for some time after rash healing. Nevertheless, it is reasonable to believe that the antiviral effect decreases with time since onset of herpes zoster, and thus education of the general population and health care personnel in recognizing and acting on symptoms of herpes zoster is essential. Complicated herpes zoster, for example, ophthalmic zoster and motor involvement, should always be treated regardless of late presentation.

The pain associated with acute herpes zoster can be excruciating and should also be treated. Several agents described in this chapter are potentially effective in this regard, justifying their use in an attempt to achieve pain control. The choice of drug should be tailored according the patient's pain severity, comorbidities, contraindications, previous experience, and preference. Treatment effects, including adverse effects, should be monitored, and patients should be made aware of the potential fluctuations in pain over time.

For mild pain, scheduled treatment with regular painkillers such as acetaminophen, nonsteroidal anti-inflammatory drugs, and/or weak opioids or tramadol may suffice, although studies on their effects are lacking. For moderate or severe pain, scheduled treatment with a strong opioid agent may be considered despite insufficient data from randomized trials demonstrating benefit. Opioid treatment should be combined with stimulant laxatives with or without stool softeners for preventing constipation, which is a common side effect. When in doubt, a pain management expert can be consulted for guidance on the treatment plan, especially if pain is severe.

If satisfactory analgesic relief is not obtained, treatment with gabapentin or pregabalin, tricyclic antidepressants, or glucocorticoids may be attempted. Gabapentin or pregabalin may be preferable because of superior side effect and interaction profiles. Pregabalin requires less frequent dosing, and titration to desired doses may be faster than for gabapentin [86], which is particularly attractive when the goal is to achieve effective pain relief rapidly. Addition of glucocorticoids to antiviral drugs may have an effect on pain and quality of life in the acute phase of herpes zoster and could thus be considered as an alternative in patients with moderate or severe pain. Glucocorticoids are also often used for intractable ocular inflammation in ophthalmic zoster and when there are symptoms of extensive nervous system inflammation, including facial paralysis, polyneuritis, meningitis, and encephalitis. As with all other drugs described herein, contraindications (e.g., hypertension, diabetes, peptic ulcer disease) should be taken into account, and advice on potential adverse events should be communicated clearly. Similarly, the use of tricyclic antidepressants for acute neuralgia is hampered by adverse events and contraindications, especially in the elderly who are at highest risk of herpes zoster and PHN. In addition, their effect has only been confirmed in one trial, which had high risk of bias [10].

Finally, in patients who are refractory to management of severe pain using the aforementioned agents, neural blockade administered by a pain specialist during hospitalization or at a specialized outpatient clinic may be required [24].

The study of prophylactic agents for chronic neuralgia is a challenging task. Large sample sizes are needed because only a minority go on to suffer from clinically significant PHN pain. Most previous trials have not been powered to specifically test for the prevention of PHN. Furthermore, methodology in existing studies is heterogeneous, complicating the compilation of evidence. In order to facilitate comparability of future studies, the Initiative on Methods, Measurement, and Pain Assessment in Clinical Trials (IMMPACT) has published recommendations to guide researchers in the design of clinical trials on chronic pain prevention [34]. In brief, the IMMPACT recommendations propose inclusion of participants within 7 days or less after rash onset with follow-up at 3 or 4 months after rash onset for presence of pain, presence of "clinically meaningful" pain (e.g., ≥3 out of 10), pain intensity and qualities, and as secondary endpoints, physical and emotional functioning. The duration of acute pain, defined as pain within 30 days or less from rash onset, should also be considered, as it may contribute to the understanding of the underlying pharmacological mechanisms. Also, it is important to track and consider the use of rescue medications, for example, by incorporating it in a secondary endpoint together with pain severity. Finally, PHN risk factors should be adjusted for in analyses. Hopefully, adoption of these guidelines and transparent reporting in accordance with the Consolidated Standards of Reporting Trials [83] will lead to publication of well-designed and comparable trials on the prevention of PHN in the near future.

References

1. Aliaga A, Armijo M, Camacho F et al (1992) A topical solution of 40 % idoxuridine in dimethyl sulfoxide compared to oral acyclovir in the treatment of herpes zoster. A double-blind multicenter clinical trial. Med Clin (Barc) 98:245–249
2. Arani RB, Soong SJ, Weiss HL et al (2001) Phase specific analysis of herpes zoster associated pain data: a new statistical approach. Stat Med 20:2429–2439
3. Attal N, Cruccu G, Baron R et al (2010) EFNS guidelines on the pharmacological treatment of neuropathic pain: 2010 revision. Eur J Neurol 17:1113–1e88
4. Balfour HH Jr, Bean B, Laskin OL et al (1983) Acyclovir halts progression of herpes zoster in immunocompromised patients. N Engl J Med 308:1448–1453
5. Bean B, Braun C, Balfour HH (1982) Acyclovir therapy for acute herpes zoster. Lancet 2:118–121
6. Benoldi D, Mirizzi S, Zucchi A, Allegra F (1991) Prevention of post-herpetic neuralgia. Evaluation of treatment with oral prednisone, oral acyclovir, and radiotherapy. Int J Dermatol 30:288–290
7. Berry JD, Petersen KL (2005) A single dose of gabapentin reduces acute pain and allodynia in patients with herpes zoster. Neurology 65:444–447
8. Beutner KR, Friedman DJ, Forszpaniak C et al (1995) Valaciclovir compared with acyclovir for improved therapy for herpes zoster in immunocompetent adults. Antimicrob Agents Chemother 39:1546–1553
9. Bodsworth NJ, Boag F, Burdge D et al (1997) Evaluation of sorivudine (BV-araU) versus acyclovir in the treatment of acute localized herpes zoster in human immunodeficiency virus-infected adults. The Multinational Sorivudine Study Group. J Infect Dis 176:103–111
10. Bowsher D (1997) The effects of pre-emptive treatment of postherpetic neuralgia with amitriptyline: a randomized, double-blind, placebo-controlled trial. J Pain Symptom Manage 13:327–331
11. Brint-Nielsen P, Christiansen LV, Damsbo N, Nørrelund N (1987) Cimetidine in herpes zoster. A double-blind study in general practice. Ugeskr Laeger 149:2138–2139
12. Chen N, Li Q, Yang J, et al (2014) Antiviral treatment for preventing postherpetic neuralgia. Cochrane Database Syst Rev (2):CD006866
13. Clemmensen OJ, Andersen KE (1984) ACTH versus prednisone and placebo in herpes zoster treatment. Clin Exp Dermatol 9:557–563
14. Cobo LM, Foulks GN, Liesegang T et al (1986) Oral acyclovir in the treatment of acute herpes zoster ophthalmicus. Ophthalmology 93:763–770
15. Cohen JI (2013) Clinical practice: Herpes zoster. N Engl J Med 369:255–263
16. Colin J, Prisant O, Cochener B et al (2000) Comparison of the efficacy and safety of valaciclovir and acyclovir for the treatment of herpes zoster ophthalmicus. Ophthalmology 107:1507–1511
17. Coplan PM, Schmader K, Nikas A et al (2004) Development of a measure of the burden of pain due to herpes zoster and postherpetic neuralgia for prevention trials: adaptation of the brief pain inventory. J Pain 5:344–356. doi:10.1016/j.jpain.2004.06.001
18. Crooks RJ, Jones DA, Fiddian AP (1991) Zoster-associated chronic pain: an overview of clinical trials with acyclovir. Scand J Infect Dis Suppl 80:62–68

19. Degreef H, Famciclovir Herpes Zoster Clinical Study Group (1994) Famciclovir, a new oral antiherpes drug: results of the first controlled clinical study demonstrating its efficacy and safety in the treatment of uncomplicated herpes zoster in immunocompetent patients. Int J Antimicrob Agents 4:241–246
20. Duschet P, Schwarz T, Soyer P et al (1988) Treatment of herpes zoster. Recombinant alpha interferon versus acyclovir. Int J Dermatol 27:193–197
21. Dworkin RH, Barbano RL, Tyring SK et al (2009) A randomized, placebo-controlled trial of oxycodone and of gabapentin for acute pain in herpes zoster. Pain 142:209–217
22. Dworkin RH, Boon RJ, Griffin DR, Phung D (1998) Postherpetic neuralgia: impact of famciclovir, age, rash severity, and acute pain in herpes zoster patients. J Infect Dis 178(Suppl 1):S76–S80
23. Dworkin RH, Johnson RW, Breuer J et al (2007) Recommendations for the management of herpes zoster. Clin Infect Dis 44(Suppl 1):S1–26. doi:10.1086/510206
24. Dworkin RH, O'Connor AB, Kent J et al (2013) Interventional management of neuropathic pain: NeuPSIG recommendations. Pain 154:2249–2261
25. Eaglstein WH, Katz R, Brown JA (1970) The effects of early corticosteroid therapy on the skin eruption and pain of herpes zoster. JAMA 211:1681–1683
26. Esmann V, Geil JP, Kroon S et al (1987) Prednisolone does not prevent post-herpetic neuralgia. Lancet 2:126–129
27. Esmann V, Ipsen J, Peterslund NA et al (1982) Therapy of acute herpes zoster with acyclovir in the nonimmunocompromised host. Am J Med 73:320–325
28. Esmann V, Wildenhoff KE (1980) Idoxuridine for herpes zoster. Lancet 2:474
29. Fields HL, Rowbotham M, Baron R (1998) Postherpetic neuralgia: irritable nociceptors and deafferentation. Neurobiol Dis 5:209–227
30. Finn R, Smith MA (1984) Oral acyclovir for herpes zoster. Lancet 2:575
31. Forbes HJ, Thomas SL, Smeeth L et al (2016) A systematic review and meta-analysis of risk factors for postherpetic neuralgia. Pain 157:30–54
32. Galbraith AW (1983) Prevention of post-herpetic neuralgia by amantadine hydrochloride (Symmetrel). Br J Clin Pract 37:304–306
33. Galbraith AW (1973) Treatment of acute herpes zoster with amantadine hydrochloride (Symmetrel). Br Med J 4:693–695
34. Gewandter JS, Dworkin RH, Turk DC et al (2015) Research design considerations for chronic pain prevention clinical trials: IMMPACT recommendations. Pain 156:1184–1197
35. Gnann JW, Crumpacker CS, Lalezari JP et al (1998) Sorivudine versus acyclovir for treatment of dermatomal herpes zoster in human immunodeficiency virus-infected patients: results from a randomized, controlled clinical trial. Collaborative Antiviral Study Group/AIDS Clinical Trials Group, Herpes Zoster Study Group. Antimicrob Agents Chemother 42:1139–1145
36. Han Y, Zhang J, Chen N, et al (2013) Corticosteroids for preventing postherpetic neuralgia. Cochrane Database Syst Rev (3):CD005582
37. Harding SP, Lipton JR, Wells JC, Campbell JA (1986) Relief of acute pain in herpes zoster ophthalmicus by stellate ganglion block. Br Med J (Clin Res Ed) 292:1428
38. Harding SP, Porter SM (1991) Oral acyclovir in herpes zoster ophthalmicus. Curr Eye Res 10 Suppl:177–182
39. Head H, Campbell AW (1900) The pathology of Herpes Zoster and its bearing on sensory localisation. Brain 23:353–362
40. Hernandez-Perez E (1980) Dehydroemetine therapy for herpes zoster. A comparison with corticosteroids. Cutis 25:424–426
41. Huff JC (1987) Oral acyclovir therapy of acute herpes zoster: a multicentre study. Res Clin Forum 9:37–43
42. Huff JC, Bean B, Balfour HH et al (1988) Therapy of herpes zoster with oral acyclovir. Am J Med 85:84–89
43. Huff JC, Drucker JL, Clemmer A et al (1993) Effect of oral acyclovir on pain resolution in herpes zoster: a reanalysis. J Med Virol Suppl 1:93–96
44. Hügler P, Siebrecht P, Hoffmann K et al (2002) Prevention of postherpetic neuralgia with varicella-zoster hyperimmune globulin. Eur J Pain 6:435–445

45. Jackson JL, Gibbons R, Meyer G, Inouye L (1997) The effect of treating herpes zoster with oral acyclovir in preventing postherpetic neuralgia. A meta-analysis. Arch Intern Med 157:909–912
46. Jensen-Dahm C, Rowbotham MC, Reda H, Petersen KL (2011) Effect of a single dose of pregabalin on herpes zoster pain. Trials 12:55
47. Johnson RW, Bouhassira D, Kassianos G et al (2010) The impact of herpes zoster and post-herpetic neuralgia on quality-of-life. BMC Med 8:37
48. Juel-Jensen BE, Khan JA, Pasvol G (1983) High-dose intravenous acyclovir in the treatment of zoster: a double-blind, placebo-controlled trial. J Infect 6:31–36
49. Juel-Jensen BE, MacCallum FO, Mackenzie AM, Pike MC (1970) Treatment of zoster with idoxuridine in dimethyl sulphoxide. Results of two double-blind controlled trials. Br Med J 4:776–780
50. Kanodia S, Seth A, Dixit A (2012) Dose related efficacy of gabapentin in acute herpetic neuralgia among geriatric patients. Indian J Dermatol 57:362
51. Kanodia SK, Singhal KC (2011) A study on efficacy of Pregabalin in acute Herpetic Neuralgia. Ann Neurol 18:148–150
52. Kawai K, Gebremeskel BG, Acosta CJ (2014) Systematic review of incidence and complications of herpes zoster: towards a global perspective. BMJ Open 10;4(6):e004833
53. Keating GM (2013) Shingles (herpes zoster) vaccine (zostavax(®)): a review of its use in the prevention of herpes zoster and postherpetic neuralgia in adults aged ≥50 years. Drugs 73:1227–1244
54. Keczkes K, Basheer AM (1980) Do corticosteroids prevent post-herpetic neuralgia? Br J Dermatol 102:551–555
55. Kernbaum S, Hauchecorne J (1981) Administration of levodopa for relief of herpes zoster pain. JAMA 246:132–134
56. Krčevski Škvarč N, Kamenik M (2010) Effects of pregabalin on acute herpetic pain and postherpetic neuralgia incidence. Wien Klin Wochenschr 122(Suppl 2):49–53
57. Lancaster T, Silagy C, Gray S (1995) Primary care management of acute herpes zoster: systematic review of evidence from randomized controlled trials. Br J Gen Pract 45:39–45
58. Lapolla W, Digiorgio C, Haitz K et al (2011) Incidence of postherpetic neuralgia after combination treatment with gabapentin and valacyclovir in patients with acute herpes zoster: open-label study. Arch Dermatol 147:901–907
59. Lee EG, Lee HJ, Hyun DJ et al (2016) Efficacy of low dose gabapentin in acute herpes zoster for preventing postherpetic neuralgia: a prospective controlled study. Dermatol Ther 29(3):184–190
60. Levy DW, Banerjee AK, Glenny HP (1985) Cimetidine in the treatment of herpes zoster. J R Coll Physicians Lond 19:96–98
61. Mandal BK, Dunbar EM, Ellis ME et al (1988) A double-masked, placebo-controlled trial of acyclovir cream in immunocompetent patients with herpes zoster. J Infect 17:57–63
62. McGill J, MacDonald DR, Fall C et al (1983) Intravenous acyclovir in acute herpes zoster infection. J Infect 6:157–161
63. McGill JI, White JE (1994) Acyclovir and post-herpetic neuralgia and ocular involvement. BMJ 309:1124
64. McKendrick MW, Care C, Burke C et al (1984) Oral acyclovir in herpes zoster. J Antimicrob Chemother 14:661–665
65. McKendrick MW, McGill JI, White JE, Wood MJ (1986) Oral acyclovir in acute herpes zoster. Br Med J (Clin Res Ed) 293(6561):1529–1532
66. McKendrick MW, McGill JI, Wood MJ (1989) Lack of effect of acyclovir on postherpetic neuralgia. BMJ 298:431
67. McKendrick MW, Wood MJ (1995) Acyclovir and post-herpetic neuralgia. Two other participating study centres report different results. BMJ 310(6985):1005
68. Merigan TC, Gallagher JG, Pollard RB, Arvin AM (1981) Short-course human leukocyte interferon in treatment of herpes zoster in patients with cancer. Antimicrob Agents Chemother 19:193–195

69. Merigan TC, Rand KH, Pollard RB et al (1978) Human leukocyte interferon for the treatment of herpes zoster in patients with cancer. N Engl J Med 298:981–987
70. Miller A, Harel D, Laor A, Lahat N (1989) Cimetidine as an immunomodulator in the treatment of herpes zoster. J Neuroimmunol 22:69–76
71. Morrison VA, Johnson GR, Schmader KE et al (2015) Long-term persistence of zoster vaccine efficacy. Clin Infect Dis 60:900–909
72. Morton P, Thomson AN (1989) Oral acyclovir in the treatment of herpes zoster in general practice. N Z Med J 102:93–95
73. Neoh C, Harding SP, Saunders D et al (1994) Comparison of topical and oral acyclovir in early herpes zoster ophthalmicus. Eye (Lond) 8(Pt 6):688–691
74. Oxman MN, Levin M, Johnson GR et al (2005) A vaccine to prevent herpes zoster and postherpetic neuralgia in older adults. N Engl J Med 352:2271–2284
75. Payne CM, Menday AP, Rogers T, Staughton RC (1989) Isoprinosine does not influence the natural history of herpes zoster or postherpetic neuralgia. Scand J Infect Dis 21:15–18
76. Peng Z, Wang S, Huang X, Xiao P (2012) Effect of hyperbaric oxygen therapy on patients with herpes zoster. Undersea Hyperb Med 39:1083–1087
77. Petersen KL, Rice FL, Farhadi M et al (2010) Natural history of cutaneous innervation following herpes zoster. Pain 150:75–82. doi:10.1016/j.pain.2010.04.002
78. Petersen KL, Rowbotham MC (2010) Natural history of sensory function after herpes zoster. Pain 150:83–92
79. Peterslund NA, Ipsen J, Schonheyder H et al (1981) Acyclovir in herpes zoster. Lancet 318:827–830
80. Reda H, Greene K, Rice FL et al (2013) Natural history of herpes zoster: late follow-up of 3.9 years (n = 43) and 7.7 years (n = 10). Pain 154:2227–2233
81. Rothman KJ, Greenland S, Lash TL (2008) Modern epidemiology, 3rd edn. Lippincott Williams & Wilkins, Philadelphia
82. Schmader KE, Studenski S (1989) Are current therapies useful for the prevention of postherpetic neuralgia? A critical analysis of the literature. J Gen Intern Med 4:83–89
83. Schulz KF, Altman DG, Moher D, CONSORT Group (2010) CONSORT 2010 statement: updated guidelines for reporting parallel group randomised trials. PLoS Med 7:e1000251
84. Sklar SH, Blue WT, Alexander EJ, Bodian CA (1985) Herpes zoster. The treatment and prevention of neuralgia with adenosine monophosphate. JAMA 253:1427–1430
85. Söltz-Szöts J, Tyring S, Andersen PL et al (1998) A randomized controlled trial of acyclovir versus netivudine for treatment of herpes zoster. International Zoster Study Group. J Antimicrob Chemother 41:549–556
86. Stacey BR, Barrett JA, Whalen E et al (2008) Pregabalin for postherpetic neuralgia: placebo-controlled trial of fixed and flexible dosing regimens on allodynia and time to onset of pain relief. J Pain 9:1006–1017
87. Stevens DA, Merigan TC (1980) Zoster immune globulin prophylaxis of disseminated zoster in compromised hosts. A randomized trial. Arch Intern Med 140:52–54
88. Thyregod HG, Rowbotham MC, Peters M et al (2007) Natural history of pain following herpes zoster. Pain 128:148–156. doi:10.1016/j.pain.2006.09.021
89. Tyring S, Barbarash RA, Nahlik JE et al (1995) Famciclovir for the treatment of acute herpes zoster: effects on acute disease and postherpetic neuralgia. A randomized, double-blind, placebo-controlled trial. Collaborative Famciclovir Herpes Zoster Study Group. Ann Intern Med 123:89–96
90. Tyring SK, Beutner KR, Tucker BA et al (2000) Antiviral therapy for herpes zoster: randomized, controlled clinical trial of valacyclovir and famciclovir therapy in immunocompetent patients 50 years and older. Arch Fam Med 9:863–869
91. Tyring SK, Plunkett S, Scribner AR et al (2012) Valomaciclovir versus valacyclovir for the treatment of acute herpes zoster in immunocompetent adults: a randomized, double-blind, active-controlled trial. J Med Virol 84:1224–1232
92. Varotti C, Rafanelli A (2001) Evaluation of efficacy and tolerance of neuramide in the treatment of herpes zoster and postherpetic neuritis. Drugs Exp Clin Res 27:199–208

93. Wassilew S, Collaborative Brivudine PHN Study Group (2005) Brivudine compared with famciclovir in the treatment of herpes zoster: effects in acute disease and chronic pain in immunocompetent patients. A randomized, double-blind, multinational study. J Eur Acad Dermatol Venereol 19:47–55
94. Wassilew SW, Reimlinger S, Nasemann T, Jones D (1987) Oral acyclovir for herpes zoster: a double-blind controlled trial in normal subjects. Br J Dermatol 117:495–501
95. Watson CP, Deck JH, Morshead C et al (1991) Post-herpetic neuralgia: further post-mortem studies of cases with and without pain. Pain 44:105–117
96. Watson CP, Vernich L, Chipman M, Reed K (1998) Nortriptyline versus amitriptyline in postherpetic neuralgia: a randomized trial. Neurology 51:1166–1171
97. Whitley RJ, Gnann JW, Hinthorn D et al (1992) Disseminated herpes zoster in the immunocompromised host: a comparative trial of acyclovir and vidarabine. The NIAID Collaborative Antiviral Study Group. J Infect Dis 165:450–455
98. Whitley RJ, Soong SJ, Dolin R et al (1982) Early vidarabine therapy to control the complications of herpes zoster in immunosuppressed patients. N Engl J Med 307:971–975
99. Whitley RJ, Weiss H, Gnann JW et al (1996) Acyclovir with and without prednisone for the treatment of herpes zoster. A randomized, placebo-controlled trial. The National Institute of Allergy and Infectious Diseases Collaborative Antiviral Study Group. Ann Intern Med 125:376–383
100. Whitley RJ, Weiss HL, Soong SJ, Gnann JW (1999) Herpes zoster: risk categories for persistent pain. J Infect Dis 179:9–15. doi:10.1086/314562
101. Wildenhoff KE, Esmann V, Ipsen J et al (1981) Treatment of trigeminal and thoracic zoster with idoxuridine. Scand J Infect Dis 13:257–262
102. Wildenhoff KE, Ipsen J, Esmann V et al (1979) Treatment of herpes zoster with idoxuridine ointment, including a multivariate analysis of symptoms and signs. Scand J Infect Dis 11:1–9
103. Wood MJ, Johnson RW, McKendrick MW et al (1994) A randomized trial of acyclovir for 7 days or 21 days with and without prednisolone for treatment of acute herpes zoster. N Engl J Med 330:896–900
104. Wood MJ, Kay R, Dworkin RH et al (1996) Oral acyclovir therapy accelerates pain resolution in patients with herpes zoster: a meta-analysis of placebo-controlled trials. Clin Infect Dis 22:341–347
105. Wood MJ, McKendrick MW, McGill JI (1987) Oral acyclovir for acute herpes zoster infections in immune-competent adults. Infection 15(Suppl 1):S9–13
106. Wood MJ, Ogan PH, McKendrick MW et al (1988) Efficacy of oral acyclovir treatment of acute herpes zoster. Am J Med 85:79–83
107. Wood MJ, Shukla S, Fiddian AP, Crooks RJ (1998) Treatment of acute herpes zoster: effect of early (<48 h) versus late (48–72 h) therapy with acyclovir and valaciclovir on prolonged pain. J Infect Dis 178(Suppl 1):S81–S84
108. United Nations, Department of Economic and Social Affairs, Population Division (2013). World Population Ageing 2013. ST/ESA/SER.A/348

Chapter 24
Herpes Zoster Vaccines

Michael N. Oxman and Ruth Harbecke

24.1 Introduction

Varicella (chickenpox) and herpes zoster (shingles) are two distinct diseases caused by the same virus, varicella-zoster virus (VZV). The differences between the two diseases are due to differences in the host's immunity to VZV and in the circumstances of infection [4, 74, 75, 179–181, 267, 268]. Varicella is a highly contagious acute febrile illness characterized by a generalized pruritic vesicular rash. It is caused by primary exogenous VZV infection of a susceptible host (Chap. 2). In temperate climates, varicella is a seasonal disease of childhood, with annual epidemics in winter and spring infecting most children before puberty (Chaps. 2 and 4). Varicella has been almost completely eliminated in countries that have introduced universal childhood varicella vaccination (Chap. 2). In contrast, herpes zoster is a localized disease of the sensory ganglion, nerve, and skin, caused by reactivation and replication of endogenous VZV that has persisted as a latent infection in sensory and autonomic neurons following an earlier episode of varicella or an inapparent primary VZV infection. Herpes zoster is a sporadic disease that occurs throughout the year [23, 106, 114, 198].

M.N. Oxman (✉)
Department of Medicine, Division of Infectious Diseases, University of California San Diego, La Jolla, California, USA

Medicine Service, Infectious Diseases Section, VA San Diego Healthcare System, San Diego, California, USA
e-mail: mnoxman@ucsd.edu

R. Harbecke
Department of Medicine, Division of Infectious Diseases, University of California San Diego, La Jolla, California, USA

Research Service, VA San Diego Healthcare System, San Diego, California, USA

C.P.N. Watson et al. (eds.), *Herpes Zoster: Postherpetic Neuralgia and Other Complications,* DOI 10.1007/978-3-319-44348-5_24

The observation that susceptible children developed varicella after contact with persons with herpes zoster [258], the recognition that the skin lesions of herpes zoster were histologically identical to those of varicella [148, 253], the demonstration that children inoculated with vesicle fluid from individuals with herpes zoster developed varicella and transmitted the disease to other susceptible children [22, 126], and the observation that epidemics of varicella initiated by contact with herpes zoster were identical in every respect to epidemics of varicella initiated by contact with varicella [105] provided compelling evidence that the two diseases were caused by the same agent. Isolation of VZV from both diseases made it possible to prove that the virus from both had identical physical, biological, and antigenic characteristics [269, 270]. Unequivocal proof of their identity was provided when molecular analysis showed that the viruses isolated from skin lesions in varicella and in a subsequent episode of herpes zoster in the same individual were identical [226] and by the isolation of the attenuated Oka vaccine strain of VZV (vOka) from cases of herpes zoster that developed in varicella vaccine recipients [69].

24.1.1 The Nature of Herpes Zoster

Herpes zoster is characterized by unilateral radicular pain and a vesicular rash that is generally limited to the dermatome innervated by a single spinal or cranial sensory ganglion [55, 83, 101, 106, 225, 262] (Chaps. 4 and 12). A prodrome of segmental neuralgia with pain and paresthesia in the involved dermatome generally precedes the herpes zoster rash by several days, occasionally by a week or more, and varies from superficial itching, tingling, or burning to severe deep boring or sharp stabbing pain. Prodromal pain may be constant or intermittent, and it is often accompanied by tenderness and subtle sensory abnormalities in the involved dermatome. It may simulate pleurisy, myocardial infarction, duodenal ulcer, cholecystitis, biliary or renal colic, appendicitis, prolapsed intervertebral disk, or early glaucoma, leading to expensive medical evaluations, misdiagnoses, and inappropriate treatments [181]. When the typical dermatomal rash appears, the diagnosis is usually obvious, although zosteriform herpes simplex is often confused with herpes zoster.

During the prodrome, the replication and spread of reactivated VZV within the sensory ganglion produces necrosis of neurons, supporting cells and fibers, intense lymphocytic inflammation, lymphocytic cuffing of small vessels, focal hemorrhage, and inflammation of the ganglion sheath [55, 101, 261, 262] (Chaps. 12 and 13). Infected neurons and satellite cells contain characteristic intranuclear inclusion bodies, virus particles, VZV antigens, and VZV nucleic acids. The peripheral nerve shows lymphocytic infiltration and focal hemorrhage with axonal degeneration and demyelination of sensory fibers, and the inflammatory process extends distally to branches innervating the affected skin. Infection and inflammation in the ganglion also extend proximally to the posterior nerve root and into adjacent regions of the spinal cord or brainstem, producing an ipsilateral segmental myelitis. There are degeneration of nerve fibers in the posterior column and inflammatory changes in the gray matter of the posterior and anterior horns [55, 101, 261, 262] (Chap. 12). Inflammation and degeneration of the anterior nerve root within the meninges and in the portion overlying the involved sensory ganglion may result in motor radiculitis. When extensive, the acute inflammatory process is followed by fibrosis of the sensory ganglion and nerve [101, 261, 262]. The acute injury to the peripheral nerve and to neurons in the sensory ganglion triggers afferent signals that are perceived as neuralgic pain during the prodrome and following onset of the herpes zoster rash. Injury to the ganglion and to the peripheral nerve also induces long-lasting changes in the physiology of second-order neurons in the spinal cord [181, 261] (Chaps. 6, 12, and 13).

Some patients experience this acute segmental neuralgia without developing a rash, a syndrome called *zoster sine herpete* [62, 80, 145, 153] (Chaps. 4 and 6).

24.1.2 Complications

Neuropathic pain, dysesthesia, hypersensitivity, sensory loss, and discomfort (e.g., allodynia, severe pruritus) are hallmarks of herpes zoster, especially in older and immunocompromised patients [272] (Chaps. 6, 12, 13, and 14). The most feared and debilitating complication of herpes zoster is postherpetic neuralgia (PHN), characterized by persistence of this neuropathic pain and dysesthesia for weeks, months, or even years after the rash has healed [41, 66, 107, 112, 122] (Chaps. 4, 6, 12, 13, and 14).

Most other complications of herpes zoster are caused by the spread of VZV from the sensory ganglia, nerve, or skin, either through the bloodstream or by direct neural extension. Dorsal root ganglia contain visceral as well as cutaneous afferents, and this explains the occurrence of visceral as well as cutaneous lesions in herpes zoster. Depending upon their location, visceral lesions may manifest clinically as laryngitis, esophagitis, gastritis, cystitis, pleuritis, or peritonitis [181]. Visceral afferents affecting gastrointestinal and bladder motility and sphincter function may also be involved. Afferents from the ganglion also innervate blood vessels in the brain and spinal cord, and extension of infection to these vessels may cause ipsilateral segmental granulomatous angiitis, which is responsible for a syndrome of ophthalmic zoster with delayed contralateral hemiplegia [18, 81, 104, 200] (Chaps. 5, 6, and 8), and a variety of other vasculopathies associated with herpes zoster, with or without rash [79, 173] (Chaps. 6, 7, and 8).

24.1.3 VZV Latency

Only humans are naturally infected by VZV, and there are no known animal reservoirs. Consequently, latency and reactivation have been essential to the survival of VZV in humans, especially in the relatively small and isolated human populations in which VZV evolved for many thousands of years. Studies of VZV latency are complicated by the absence of laboratory models that fully replicate VZV latency and reactivation in the human host. In vitro models using human stem cell-derived neurons appear to recapitulate VZV latency and reactivation and promise to shed new light on these processes [158, 205]. Nevertheless, VZV latency and reactivation remain undefined areas of continued controversy [120].

During primary infection, VZV gains access to sensory neurons in cranial, dorsal root and autonomic ganglia by centripetal axonal transport from mucocutaneous lesions, or hematogenously in infected tonsillar T cells, in which expression of surface proteins may be altered by VZV infection to promote trafficking to the skin and, possibly, to neurons [124, 125, 218]. The axonal route of infection is supported by the correlation between the dermatomal distribution of cases of herpes zoster and the distribution of varicella skin lesions [60, 71, 106, 181, 223], as well as by the observation that in vaccinated children most cases of herpes zoster caused by the vOka strain occur in the extremity where the vaccine had been administered, whereas cases of herpes zoster caused by wild-type strains of VZV do not show such localization [69, 72, 94, 266] (Chap. 2). The hematogenous route is supported by the occurrence of latent VZV infection in autonomic neurons, as well as in sensory neurons in individuals in whom primary VZV infection or vaccination was not accompanied by a rash [4, 72, 284]. VZV also establishes latency in ganglia in the enteric nervous system, which may be infected by viremia or by axonal transport from neurons in sensory ganglia [32].

In latently infected sensory ganglia, 1–7 % of sensory neurons contain small numbers of VZV genomes (e.g., 2–10) in the form of circular episomes, with little or no VZV gene expression due, at least in part, to epigenetic silencing [40, 46, 71, 118, 128, 138, 191, 226, 259]. VZV gene expression during latency has been studied in human cranial and dorsal root ganglia obtained at autopsy, and the interval between death and autopsy, as well as stress associated with terminal illness, are likely to alter regulation of VZV gene expression. Transcripts of as many as 12 VZV genes have been reported in latently infected cadaver ganglia [42–44, 47, 50, 117–119, 162, 172]. However, Ouwendijk et al. found only transcripts from ORF63 in trigeminal ganglia obtained <9 h post-mortem, whereas they detected transcripts from a number of additional VZV genes in trigeminal ganglia obtained >9 h post-mortem [178]. Latently infected autonomic ganglia removed at surgery and analyzed without delay exhibited more extensive VZV gene expression [71]. While anesthesia and surgery are still stressful events, these observations suggest that latency in autonomic neurons may be characterized by more extensive VZV gene expression than observed in latently infected neurons in sensory ganglia.

24.1.4 Reactivation of Latent VZV

Reactivation of latent VZV results in expression of the full complement of VZV genes and a productive lytic infection that spreads to adjacent neurons and supporting cells within the ganglion. Although it is not known whether VZV-specific cell-mediated immunity (VZV CMI) prevents reactivation of latent VZV, it is likely to limit the subsequent multiplication and spread of the reactivated virus, reducing the number of reactivations that result in episodes of herpes zoster and the severity of

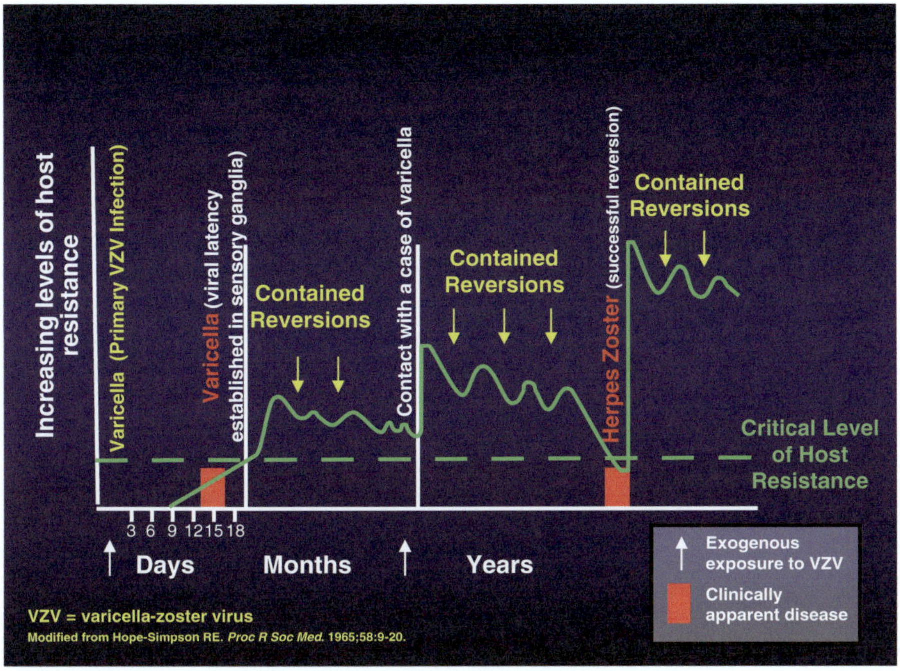

Fig. 24.1 Pathogenesis of herpes zoster (Modified from Fig. 7 in Hope-Simpson [106])

cases of herpes zoster that do occur. When latent VZV reactivates in a sensory neuron, but VZV CMI prevents the subsequent development of herpes zoster, the result is a "contained reversion" (Sect. 24.2 and Fig. 24.1).

24.2 The Central Role of Immunity to VZV in the Pathogenesis of Herpes Zoster

Clinical observations by the British physician R. Edgar Hope-Simpson, reported in a landmark publication in 1965 [106], provided the rationale for the development of a vaccine to prevent herpes zoster and its complications. Carefully following every case of varicella and herpes zoster in his clinical practice over a period of 16 years, Hope-Simpson observed that the incidence and severity of herpes zoster and PHN increased with increasing age. He hypothesized that primary infection with VZV (i.e., varicella) establishes lifelong latent VZV infections in sensory neurons and also induces immunity to VZV that limits the ability of the latent virus to reactivate and multiply to cause herpes zoster (Fig. 24.1). He further postulated that this immunity to VZV gradually decreases over time until it falls below a critical threshold, permitting latent VZV to reactivate, multiply, and spread, resulting in herpes zoster. Hope-Simpson also proposed that subclinical VZV infection caused by exogenous exposure to varicella and spontaneous endogenous ganglionic reactivations that are contained by host immunity to VZV before herpes zoster can develop ("contained reversions") both stimulate the host's immunity to VZV, delaying its decline to levels permitting latent VZV to reactivate, replicate, and reemerge as herpes zoster (Fig. 24.1). Finally, Hope-Simpson observed that second episodes of herpes zoster were uncommon and hypothesized that the large amount of VZV produced during an episode of herpes zoster boosted immunity to VZV, essentially immunizing the afflicted individual against another episode of herpes zoster (Fig. 24.1).

During the past 50 years, investigators have confirmed every aspect of Hope-Simpson's proposed pathophysiology of herpes zoster and demonstrated unequivocally that it is VZV CMI rather than antibody to VZV that limits the capacity of latent VZV to reactivate and cause herpes zoster [2, 12, 24, 135, 163, 180, 182, 206, 265]. Results from a community-based prospective study on the incidence of herpes zoster and PHN in Shozu County, Japan (SHEZ Study) [235], demonstrated that the reaction to an intradermally injected VZV antigen is inversely correlated with the risk of herpes zoster [176], with herpes zoster severity [6], and with development of PHN [108]. Conversely, levels of VZV antibody, which are similar over a wide age range, and do not decline with increasing age [142, 206, 232, 236], are not predictive of protection against herpes zoster or PHN [183, 265] and did not correlate with herpes zoster severity or with PHN in SHEZ Study participants who developed herpes zoster [6, 108].

A number of epidemiologic studies have shown that the age-specific incidence of herpes zoster has increased steadily over the past six decades in the United States, with no change in the rate of increase before and after the introduction of varicella immunization, which has virtually eliminated varicella and, consequently, the boost in immunity to VZV in older adults that results from exogenous exposure to VZV [114] (Chaps. 2 and 4) (Fig. 24.1). This and the observation of similar increases in the age-specific incidence of herpes zoster in Canada, the United Kingdom, Spain, Taiwan, Japan, and Australia [21, 27, 113, 188, 204, 240], irrespective of the presence or absence of varicella vaccination programs, indicate that endogenous boosting by "contained reversions," rather than exogenous exposure to VZV, is primarily responsible for maintaining levels of VZV CMI in older adults. This conclusion is supported by direct evidence that the absence of exogenous exposure to VZV is not associated with an increase in the age-specific incidence of herpes zoster among older adults [51, 67, 91, 134, 281].

When CMI temporarily declines due to unusually stressful situations, subclinical reactivation can lead to asymptomatic shedding of VZV DNA in blood and saliva [136, 161]. Detection of VZV DNA in blood and saliva has been reported in persons with herpes zoster, recipients of zoster vaccine, and a small proportion of highly stressed individuals such as astronauts during and after (but not before) space flight [45, 136]. However, the presence of VZV DNA does not necessarily equate to the presence of infectious virus. This is obvious when one considers that a single infected cell contains thousands of copies of the VZV genome [93] and that this VZV DNA may be cleared very slowly from sites of infection, such as affected sensory ganglia that contain large numbers of infected cells and inflammatory infiltrates following herpes zoster. In addition, "contained reversions" [106] (Fig. 24.1) provide a fresh source of infectious virus and VZV DNA in healthy asymptomatic persons. These considerations present a problem in seeking to attribute complications like PHN to continuing VZV replication (Chaps. 6, 12, 13, and 14), or in asserting that vasculopathy without rash is a manifestation of herpes zoster on the basis of finding VZV DNA in blood or saliva (Chaps. 6 and 8).

The absence of credible epidemiological evidence of transmission of VZV from persons without symptomatic VZV infection (in contrast to herpes simplex virus where most new infections are acquired from someone shedding virus asymptomatically) suggests that VZV DNA in saliva is rarely indicative of the presence of infectious VZV. This conclusion is supported by the observation that index cases in outbreaks of varicella are invariably symptomatic or prodromal, whereas asymptomatic persons who are not incubating varicella are not infectious despite the frequent reports of VZV DNA in their saliva.

Nevertheless, VZV DNA in blood or saliva may be useful in diagnosing inapparent or atypical herpes zoster [73] and, if detected early in the prodrome, may permit initiation of antiviral therapy sufficiently early to reduce the extent of ganglionic pathology and thereby reduce the occurrence or severity of PHN.

24.3 Live Attenuated Oka VZV Vaccine

The Oka strain of VZV was isolated from a healthy Japanese child with varicella and attenuated by serial passage at 34 °C in human and guinea pig cells, producing the live attenuated Oka vaccine strain of VZV (vOka), which is temperature sensitive [182, 230, 233, 234] (Chap. 2). Clinical studies in Japan and subsequently in the United States demonstrated the safety, immunogenicity, and clinical efficacy of an Oka varicella vaccine in immunocompetent and immunocompromised children and adults [1, 7, 13, 70, 71, 76–78, 137, 140, 141, 229, 231, 232, 241, 263] (Chap. 2).

Live, attenuated Oka vaccines are not monoclonal, but consist of a mixture of distinct VZV genotypes. Forty-two single-nucleotide polymorphisms (SNPs), 20 of which specify amino acid changes, distinguish vOka from the wild-type Oka parent strain of VZV [87]. Although each is genotypically unique, all of the genotypes in Oka vaccine preparations share a subset of the 42 SNPs in variable proportions. Six fixed or nearly fixed SNPs at position 560 in a noncoding region of open reading frame (ORF) 0 and at positions 105544, 105705, 106262, 107252, and 108111 in ORF 62, which encodes a transactivator of viral genes required for VZV replication, generally distinguish vOka from wild-type strains of VZV [20, 56, 57, 84–87, 182, 195, 197, 216, 248, 278].

Although the vOka strain is attenuated for replication in the skin [3, 165], 2–3 % of healthy recipients of varicella vaccine develop vesicular rashes. In addition, vOka can establish latent infections in sensory neurons and can reactivate months or years later to cause herpes zoster [39]. Consequently, it is important to carefully characterize the genotypes responsible for vaccine-associated rashes and episodes of herpes zoster in varicella vaccine recipients to determine whether they might represent more pathogenic components of the vaccine itself or new genotypes resulting from reversion of specific vaccine SNPs to the wild-type allele or even from recombination with wild-type VZV. A study by Depledge et al. [56] suggests that viruses responsible for vaccine rashes and vaccine-associated herpes zoster are not the result of selection of genotypes with increased pathogenicity present in the Oka vaccine or the production of new more pathogenic strains of vOka by recombination but, instead, represent selection in vaccine recipients against alleles in the vaccine that are essentially neutral [56]. The observation that there are differences in strain content among Oka vaccines produced by different manufacturers and even between different batches from the same manufacturer [56, 194, 196, 211, 237, 238] underlines the importance of genotyping VZV strains isolated from vaccine-associated illnesses [56, 139, 151, 177, 197, 210] and of regular surveillance of Oka vaccine batches during manufacturing and post-production to identify potentially pathogenic alleles and to monitor for reversions to wild-type alleles or recombination with wild-type VZV.

24.4 The Challenge of Vaccinating Against Herpes Zoster

Varicella vaccine, like other vaccines against common childhood viral diseases such as measles, mumps, and rubella, is administered to susceptible persons *prior to* exogenous exposure to the virus, inducing immunity that prevents primary infection and disease. We expect such vaccines to have >95 % efficacy and induce herd immunity.

Vaccination against herpes zoster is directed at persons who have already experienced primary VZV infection and thus already have solid immunity against varicella, but are harboring latent VZV that can reactivate and cause herpes zoster [180, 182]. To be effective, zoster vaccine must function as a "therapeutic vaccine" and alter the host-virus relationship to prevent or modify a disease caused by a pathogen with which the vaccine recipient is already infected and to which clinically significant immunity already exists. Zoster vaccine acts by boosting declining levels of preexisting CMI to VZV in older adults, thereby reducing the frequency and severity of a disease caused by reactivation and multiplication of endogenous latent VZV. We do not expect vaccines against such endogenous infections (i.e., "therapeutic vaccines") to approach 95 % efficacy or to induce herd immunity. However, the natural history of herpes zoster described by Hope-Simpson [106], where one episode of herpes zoster protects against a second episode, provided a model for successful vaccination of older adults against herpes zoster. There is no comparable natural resistance to recurrent herpes simplex, which recurs repeatedly in immunocompetent persons [182].

24.5 The Shingles Prevention Study (SPS)

The development of the Oka strain of live attenuated VZV by Takahashi and his colleagues and the demonstration of its safety and immunogenicity in children and older adults [74, 229, 230, 234] (Chap. 2) permitted initiation of a large clinical trial, Department of Veterans Affairs (VA) Cooperative Study #403: The Shingles

Prevention Study (SPS), to test Hope-Simpson's hypothesis that boosting declining VZV CMI in older adults would protect them from herpes zoster and its complications, particularly PHN [180, 184, 185].

24.5.1 Description of the Study

The SPS was a double-blind, placebo-controlled clinical trial in which 38,546 adults ≥60 years of age were randomized to receive a single dose of high-potency live attenuated Oka/Merck VZV vaccine (zoster vaccine) or placebo at 22 study sites across the continental United States [184, 185]. Randomization was stratified by study site and age group (60–69 and ≥70 years of age). There were two co-primary end points: (1) the *burden of illness due to herpes zoster*, a severity-by-duration measure representing the total pain and discomfort due to herpes zoster in a population of subjects (e.g., all 60–69-year-old vaccine recipients), and (2) the *incidence of clinically significant PHN*, defined as pain and/or discomfort (e.g., allodynia, severe pruritus) due to herpes zoster, scored as ≥3 on a 0–10 Likert scale, that persisted for more than 90 days after rash onset [48, 184]. The *incidence of herpes zoster* was also determined. Zoster vaccine was at least 14 times more potent than the licensed varicella vaccine because preliminary studies indicated that potency of at least this magnitude was required to induce a significant increase in VZV CMI among older adults already immune to varicella [184]. There is no evidence that varicella vaccine is a suitable substitute for zoster vaccine.

A total of 19,270 subjects received zoster vaccine and 19,276 received placebo. The median age in both groups was 69 years; 6.6 % of vaccine recipients and 6.9 % of placebo recipients were ≥80 years of age. Forty-one percent were female. The mean duration of surveillance for herpes zoster was 3.13 years (median, 3.12 years; range, 1 day to 4.90 years) with no difference between the vaccine and placebo groups. More than 95 % of enrolled subjects were actively followed to the end of the study; only 0.6 % withdrew or were lost to follow-up and 4.1 % died before the study ended. None of these figures differed between vaccine and placebo recipients [184].

24.5.2 Evaluation of Suspected Cases of Herpes Zoster

Subjects with a new rash and/or new unilateral pain were evaluated by study personnel as soon as possible (a median of 5 days after rash onset). Subjects with unilateral rashes and no clear alternative diagnosis were classified as having "suspected herpes zoster", actively followed according to the study protocol, and offered famciclovir without charge and standard-of-care treatment for pain [184]. Herpes zoster-associated pain and discomfort were measured with the Zoster Brief Pain Inventory, a questionnaire completed by the subject that was specifically designed to measure pain and discomfort in herpes zoster [48]. This questionnaire and others [19, 260] were used to assess the effect of herpes zoster on subjects' activities of daily living, quality of life, and general health status. Characteristics of the rash, associated complications, and medication use were also recorded. These evaluations were repeated over a period of at least 182 days, according to a schedule specified by the study protocol [184]. Digital photographs and specimens for laboratory diagnosis were also obtained.

24.5.3 Diagnosis of Herpes Zoster

Valid assessment of vaccine efficacy must be carried out in proven cases of the targeted disease. More than 3500 rashes in subjects in each treatment group were evaluated clinically but were not considered to be suspected cases of herpes zoster. A total of 1308 suspected cases of herpes zoster were evaluated. The threshold for considering a subject to have a suspected case of herpes zoster was intentionally set very low to capture even mild and atypical cases of herpes zoster, to avoid missing potential vaccine-modified cases, and to describe the natural history of herpes zoster among the placebo recipients. A real-time PCR assay was designed and validated for the SPS to detect and discriminate among DNA from wild-type and vOka strains of VZV and herpes simplex virus. Primers and probes for each viral DNA were multiplexed with primers and probe for the human β-globin gene to assess specimen adequacy [93]. Every suspected case of herpes zoster was also adjudicated by an expert Clinical Evaluation Committee blinded to treatment and diagnostic laboratory results although, when available, the PCR assay results were determinative [184]. The results of the PCR assay correlated well with the diagnoses of the Clinical Evaluation Committee [93]. Of the 1308 suspected cases of herpes zoster, 317 were determined not to be herpes zoster (156 in the vaccine group, 161 in the placebo group); 49 of these (24 in vaccine recipients, 25 in placebo recipients) were caused by herpes simplex virus, confirmed by PCR assay or virus culture. An additional 7 cases (3 in the vaccine and 4 in the placebo group) were excluded because they were not seen until after rash crusting; 984 (75.2 %) were confirmed cases of herpes zoster. Of these, 24 were excluded from the primary efficacy analysis per protocol because they occurred within 30 days of vaccination (6 in vaccine recipients, 18 in placebo recipients), and 3 were excluded because they were the subject's second case of herpes zoster (1 in a vaccine recipient and 2 in placebo recipients). The remaining 957 cases of herpes zoster (315 in vaccine recipients, 642 in placebo recipients) constituted the end points for the primary efficacy analysis. In each group, >93 % of the subjects with confirmed herpes zoster were positive for wild-type VZV DNA by PCR assay. Vaccine virus (vOka) was never detected in any subject [93, 184].

24.5.4 Efficacy End Points

The primary end point was the *burden of illness due to herpes zoster* [26, 48, 154]. For each confirmed case of herpes zoster, responses to the "worst pain" question in the Zoster Brief Pain Inventory (ZBPI) were used to calculate a *herpes zoster severity-of-illness score*, defined as the area under the curve (AUC) of herpes zoster pain severity plotted against time during the 182-day period after zoster rash onset. Subjects with herpes zoster had *herpes zoster severity-of-illness scores* ranging from 0 to 1813. Increasing mean scores are highly correlated with a decrease in the health-related quality of life and in functional status among older adults [48, 54]. A score of 0 was recorded for subjects in whom herpes zoster did not develop during the study period. The *burden of illness due to herpes zoster* represented the average severity of illness among all subjects in the group (e.g., all placebo recipients or all 60–69-year-old placebo recipients); it was calculated as the sum of the *herpes zoster severity-of-illness scores* of all members of a group divided by the total number of subjects in the group.

The second (co-primary) study end point was the *incidence of clinically significant PHN* [107], defined as pain and discomfort due to herpes zoster rated as ≥3 on the ZBPI Likert scale ranging from 0 ("no pain") to 10 ("pain as bad as you can imagine"), persisting or appearing more than 90 days after zoster rash onset. (Scores lower than 3 are not associated with significant decrements in the quality of life or the ability to carry out activities of daily living and were therefore not considered to represent PHN [48]).

An additional end point was the *incidence of herpes zoster*.

24.5.5 Results: Vaccine Efficacy

Zoster vaccine efficacy for the *herpes zoster burden of illness* was 61.1 % (95 % CI, 51.1–69.1 %; $P < 0.001$), 65.5 % in subjects aged 60–69, and 55.4 % in subjects aged ≥70 (Fig. 24.2). There were no significant differences in vaccine efficacy by gender or age stratum, although vaccine efficacy appeared to be slightly lower in the older subjects. The median duration of pain and discomfort among subjects with herpes zoster was significantly shorter in vaccine than placebo recipients (21 days vs. 24 days, $P = 0.03$), and the mean *herpes zoster severity-of-illness score* among subjects with herpes zoster was significantly lower in vaccine than placebo recipients (141.2 vs. 180.5, $P = 0.008$). For virtually every level of *herpes zoster severity-of-illness score*, fewer cases were seen in the vaccine group than in the placebo group; this was especially notable for cases with higher scores – that is, cases with more painful and protracted disease [184, 185]. For example, *herpes zoster severity-of-illness scores* of 1600, equivalent to 160 days of "the worst pain imaginable," were observed in 11 zoster vaccine recipients compared with 40 placebo recipients, a 73 % reduction in the vaccine recipients.

Zoster vaccine efficacy for the *incidence of clinically significant PHN* was 66.5 % (95 % CI, 47.5–79.2 %; $P < 0.001$), 65.7 % in subjects aged 60–69, and 66.8 % in subjects aged ≥70 (Fig. 24.3). Zoster vaccine also reduced the proportion of subjects with herpes zoster who developed PHN by more than 31 %, with most of this benefit in the ≥70-year-old age group who had the highest risk of developing this complication [184]. In addition, the mean *herpes zoster severity-of-illness score* was lower in subjects with PHN who had received zoster vaccine than in those who had received placebo. Among placebo recipients, the *incidence of clinically significant PHN* was

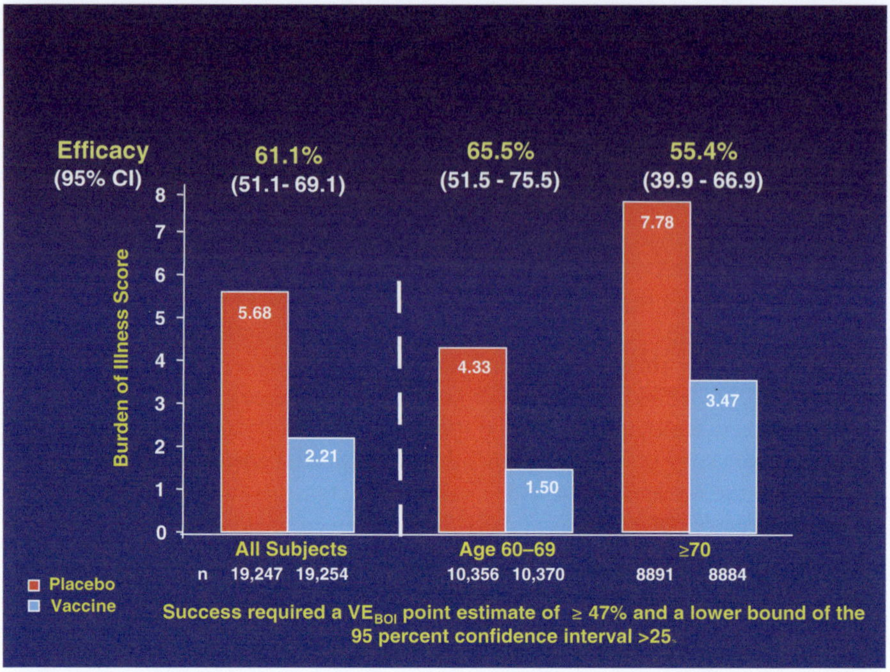

Fig. 24.2 Zoster vaccine efficacy for the *herpes zoster burden of illness* (VEBOI) in the SPS. The *herpes zoster burden of illness* is a severity-by-duration measure of the total pain and discomfort caused by herpes zoster in a specific group or population. For each case of herpes zoster, Zoster Brief Pain Inventory (ZBPI) data were used to calculate a *herpes zoster severity-of-illness score*, defined as the area under the curve of severity of pain and discomfort due to herpes zoster plotted against time during the 182-day period after herpes zoster rash onset. A *herpes zoster severity-of-illness score* of 0 was recorded for subjects who did not develop herpes zoster. The *herpes zoster burden-of-illness score* represents the average *herpes zoster severity-of-illness score* among all subjects in a group (e.g., all ≥70-year-old vaccine recipients) calculated as the sum of the *herpes zoster severity-of-illness scores* of all members of the group divided by the total number of subjects in the group. The figure within each bar is the average *herpes zoster burden-of-illness score* (Based upon data published in Oxman et al. [184]: Table 2)

lower in females than in males, whereas vaccine efficacy for *incidence of clinically significant PHN* was higher in females than males, but these differences were not statistically significant [184].

Zoster vaccine efficacy for the *incidence of herpes zoster* was 51.3 % (95 % CI, 44.2–57.6 %; $P < 0.001$), 63.9 % in subjects ages 60–69, but only 37.6 % in subjects aged ≥70 (Fig. 24.4). In a time-to-event analysis, the cumulative incidence of herpes zoster was significantly lower in the vaccine recipients than in the placebo recipients ($P < 0.001$) (Fig. 2 in Oxman et al. [184]). Among placebo recipients, the incidence of herpes zoster was slightly higher in females than males (11.79 versus 10.65 per 1000 person years), but vaccine efficacy for incidence of herpes zoster did not differ by gender [184].

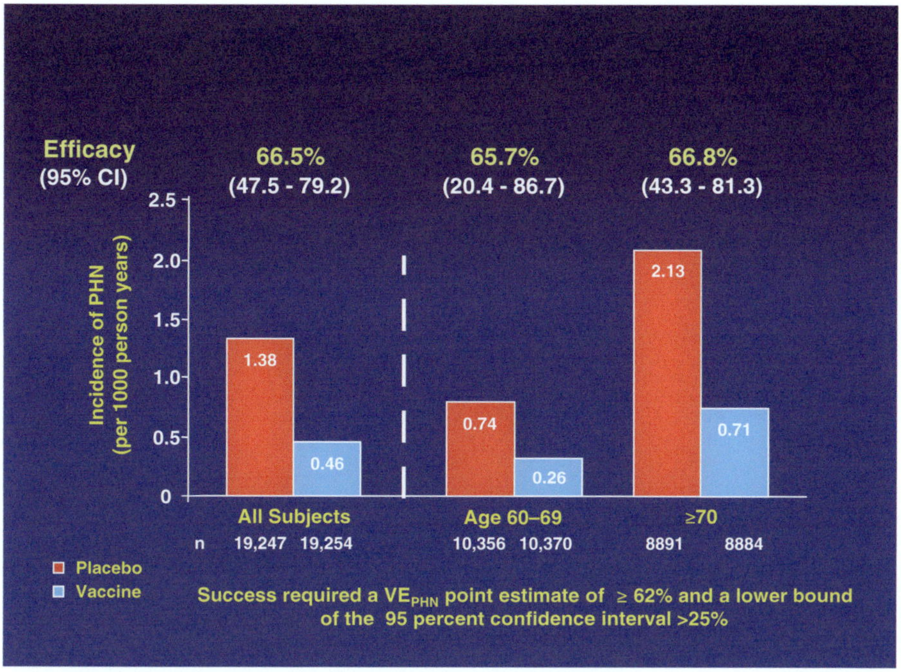

Fig. 24.3 Zoster vaccine efficacy for the *incidence of postherpetic neuralgia* (VE_{PHN}) in the SPS. Zoster vaccine reduced the incidence of clinically significant postherpetic neuralgia (PHN) by approximately two-thirds in all subjects and in both age strata. Clinically significant PHN was defined as pain and discomfort caused by herpes zoster scored as 3 or more on a scale ranging from 0 ("no pain") to 10 ("pain as bad as you can imagine") persisting or appearing more than 90 days after herpes zoster rash onset. The figure within each bar is the incidence of PHN per 1000 person years. It is important to note that this reduction is calculated for all subjects and not just those with herpes zoster (Based upon data published in Oxman et al. [184]: Table 3)

These results suggest that the zoster vaccine-induced increase in VZV CMI, though not always sufficient to prevent herpes zoster, reduced the severity and duration of the episodes that did occur. This was reflected by a reduction in the *herpes zoster severity-of-illness score*, in the *incidence of clinically significant PHN*, and in the duration and severity of the cases of PHN that did occur in vaccine recipients.

The effects of zoster vaccine differed among the younger and older subjects. Among the 60–69-year-olds, much of the reduction in the burden of illness and incidence of PHN was due to the prevention of herpes zoster. However, among the ≥70-year-old subjects, much of the reduction in the burden of illness and incidence of PHN resulted from a significant reduction in the severity of the disease [184].

There were only two second episodes of herpes zoster among 660 placebo recipients who developed documented herpes zoster and one second episode among 321 vaccine recipients who developed documented herpes zoster [184]; none of the

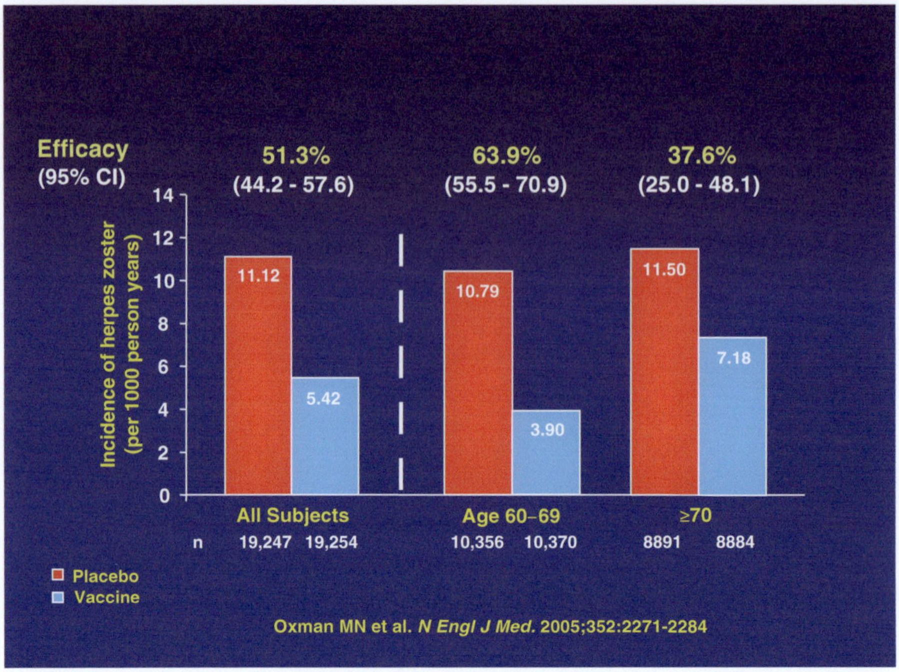

Fig. 24.4 Zoster vaccine efficacy (VE) for the *incidence of herpes zoster* in the SPS. Zoster vaccine efficacy for the incidence of herpes zoster was 51.3 % overall; 63.9 % in subjects 60–69 years of age but only 37.6 % in subjects ≥70 years of age. The figure within each bar is the incidence of herpes zoster per 1000 person years (Based upon data published in Oxman et al. [184]: Table 3)

three second episodes occurred in a subject who was immunosuppressed. This indicates that in the SPS the risk of developing a second episode of herpes zoster was at least tenfold lower than the risk of developing a first episode (i.e., ≤0.1 % per year vs. approximately 1.1 % per year in placebo recipients) [184]. Retrospective analysis of large databases report higher rates of recurrent herpes zoster [114, 282], but they included immunocompromised persons, and few of the cases of herpes zoster were proven by virus culture or PCR. The reported incidence of second episodes of herpes zoster is influenced by the accuracy of diagnosis and by the duration of follow-up.

The distribution of herpes zoster by dermatome was similar in vaccine and placebo recipients, and the proportion of cases of ophthalmic zoster, approximately 11 %, was also comparable [187]. In addition, among placebo recipients, there was no significant difference between the mean *herpes zoster severity-of-illness score* of ophthalmic zoster and that of zoster in other dermatomes, and zoster vaccine efficacy was comparable in ophthalmic zoster and zoster in other dermatomes [187]. It should be noted that the early initiation of effective antiviral therapy in both vaccine and placebo recipients with ophthalmic herpes zoster is likely to have prevented most complications other than PHN (Chap. 5).

24.5.6 Results: Vaccine Efficacy for Zoster-Related Interference with Functional Status and Quality of Life

Assessments of subjects included evaluation of the impact of herpes zoster on health-related quality of life and capacity to carry out activities of daily living [48]. Detailed analyses of these data provided further evidence of the severe impact of acute herpes zoster pain and discomfort and of PHN on these aspects of patients' well-being [184, 212, 215] and demonstrated that the efficacy of zoster vaccine in reducing the decrement in quality of life and in the capacity to carry out activities of daily living caused by herpes zoster was similar to zoster vaccine's efficacy for the *burden of illness due to herpes zoster* (Fig. 24.5) [212, 215]. These results are consistent with the results of a large prospective study of the impact of herpes zoster in adults in Canada [61].

24.5.7 Safety of Zoster Vaccine

Rates of serious adverse events, systemic adverse events, hospitalizations, and deaths were low among zoster vaccine recipients and comparable to those among placebo recipients [184, 220]. Local reactions at the injection site occurred in 48.3 % of vaccine recipients compared with 16.6 % of placebo recipients, but they were generally mild and transient [184]. Vesicular rashes at the injection site were more common in vaccine than placebo recipients, but they were infrequent and

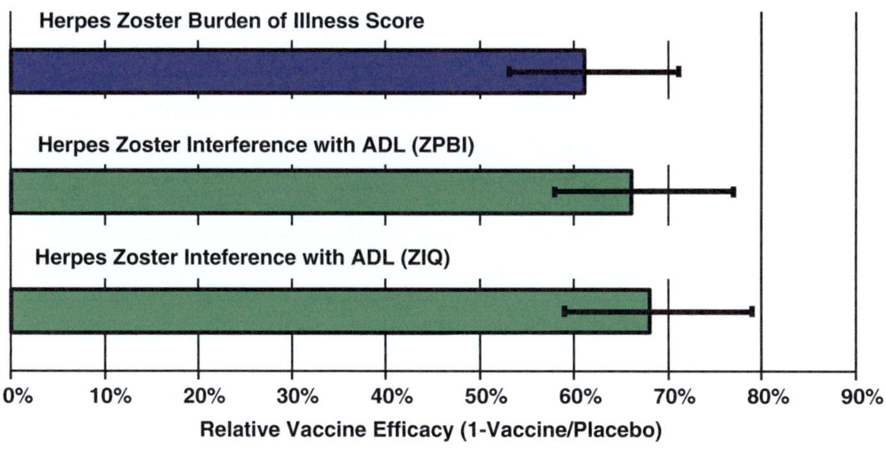

Fig. 24.5 Zoster vaccine efficacy for the herpes zoster burden of illness and interference with activities of daily living (ADL) end points in the SPS. Zoster vaccine efficacy for the *herpes zoster burden-of-illness score* compared to zoster vaccine efficacy for interference with *activities of daily living* assessed with the Zoster Brief Pain Inventory (*ZBPI*) and the Zoster Impact Questionnaire (*ZIQ*) (From Fig. 2 in Oxman [182], based, in part, upon data in Schmader et al. [212])

limited in extent and duration [184]. Neither wild-type VZV nor vOka was demonstrated in these injection site lesions, either by culture or PCR assay [93], and there were no documented episodes of disseminated vesicular disease caused by vOka [184]. Of particular note were the observations that during the first 42 days after vaccination, there were 24 cases of herpes zoster in placebo recipients and only 7 cases in vaccine recipients and that no cases of herpes zoster at any time during the study were caused by vOka [93]. This demonstrated that zoster vaccine neither caused herpes zoster nor induced herpes zoster caused by reactivation of endogenous wild-type VZV [93, 184, 185]. A more detailed analysis provided additional data indicating the safety of zoster vaccine in SPS subjects [220].

24.5.8 Immune Responses to Zoster Vaccine

Immune responses to VZV were assessed in a representative subset of 1395 SPS subjects at enrollment (691 vaccine and 704 placebo recipients), as well as in all suspected cases of herpes zoster throughout the study. VZV CMI was measured by VZV responder cell frequency and interferon-γ ELISPOT assays and antibody to VZV by glycoprotein (gp) ELISA [142, 265]. Baseline measurements prior to vaccination demonstrated that the progressive decline in VZV CMI with increasing age [2, 12, 24, 163, 206] continues in adults into the eighth and ninth decades of life [142]. In contrast, antibody to VZV persists at high levels with little or no decline in the aged [143, 163, 206]. VZV CMI and antibody responses were significantly increased at 6 weeks after immunization in vaccine recipients compared to placebo recipients and compared to baseline values. The vaccine-induced increases in VZV CMI persisted during the 3 years of follow-up, although their magnitude decreased over time, with the greatest decrease between 6 weeks and 1 year after vaccination [142]. The magnitude of VZV CMI responses to zoster vaccine decreased with age and was significantly higher in 60–69-year-olds than in vaccine recipients ≥70 years of age. In contrast, there was no significant difference in VZV antibody titers between the two age strata. The maximum levels in VZV CMI in the first year post-vaccination are reflected by the observation that the highest levels of zoster vaccine efficacy for all three study end points occurred during the first year post-vaccination [167, 214]. This indicates that results of studies of vaccine effectiveness will be affected by the duration of post-vaccination observation.

Immunologic assessments were also carried out in 981 subjects (321 vaccine and 660 placebo recipients) who developed herpes zoster and 1362 subjects (682 vaccine and 680 placebo recipients) who did not, using the same assays for VZV CMI and antibody to VZV described above [265]. High levels of VZV CMI at onset of herpes zoster were correlated with reduced *herpes zoster severity-of-illness scores* and a significantly lower risk of developing PHN. In contrast, higher levels of antibody to VZV after rash onset were associated with increased *herpes zoster severity-of-illness scores* and increased occurrence of PHN [265], probably reflecting more extensive VZV replication. Of note, the magnitude and duration of the increase in VZV CMI induced by zoster vaccine and by an episode of herpes zoster were similar, despite exposure to orders of magnitude more VZV during an episode of herpes zoster [265].

24.6 Persistence of Zoster Vaccine Efficacy

The SPS demonstrated persistence of zoster vaccine efficacy for the study end points through 4 years post-vaccination [184]. A Short-Term Persistence Substudy of SPS subjects assessed the duration of vaccine efficacy through 7 years post-vaccination [214]. In the Short-Term Persistence Substudy, compared to the SPS, vaccine efficacy for the *herpes zoster burden of illness* decreased from 61.1 to 50.1 %, for the *incidence of PHN* from 66.5 to 60.1 %, and for the *incidence of herpes zoster* from 51.3 to 39.6 %. Analysis of vaccine efficacy using pooled SPS and Short-Term Persistence Substudy data for each year post-vaccination showed that vaccine efficacy for all three study end points was greatest in year 1 post-vaccination and declined thereafter, and that vaccine efficacy for the *herpes zoster burden of illness* and for the *incidence of herpes zoster* were statistically significant through year 5 post-vaccination [214]. A Long-Term Persistence Substudy assessed the duration of vaccine efficacy for up to 11 years post-vaccination [167]. Because, per protocol, placebo recipients were offered zoster vaccine without charge after completion of the SPS and most accepted, the Long-Term Persistence Substudy was complicated by the absence of a placebo group. This necessitated the use of SPS and Short-Term Persistence Substudy placebo results to model reference placebo groups. Compared to the SPS, estimated vaccine efficacy in the Long-Term Persistence Substudy decreased from 61.1 to 37.3 % for the *herpes zoster burden of illness*, from 66.5 to 35.4 % for the *incidence of PHN*, and from 51.3 to 21.1 % for the *incidence of herpes zoster*, and declined for all three outcome measures from 7 through 11 years post-vaccination. Statistically significant estimates of vaccine efficacy persisted into year 10 post-vaccination for *burden of illness due to herpes zoster* and through year 8 for the *incidence of herpes zoster* [167].

Life expectancy in the United Kingdom in 1965 was only about 70 years, whereas it is now substantially longer in most developed countries and likely to increase further in the future. Thus, the lifetime risk of second episodes of herpes zoster observed by Hope-Simpson [106] will be substantially greater in the present or in the future. Consequently, it will be necessary to utilize more effective vaccines and/or booster doses to protect persons against recurrences of herpes zoster in their older years.

24.7 Immune Response to a Booster Dose of Zoster Vaccine

The observation that the efficacy of zoster vaccine declined substantially during the decade following administration and the lack of definitive information on the actual duration of clinical efficacy [167, 214] present a challenge for those responsible for public health recommendations [92, 186, 271].

VZV CMI (interferon-γ ELISPOT) and antibody (gpELISA) responses were evaluated in 210 adults ages 60 years and older randomized to receive two doses of live attenuated Oka/Merck zoster vaccine or placebo separated by 6 weeks [256]. The second dose of zoster vaccine was well tolerated and immunogenic, but correlation between the VZV CMI and antibody assays was poor, and the second dose did not boost VZV-specific immunity beyond the levels achieved with the first dose.

A second study evaluated VZV antibody responses (by gpELISA) in 616 adults aged 70 years and older randomized to receive one or two doses of zoster vaccine separated by 1 or by 3 months [257]. The vaccine was well tolerated, with no increase in the incidence of adverse events following the second dose. However, neither two-dose regimen increased the VZV antibody response compared to a single dose.

In a third study, Levin et al. measured VZV CMI (by dual-color interferon-γ and interleukin-2 ELISPOT) and antibody (by gpELISA) at baseline and 1, 6, and 52 weeks following administration of zoster vaccine to ~200 subjects ≥70 years of age who had received zoster vaccine ≥10 years earlier, as well as ~100 subjects aged 50–59, ~100 subjects aged 60–69, and ~200 subjects ≥70 years of age who were all receiving their first dose of zoster vaccine [143]. VZV antibody responses were similar at baseline and after vaccination in all four groups, with no difference between the ≥70-year-olds receiving a booster dose and those receiving their first dose of zoster vaccine. The VZV CMI responses were lower in the ≥70-year-olds receiving their first dose of zoster vaccine than in the 60–69-year-olds, but the VZV CMI responses of the boosted ≥70-year-olds were similar to those of the vaccinated 60–69-year-olds, and the responses of the 50–59-year-olds were even higher. The higher VZV CMI responses in the boosted ≥70-year-olds than in the ≥70-year-olds receiving their first dose of zoster vaccine suggest that some residual effect of zoster vaccine on VZV CMI persisted for ≥10 years after vaccination and was enhanced by the booster dose.

24.7.1 Intramuscular Versus Subcutaneous Administration

In a recent clinical trial, 354 subjects ≥50 years of age were randomized 1:1 to receive Zostavax® by intramuscular or subcutaneous injection [59]. Immunogenicity was measured by antibody titers using gpELISA and, in a subset of participants ($N = 228$), by interferon-γ ELISPOT. Subjects vaccinated intramuscularly reported fewer adverse events than those vaccinated subcutaneously, with injection site reactions observed in 34.1 % and 64.4 %, respectively. The number of systemic adverse events and vaccine-related systemic adverse events were comparable in both treatment groups. Immunogenicity measured by geometric mean VZV antibody titer and fold-rise and by geometric mean ELISPOT count and fold-rise was comparable in both groups.

24.8 Efficacy and Safety of Zoster Vaccine in Persons 50–59 Years of Age

A randomized double-blind placebo-controlled trial, similar in design to the SPS, was carried out in 22,439 subjects 50–59 years of age at 105 study sites in Europe and North America [213]. Zoster vaccine was shown to reduce the incidence of herpes zoster by approximately 70 % in subjects followed for a mean of 1.3 years. The safety profile was similar to that observed in the SPS, except that the proportion of injection site reactions was higher (49.5 %), probably reflecting the more robust immune response of younger persons to zoster vaccine [135, 142, 184, 220]. The total burden of acute pain due to herpes zoster was significantly lower in zoster vaccine than placebo recipients, extending the trend observed in the younger age stratum in the SPS; most of the benefit of zoster vaccine in 50–59-year-olds was due to the reduction in incidence of herpes zoster; the burden of acute pain in subjects with herpes zoster was similar in vaccine and placebo recipients [213].

24.9 Other Oka Zoster Vaccines

24.9.1 Refrigerator-Stable Zoster Vaccine

In a double-blind, randomized safety and immunogenicity study, refrigerator-stable and frozen formulations of Zostavax® were compared in 367 immunocompetent subjects ≥50 years of age [82]. During 28 days of safety follow-up, more injection site adverse events were reported in recipients of the frozen than the refrigerator-stable formulation (46.4 % versus 35.6 %), while no difference in the frequency of systemic adverse events was observed between the two formulations. No vaccine-related serious adverse events were reported in either group. Immunogenicity was assessed by measuring VZV antibody titers by gpELISA at baseline (before vaccination) and on day 28 post-vaccination. Based on the subjects' antibody responses, the immunogenicity of the two formulations of Zostavax® was judged to be similar.

The safety and immunogenicity of refrigerator-stable Zostavax® was evaluated in an open-label single-arm trial in 150 Taiwanese subjects ≥50 years of age [280]. The vaccine was immunogenic, based on geometric mean fold-rise in VZV antibody titers (by gpELISA), from before immunization to 4 weeks after vaccination. Adverse events were generally mild, with 36.0 % of subjects reporting one or more vaccine-related adverse events. The most frequently reported vaccine-related adverse event was injection site pain (26.7 %). No serious adverse events were reported during the 28 day safety follow-up.

Since the need to keep zoster vaccine frozen prior to reconstitution is an impediment to its widespread use, a refrigerator-stable formulation would be very advantageous. However, since VZV CMI rather than antibody to VZV is responsible for the zoster vaccine-induced reduction in the incidence and severity of herpes zoster and PHN observed in the SPS, the reported equivalent immunogenicity of the refrigerator-stable and frozen formulations of Zostavax® is open to question [186, 265].

The refrigerator-stable formulation of Zostavax® is used in the United Kingdom and other European countries, and it may be administered subcutaneously or intramuscularly [64, 193].

24.9.2 Heat-Inactivated Zoster Vaccine

Early studies in healthy VZV-seropositive adults ≥55 years of age demonstrated comparable safety and immunogenicity of live attenuated and heat-inactivated Oka vaccines, although efficacy was not determined [100, 144].

Redman et al. administered heat-inactivated varicella vaccine (zoster vaccine was not yet available) to bone marrow transplant recipients beginning 1 month after transplantation in an attempt to prevent herpes zoster. VZV CMI was increased significantly among 14 patients given a single dose compared with 14 unvaccinated patients, but their incidence and severity of herpes zoster was unaltered. VZV CMI was also increased in 24 patients vaccinated with three doses of heat-inactivated varicella vaccine at 1, 2, and 3 months after transplantation compared with 23 unvaccinated patients. Although the incidence of herpes zoster was not altered in vaccine recipients, the severity of the disease was decreased dramatically [199].

Hata et al. administered heat-inactivated varicella vaccine to patients receiving autologous hematopoietic cell transplantation for non-Hodgkin's or Hodgkin's lymphoma in an attempt to prevent herpes zoster. The incidence of herpes zoster during 1 year after transplantation was reduced from 30 % among 58 placebo recipients to 13 % among 53 patients who received the vaccine 1 month before and 1, 2, and 3 months after transplantation. Protection was correlated with reconstitution of T cell immunity to VZV [96].

A safe and effective zoster vaccine is needed for immunocompromised patients in whom the risk and severity of herpes zoster is markedly increased. With this in mind, Merck and Co., Inc., has developed a heat-inactivated zoster vaccine. Encouraged by the promising early proof of concept trials described above [96, 199], Mullane et al. conducted a randomized double-blind placebo-controlled phase I trial in 341 adults with either solid tumors, hematologic malignancies, HIV infection with ≤200 CD4+ T cells, or autologous and allogeneic hematopoietic stem cell transplants, stratified by disease, with 262 receiving four doses of vaccine at 30-day intervals and 79 receiving placebo. Safety follow-up through day 28 after dose 4 revealed injection site reactions ranging from 3.3 % among subjects with HIV infection to 36.8 % among subjects with solid tumors. The heat-inactivated zoster vaccine induced a significant increase in interferon-γ ELISPOT counts after the fourth dose in all patient groups, except the allogeneic stem cell transplant recipients. VZV antibody titers were also increased, but not in either group of stem cell transplant recipients [169].

24.10 Safety and Effectiveness of Live Attenuated Zoster Vaccine

Zoster vaccine was licensed by the FDA in May 2006 and recommended by the Centers for Disease Control and Prevention (CDC) Advisory Committee on Immunization Practices (ACIP) in October 2006 for routine immunization of all adults age 60 years or older without contraindications for prevention of herpes zoster and its complications, principally PHN [95]. Zoster vaccine has now been licensed in more than 60 countries, and post-licensure studies have provided additional evidence of the safety of zoster vaccine and of its effectiveness against herpes zoster, ophthalmic zoster, and PHN in diverse populations, including persons with a variety of comorbidities [10, 28, 59, 115, 116, 131, 132, 155, 157, 164, 168, 170, 203, 228, 244–247, 249, 251, 280, 285, 286].

24.10.1 Safety of Zoster Vaccine in the General Population

A 1:1 placebo-controlled study evaluated the safety of zoster vaccine in 11,980 adults ≥60 years of age by assessing the rate of serious adverse events [170]. During the first 42 days after vaccination, 1.4 % of vaccine and 1.1 % of placebo recipients reported serious adverse events. Serious adverse events reported during the entire 182-day follow-up period were comparable, with 5.7 % in vaccine and 5.0 % in placebo recipients. Types of serious adverse events were also comparable in the two groups. Two subjects in the vaccine group and one subject in the placebo group reported serious adverse events that were deemed vaccine-related. The overall safety profile was similar for both groups and comparable to vaccine safety reported in the SPS, which observed a serious adverse events rate of 1.4 % in both vaccine and placebo recipients [184, 220].

A retrospective study by Tseng et al. [246] examined Vaccine Safety Datalink data from 193,083 zoster vaccine recipients ≥50 years of age. The most common adverse events following zoster vaccination were injection site reactions consisting of pain, erythema, and/or swelling occurring within 1–7 days after vaccination. This adverse event was generally mild to moderate and self-limited and more common in younger than in older vaccinees. These results are similar to safety data from the SPS, which reported that injection site reactions were more common in 60–69-year-olds than in ≥70-year-olds. The difference was attributed to the inverse relationship between age at vaccination and the magnitude of the VZV CMI response; it was suggested that local injection site reactions reflect the CMI response to zoster vaccine [220]. Tseng et al. [246] found no significant increase in the risk of cerebrovascular or cardiovascular adverse events in zoster vaccine recipients within 42 days of vaccination.

Kaiser Permanente Northern California performed a retrospective cohort study of approximately 29,000 persons ≥60 years of age to assess hospitalizations or emergency department visits in 42 days after administration of zoster vaccine. The study did not reveal any short-term safety concerns, such as acute myocardial infarction or stroke, linked to vaccination [10].

In 2014, Tseng et al. [248] reported the first laboratory-documented case of herpes zoster caused by the vaccine strain (vOka). The case was identified as part of a prospective herpes zoster study at Kaiser Permanente Southern California (KPSC), in which specimens from patients with clinically diagnosed herpes zoster are submitted to the National VZV Laboratory at CDC for genotyping. A 68-year-old healthy female, who had received Zostavax®, was diagnosed with herpes zoster 10 months after vaccination. The rash specimen contained vOka, identified by the presence of 15 of 42 recognized vaccine strain SNPs, including five of the fixed vaccine markers (see Sect. 24.3). While the patient had spent her entire life in the United States and had likely experienced varicella, past primary infection with VZV was not confirmed. This report may represent a rare case of herpes zoster caused by the Oka vaccine strain in a person who was VZV-naïve when she received Zostavax®. Almost all long-term US residents ≥60 years of age are seropositive for VZV as a result of childhood varicella. For example, none of the 1395 subjects in the SPS Immunology Substudy were seronegative at baseline [142]. Approximately 30 % of the eligible immunocompetent adults ≥60 years of age in the United States have now been vaccinated [152, 273] without serologic screening. Even if only 0.1 % of these ≥60-year-old adults were VZV seronegative, this would equate to some 20,000 VZV-seronegative adults who have received zoster vaccine without any reported serious adverse events.

Macaladad et al. [155] evaluated safety and immunogenicity of zoster vaccine in a double-blind, placebo-controlled randomized study with 21 healthy VZV-seronegative and low seropositive adults ≥40 years of age (18 vaccines and 3 placebo recipients). All 4 seronegative subjects who received vaccine seroconverted by day 42 and antibody titers increased significantly in seropositive vaccine recipients, demonstrating immunogenicity. No vaccine-related serious adverse events and no varicella or varicella-like rashes were reported, indicating that zoster vaccine is both safe and immunogenic in VZV-seronegative or low seropositive persons.

After completion of the SPS, 13,681 placebo participants, including 420 subjects who had previously developed confirmed herpes zoster 5 to 85 months earlier, received zoster vaccine and were followed for 28 days after vaccination [168]. The proportion of persons with prior herpes zoster reporting ≥1 serious adverse event was not significantly different from that of vaccinated placebo recipients who had not had herpes zoster. The study demonstrated that zoster vaccine safety was unaltered by a recent history of herpes zoster.

A study with 101 persons ≥50 years of age with a prior history of herpes zoster who received zoster vaccine and were followed for 28 days after vaccination also demonstrated that zoster vaccine was well tolerated and safe [164]. These studies support the ACIP recommendation that zoster vaccine should be administered to persons ≥60 years of age regardless of a history of herpes zoster [95].

24.10.2 Zoster Vaccine Effectiveness in the General Population

Effectiveness of zoster vaccine against herpes zoster, ophthalmic zoster, and PHN in the general population and in adults with a variety of medical conditions, including moderate immunosuppression resulting from disease and/or treatment, has been demonstrated by retrospective cohort studies using claims data from KPSC [244, 245, 247, 249], Medicare [131, 132, 286], or Aetna Healthcare [285]. In addition, Marin et al. [157] utilized active surveillance for herpes zoster in Olmsted County, Minnesota, and the Rochester Epidemiology Report.

Four studies examined zoster vaccine effectiveness for the incidence of herpes zoster in the general population (Table 24.1). Three of these studies included persons ≥60 years of age [157, 244, 249], and one study included persons ≥65 years of age [131]. Additional measures of vaccine effectiveness were the incidence of prodromal pain [157], PHN [131, 157], ophthalmic zoster [249], and herpes zoster-associated hospitalizations [249]. The study by Langan et al. [131] included both immunocompetent and immunosuppressed persons.

Tseng et al. [249] carried out a retrospective cohort study in 75,761 zoster vaccine recipients and 227,283 matched unvaccinated controls, using KPSC claims data (Table 24.1). Zoster vaccine was found to reduce the incidence of herpes zoster by 55 %, the incidence of ophthalmic zoster by 63 %, and zoster-associated hospitalizations by 65 % over a mean follow-up period of 1.72 years. Effectiveness for incidence of herpes zoster was similar to the 51 % efficacy observed in the SPS, but, in contrast to the SPS [184], effectiveness was retained across all age strata [249]. To control for bias, the adjusted rate ratios of 13 acute conditions unrelated to herpes zoster in vaccine and placebo recipients were determined. These ranged from 0.76 to 1.36 with a mean of 1.05; none were as low as the 0.45 rate ratio observed for herpes zoster.

Langan et al. reported an overall vaccine effectiveness for the incidence of herpes zoster of 48 % in persons ≥65 years of age, using a 5 % random sample of Medicare claims (Table 24.1). Zoster vaccine reduced the incidence of herpes zoster in both immunocompetent and immunosuppressed persons. The incidence of PHN with pain beyond 90 days after zoster rash onset was reduced by 59 % [131].

Marin et al. [157] determined zoster vaccine effectiveness by means of a case-control study in 266 patients ≥60 years of age with herpes zoster and 362 matched controls in Olmsted County, Minnesota, identified by active surveillance and verified by medical record review. Zoster vaccine effectiveness for prevention of herpes zoster prodrome, medically attended prodrome, and PHN was determined by analyzing a subset of herpes zoster case patients with the respective outcome of interest and their matched controls. The overall effectiveness of zoster vaccine for preventing herpes zoster was 54.2 % during 3 years of follow-up; 67.1 % in persons vaccinated before age 70 years and 38.3 % with vaccination after age 70 years. These results are almost identical to zoster vaccine efficacy in the SPS, where overall efficacy for incidence of herpes zoster was 51.3, 63.9 % in 60–69-year-olds and 37.6 % in subjects ≥70 years of age [184]. A separate analysis of herpes zoster characteristics by vaccination status found that the risk, duration, and severity of prodromal symptoms, as well as the pro-

Table 24.1 Effectiveness of live attenuated Oka/Merck zoster vaccine in general populations

Vaccine effectiveness (% reduction of HZ incidence[a])	Population	Mean duration of follow-up	Reduction in other measures of effectiveness	Data source	Reference
55 %	General population ≥60 yoa, immunocompetent; 75,761 vaccinated 227,283 unvaccinated	Vaccinated: 1.72 years Unvaccinated: 1.56 years	Ophthalmic zoster: 63 % Hospitalizations: 65 %	KPSC	[249]
48 % overall [37 % in immunosuppressed]	General population ≥65 yoa; 29,785 vaccinated 736,545 unvaccinated this included: immunosuppressed 4469 vaccinated 14,0925 unvaccinated	Vaccinated: 0.95 years Unvaccinated: 1.75 years	PHN (30 days): 62 % PHN (90 days): 59 %	US Medicare (5 % random sample)	[131]
54.2 % overall 67.1 % in <70 yoa 38.3 % in ≥70 yoa	General population ≥60 yoa; HZ cases ($N = 266$); confirmatory cohort ($N = 362$)	3 years	Prodrome: 58.0 %, medically attended prodrome: 70.0 % PHN (30 days): 60.5 %; PHN (60 days): 69.1 %; PHN (90 days): 55.2 %	Active surveillance; REP	[157]
49 % overall 68.7 % – Year 1 49.5 % – Year 2 39.1 % – Year 3 35.2 % – Year 4 37.1 % – Year 5 32.9 % – Year 6 16.5 % – Year 7 4.2 % – Year 8	General population ≥60 yoa, immunocompetent; 176,078 vaccinated 528,234 unvaccinated	Vaccinated: 3.71 years Unvaccinated: 2.65 years	Not done	KPSC	[244]

Abbreviations: HZ herpes zoster, *KPSC* Kaiser Permanente Southern California, *PHN* postherpetic neuralgia, *REP* Rochester Epidemiology Report, *yoa* years of age

[a]Percent effectiveness is based on adjusted vaccine efficacy or adjusted hazard ratios published in the respective references

portion of patients who sought medical care for prodromal symptoms, were lower in vaccinated than unvaccinated patients with herpes zoster (Table 24.1). However, as in the SPS, there was no difference in the dermatomal distribution of cases of herpes zoster in patients who did and did not receive zoster vaccine [157, 187].

These effectiveness studies in the general population have shown that zoster vaccine reduces the incidence of herpes zoster by approximately 50 % (Table 24.1), similar to the 51.3 % reduction observed in the SPS [184].

Due to their short follow-up times, most effectiveness studies cannot address long-term persistence of vaccine effectiveness. However, a recent study by Tseng et al. [244], which followed KPSC zoster vaccine recipients for up to 8 years, observed that vaccine effectiveness declined from 68.7 % in year 1 to 4.2 % in year 8 post-vaccination (Table 24.1). These observations are similar to the results of the SPS Persistence Substudies [167, 214].

24.10.3 Additional Effectiveness Studies

A study among KPSC patients showed that the short-term risk of herpes zoster recurrence was low (0.2 %/year) in adults with a recent episode of herpes zoster and was not further reduced by administration of zoster vaccine [243]. Assessment of 13 unrelated medical conditions, as described by Tseng et al. [249] was included to control for bias.

In a retrospective review of KPSC medical records, Tseng et al. [245] compared 1155 vaccinated adults ≥60 years of age who developed herpes zoster with a matched group of unvaccinated adults with herpes zoster to assess vaccine effectiveness for the incidence of PHN [245]. While vaccination was associated with a lower risk of PHN in women, no effect of the vaccine on PHN was observed in men. However, the number of persons experiencing PHN was small in both groups (56 in vaccine recipients, 100 in unvaccinated controls). Furthermore, since there is no ICD code for PHN, diagnosis in many cases is based upon circumstantial evidence, e.g., prescriptions for pain medication that coincide with episodes of herpes zoster. Thus, identification of PHN in electronic health records is difficult.

It is important to note that zoster vaccine is not effective for treatment of PHN.

Safety, immunogenicity, and effectiveness of concomitant administration of Zostavax® with other vaccines is described in Sect. 24.11.

24.10.4 Safety and Effectiveness of Zoster Vaccine in Persons with Comorbidities

Retrospective reviews have identified a number of comorbidities that are associated with increased incidence of herpes zoster [25, 29, 30, 63, 65, 88, 89, 99, 103, 111, 127, 132, 146, 150, 159, 171, 221, 224, 227, 242, 255, 274, 277, 279, 283] (Chap. 4).

The increase in age-specific risk of herpes zoster ranges from 1.2- to threefold for such diseases as chronic obstructive pulmonary disease, depression, inflammatory bowel disease, rheumatoid arthritis, psoriasis, asthma, multiple sclerosis, and diabetes mellitus. The increased risk may reflect inhibition of VZV CMI by one or more medications, such as corticosteroids, but the disease itself may also adversely affect host immune responses. The addition of corticosteroids to other therapies in patients with rheumatoid arthritis and other immune-mediated diseases results in a twofold increase in the incidence of herpes zoster [224]. The use of tofacitinib (a small-molecule inhibitor of janus kinase 1 and janus kinase 3) in patients with rheumatoid arthritis was associated with a twofold increase in the risk of herpes zoster compared to that in recipients of anti-TNF and other biologics [53].

In a preliminary study for the SPS, the investigational live attenuated zoster vaccine was determined to be safe and immunogenic in subjects ≥60 years of age with diabetes mellitus and with chronic obstructive pulmonary disease (COPD) receiving inhaled corticosteroids at a total dose equivalent to <20 mg/day of prednisone. Consequently, subjects with diabetes mellitus and COPD were enrolled in the SPS.

A double-blind placebo-controlled study evaluated the safety, tolerability, and immunogenicity of zoster vaccine in 309 subjects (207 vaccine and 102 placebo recipients) receiving chronic/maintenance corticosteroid therapy (5–20 mg prednisone/day) for ≥2 weeks pre- and ≥6 weeks post-vaccination [203]. Injection site adverse events were more frequent in the vaccine than in the placebo group (22 % versus 12 %); frequency of systemic adverse events was 28 % and 24 %, respectively, in vaccine and placebo recipients, similar to systemic adverse events in the SPS. Vaccine-related systemic adverse events were also more frequent in vaccine than placebo recipients (4.6 % versus 2.8 %). Zoster vaccine was immunogenic based on VZV gpELISA antibody titer and geometric mean fold-rise from prevaccination to 6 weeks after vaccination.

A retrospective study using Vaccine Safety Datalink data assessed the risks associated with zoster vaccine in 14,554 persons taking immunosuppressive medications, including oral corticosteroids, non-biological disease-modifying antirheumatic drugs (DMARDs), and oral antirejection drugs at the time of vaccination [28]. Patients who had immunosuppressive drugs dispensed between 30 days before and 5 days after vaccination and with medication supply crossing into this time period (defined as "current use," 4826 persons) were compared to patients using these drugs in the 365 days before but not extending into the current time window (defined as "remote use," 9728 persons). A modest increase in the incidence of herpes zoster was observed in the first 42 days after zoster vaccination in patients with current immunosuppressive drug use when compared to patients with remote immunosuppressant drug use. Patients with current immunosuppressive drug uses constituted 1.7 % of the total eligible population (4826/277,358 persons) that had received zoster vaccine during the study period, and only 0.2 % (550/277,358) were receiving high-dose corticosteroids (equivalent to ≥20 mg/day of prednisone) at the time of vaccination, which is contraindicated.

A Japanese pilot study administered investigational BIKEN varicella vaccine with a mean titer of 50,000 plaque-forming units of VZV (comparable to Zostavax®)

to 10 diabetic patients (with 6 % to 9.5 % glycosylated hemoglobin) and 10 healthy adults aged 60–70 years [98]. The vaccine was found to be safe and immunogenic.

In a pilot study, 10 moderately immunosuppressed subjects with systemic lupus erythematosus (SLE) and 10 control subjects ≥50 years of age received Zostavax® and were followed for 12 weeks. Three subjects in each group experienced mild adverse events at the injection site. No serious adverse events or SLE flares were reported during the study [90].

Patients with rheumatoid arthritis are at a twofold higher risk of developing herpes zoster, compared with the general population [221, 224, 274, 283]. This increased risk may be due, in part, to their underlying disease and in part to immunosuppressive therapies. Relatively few of such patients have received zoster vaccine because of safety concerns. Current ACIP guidelines [95] recommend zoster vaccine for patients who use methotrexate at <0.4 mg/kg/week or low to moderate doses of corticosteroids (prednisone equivalent of <20 mg/day), but the vaccine is contraindicated in patients using antitumor necrosis factor-α (TNF) and other biologics [285].

Zoster vaccine was administered to 152 patients with rheumatoid arthritis, ankylosing spondylitis, or psoriatic arthritis receiving intravenous (110 patients) or subcutaneous (42 patients) anti-TNF and other biologic therapies, at the time of a scheduled dose of biologic, which was withheld [147]. Zoster vaccine appeared to be safe, with no increase in herpes zoster incidence or disseminated herpes zoster. A safety and effectiveness study of zoster vaccine in adults on anti-TNF therapy, which is currently underway (https://clinicaltrials.gov/ct2/show/NCT02538757), has not yet reported any safety concerns [283].

Using Aetna Healthcare administrative claims data, Zhang et al. [285] examined zoster vaccine uptake and incidence of herpes zoster in persons ≥50 years of age with immune-mediated diseases (rheumatoid arthritis, psoriasis, psoriatic arthritis, ankylosing spondylitis, and/or inflammatory bowel disease) (Table 24.2). The incidence rates of herpes zoster were similar in vaccinated and unvaccinated patients. However, only 551 (1.2 %) of the 44,115 patients with immune-mediated diseases had received zoster vaccine, and most of these were not receiving immunosuppressive medications at time of vaccination. No cases of herpes zoster were observed among 47 vaccine recipients who received anti-TNF or other biologics within 30 days before or after vaccination. Thus, although the numbers are small, likely reflecting physicians' hesitation to administer a live virus vaccine to patients with autoimmune diseases, the results suggest that administration of zoster vaccine is relatively safe in these patients.

Zhang et al. [286] reported that zoster vaccine reduced the incidence of herpes zoster in Medicare recipients ≥60 years of age with immune-mediated diseases (rheumatoid arthritis, psoriatic arthritis, psoriasis, inflammatory bowel disease, ankylosing spondylitis) by 39 % over a median of 2 years of follow-up (Table 24.2). No cases of varicella or herpes zoster occurred in the first 42 days following vaccinations in 663 patients treated with biologics (in whom Zostavax® is contraindicated), including 551 receiving anti-TNF biologics, suggesting that zoster vaccine is safe in these patients.

Table 24.2 Effectiveness of live attenuated Oka/Merck zoster vaccine in populations with comorbidities

Vaccine effectiveness (% reduction of HZ incidence[a])	Population	Mean duration of follow-up	Data source	Reference
None	Patients ≥50 yoa with immune-mediated diseases; 551 vaccinated 43,564 unvaccinated	Vaccinated: 0.91 years Unvaccinated: 2.02 years	Aetna Healthcare	[285]
39 %	Patients ≥60 yoa with immune-mediated diseases; 18,683 vaccinated 444,858 unvaccinated	Vaccinated: 1.15 years[b] Unvaccinated: 1.92 years[b]	US Medicare	[286]
42 %	Patients ≥60 yoa who received myelosuppressive chemotherapy after vaccination; 4710 vaccinated 16,766 unvaccinated	Vaccinated: 1.50 years Unvaccinated: 1.58 years	KPSC	[251]
CKD: 49 % DM: 50 % CKD + DM: 46 %	Patients ≥65 yoa with CKD or DM or CKD and DM; 29,785 vaccinated 736,545 unvaccinated	Vaccinated: 0.95 years Unvaccinated: 1.75 years	US Medicare (5 % random sample)	[132]
51 %	End-stage renal disease patients ≥60 yoa; 582 vaccinated 2910 unvaccinated	Vaccinated: 2.36 years Unvaccinated: 1.94 years	KPSC	[247]

Abbreviations: CKD chronic kidney disease, *DM* diabetes mellitus, *HZ* herpes zoster, *KPSC* Kaiser Permanente Southern California, *yoa* years of age

[a]Percent effectiveness is based on adjusted vaccine efficacy or adjusted hazard ratios published in the respective references

[b]Based on person years in Table 4 in [286]

A cohort study by Tseng et al. [251] using KPSC electronic records demonstrated zoster vaccine effectiveness in persons ≥60 years of age who underwent chemotherapy following vaccination (Table 24.2). Zoster vaccine reduced the incidence of herpes zoster compared to unvaccinated chemotherapy recipients; vaccine effectiveness, 42 %, was comparable to that reported in a vaccinated immunocompetent cohort [249].

A population-based cohort study in a 5 % sample of Medicare patients demonstrated that zoster vaccine is effective in reducing the incidence of herpes zoster in persons ≥65 years of age with chronic kidney disease, diabetes, or both (Table 24.2). While herpes zoster incidence rates are higher in these patients, vaccine effectiveness is comparable to that in vaccine recipients without these comorbidities [132].

A retrospective study using KPSC claims data demonstrated that zoster vaccine reduced the incidence of herpes zoster by 51 % in patients with end-stage renal disease [247] (Table 24.2). The authors recommend that zoster vaccine be administrated before or shortly after initiation of dialysis to mitigate immunosuppression that might be associated with uremia, systemic inflammation, or hemodialysis.

24.10.5 Safety and Immunogenicity of Live Attenuated Oka Vaccine in Persons with HIV Infection

HIV-infected persons are at much higher risk of herpes zoster, recurrent herpes zoster, and complications of herpes zoster than the general population [68]. Weinberg et al. [264] administered two doses of varicella vaccine, 12 weeks apart, to 33 VZV-seropositive, HIV-infected adults with ≥400 CD4+ T cells per mm^3 to boost immunity against VZV and to protect against herpes zoster. The vaccine was well tolerated and safe, but only modestly immunogenic [264].

Two doses of Zostavax®, given 6 weeks apart to 295 VZV-seropositive HIV-infected adults ≥18 years of age on antiretroviral therapy (ART) with ≥200 CD4+ T cells per mm^3 and a viral load of <75 copies of HIV RNA per ml, were safe and immunogenic [11]. VZV antibody (gpELISA) titers after the second dose of zoster vaccine were comparable to those after the first dose.

A retrospective chart review at the Cleveland Clinic assessed zoster vaccine safety in 38 HIV-infected patients ≥50 years of age with CD4+ T cell counts of ≥350 per mm^3 [16]. The median CD4+ T cell count was 610 per mm^3, and safety follow-up was 42 days. Two of the 38 patients experienced adverse events that prevented normal everyday activities (grade 3): one developed a grade 3 injection site reaction; the other, who had a CD4+ T cell count of 6 per mm^3 (HIV infection was not recognized at the time of vaccination) developed grade 3 pruritus. Although the number of vaccine recipients was small, zoster vaccine appeared to be safe for this population.

There are currently no efficacy trials underway with live attenuated zoster vaccine in HIV-infected adults.

A double-blind placebo-controlled phase I/IIa clinical trial administered three doses of a liposome-based adjuvanted VZV glycoprotein E subunit vaccine (see Sect. 24.12) to 74 HIV-infected subjects ≥18 years of age on active ART with high (≥200 per mm^3) and low (50–199 per mm^3) CD4+ T cell counts and to ART-naïve adults with high (≥500 per mm^3) CD4+ T cell counts [14]. The vaccine was immunogenic with humoral and cell-mediated immune responses that persisted through 18 months of follow-up. There were no vaccine-related serious adverse events and no impact on HIV viral load or CD4+ T cell counts.

24.10.6 *Safety and Immunogenicity of Zoster Vaccine in Immunocompromised Patients*

Live attenuated Oka VZV vaccines are the most attenuated of all live viral vaccines. In addition, with respect to zoster vaccine, preexisting immunity to VZV resulting from prior varicella provides an added margin of safety in most adults. For example, even untreated HIV-infected patients with <50 CD4+ T cells per mm^3, who are VZV seropositive as a result of prior varicella, do not develop another episode of varicella when exposed to exogenous VZV. Finally, the vaccine strain of VZV (vOka) is temperature sensitive and is susceptible to antiviral drugs, including acyclovir, famciclovir, valacyclovir, and foscarnet (Chaps. 2 and 8). Nevertheless, the ACIP and other expert advisory groups [95, 201, 202, 102, 192, 193, 276] have recommended that zoster vaccine not be administered to persons with a variety of diseases and treatments that result in severely compromised CMI, such as leukemias, lymphomas, and other malignant neoplasms affecting the bone marrow or lymphatic system; persons with clinical manifestations of HIV infection, including with CD4+ T lymphocyte counts ≤200 per mm^3; persons on immunosuppressive therapy, including high-dose corticosteroids (≥20 mg/day of prednisone or equivalent) for ≥2 weeks; and persons undergoing hematopoietic stem cell or bone marrow transplantation unless they are at least 2 years post-transplantation, are not receiving immunosuppressive therapy, and are free of graft-versus-host disease. However, despite recommendations to the contrary, zoster vaccine has been administered to a number of immunosuppressed adults, both in clinical trials and in the course of general medical management, with few severe adverse events [110, 132, 150, 174, 190, 219, 252, 285, 286].

Retrospective reviews of electronic records from large integrated healthcare organizations and large administrative medical claims databases have identified a number of diseases and treatments associated with increased incidence and severity of herpes zoster and PHN (Chap. 4). The affected patients have immune-mediated disorders resulting in decreased VZV CMI, and the magnitude of the increase in risk is correlated with the degree of immunosuppression. The increase in age-specific risk of herpes zoster ranges from 1.5- to threefold for such diseases as rheumatoid arthritis, psoriasis, asthma, multiple sclerosis, and type II diabetes mellitus to 10- to more than 40-fold for patients with bone marrow and stem cell transplants, solid organ transplants, HIV infection, multiple myeloma, and systemic lupus erythematosus [17, 29, 33, 88, 99, 110, 159, 189, 224, 286] (Chap. 4).

A retrospective cohort study assessing vaccine effectiveness against the incidence of herpes zoster in the general population [131] (see Sect. 24.10.2) also captured zoster vaccine effectiveness in immunosuppressed persons ≥65 years of age (patients with leukemia, lymphoma, and HIV infection) and found a 37 % reduction in the incidence of herpes zoster in vaccinated versus unvaccinated immunocompromised patients.

In a study by Chou et al. [38], live attenuated Oka varicella vaccine was administered to 44 VZV-seronegative children <20 years of age who had received allogeneic hematopoietic cell transplants ≥24 months earlier, were off immunosuppressive therapy, had no evidence of ongoing chronic graft-versus-host disease, had >200 CD4+ T cells per mm^3, and had developed a specific antibody response to ≥1 post-transplantation vaccines. Three of the 44 developed self-limited varicella-like rashes within 2.5 weeks of vaccination that resolved without treatment. There were no subsequent varicella-like rashes and no cases of herpes zoster during a median follow-up of 29.1 months. Twenty-eight patients seroconverted after one injection; 10 of 14 who did not respond to the initial dose of varicella vaccine responded to a second dose. This study demonstrated the safety and immunogenicity of live attenuated Oka varicella vaccine in VZV-seronegative pediatric hematopoietic cell transplant recipients. Zoster vaccine contains a larger quantity of the same live attenuated vOka as varicella vaccine [184] but is administered to older adults, almost all of whom are immune to varicella. Thus, the safety and immunogenicity demonstrated in this study in children lacking any prior exposure to VZV are relevant to the use of zoster vaccine in hematopoietic cell transplant recipients.

Two doses of varicella vaccine, given to VZV-seropositive adults 4.5 and 6.5 months after autologous hematopoietic cell transplantation, were safe but poorly immunogenic [209].

Issa et al. [109] administered zoster vaccine to 110 adult autologous (52) and allogeneic (58) hematopoietic stem cell transplant recipients who were approximately 2 years post-transplantation, free of graft-versus-host disease, on no immunosuppressive medication, and not receiving prophylactic antivirals. One hundred eight vaccine recipients had no clinically apparent adverse events during a median of 9.5 months of follow-up. Two (1.8 %) developed a skin rash within 42 days post-vaccination that resolved with antiviral therapy. One of these rashes was zoster-like with associated pain that was treated with oral acyclovir. This patient developed PHN which had improved but not disappeared by 6 months after rash onset. The second patient, who had also received measles-mumps-rubella (MMR) vaccine on the same day as zoster vaccine, developed a vesicular skin rash on her upper extremities and trunk 24 days post-vaccination. She was treated with valacyclovir with complete resolution of her rash, which had been diagnosed as multidermatomal herpes zoster. As no diagnostic specimens were obtained, it is not known whether wild-type VZV or vOka caused the rash in either patient [109].

Review of pharmacy records from the UCSF Hematology Clinic identified 62 patients >18 years of age with hematologic malignancies who received hematopoietic cell transplants and, subsequently, Zostavax® at their provider's discretion [174]. Of these, 26 had received autologous and 5 allogeneic hematopoietic cell transplants at a median time of 482 days and 1323 days, respectively, prior to vaccination. No adverse events were reported during a median of 268.5 days of follow-up. One patient developed trigeminal herpes zoster 3 weeks after vaccination, but as no rash specimens were obtained, it could not be determined whether the episode of herpes zoster was caused by wild-type VZV or vOka [174].

These and other reports indicate that zoster vaccine can be safely administered to carefully selected patients following hematopoietic cell transplantation or chemotherapy for hematologic malignancies; however, several case reports demonstrate that this is not always the case. Costa et al. [49] reported fatal disseminated vOka following administration of zoster vaccine to a 79-year-old man with treated chronic lymphocytic leukemia who had not received immunosuppressive therapy for 6 months prior to vaccination [49].

Bhalla et al. [15] described a 67-year-old HIV-negative, childhood varicella-positive man successfully treated for relapse of non-Hodgkin lymphoma, who had received chemotherapy and autologous stem cell transplantation for diffuse large B cell lymphoma. Measles, mumps, rubella, hepatitis, and varicella vaccines were administered 4 years post-transplantation, at a time when he had diffuse large B cell lymphoma that was localized to a few lymph nodes and quiescent. Three months after vaccination, he developed a zosteriform rash on his forehead that was treated with valacyclovir. Zosteriform rashes recurred, progressed, and disseminated despite aggressive treatment with intravenous acyclovir. Skin and liver biopsies revealed granulomas containing VZV, which was shown by PCR and selective sequencing to be vOka. Sequencing of the VZV thymidine kinase gene revealed a mutation that resulted in acyclovir resistance. These rare cases emphasize the importance of avoiding administration of Oka vaccines to patients where significant deficiencies in CMI are suspected.

24.11 Concomitant Administration

An important impediment to administering the recommended vaccines to older adults is the low frequency of encounters with healthcare providers. Thus, it is essential to administer more than one vaccine during a single visit. The FDA and ACIP have routinely endorsed the safety and efficacy of concomitant administration of live attenuated and inactivated vaccines at separate injection sites [123]. However, in 2009, a clinical trial [156] demonstrated that concomitant administration of Zostavax® and Pneumovax®23 resulted in lower VZV antibody titers than observed when Zostavax® was administered 4 weeks after Pneumovax®23. Based upon these results, a sentence was added to the Zostavax® prescribing information [package insert] advising against concomitant administration of the two vaccines.

Subsequently a large observational study [250] compared zoster vaccine effectiveness against the incidence of herpes zoster in eligible adults receiving concomitant versus sequential vaccination with Pneumovax®23 and Zostavax® and demonstrated that the effectiveness of zoster vaccine was not reduced by concomitant administration. However, the prescribing information [package insert] continues to caution that concomitant administration resulted in lower antibody titers to VZV without citing the observational study (ZOSTAVAX prescribing information [package insert]; Merck & Co., Inc., Whitehouse Station, NJ, 02-2016 https://www.merck.com/product/usa/pi_circulars/z/zostavax/zostavax_pi2.pdf - accessed 06-09-2016). Retention of the cautionary statement regarding concomitant administration of the two vaccines, which likely results in many older adults failing to receive one or the other of these essential vaccines, emphasizes the importance of basing public health policy recommendations on clinically relevant observations – in this case, the recognition that it is VZV CMI and not VZV antibody that protects older adults from herpes zoster and PHN [142, 182, 183, 265].

A small placebo-controlled Japanese study [97] assessed the immunogenicity of zoster vaccine in elderly subjects with diabetes mellitus who received investigational zoster vaccine (BIKEN) containing approximately 50,000 plaque-forming units or placebo concomitantly with one dose of a 23-valent pneumococcal polysaccharide vaccine (PPSV23). Immunogenicity of zoster vaccine was assessed by VZV skin test [8], VZV interferon-γ ELISPOT, and VZV antibody assays. Zoster vaccine was safe but did not boost VZV CMI or antibody levels in 25 persons who received the vaccine, possibly due to concomitant PPSV23 administration. However, the study was small, and most subjects used dipeptidyl peptidase-4 inhibitors as antidiabetic drugs, which might have interfered with VZV CMI responses [31, 287].

The safety profile in persons ≥50 years of age receiving Zostavax® and inactivated influenza vaccine concomitantly was similar to that in persons who received the two vaccines ≥4 weeks apart [121], with injection site reactions observed in 44.7 % and 38.3 %, respectively. Systemic adverse events during 28 days following administration of zoster vaccine were reported in 4.5 % and 4.8 % of vaccinees, respectively. No serious adverse events were reported during the study.

24.12 New Improved Vaccines

Identification of VZV genes responsible for aspects of VZV pathogenicity, such as an ORF7-encoded neurotropic factor [217], and VZV epitopes with the capacity to induce protective immune responses [34] will facilitate the production of a new generation of VZV vaccines with improved safety and efficacy.

The development by GlaxoSmithKline of a liposome-based adjuvanted VZV glycoprotein E (gE) subunit vaccine (HZ/su) promises to radically change the prospects for immunization against herpes zoster and its complications. The liposome-based $AS01_B$ adjuvant system contains two immunostimulants: MPL, a nontoxic derivative of *Salmonella minnesota* lipopolysaccharide that is a TLR4 agonist that can stimulate NF-κB transcription and cytokine production and activate antigen-presenting cells, and QS-21, a natural saponin from the bark of the South American tree *Quillaja saponaria Molina* that promotes antigen-specific antibody and CD8+ T cell responses [58, 254]. VZV gE is an excellent candidate for a subunit antigen because it is a major component of the virion envelope and the most abundant glycoprotein in VZV virions and infected cells, it is essential for virus replication and cell-to-cell spread, and it is a major target for VZV-specific CD4+ T cell responses [5, 133].

A number of phase I, I/II, and II studies [35, 37, 130, 133] established that two doses of HZ/su containing 50 μg of recombinant VZV gE and the liposome-based $AS01_B$ adjuvant system containing 50 μg of MPL and 50 μg of QS21, administered one or 2 months apart, were well tolerated and induced much more robust

VZV-specific and VZV gE-specific CD4+ T cell and antibody responses than live attenuated zoster vaccine in normal younger and older adults. These VZV-specific immune responses have remained elevated for ≥6 years after vaccination [36, 133]. Administration of live attenuated zoster vaccine at the same time as HZ/su did not increase the immune responses induced by HZ/su. HZ/su had an acceptable safety profile, although non-serious adverse events were more frequent than those induced by live attenuated zoster vaccine, particularly injection site reactions during the first 7 days after vaccination. Small studies have demonstrated the immunogenicity and safety of HZ/su in two groups of immunocompromised patients: HIV-infected adults [14] and autologous hematopoietic cell transplant recipients [222].

A large randomized, placebo-controlled phase III efficacy trial of HZ/su in adults ≥50 years of age (ZOE-50) was carried out in 18 countries, in 15,411 subjects stratified by age, randomized to receive two intramuscular doses of HZ/su or placebo at months 0 and 2, and actively followed for herpes zoster for a mean of 3.2 years, beginning 1 month after the second dose [129]. Case determination was similar to that of the SPS (PCR for VZV DNA and for β-actin DNA to assess specimen adequacy [166]; all suspected cases of herpes zoster were adjudicated by a 5-member ascertainment committee independent of GSK and blinded to treatment and diagnostic laboratory results, with unanimity required for a positive assignment; PCR results always took precedence). Herpes zoster was confirmed in 6 HZ/su recipients and 210 placebo recipients, demonstrating an overall HZ/su vaccine efficacy for the incidence of herpes zoster of 97.2 %. Vaccine efficacy was comparable in all age groups.

HZ/su was more reactogenic than placebo; solicited or unsolicited adverse events within 7 days after vaccination were reported in 84.4 % of HZ/su recipients and 37.8 % of placebo recipients. Among HZ/su recipients, 81.5 % had injection site reactions, primarily pain, and 66.1 % had systemic reactions, primarily myalgia and fatigue, compared to 11.9 % and 29.5 %, respectively, in placebo recipients. Reactions were transient and mostly of mild-to-moderate intensity, but 17.0 % of HZ/su recipients and 3.2 % of placebo recipients had symptoms that prevented normal everyday activities (i.e., grade 3). Nevertheless, approximately 96 % of the participants in both the HZ/su and placebo groups received two doses of the study vaccine, indicating that reactogenicity did not greatly affect participants' willingness to receive the second dose. There were no significant efficacy differences according to age, in contrast to results of the SPS where vaccine efficacy for incidence of herpes zoster declined markedly in subjects ≥70 years of age.

To provide definitive data on the efficacy and safety of HZ/su in persons ≥70 years of age, a second study, ZOE-70, which had the same design, was performed concurrently with ZOE-50 at the same sites. In ZOE-70, 13,900 adults ≥70 years of age were randomized to receive HZ/su or placebo and followed for a mean of 3.7 years. Herpes zoster was confirmed in 23 HZ/su recipients and 223 placebo recipients (0.9 versus 9.2 cases per 1000 patient years); vaccine efficacy for incidence of herpes zoster was 89.8 %, $P < 0.001$, and was similar in 70–79-year-old (90.0 %) and ≥80-year-old (89.1 %) HZ/su recipients [52].

In a pooled analysis of data from subjects ≥70 years of age in ZOE-50 and ZOE-70 ($N = 16{,}596$), vaccine efficacy against herpes zoster was 91.3 %, $P < 0.001$. Four of 32 HZ/su recipients with herpes zoster and 46 of 477 placebo recipients with herpes zoster, all >70 years of age, had PHN, for a vaccine efficacy against PHN of 88.8 %, $P < 0.001$. Solicited injection site and systemic reactions were more frequent in HZ/su than placebo recipients (79.0 % versus 29.5 %), but serious adverse events, potential immune-mediated diseases, and deaths occurred with similar frequencies in the HZ/su and placebo recipients [52].

The HZ/su zoster vaccine should be safe and effective in the growing population of immunocompromised patients for whom live attenuated Oka zoster vaccines are contraindicated. The inclusion of a potent adjuvant system is of some concern with regard to widespread use in immunocompetent individuals because of the theoretical potential for inducing autoimmune diseases such as multiple sclerosis, systemic lupus erythematosus, and rheumatoid arthritis. The increasing frequency of these diseases and their potentially long incubation periods make the design and execution of long-term safety studies challenging. However, the absence of potential immune-mediated diseases in the large HZ/su efficacy trials conducted to date and in the relatively small numbers of subjects vaccinated almost a decade ago in phase I and I/II studies is encouraging.

Widespread use of new vaccines with gE as the sole VZV antigen may exert selective pressure in favor of gE escape mutants. However, the important functions of gE in VZV replication suggest that many such escape mutants would be less efficient than wild-type VZV in replication and spread.

Missense mutations in VZV gE that lead to loss of a B cell epitope and abrogate the binding of an antibody frequently used to diagnose VZV infection have been identified [207, 208, 239, 275]. These variants are suspected of having increased virulence: VZV-MSP [207, 208] exhibits accelerated cell-to-cell spread in tissue culture and in the SCID-hu mouse model, as well as a different egress pattern, when compared to wild-type VZV [207]. The variant BC, independently isolated from a patient with severe HZ lesions, carries the same missense mutation as VZV-MSP [239]. Two additional VZV gE variants with a different mutation in the same epitope were isolated from patients in Sweden [275], and a variant with another mutation in the this epitope was isolated from a fatal case of varicella [175]. If these or similar mutations alter targets for HZ/su-induced immunity to VZV, they might function as escape mutants. The potential for selection of gE escape mutants would likely be much greater if a gE subunit vaccine were used for primary vaccination (i.e., as a varicella vaccine) than to prevent herpes zoster and its complications in VZV-seropositive adults with latent VZV infection.

24.13 Barriers to Uptake of Zoster Vaccine

Zoster vaccine has been recommended for routine administration to persons ≥60 years of age for a decade in the United States, but only about 30 % of those for whom it has been recommended have been vaccinated [152, 273]. This reflects the existence of multiple barriers at the patient, provider, and system levels [9, 160, 182]. Many patients remain unaware of the risk and severity of herpes zoster and its complications, particularly PHN, of the fact that antiviral treatment does not prevent

PHN, and of the lack of effective treatment for this debilitating complication (Chaps. 4, 9, 16, 17, 18, 19, and 20). In addition, patients may not be aware of the availability of zoster vaccine or may not believe it to be effective (perhaps because, like most vaccines recommended for older adults, zoster vaccine is only partially protective, in contrast to the almost complete protection provided by most vaccines administered to children). Fear of side effects, stoked by a small but vociferous anti-vaccine establishment that has garnered media coverage, and the lack of effective rebuttal may lead to patients' refusal even when vaccination is offered. Low levels of education and the absence of strong recommendations by their healthcare providers are added disincentives, as are busy schedules and competing demands.

Barriers at the provider level include failure to initiate discussion of the need for zoster vaccine, to educate the patient, and to strongly recommend vaccination. This may reflect inadequate knowledge of the disease and the severity of its impact on elderly patients and/or lack of familiarity with zoster vaccine by the specialists who often provide care to older adults. It may also reflect the provider's skepticism about the need for zoster vaccine, concerns about its efficacy and safety, or uncertainty about the duration of protection. Insufficient knowledge about contraindicated medications and drug doses may lead to additional hesitation when considering zoster vaccine for patients with immune-mediated diseases or other comorbidities [9]. Concern about reduced immunogenicity and efficacy in older adults should be offset by recognition that the incidence and severity of herpes zoster, and especially of PHN, are markedly increased in older persons, so that even if effectiveness is somewhat lower, the net benefit of vaccination is greatest in the oldest patients. There is also a lack of incentives for physicians to administer recommended vaccines. This is compounded by inadequate time with the patient, especially since the primary purpose of most encounters is to address existing disease(s).

There are also major barriers at the system level, including lack of reimbursement for vaccine administration and concerns about out-of-pocket costs for patients, both related to coverage by Medicare Part D. The cost of zoster vaccine represents another barrier. Additional barriers include the lack of a tracking and reminder system, the need for frozen storage of zoster vaccine, and advice against concomitant administration of zoster vaccine and Pneumovax® 23 (see Sect. 24.11). Improving uptake of zoster vaccine will depend upon overcoming these obstacles and require strong, knowledgeable, and authoritative advocacy by the patients' healthcare providers. The single most important determinant of patients' acceptance of vaccination is a strong and reasoned recommendation by their healthcare provider.

24.14 Future Directions

Additional studies utilizing electronic medical records and medical claims databases will likely demonstrate the safety and efficacy of currently licensed live attenuated zoster vaccine in diverse populations of moderately immunocompromised patients, such as patients with immune-mediated diseases treated with anti-TNF biologics who are currently the subject of a prospective clinical trial (NCT02538757).

Studies with longer periods of post-vaccination follow-up will provide more precise data on the durability of protection and the need for and timing of booster vaccination. A major question is whether long-lasting immunity to VZV (and other herpesviruses) depends upon the vaccine virus's capacity to establish latency and repeatedly reactivate subclinically.

Evaluation and licensure of inactivated zoster vaccines and non-replication subunit vaccines, with and without adjuvants, will extend the benefits of zoster vaccination to the growing population of immunocompromised patients who are not candidates for currently licensed live attenuated zoster vaccines. Advances in molecular virology and vaccinology will lead to the introduction of an increasing number of new vaccines, new vaccine platforms, adjuvants, and vaccine combinations. These include new therapeutic vaccines targeting other human herpesviruses (e.g., herpes simplex virus, cytomegalovirus, and Epstein-Barr virus) and other pathogens like HIV and hepatitis C virus that cause latent or persistent infections.

The plethora of new investigational vaccines will make it impossible to base approval of vaccines and recommendations for vaccine usage, including concomitant administration, on the results of large randomized double-blind placebo-controlled clinical trials such as the SPS. Since elements of CMI are the host defenses of primary importance for many of the agents targeted by these new vaccines, it will be important to develop and utilize validated assays of virus-specific CMI that can provide reliable laboratory correlates of clinically significant immunity to the disease targeted by the vaccine in question. The selection and evaluation of such clinically valid and reliable assays, as well as the design of suitable phase IV clinical studies to establish safety and effectiveness, will need to be accomplished in advance to ensure that recommendations on vaccine usage are evidence-based.

Acknowledgments MNO and RH are grateful to Ms. Donna Mussatto for her expertise and patience in preparing the manuscript and to the James M. and Jesse V. Scott Fund for Shingles Research for continued support. We also thank Dr. Mona Marin for her kind help in clarifying the details of her recent studies.

References

1. Arbeter AM, Starr SE, Plotkin SA (1986) Varicella vaccine studies in healthy children and adults. Pediatrics 78(4 Pt 2):748–756
2. Arvin AM (1992) Cell-mediated immunity to varicella-zoster virus. J Infect Dis 166(Suppl 1):S35–S41
3. Arvin AM (2001) Varicella-zoster virus: molecular virology and virus-host interactions. Curr Opin Microbiol 4(4):442–449
4. Arvin AM, Gilden D (2013) Varicella-zoster virus. In: Knipe DM, Howley PM (eds) Fields virology, vol 2. 6thth edn. Lippincott Williams & Wilkins, Philadelphia, pp 2015-2057
5. Arvin AM, Oliver S, Reichelt M, Moffat JF, Sommer M, Zerboni L, Berarducci B (2010) Analysis of the functions of glycoproteins E and I and their promoters during VZV replication in vitro and in skin and T-cell xenografts in the SCID mouse model of VZV pathogenesis. In: Abendroth A, Arvin AM, Moffat JF (eds) Current topics in microbiology and immunology, Varicella-zoster virus, vol 342. Springer, Berlin/Heidelberg, pp. 129–146

6. Asada H, Nagayama K, Okazaki A, Mori Y, Okuno Y, Takao Y, Miyazaki Y, Onishi F, Okeda M, Yano S, Kumihashi H, Gomi Y, Maeda K, Ishikawa T, Iso H, Yamanishi K (2013) An inverse correlation of VZV skin-test reaction, but not antibody, with severity of herpes zoster skin symptoms and zoster-associated pain. J Dermatol Sci 69(3):243–249. doi:10.1016/j.jdermsci.2012.10.015
7. Asano Y, Nagai T, Miyata T, Yazaki T, Ito S, Yamanishi K, Takahashi M (1985) Long-term protective immunity of recipients of the OKA strain of live varicella vaccine. Pediatrics 75(4):667–671
8. Asano Y, Shiraki K, Takahashi M, Nagai H, Ozaki T, Yazaki T (1981) Soluble skin test antigen of varicella-zoster virus prepared from the fluid of infected cultures. J Infect Dis 143(5):684–692
9. Aziz M, Kessler H, Huhn G (2013) Providers' lack of knowledge about herpes zoster in HIV-infected patients is among barriers to herpes zoster vaccination. Int J STD AIDS 24(6):433–439. doi:10.1177/0956462412472461
10. Baxter R, Tran TN, Hansen J, Emery M, Fireman B, Bartlett J, Lewis N, Saddier P (2012) Safety of Zostavax – a cohort study in a managed care organization. Vaccine 30(47):6636–6641. doi:10.1016/j.vaccine.2012.08.070
11. Benson C, Hua L, Andersen J, Jiang J, Bozzolo D, Annunziato P, Read S, Pollard R, Rusin D, Lennox J (2012) ZOSTAVAX is generally safe and immunogenic in HIV+ adults virologically suppressed on ART: results of a Phase 2, randomized, double-blind, placebo-controlled trial. Paper presented at the 19th Conference on Retroviruses and Opportunistic Infections, Seattle, WA, 5–8 Mar 2012
12. Berger R, Florent G, Just M (1981) Decrease of the lymphoproliferative response to varicella-zoster virus antigen in the aged. Infect Immun 32(1):24–27
13. Berger R, Luescher D, Just M (1984) Enhancement of varicella-zoster-specific immune responses in the elderly by boosting with varicella vaccine. J Infect Dis 149(4):647–647
14. Berkowitz EM, Moyle G, Stellbrink HJ, Schürmann D, Kegg S, Stoll M, El Idrissi M, Oostvogels L, Heineman TC, Group Z-HsS (2015) Safety and immunogenicity of an adjuvanted herpes zoster subunit candidate vaccine in HIV-infected adults: a phase 1/2a randomized, placebo-controlled study. J Infect Dis 211(8):1279–1287. doi:10.1093/infdis/jiu606
15. Bhalla P, Forrest GN, Gershon M, Zhou Y, Chen J, LaRussa P, Steinberg S, Gershon AA (2015) Disseminated, persistent, and fatal infection due to the vaccine strain of varicella-zoster virus in an adult following stem cell transplantation. Clin Infect Dis 60(7):1068–1074. doi:10.1093/cid/ciu970
16. Bombatch C, Pallotta A, Neuner EA, Taege AJ (2016) Evaluation of herpes zoster vaccination in HIV-infected patients 50 years of age and older. Ann Pharmacother 50(4):326–327. doi:10.1177/1060028016632262
17. Borba EF, Ribeiro AC, Martin P, Costa LP, Guedes LK, Bonfa E (2010) Incidence, risk factors, and outcome of herpes zoster in systemic lupus erythematosus. J Clin Rheumatol 16(3):119–122. doi:10.1097/RHU.0b013e3181d52ed7
18. Bourdette DN, Rosenberg NL, Yatsu FM (1983) Herpes zoster ophthalmicus and delayed ipsilateral cerebral infarction. Neurology 33(11):1428–1432
19. Brazier J, Jones N, Kind P (1993) Testing the validity of the Euroqol and comparing it with the SF-36 health survey questionnaire. Qual Life Res 2(3):169–180
20. Breuer J, Schmid DS (2008) Vaccine Oka variants and sequence variability in vaccine-related skin lesions. J Infect Dis 197(Suppl 2):S54–S57. doi:10.1086/522140
21. Brisson M, Edmunds WJ, Law B, Gay NJ, Walld R, Brownell M, Roos L, De Serres G (2001) Epidemiology of varicella zoster virus infection in Canada and the United Kingdom. Epidemiol Infect 127(2):305–314
22. Bruusgaard E (1932) The mutual relation between zoster and varicella. Br J Dermatol Syphilol 44:1–24
23. Burgoon CF Jr, Burgoon JS, Baldridge GD (1957) The natural history of herpes zoster. JAMA 164(3):265–269

24. Burke BL, Steele RW, Beard OW, Wood JS, Cain TD, Marmer DJ (1982) Immune responses to varicella-zoster in the aged. Arch Intern Med 142(2):291–293
25. Chakravarty EF, Michaud K, Katz R, Wolfe F (2013) Increased incidence of herpes zoster among patients with systemic lupus erythematosus. Lupus 22(3):238–244. doi:10.1177/0961203312470186
26. Chang MN, Guess HA, Heyse JF (1994) Reduction in burden of illness: a new efficacy measure for prevention trials. Stat Med 13(18):1807–1814
27. Chao DY, Chien YZ, Yeh YP, Hsu PS, Lian IB (2012) The incidence of varicella and herpes zoster in Taiwan during a period of increasing varicella vaccine coverage, 2000-2008. Epidemiol Infect 140(6):1131–1140. doi:10.1017/S0950268811001786
28. Cheetham TC, Marcy SM, Tseng HF, Sy LS, Liu IL, Bixler F, Baxter R, Donahue JG, Naleway AL, Jacobsen SJ (2015) Risk of herpes zoster and disseminated varicella zoster in patients taking immunosuppressant drugs at the time of zoster vaccination. Mayo Clin Proc 90(7):865–873. doi:10.1016/j.mayocp.2015.04.021
29. Chen HH, Chen YM, Chen TJ, Lan JL, Lin CH, Chen DY (2011) Risk of herpes zoster in patients with systemic lupus erythematosus: a three-year follow-up study using a nationwide population-based cohort. Clinics (Sao Paulo) 66(7):1177–1182
30. Chen HH, Lin CL, Yeh SY, Kao CH (2016b) Short-term dipeptidyl peptidase-4 inhibitor use increases the risk of herpes zoster infection in Asian patients with diabetes. QJM 109(2):91–95. doi:10.1093/qjmed/hcv096
31. Chen HH, Lin IC, Chen HJ, Yeh SY, Kao CH (2016a) Association of herpes zoster and type 1 diabetes mellitus. PLoS One 11(5):e0155175. doi:10.1371/journal.pone.0155175
32. Chen JJ, Gershon AA, Zhishan L, Cowles RA, Gershon MD (2011) Varicella zoster virus (VZV) infects and establishes latency in enteric neurons. J Neurovirol 17(6):578–589. doi:10.1007/s13365-011-0070-1
33. Chen SY, Suaya JA, Li Q, Galindo CM, Misurski D, Burstin S, Levin MJ (2014) Incidence of herpes zoster in patients with altered immune function. Infection 42(2):325–334. doi:10.1007/s15010-013-0550-8
34. Chiu C, McCausland M, Sidney J, Duh FM, Rouphael N, Mehta A, Mulligan M, Carrington M, Wieland A, Sullivan NL, Weinberg A, Levin MJ, Pulendran B, Peters B, Sette A, Ahmed R (2014) Broadly reactive human CD8 T cells that recognize an epitope conserved between VZV, HSV and EBV. PLoS Pathogens 10(3):e1004008. doi:10.1371/journal.ppat.1004008
35. Chlibek R, Bayas JM, Collins H, de la Pinta ML, Ledent E, Mols JF, Heineman TC (2013) Safety and immunogenicity of an AS01-adjuvanted varicella-zoster virus subunit candidate vaccine against herpes zoster in adults >=50 years of age. J Infect Dis 208(12):1953–1961. doi:10.1093/infdis/jit365
36. Chlibek R, Pauksens K, Rombo L, van Rijckevorsel G, Richardus JH, Plassmann G, Schwarz TF, Catteau G, Lal H, Heineman TC (2016) Long-term immunogenicity and safety of an investigational herpes zoster subunit vaccine in older adults. Vaccine 34(6):863–868. doi:10.1016/j.vaccine.2015.09.073
37. Chlibek R, Smetana J, Pauksens K, Rombo L, Van den Hoek JA, Richardus JH, Plassmann G, Schwarz TF, Ledent E, Heineman TC (2014) Safety and immunogenicity of three different formulations of an adjuvanted varicella-zoster virus subunit candidate vaccine in older adults: a phase II, randomized, controlled study. Vaccine 32(15):1745–1753. doi:10.1016/j.vaccine.2014.01.019
38. Chou JF, Kernan NA, Prockop S, Heller G, Scaradavou A, Kobos R, Knowles MA, Papadopoulos EB, Casson A, Copeland C, Torok-Castanza J, Zakak N, Ruggiero J, Small TN (2011) Safety and immunogenicity of the live attenuated varicella vaccine following T replete or T cell-depleted related and unrelated allogeneic hematopoietic cell transplantation (allo-HCT). Biol Blood Marrow Transplant 17(11):1708–1713. doi:10.1016/j.bbmt.2011.05.006
39. Civen R, Chaves SS, Jumaan A, Wu H, Mascola L, Gargiullo P, Seward JF (2009) The incidence and clinical characteristics of herpes zoster among children and adolescents after implementation of varicella vaccination. Pediatr Infect Dis J 28(11):954–959. doi:10.1097/INF.0b013e3181a90b16

40. Clarke P, Beer T, Cohrs R, Gilden DH (1995) Configuration of latent varicella-zoster virus DNA. J Virol 69(12):8151–8154
41. Cohen JI (2013) Clinical practice: herpes zoster. N Engl J Med 369(3):255–263. doi:10.1056/NEJMcp1302674
42. Cohrs RJ, Barbour M, Gilden DH (1996) Varicella-zoster virus (VZV) transcription during latency in human ganglia: detection of transcripts mapping to genes 21, 29, 62, and 63 in a cDNA library enriched for VZV RNA. J Virol 70(5):2789–2796
43. Cohrs RJ, Gilden DH (2007) Prevalence and abundance of latently transcribed varicella-zoster virus genes in human ganglia. J Virol 81(6):2950–2956. doi:10.1128/JVI.02745-06
44. Cohrs RJ, Gilden DH, Kinchington PR, Grinfeld E, Kennedy PG (2003) Varicella-zoster virus gene 66 transcription and translation in latently infected human Ganglia. J Virol 77(12):6660–6665
45. Cohrs RJ, Mehta SK, Schmid DS, Gilden DH, Pierson DL (2008) Asymptomatic reactivation and shed of infectious varicella zoster virus in astronauts. J Med Virol 80(6):1116–1122. doi:10.1002/jmv.21173
46. Cohrs RJ, Randall J, Smith J, Gilden DH, Dabrowski C, van Der Keyl H, Tal-Singer R (2000) Analysis of individual human trigeminal ganglia for latent herpes simplex virus type 1 and varicella-zoster virus nucleic acids using real-time PCR. J Virol 74(24):11464–11471
47. Cohrs RJ, Srock K, Barbour MB, Owens G, Mahalingam R, Devlin ME, Wellish M, Gilden DH (1994) Varicella-zoster virus (VZV) transcription during latency in human ganglia: construction of a cDNA library from latently infected human trigeminal ganglia and detection of a VZV transcript. J Virol 68(12):7900–7908
48. Coplan PM, Schmader K, Nikas A, Chan ISF, Choo P, Levin MJ, Johnson G, Bauer M, Williams HM, Kaplan KM, Guess HA, Oxman MN (2004) Development of a measure of the burden of pain due to herpes zoster and postherpetic neuralgia for prevention trials: adaptation of the brief pain inventory. J Pain 5(6):344–356
49. Costa E, Buxton J, Brown J, Templeton KE, Breuer J (2016) Johannessen I (2016) Fatal disseminated varicella zoster infection following zoster vaccination in an immunocompromised patient. BMJ Case Rep. doi:10.1136/bcr-2015-212688
50. Croen KD, Ostrove JM, Dragovic LJ, Straus SE (1988) Patterns of gene expression and sites of latency in human nerve ganglia are different for varicella-zoster and herpes simplex viruses. Proc Natl Acad Sci U S A 85(24):9773–9777
51. Crumpacker CS (2011) Absence of exposure to varicella does not increase the risk of zoster. Clin Infect Dis 53(5):411–412. doi:10.1093/cid/cir439
52. Cunningham AL, Lal H, Kovac M, Chlibek R, Hwang SJ, Díez-Domingo J, Godeaux O, Levin MJ, McElhaney JE, Puig-Barberà J, Vanden Abeele C, Vesikari T, Watanabe D, Zahaf T, Ahonen A, Athan E, Barba-Gomez JF, Campora L, de Looze F, Downey HJ, Ghesquiere W, Gorfinkel I, Korhonen T, Leung E, McNeil SA, Oostvogels L, Rombo L, Smetana J, Weckx L, Yeo W, Heineman TC, ZOE-70 Study Group (2016) Efficacy of the herpes zoster subunit vaccine in adults 70 years of age or older. N Engl J Med 375(11):1019–1032. doi:10.1056/NEJMoa1603800
53. Curtis JR, Xie F, Yun H, Bernatsky S, Winthrop KL (2016) Real-world comparative risks of herpes virus infections in tofacitinib and biologic-treated patients with rheumatoid arthritis. Ann Rheum Dis 75(10):1–5. doi:10.1136/annrheumdis-2016-209131
54. Daut RL, Cleeland CS, Flanery RC (1983) Development of the Wisconsin brief pain questionnaire to assess pain in cancer and other diseases. Pain 17(2):197–210
55. Denny-Brown D, Adams RD, Fitzgerald PJ (1944) Pathologic features of herpes zoster: a note on "geniculate herpes". Arch Neurol Psychiatry 51(3):216–231
56. Depledge DP, Kundu S, Jensen NJ, Gray ER, Jones M, Steinberg S, Gershon A, Kinchington PR, Schmid DS, Balloux F, Nichols RA, Breuer J (2014) Deep sequencing of viral genomes provides insight into the evolution and pathogenesis of varicella zoster virus and its vaccine in humans. Mol Biol Evol 31(2):397–409. doi:10.1093/molbev/mst210
57. Depledge DP, Yamanishi K, Gomi Y, Gershon AA, Breuer J (2016) Deep sequencing of distinct preparations of the live attenuated varicella-zoster virus vaccine reveals a conserved core

of attenuating single-nucleotide polymorphisms. J Virol 90(19):8698–8704. doi:10.1128/JVI.00998-16
58. Didierlaurent AM, Collignon C, Bourguignon P, Wouters S, Fierens K, Fochesato M, Dendouga N, Langlet C, Malissen B, Lambrecht BN, Garçon N, Van Mechelen M, Morel S (2014) Enhancement of adaptive immunity by the human vaccine adjuvant AS01 depends on activated dendritic cells. J Immunol 193(4):1920–1930. doi:10.4049/jimmunol.1400948
59. Diez-Domingo J, Weinke T, Garcia de Lomas J, Meyer CU, Bertrand I, Eymin C, Thomas S, Sadorge C (2015) Comparison of intramuscular and subcutaneous administration of a herpes zoster live-attenuated vaccine in adults aged ≥50 years: a randomised non-inferiority clinical trial. Vaccine 33(6):789–795. doi:10.1016/j.vaccine.2014.12.024
60. Donahue JG, Choo PW, Manson JE, Platt R (1995) The incidence of herpes zoster. Arch Intern Med 155(15):1605–1609
61. Drolet M, Brisson M, Schmader KE, Levin MJ, Johnson R, Oxman MN, Patrick D, Blanchette C, Mansi JA (2010) The impact of herpes zoster and postherpetic neuralgia on health-related quality of life: a prospective study. CMAJ 182(16):1731–1736. doi:10.1503/cmaj.091711
62. Easton HG (1970) Zoster sine herpete causing acute trigeminal neuralgia. Lancet 2(7682):1065–1066
63. Esteban-Vasallo MD, Domínguez-Berjón MF, Gil-Prieto R, Astray-Mochales J, Gil de Miguel A (2014) Sociodemographic characteristics and chronic medical conditions as risk factors for herpes zoster: a population-based study from primary care in Madrid (Spain). Hum Vaccin Immunother 10(6):1650–1660. doi:10.4161/hv.28620
64. European Medicines Agency. (2016) Annex 1 summary of product characteristics. http://wwwemaeuropaeu/docs/en_GB/document_library/EPAR_-_Product_Information/human/000674/WC500053462pdf. Accessed 18 June 2016
65. Forbes HJ, Bhaskaran K, Thomas SL, Smeeth L, Clayton T, Langan SM (2014) Quantification of risk factors for herpes zoster: population based case-control study. BMJ 348:g2911. doi:10.1136/bmj.g2911 (Clinical research ed)
66. Forbes HJ, Thomas SL, Smeeth L, Clayton T, Farmer R, Bhaskaran K, Langan SM (2016) A systematic review and meta-analysis of risk factors for postherpetic neuralgia. Pain 157(1):30–54. doi:10.1097/j.pain.0000000000000307
67. Gaillat J, Gajdos V, Launay O, Malvy D, Demoures B, Lewden L, Pinchinat S, Derrough T, Sana C, Caulin E, Soubeyrand B (2011) Does monastic life predispose to the risk of Saint Anthony's fire (herpes zoster)? Clin Infect Dis 53(5):405–410
68. Gebo KA, Kalyani R, Moore RD, Polydefkis MJ (2005) The incidence of, risk factors for, and sequelae of herpes zoster among HIV patients in the highly active antiretroviral therapy era. J Acquir Immune Defic Syndr 40(2):169–174
69. Gelb LD, Dohner DE, Gershon AA, Steinberg SP, Waner JL, Takahashi M, Dennehy PH, Brown AE (1987) Molecular epidemiology of live, attenuated varicella virus vaccine in children with leukemia and in normal adults. J Infect Dis 155(4):633–640
70. Gershon AA (2001) Live-attenuated varicella vaccine. Infect Dis Clin North Am 15(1):65–81 , viii
71. Gershon AA, Breuer J, Cohen JI, Cohrs RJ, Gershon MD, Gilden D, Grose C, Hambleton S, Kennedy PG, Oxman MN, Seward JF, Yamanishi K (2015a) Varicella zoster virus infection. Nat Rev Dis Primers 1:1–18. doi:10.1038/nrdp.2015.16
72. Gershon AA, Chen J, Davis L, Krinsky C, Cowles R, Reichard R, Gershon M (2012) Latency of varicella zoster virus in dorsal root, cranial, and enteric ganglia in vaccinated children. Trans Am Clin Climatol Assoc 123:17–33 ; discussion 33-35
73. Gershon AA, Chen J, Gershon MD (2015b) Use of saliva to identify varicella zoster virus infection of the gut. Clin Infect Dis 61(4):536–544. doi:10.1093/cid/civ320
74. Gershon AA, Gershon MD (2013) Pathogenesis and current approaches to control of varicella-zoster virus infections. Clin Microbiol Rev 26(4):728–743. doi:10.1128/CMR.00052-13
75. Gershon AA, Gershon MD, Breuer J, Levin MJ, Oaklander AL, Griffiths PD (2010) Advances in the understanding of the pathogenesis and epidemiology of herpes zoster. J Clin Virol 48(Suppl 1):S2–S7. doi:10.1016/S1386-6532(10)70002-0

76. Gershon AA, LaRussa P, Steinberg S, Mervish N, Lo SH, Meier P (1996) The protective effect of immunologic boosting against zoster: an analysis in leukemic children who were vaccinated against chickenpox. J Infect Dis 173(2):450–453
77. Gershon AA, Levin MJ, Weinberg A, Song LY, LaRussa PS, Steinberg SP, Bartlett P, Pediatric ACTGT (2009) A phase I-II study of live attenuated varicella-zoster virus vaccine to boost immunity in human immunodeficiency virus-infected children with previous varicella. Pediatr Infect Dis J 28(7):653–655. doi:10.1097/INF.0b013e3181998f06
78. Gershon AA, Steinberg SP, LaRussa P, Ferrara A, Hammerschlag M, Gelb L (1988) Immunization of healthy adults with live attenuated varicella vaccine. J Infect Dis 158(1):132–137
79. Gilden D, Cohrs RJ, Mahalingam R, Nagel MA (2010) Neurological disease produced by varicella zoster virus reactivation without rash. Curr Top Microbiol Immunol 342:243–253. doi:10.1007/82_2009_3
80. Gilden DH, Dueland AN, Devlin ME, Mahalingam R, Cohrs R (1992) Varicella-zoster virus reactivation without rash. J Infect Dis 166(Suppl 1):S30–S34
81. Gilden DH, Kleinschmidt-DeMasters BK, Wellish M, Hedley-Whyte ET, Rentier B, Mahalingam R (1996) Varicella zoster virus, a cause of waxing and waning vasculitis: the New England Journal of Medicine case 5-1995 revisited. Neurology 47(6):1441–1446
82. Gilderman LI, Lawless JF, Nolen TM, Sterling T, Rutledge RZ, Fernsler DA, Azrolan N, Sutradhar SC, Wang WW, Chan IS, Schlienger K, Schodel F, Silber JL, Zostavax Protocol 010 Study G (2008) A double-blind, randomized, controlled, multicenter safety and immunogenicity study of a refrigerator-stable formulation of Zostavax. Clin Vaccine Immunol 15(2):314–319. doi:10.1128/CVI.00310-07
83. Gnann JW Jr, Whitley RJ (2002) Clinical practice. Herpes zoster. N Engl J Med 347(5):340–346. doi:10.1056/NEJMcp013211
84. Gomi Y, Imagawa T, Takahashi M, Yamanishi K (2000) Oka varicella vaccine is distinguishable from its parental virus in DNA sequence of open reading frame 62 and its transactivation activity. J Med Virol 61(4):497–503
85. Gomi Y, Imagawa T, Takahashi M, Yamanishi K (2001) Comparison of DNA sequence and transactivation activity of open reading frame 62 of Oka varicella vaccine and its parental viruses. Arch Virol Suppl 17(17):49–56
86. Gomi Y, Ozaki T, Nishimura N, Narita A, Suzuki M, Ahn J, Watanabe N, Koyama N, Ushida H, Yasuda N, Nakane K, Funahashi K, Fuke I, Takamizawa A, Ishikawa T, Yamanishi K, Takahashi M (2008) DNA sequence analysis of varicella-zoster virus gene 62 from subclinical infections in healthy children immunized with the Oka varicella vaccine. Vaccine 26(44):5627–5632
87. Gomi Y, Sunamachi H, Mori Y, Nagaike K, Takahashi M, Yamanishi K (2002) Comparison of the complete DNA sequences of the Oka varicella vaccine and its parental virus. J Virol 76(22):11447–11459
88. Guignard AP, Greenberg M, Lu C, Rosillon D, Vannappagari V (2014) Risk of herpes zoster among diabetics: a matched cohort study in a US insurance claim database before introduction of vaccination, 1997-2006. Infection 42(4):729–735. doi:10.1007/s15010-014-0645-x
89. Gupta G, Lautenbach E, Lewis JD (2006) Incidence and risk factors for herpes zoster among patients with inflammatory bowel disease. Clin Gastroenterol Hepatol 4(12):1483–1490. doi:10.1016/j.cgh.2006.09.019
90. Guthridge JM, Cogman A, Merrill JT, Macwana S, Bean KM, Powe T, Roberts V, James JA, Chakravarty EF (2013) Herpes zoster vaccination in SLE: a pilot study of immunogenicity. J Rheumatol 40(11):1875–1880. doi:10.3899/jrheum.130170
91. Hales CM, Harpaz R, Joesoef MR, Bialek SR (2013) Examination of links between herpes zoster incidence and childhood varicella vaccination. Ann Intern Med 159(11):739–745. doi:10.7326/0003-4819-159-11-201312030-00006
92. Hales CM, Harpaz R, Ortega-Sanchez I, Bialek SR, Centers for Disease Control and P (2014) Update on recommendations for use of herpes zoster vaccine. MMWR 63(33):729–731
93. Harbecke R, Oxman MN, Arnold BA, Ip C, Johnson GR, Levin MJ, Gelb LD, Schmader KE, Straus SE, Wang H, Wright PF, Pachucki CT, Gershon AA, Arbeit RD, Davis LE, Simberkoff

MS, Weinberg A, Williams HM, Cheney C, Petrukhin L, Abraham KG, Shaw A, Manoff S, Antonello JM, Green T, Wang Y, Tan C, Keller PM, Shingles Prevention SG (2009) A real-time PCR assay to identify and discriminate among wild-type and vaccine strains of varicella-zoster virus and herpes simplex virus in clinical specimens, and comparison with the clinical diagnoses. J Med Virol 81(7):1310–1322. doi:10.1002/jmv.21506

94. Hardy I, Gershon AA, Steinberg SP, LaRussa P (1991) The incidence of zoster after immunization with live attenuated varicella vaccine. A study in children with leukemia. Varicella Vaccine Collaborative Study Group. N Engl J Med 325(22):1545–1550
95. Harpaz R, Ortega-Sanchez IR, Seward JF, Advisory Committee on Immunization Practices Centers for Disease Control and P (2008) Prevention of herpes zoster: recommendations of the Advisory Committee on Immunization Practices (ACIP). MMWR Recomm Rep 57(RR-5):1–30 ; quiz CE32-34. doi:rr5705a1[pii]
96. Hata A, Asanuma H, Rinki M, Sharp M, Wong RM, Blume K, Arvin AM (2002) Use of an inactivated varicella vaccine in recipients of hematopoietic-cell transplants. N Engl J Med 347(1):26–34. doi:10.1056/NEJMoa013441
97. Hata A, Inoue F, Hamamoto Y, Yamasaki M, Fujikawa J, Kawahara H, Kawasaki Y, Honjo S, Koshiyama H, Moriishi E, Mori Y, Ohkubo T (2015) Efficacy and safety of live varicella zoster vaccine in diabetes: a randomized, double-blind, placebo-controlled trial. Diabet Med 33(8):1094–1101. doi:10.1111/dme.13038
98. Hata A, Inoue F, Yamasaki M, Fujikawa J, Kawasaki Y, Hamamoto Y, Honjo S, Moriishi E, Mori Y, Koshiyama H (2013) Safety, humoral and cell-mediated immune responses to herpes zoster vaccine in subjects with diabetes mellitus. J Infect 67(3):215–219. doi:10.1016/j.jinf.2013.04.010
99. Hata A, Kuniyoshi M, Ohkusa Y (2011) Risk of herpes zoster in patients with underlying diseases: a retrospective hospital-based cohort study. Infection 39(6):537–544. doi:10.1007/s15010-011-0162-0
100. Hayward AR, Buda K, Levin MJ (1994) Immune response to secondary immunization with live or inactivated VZV vaccine in elderly adults. Viral Immunol 7(1):31–36. doi:10.1089/vim.1994.7.31
101. Head H, Campbell AW (1900) The pathology of herpes zoster and its bearing on sensory localisation. Brain 23:353–523
102. Health Protection Scotland (2016) Patient group direction template: administration of shingles (herpes zoster) vaccine (live) Zostavax® Version 5.4. http://www.documents.hps.scot.nhs.uk/immunisation/patient-group-directions/pgd-zostavax-v5.4.pdf. Accessed 15 June 2016
103. Heymann AD, Chodick G, Karpati T, Kamer L, Kremer E, Green MS, Kokia E, Shalev V (2008) Diabetes as a risk factor for herpes zoster infection: results of a population-based study in Israel. Infection 36(3):226–230. doi:10.1007/s15010-007-6347-x
104. Hilt DC, Buchholz D, Krumholz A, Weiss H, Wolinsky JS (1983) Herpes zoster ophthalmicus and delayed contralateral hemiparesis caused by cerebral angiitis: diagnosis and management approaches. Ann Neurol 14(5):543–553. doi:10.1002/ana.410140509
105. Hope-Simpson RE (1954) Studies on Shingles: Is the virus ordinary chickenpox virus? Lancet 2:1299–1302
106. Hope-Simpson RE (1965) The nature of herpes zoster: a long-term study and a new hypothesis. Proc R Soc Med 58:9–20
107. Hope-Simpson RE (1975) Postherpetic neuralgia. J R Coll Gen Pract 25(157):571–575
108. Imoto K, Okazaki A, Onishi F, Miyazaki Y, Okeda M, Yano S, Takao Y, Gomi Y, Ishikawa T, Okuno Y, Mori Y, Iso H, Yamanishi K, Asada H (2015) VZV skin-test reaction, but not antibody, is an important predictive factor for postherpetic neuralgia. J Dermatol Sci 79(3):235–240. doi:10.1016/j.jdermsci.2015.05.011
109. Issa NC, Marty FM, Leblebjian H, Galar A, Shea MM, Antin JH, Soiffer RJ, Baden LR (2014) Live attenuated varicella-zoster vaccine in hematopoietic stem cell transplantation recipients. Biol Blood Marrow Transplant 20(2):285–287. doi:10.1016/j.bbmt.2013.11.013
110. Jansen K, Haastert B, Michalik C, Guignard A, Esser S, Dupke S, Plettenberg A, Skaletz-Rorowski A, Brockmeyer NH (2013) Incidence and risk factors of herpes zoster among

hiv-positive patients in the german competence network for HIV/AIDS (KompNet): a cohort study analysis. BMC Infect Dis 13:372. doi:10.1186/1471-2334-13-372

111. Joesoef RM, Harpaz R, Leung J, Bialek SR (2012) Chronic medical conditions as risk factors for herpes zoster. Mayo Clin Proc 87(10):961–967. doi:10.1016/j.mayocp.2012.05.021
112. Johnson RW, Rice AS (2014) Clinical practice. Postherpetic neuralgia. N Engl J Med 371(16):1526–1533. doi:10.1056/NEJMcp1403062
113. Kawai K, Gebremeskel BG, Acosta CJ (2014) Systematic review of incidence and complications of herpes zoster: towards a global perspective. BMJ Open 4(6):e004833. doi:10.1136/bmjopen-2014-004833
114. Kawai K, Yawn BP, Wollan P, Harpaz R (2016) Increasing incidence of herpes zoster over a 60-year period from a population-based study. Clin Infect Dis 63(2):221–226. doi:10.1093/cid/ciw296
115. Keating GM (2013) Shingles (herpes zoster) vaccine (zostavax((R))): a review of its use in the prevention of herpes zoster and postherpetic neuralgia in adults aged >/=50 years. Drugs 73(11):1227–1244. doi:10.1007/s40265-013-0088-1
116. Keating GM (2016) Shingles (herpes zoster) vaccine (Zostavax(®)): a review in the prevention of herpes zoster and postherpetic neuralgia. BioDrugs 30(3):243–254. doi:10.1007/s40259-016-0180-7
117. Kennedy PG, Grinfeld E, Bell JE (2000) Varicella-zoster virus gene expression in latently infected and explanted human ganglia. J Virol 74(24):11893–11898
118. Kennedy PG, Grinfeld E, Gow JW (1998) Latent varicella-zoster virus is located predominantly in neurons in human trigeminal ganglia. Proc Natl Acad Sci U S A 95(8):4658–4662
119. Kennedy PG, Grinfeld E, Gow JW (1999) Latent Varicella-zoster virus in human dorsal root ganglia. Virology 258(2):451–454. doi:10.1006/viro.1999.9745
120. Kennedy PG, Rovnak J, Badani H, Cohrs RJ (2015) A comparison of herpes simplex virus type 1 and varicella-zoster virus latency and reactivation. J Gen Virol 96(7):1581–1602. doi:10.1099/vir.0.000128
121. Kerzner B, Murray AV, Cheng E, Ifle R, Harvey PR, Tomlinson M, Barben JL, Rarrick K, Stek JE, Chung MO, Schödel FP, Wang WW, Xu J, Chan IS, Silber JL, Schlienger K (2007) Safety and immunogenicity profile of the concomitant administration of ZOSTAVAX and inactivated influenza vaccine in adults aged 50 and older. J Am Geriatr Soc 55(10):1499–1507. doi:10.1111/j.1532-5415.2007.01397.x
122. Kost RG, Straus SE (1996) Postherpetic neuralgia – pathogenesis, treatment, and prevention. N Engl J Med 335(1):32–42
123. Kroger AT, Atkinson WL, Marcuse EK, Pickering LK (2006) General recommendations on immunization: recommendations of the Advisory Committee on Immunization Practices (ACIP). MMWR Recomm Rep 55(RR-15):1–48
124. Ku CC, Padilla JA, Grose C, Butcher EC, Arvin AM (2002) Tropism of varicella-zoster virus for human tonsillar CD4(+) T lymphocytes that express activation, memory, and skin homing markers. J Virol 76(22):11425–11433
125. Ku CC, Zerboni L, Ito H, Graham BS, Wallace M, Arvin AM (2004) Varicella-zoster virus transfer to skin by T Cells and modulation of viral replication by epidermal cell interferon-alpha. J Exp Med 200(7):917–925. doi:10.1084/jem.20040634
126. Kundratitz K (1925) Experimentelle Ubertragung von herpes zoster auf den Menschen und die Beziehungen von herpes zoster zu varicellen. Monatsschr Kinderheilkd 29:516–523
127. Kwon HJ, Bang DW, Kim EN, Wi CI, Yawn BP, Wollan PC, Lahr BD, Ryu E, Juhn YJ (2016) Asthma as a risk factor for zoster in adults: a population-based case-control study. J Allergy Clin Immunol 137(5):1406–1412. doi:10.1016/j.jaci.2015.10.032
128. LaGuardia JJ, Cohrs RJ, Gilden DH (1999) Prevalence of varicella-zoster virus DNA in dissociated human trigeminal ganglion neurons and nonneuronal cells. J Virol 73(10):8571–8577
129. Lal H, Cunningham AL, Godeaux O, Chlibek R, Diez-Domingo J, Hwang SJ, Levin MJ, McElhaney JE, Poder A, Puig-Barberà J, Vesikari T, Watanabe D, Weckx L, Zahaf T, Heineman TC, Group Z-S (2015) Efficacy of an adjuvanted herpes zoster subunit vaccine in older adults. N Engl J Med 372(22):2087–2096. doi:10.1056/NEJMoa1501184

130. Lal H, Zahaf T, Heineman TC (2013) Safety and immunogenicity of an AS01-adjuvanted varicella zoster virus subunit candidate vaccine (HZ/su): a phase-I, open-label study in Japanese adults. Hum Vaccin Immunother 9(7):1425–1429. doi:10.4161/hv.24269
131. Langan SM, Smeeth L, Margolis DJ, Thomas SL (2013) Herpes zoster vaccine effectiveness against incident herpes zoster and post-herpetic neuralgia in an older US population: a cohort study. PLoS Med 10(4):e1001420. doi:10.1371/journal.pmed.1001420
132. Langan SM, Thomas SL, Smeeth L, Margolis DJ, Nitsch D (2016) Zoster vaccination is associated with a reduction of zoster in elderly patients with chronic kidney disease. Nephrol Dial Transplant. doi:10.1093/ndt/gfv432
133. Leroux-Roels I, Leroux-Roels G, Clement F, Vandepapelière P, Vassilev V, Ledent E, Heineman TC (2012) A phase 1/2 clinical trial evaluating safety and immunogenicity of a varicella zoster glycoprotein e subunit vaccine candidate in young and older adults. J Infect Dis 206(8):1280–1290. doi:10.1093/infdis/jis497
134. Leung J, Harpaz R, Molinari NA, Jumaan A, Zhou F (2011) Herpes zoster incidence among insured persons in the United States, 1993-2006: evaluation of impact of varicella vaccination. Clin Infect Dis 52(3):332–340. doi:10.1093/cid/ciq077
135. Levin MJ (2012) Immune senescence and vaccines to prevent herpes zoster in older persons. Curr Opin Immunol 24(4):494–500. doi:10.1016/j.coi.2012.06.002
136. Levin MJ (2014) Varicella-zoster virus and virus DNA in the blood and oropharynx of people with latent or active varicella-zoster virus infections. J Clin Virol 61(4):487–495. doi:10.1016/j.jcv.2014.09.012
137. Levin MJ, Barber D, Goldblatt E, Jones M, LaFleur B, Chan C, Stinson D, Zerbe GO, Hayward AR (1998) Use of a live attenuated varicella vaccine to boost varicella-specific immune responses in seropositive people 55 years of age and older: duration of booster effect. J Infect Dis 178(Suppl 1):109–112
138. Levin MJ, Cai GY, Manchak MD, Pizer LI (2003) Varicella-zoster virus DNA in cells isolated from human trigeminal ganglia. J Virol 77(12):6979–6987
139. Levin MJ, DeBiasi RL, Bostik V, Schmid DS (2008a) Herpes zoster with skin lesions and meningitis caused by 2 different genotypes of the Oka varicella-zoster virus vaccine. J Infect Dis 198(10):1444–1447
140. Levin MJ, Gershon AA, Weinberg A, Song L-Y, Fentin T, Nowak B (2006) Administration of live varicella vaccine to HIV-infected children with current or past significant depression of CD4(+) T cells. J Infect Dis 194(2):247–255
141. Levin MJ, Murray M, Rotbart HA, Zerbe GO, White CJ, Hayward AR (1992) Immune response of elderly individuals to a live attenuated varicella vaccine. J Infect Dis 166(2):253–259
142. Levin MJ, Oxman MN, Zhang JH, Johnson GR, Stanley H, Hayward AR, Caulfield MJ, Irwin MR, Smith JG, Clair J, Chan ISF, Williams H, Harbecke R, Marchese R, Straus SE, Gershon A, Weinberg A (2008b) Varicella-zoster virus-specific immune responses in elderly recipients of a herpes zoster vaccine. J Infect Dis 197(6):825–835
143. Levin MJ, Schmader KE, Pang L, Williams-Diaz A, Zerbe G, Canniff J, Johnson MJ, Caldas Y, Cho A, Lang N, Su SC, Parrino J, Popmihajlov Z, Weinberg A (2016) Cellular and humoral responses to a second dose of herpes zoster vaccine administered 10 years after the first dose among older adults. J Infect Dis 213(1537-6613 (Electronic)):14–22
144. Levine MJ, Ellison MC, Zerbe GO, Barber D, Chan C, Stinson D, Jones M, Hayward AR (2000) Comparison of a live attenuated and an inactivated varicella vaccine to boost the varicella-specific immune response in seropositive people 55 years of age and older. Vaccine 18(25):2915–2920
145. Lewis GW (1958) Zoster sine herpete. Br Med J 2(5093):418–421
146. Lin SY, Liu JH, Lin CL, Tsai IJ, Chen PC, Chung CJ, Liu YL, Wang IK, Lin HH, Huang CC (2012) A comparison of herpes zoster incidence across the spectrum of chronic kidney disease, dialysis and transplantation. Am J Nephrol 36(1):27–33. doi:10.1159/000339004
147. Lindsey S (2014) Safety of zoster vaccination administration in rheumatic patients on current biologic therapy. Arthritis Rheumatol 66(10 (Supplement)):S806

148. Lipschutz B (1921) Untersuchungen uber die atiologie der krankheiten der herpesgruppe (herpes zoster, herpes genitalis, herpes febrilis). Arch Dermatol Syphilol (Berl) 136:428–482
149. Ljungman P, Cordonnier C, Einsele H, Englund J, Machado CM, Storek J, Small T, Research CfIBaMT, Program NMD, Group EBaMT, Transplantation ASoBaM, Group CBaMT, America IDSo, America SfHEo, Canada AoMMaID, Prevention CfDCa (2009) Vaccination of hematopoietic cell transplant recipients. Bone Marrow Transplant 44 (8):521-526. doi:10.1038/bmt.2009.263
150. Long MD, Martin C, Sandler RS, Kappelman MD (2013) Increased risk of herpes zoster among 108 604 patients with inflammatory bowel disease. Aliment Pharmacol Ther 37(4):420–429. doi:10.1111/apt.12182
151. Loparev VN, Rubtcova E, Seward JF, Levin MJ, Schmid DS (2007) DNA sequence variability in isolates recovered from patients with postvaccination rash or herpes zoster caused by Oka varicella vaccine. J Infect Dis 195(4):502–510
152. Lu PJ, O'Halloran A, Williams WW, Harpaz R (2017) National and state-specific shingles vaccination among adults aged ≥60 years. Am J Prev Med 52(3):362–372. doi:10.1016/j.amepre.2016.08.031
153. Luby JP, Ramierez-Ronda C, Rinner S (1977) A longitudinal study of varicella-zoster virus infections in renal transplant recipients. J Infect Dis 135(4):659–663
154. Lydick E, Epstein RS, Himmelberger D, White CJ (1995) Area under the curve: a metric for patient subjective responses in episodic diseases. Qual Life Res 4(1):41–45
155. Macaladad N, Marcano T, Guzman M, Moya J, Jurado F, Thompson M, Meechan C, Li D, Schlienger K, Chan I, Sadoff J, Schodel F, Silber JL (2007) Safety and immunogenicity of a zoster vaccine in varicella-zoster virus seronegative and low-seropositive healthy adults. Vaccine 25(11):2139–2144. doi:10.1016/j.vaccine.2006.11.011
156. MacIntyre CR, Egerton T, McCaughey M, Parrino J, Campbell BV, Su SC, Pagnoni MF, Stek JE, Xu J, Annunziato PW, Chan IS, Silber JL (2010) Concomitant administration of zoster and pneumococcal vaccines in adults >/=60 years old. Hum Vaccin 6(11):894–902
157. Marin M, Yawn BP, Hales CM, Wollan PC, Bialek SR, Zhang J, Kurland MJ, Harpaz R (2015) Herpes zoster vaccine effectiveness and manifestations of herpes zoster and associated pain by vaccination status. Hum Vaccin Immunother 11(5):1157–1164. doi:10.1080/21645515.2015.1016681
158. Markus A, Lebenthal-Loinger I, Yang IH, Kinchington PR, Goldstein RS (2015) An in vitro model of latency and reactivation of varicella zoster virus in human stem cell-derived neurons. PLoS Pathog 11(6):e1004885. doi:10.1371/journal.ppat.1004885
159. McDonald JR, Zeringue AL, Caplan L, Ranganathan P, Xian H, Burroughs TE, Fraser VJ, Cunningham F, Eisen SA (2009) Herpes zoster risk factors in a national cohort of veterans with rheumatoid arthritis. Clin Infect Dis 48(10):1364–1371. doi:10.1086/598331
160. McLaughlin JM, McGinnis JJ, Tan L, Mercatante A, Fortuna J (2015) Estimated human and economic burden of four major adult vaccine-preventable diseases in the United States, 2013. J Prim Prev 36(4):259–273. doi:10.1007/s10935-015-0394-3
161. Mehta SK, Cohrs RJ, Forghani B, Zerbe G, Gilden DH, Pierson DL (2004) Stress-induced subclinical reactivation of varicella zoster virus in astronauts. J Med Virol 72(1):174–179. doi:10.1002/jmv.10555
162. Meier JL, Holman RP, Croen KD, Smialek JE, Straus SE (1993) Varicella-zoster virus transcription in human trigeminal ganglia. Virology 193(1):193–200. doi:10.1006/viro.1993.1115
163. Miller AE (1980) Selective decline in cellular immune response to varicella-zoster in the elderly. Neurology 30(6):582–587
164. Mills R, Tyring SK, Levin MJ, Parrino J, Li X, Coll KE, Stek JE, Schlienger K, Chan IS, Silber JL (2010) Safety, tolerability, and immunogenicity of zoster vaccine in subjects with a history of herpes zoster. Vaccine 28(25):4204–4209. doi:10.1016/j.vaccine.2010.04.003
165. Moffat JF, Zerboni L, Kinchington PR, Grose C, Kaneshima H, Arvin AM (1998) Attenuation of the vaccine Oka strain of varicella-zoster virus and role of glycoprotein C in alphaherpesvirus virulence demonstrated in the SCID-hu mouse. J Virol 72(2):965–974

166. Mols JF, Ledent E, Heineman TC (2013) Sampling of herpes zoster skin lesion types and the impact on viral DNA detection. J Virol Methods 188(1-2):145–147. doi:10.1016/j.jviromet.2012.12.013
167. Morrison VA, Johnson GR, Schmader KE, Levin MJ, Zhang JH, Looney DJ, Betts R, Gelb L, Guatelli JC, Harbecke R, Pachucki C, Keay S, Menzies B, Griffin MR, Kauffman CA, Marques A, Toney J, Boardman K, Su SC, Li X, Chan IS, Parrino J, Annunziato P, Oxman MN, Group SPS (2015) Long-term persistence of zoster vaccine efficacy. Clin Infect Dis 60(6):900–909. doi:10.1093/cid/ciu918
168. Morrison VA, Oxman MN, Levin MJ, Schmader KE, Guatelli JC, Betts RF, Gelb LD, Pachucki CT, Keay SK, Menzies B, Griffin MR, Kauffman CA, Marques AR, Toney JF, Simberkoff MS, Serrao R, Arbeit RD, Gnann JW, Greenberg RN, Holodniy M, Keitel WA, Yeh SS, Davis LE, Crawford GE, Neuzil KM, Johnson GR, Zhang JH, Harbecke R, Chan IS, Keller PM, Williams HM, Boardman KD, Silber JL, Annunziato PW, Shingles Prevention Study G (2013) Safety of zoster vaccine in elderly adults following documented herpes zoster. J Infect Dis 208(4):559–563. doi:10.1093/infdis/jit182
169. Mullane KM, Winston DJ, Wertheim MS, Betts RF, Poretz DM, Camacho LH, Pergam SA, Mullane MR, Stek JE, Sterling TM, Zhao Y, Manoff SB, Annunziato PW (2013) Safety and immunogenicity of heat-treated zoster vaccine (ZVHT) in immunocompromised adults. J Infect Dis 208(9):1375–1385. doi:10.1093/infdis/jit344
170. Murray AV, Reisinger KS, Kerzner B, Stek JE, Sausser TA, Xu J, Wang WW, Chan IS, Annunziato PW, Parrino J (2011) Safety and tolerability of zoster vaccine in adults >/=60 years old. Hum Vaccine 7(11):1130–1136. doi:10.4161/hv.7.11.17982
171. Nagasawa K, Yamauchi Y, Tada Y, Kusaba T, Niho Y, Yoshikawa H (1990) High incidence of herpes zoster in patients with systemic lupus erythematosus: an immunological analysis. Ann Rheum Dis 49(8):630–633
172. Nagel MA, Choe A, Traktinskiy I, Cordery-Cotter R, Gilden D, Cohrs RJ (2011) Varicella-zoster virus transcriptome in latently infected human ganglia. J Virol 85(5):2276–2287. doi:10.1128/JVI.01862-10
173. Nagel MA, Gilden D (2014) Neurological complications of varicella zoster virus reactivation. Curr Opin Neurol 27(3):356–360. doi:10.1097/WCO.0000000000000092
174. Naidus E, Damon L, Schwartz BS, Breed C, Liu C (2012) Experience with use of Zostavax((R)) in patients with hematologic malignancy and hematopoietic cell transplant recipients. Am J Hematol 87(1):123–125. doi:10.1002/ajh.22196
175. Natoli S, Ciotti M, Paba P, Testore GP, Palmieri G, Orlandi A, Sabato AF, Leonardis F (2006) A novel mutation of varicella-zoster virus associated to fatal hepatitis. J Clin Virol 37(1):72–74. doi:10.1016/j.jcv.2006.06.004
176. Okuno Y, Takao Y, Miyazaki Y, Ohnishi F, Okeda M, Yano S, Kumihashi H, Gomi Y, Maeda K, Ishikawa T, Mori Y, Asada H, Iso H, Yamanishi K (2013) Assessment of skin test with varicella-zoster virus antigen for predicting the risk of herpes zoster. Epidemiol Infect 141(4):706–713. doi:10.1017/s0950268812002671
177. Otsuka T, Gomi Y, Inoue N, Uchiyama M (2009) Transmission of varicella vaccine virus, Japan. Emerg Infect Dis 15(10):1702–1703. doi:10.3201/eid1510.090597
178. Ouwendijk WJ, Choe A, Nagel MA, Gilden D, Osterhaus AD, Cohrs RJ, Verjans GM (2012) Restricted varicella-zoster virus transcription in human trigeminal ganglia obtained soon after death. J Virol 86(18):10203–10206. doi:10.1128/JVI.01331-12
179. Oxman MN (1987) Varicella and herpes zoster. In: Fitzpatrick TB (ed) Dermatology in general medicine, 3rd edn. McGraw-Hill, Boston, pp. 2314–2340
180. Oxman MN (1995) Immunization to reduce the frequency and severity of herpes zoster and its complications. Neurology 45(12):41–46
181. Oxman MN (2000) Clinical manifestations of herpes zoster. In: Arvin AM, Gershon AA (eds) Varicella-zoster virus: virology and clinical management. Cambridge University Press, Cambridge, pp. 246–275

182. Oxman MN (2010) Zoster vaccine: current status and future prospects. Clin Infect Dis 51(2):197–213. doi:10.1086/653605
183. Oxman MN, Gershon AA, Poland GA (2011) Zoster vaccine recommendations: the importance of using a clinically valid correlate of protection. Vaccine 29(20):3625–3627. doi:10.1016/j.vaccine.2011.04.019
184. Oxman MN, Levin MJ, Johnson GR, Schmader KE, Straus SE, Gelb LD, Arbeit RD, Simberkoff MS, Gershon AA, Davis LE, Weinberg A, Boardman KD, Williams HM, Zhang JH, Peduzzi PN, Beisel CE, Morrison VA, Guatelli JC, Brooks PA, Kauffman CA, Pachucki CT, Neuzil KM, Betts RF, Wright PF, Griffin MR, Brunell P, Soto NE, Marques AR, Keay SK, Goodman RP, Cotton DJ, Gnann JW, Loutit J, Holodniy M, Keitel WA, Crawford GE, Yeh SS, Lobo Z, Toney JF, Greenberg RN, Keller PM, Harbecke R, Hayward AR, Irwin MR, Kyriakides TC, Chan CY, Chan ISF, Wang WWB, Annunziato PW, Silber JL (2005a) A vaccine to prevent herpes zoster and postherpetic neuralgia in older adults. N Engl J Med 352(22):2271–2284
185. Oxman MN, Levin MJ, Shingles Prevention Study G (2008) Vaccination against herpes zoster and postherpetic neuralgia. J Infect Dis 197(Suppl 2):S228–S236. doi:10.1086/522159
186. Oxman MN, Schmader KE (2014) Editorial commentary: zoster vaccine in immunocompromised patients: time to reconsider current recommendations. Clin Infect Dis 59(7):920–922. doi:10.1093/cid/ciu501
187. Oxman MN, Williams HM, Levin MJ, Johnson GR, Zhang JH, Schmader KE, Straus SE, Gelb LD, Arbeit RD, Simberkoff MS, Annunziato PW, Chan ISF, Silber JL, Wang WWB (2005b) Efficacy of zoster vaccine according to dermatome region. Paper presented at the 45th Interscience Conference of Antimicrobial Agents and Chemotherapy (ICAAC), Washington, DC, 16–19 Dec
188. Perez-Farinos N, Ordobas M, Garcia-Fernandez C, Garcia-Comas L, Canellas S, Rodero I, Gutierrez-Rodriguez A, Garcia-Gutierrez J, Ramirez R (2007) Varicella and herpes zoster in Madrid, based on the Sentinel General Practitioner Network: 1997-2004. BMC Infect Dis 7:59. doi:10.1186/1471-2334-7-59
189. Pergam SA, Forsberg CW, Boeckh MJ, Maynard C, Limaye AP, Wald A, Smith NL, Young BA (2011) Herpes zoster incidence in a multicenter cohort of solid organ transplant recipients. Transpl Infect Dis 13(1):15–23. doi:10.1111/j.1399-3062.2010.00547.x
190. Perry LM, Winthrop KL, Curtis JR (2014) Vaccinations for rheumatoid arthritis. Curr Rheumatol Rep 16(8):431. doi:10.1007/s11926-014-0431-x
191. Pevenstein SR, Williams RK, McChesney D, Mont EK, Smialek JE, Straus SE (1999) Quantitation of latent varicella-zoster virus and herpes simplex virus genomes in human trigeminal ganglia. J Virol 73(12):10514–10518
192. Public Health England (2015) Herpes zoster (shingles) immunization programme 2014/2015: Report for England. https://wwwgovuk/government/uploads/system/uploads/attachment_data/file/480254/Shingles_annual_1415pdf Accessed 14 June 2016
193. Public Health England (2016) Shingles (herpes zoster). The Green Book Chapter 28a v3_0 https://wwwgovuk/government/uploads/system/uploads/attachment_data/file/503773/2905109_Green_Book_Chapter_28a_v3_0WPDF. Accessed 10 June 2016
194. Quinlivan M, Breuer J, Schmid DS (2011) Molecular studies of the Oka varicella vaccine. Expert Rev Vaccines 10(9):1321–1336
195. Quinlivan M, Gershon AA, Steinberg SP, Breuer J (2005) An evaluation of single nucleotide polymorphisms used to differentiate vaccine and wild type strains of varicella-zoster virus. J Med Virol 75(1):174–180
196. Quinlivan ML, Gershon AA, Steinberg SP, Breuer J (2004) Rashes occurring after immunization with a mixture of viruses in the Oka vaccine are derived from single clones of virus. J Infect Dis 190(4):793–796
197. Quinlivan ML, Jensen NJ, Radford KW, Schmid DS (2012) Novel genetic variation identified at fixed loci in ORF62 of the Oka varicella vaccine and in a case of vaccine-associated herpes zoster. J Clin Microbiol 50(5):1533–1538. doi:10.1128/JCM.06630-11
198. Ragozzino MW, Melton LJ 3rd, Kurland LT, Chu CP, Perry HO (1982) Population-based study of herpes zoster and its sequelae. Medicine 61(5):310–316

199. Redman RL, Nader S, Zerboni L, Liu C, Wong RM, Brown BW, Arvin AM (1997) Early reconstitution of immunity and decreased severity of herpes zoster in bone marrow transplant recipients immunized with inactivated varicella vaccine. J Infect Dis 176(3):578–585
200. Rosenblum WI, Hadfield MG (1972) Granulomatous angiitis of the nervous system in cases of herpes zoster and lymphosarcoma. Neurology 22(4):348–354
201. Rubin LG, Levin MJ, Ljungman P, Davies EG, Avery R, Tomblyn M, Bousvaros A, Dhanireddy S, Sung L, Keyserling H, Kang I, America IDSo (2014a) 2013 IDSA clinical practice guideline for vaccination of the immunocompromised host. Clin Infect Dis 58(3):e44–100. doi:10.1093/cid/cit684
202. Rubin LG, Levin MJ, Ljungman P, Davies EG, Avery R, Tomblyn M, Bousvaros A, Dhanireddy S, Sung L, Keyserling H, Kang I, America IDSo (2014b) 2013 IDSA clinical practice guideline for vaccination of the immunocompromised host. Clin Infect Dis 58(3):309–318. doi:10.1093/cid/cit816
203. Russell AF, Parrino J, Fisher CL, Spieler W, Stek JE, Coll KE, Su SC, Xu J, Li X, Schlienger K, Silber JL (2015) Safety, tolerability, and immunogenicity of zoster vaccine in subjects on chronic/maintenance corticosteroids. Vaccine 33(27):3129–3134. doi:10.1016/j.vaccine.2015.04.090
204. Russell ML, Schopflocher DP, Svenson L, Virani SN (2007) Secular trends in the epidemiology of shingles in Alberta. Epidemiol Infect 135(6):908–913. doi:10.1017/S0950268807007893
205. Sadaoka T, Depledge DP, Rajbhandari L, Venkatesan A, Breuer J, Cohen JI (2016) In vitro system using human neurons demonstrates that varicella-zoster vaccine virus is impaired for reactivation, but not latency. Proc Natl Acad Sci U S A 113(17):E2403–E2412. doi:10.1073/pnas.1522575113
206. Sadaoka K, Okamoto S, Gomi Y, Tanimoto T, Ishikawa T, Yoshikawa T, Asano Y, Yamanishi K, Mori Y (2008) Measurement of varicella-zoster virus (VZV)-specific cell-mediated immunity: comparison between VZV skin test and interferon-gamma enzyme-linked immunospot assay. J Infect Dis 198(9):1327–1333. doi:10.1086/592219
207. Santos RA, Hatfield CC, Cole NL, Padilla JA, Moffat JF, Arvin AM, Ruyechan WT, Hay J, Grose C (2000) Varicella-zoster virus gE escape mutant VZV-MSP exhibits an accelerated cell-to-cell spread phenotype in both infected cell cultures and SCID-hu mice. Virology 275(2):306–317. doi:10.1006/viro.2000.0507
208. Santos RA, Padilla JA, Hatfield C, Grose C (1998) Antigenic variation of varicella zoster virus Fc receptor gE: loss of a major B cell epitope in the ectodomain. Virology 249(1):21–31. doi:10.1006/viro.1998.9313
209. Sasadeusz J, Prince HM, Schwarer A, Szer J, Stork A, Bock HL, Povey M, Nicholson O, Innis BL (2014) Immunogenicity and safety of a two-dose live attenuated varicella vaccine given to adults following autologous hematopoietic stem cell transplantation. Transpl Infect Dis 16(6):1024–1031. doi:10.1111/tid.12295
210. Sauerbrei A, Rubtcova E, Wutzler P, Schmid DS, Loparev VN (2004) Genetic profile of an Oka varicella vaccine virus variant isolated from an infant with zoster. J Clin Microbiol 42(12):5604–5608
211. Sauerbrei A, Zell R, Harder M, Wutzler P (2006) Genotyping of different varicella vaccine strains. J Clin Virol 37(2):109–117. doi:10.1016/j.jcv.2006.07.002
212. Schmader KE, Johnson GR, Saddier P, Ciarleglio M, Wang WW, Zhang JH, Chan IS, Yeh SS, Levin MJ, Harbecke RM, Oxman MN, Shingles Prevention Study G (2010) Effect of a zoster vaccine on herpes zoster-related interference with functional status and health-related quality-of-life measures in older adults. J Am Geriatr Soc 58(9):1634–1641. doi:10.1111/j.1532-5415.2010.03021.x
213. Schmader KE, Levin MJ, Gnann JW Jr, McNeil SA, Vesikari T, Betts RF, Keay S, Stek JE, Bundick ND, Su SC, Zhao Y, Li X, Chan IS, Annunziato PW, Parrino J (2012a) Efficacy, safety, and tolerability of herpes zoster vaccine in persons aged 50-59 years. Clin Infect Dis 54(7):922–928. doi:10.1093/cid/cir970
214. Schmader KE, Oxman MN, Levin MJ, Johnson G, Zhang JH, Betts R, Morrison VA, Gelb L, Guatelli JC, Harbecke R, Pachucki C, Keay S, Menzies B, Griffin MR, Kauffman C, Marques

A, Toney J, Keller PM, Li X, Chan IS, Annunziato P, Shingles Prevention Study G (2012b) Persistence of the efficacy of zoster vaccine in the shingles prevention study and the short-term persistence substudy. Clin Infect Dis 55(10):1320–1328. doi:10.1093/cid/cis638

215. Schmader KE, Sloane R, Pieper C, Coplan PM, Nikas A, Saddier P, Chan IS, Choo P, Levin MJ, Johnson G, Williams HM, Oxman MN (2007) The impact of acute herpes zoster pain and discomfort on functional status and quality of life in older adults. Clin J Pain 23(6):490–496. doi:10.1097/AJP.0b013e318065b6c9
216. Schmid DS (2010) Varicella-zoster virus vaccine: molecular genetics. In: Current Topics in Microbiology and Immunology, Varicella-zoster virus, vol 342. Springer, Heidelberg, pp. 323–340. doi:10.1007/82_2010_14
217. Selariu A, Cheng T, Tang Q, Silver B, Yang L, Liu C, Ye X, Markus A, Goldstein RS, Cruz-Cosme RS, Lin Y, Wen L, Qian H, Han J, Dulal K, Huang Y, Li Y, Xia N, Zhu H (2012) ORF7 of varicella-zoster virus is a neurotropic factor. J Virol 86(16):8614–8624. doi:10.1128/JVI.00128-12
218. Sen N, Mukherjee G, Sen A, Bendall SC, Sung P, Nolan GP, Arvin AM (2014) Single-cell mass cytometry analysis of human tonsil T cell remodeling by varicella zoster virus. Cell Rep 8(2):633–645
219. Shafran SD (2016) Live attenuated herpes zoster vaccine for HIV-infected adults. HIV Med 17(4):305–310. doi:10.1111/hiv.12311
220. Simberkoff MS, Arbeit RD, Johnson GR, Oxman MN, Boardman KD, Williams HM, Levin MJ, Schmader KE, Gelb LD, Keay S, Neuzil K, Greenberg RN, Griffin MR, Davis LE, Morrison VA, Annunziato PW, Shingles Prevention Study G (2010) Safety of herpes zoster vaccine in the shingles prevention study: a randomized trial. Ann Intern Med 152(9):545–554. doi:10.1059/0003-4819-152-9-201005040-00004
221. Smitten AL, Choi HK, Hochberg MC, Suissa S, Simon TA, Testa MA, Chan KA (2007) The risk of herpes zoster in patients with rheumatoid arthritis in the United States and the United Kingdom. Arthritis Rheum 57(8):1431–1438. doi:10.1002/art.23112
222. Stadtmauer EA, Sullivan KM, Marty FM, Dadwal SS, Papanicolaou GA, Shea TC, Mossad SB, Andreadis C, Young JA, Buadi FK, El Idrissi M, Heineman TC, Berkowitz EM (2014) A phase 1/2 study of an adjuvanted varicella-zoster virus subunit vaccine in autologous hematopoietic cell transplant recipients. Blood 124(19):2921–2929. doi:10.1182/blood-2014-04-573048
223. Stern ES (1937) The mechanism of herpes zoster and its relation to chicken-pox. Br J Dermatol 49:263–271
224. Strangfeld A, Listing J, Herzer P, Liebhaber A, Rockwitz K, Richter C, Zink A (2009) Risk of herpes zoster in patients with rheumatoid arthritis treated with anti-TNF-alpha agents. JAMA 301(7):737–744. doi:10.1001/jama.2009.146
225. Straus SE, Oxman MN (1999) Varicella and herpes zoster. In: Fitzpatrick TB et al (eds) Dermatology in general medicine, 5th edn. McGraw-Hill Book Company, New York, pp. 2427–2450
226. Straus SE, Reinhold W, Smith HA, Ruyechan WT, Henderson DK, Blaese RM, Hay J (1984) Endonuclease analysis of viral DNA from varicella and subsequent zoster infections in the same patient. N Engl J Med 311(21):1362–1364
227. Suaya JA, Chen SY, Li Q, Burstin SJ, Levin MJ (2014) Incidence of herpes zoster and persistent post-zoster pain in adults with or without diabetes in the United States. Open Forum Infect Dis 1(2):ofu049. doi:10.1093/ofid/ofu049
228. Sutradhar SC, Wang WW, Schlienger K, Stek JE, Xu J, Chan IS, Silber JL (2009) Comparison of the levels of immunogenicity and safety of Zostavax in adults 50 to 59 years old and in adults 60 years old or older. Clin Vaccine Immunol 16(5):646–652. doi:10.1128/CVI.00407-08
229. Takahashi M (1986) Clinical overview of varicella vaccine: development and early studies. Pediatrics 78(4 Pt 2):736–741
230. Takahashi M, Asano Y, Kamiya H, Baba K, Ozaki T, Otsuka T, Yamanishi K (2008) Development of varicella vaccine. J Infect Dis 197(Suppl 2):41–44

231. Takahashi M, Iketani T, Sasada K, Hara J, Kamiya H, Asano Y, Baba K, Shiraki K (1992) Immunization of the elderly and patients with collagen vascular diseases with live varicella vaccine and use of varicella skin antigen. J Infect Dis 166(Suppl 1):58–62
232. Takahashi M, Okada S, Miyagawa H, Amo K, Yoshikawa K, Asada H, Kamiya H, Torigoe S, Asano Y, Ozaki T, Terada K, Muraki R, Higa K, Iwasaki H, Akiyama M, Takamizawa A, Shiraki K, Yanagi K, Yamanishi K (2003) Enhancement of immunity against VZV by giving live varicella vaccine to the elderly assessed by VZV skin test and IAHA, gpELISA antibody assay. Vaccine 21(25-26):3845–3853
233. Takahashi M, Okuno Y, Otsuka T, Osame J, Takamizawa A (1975) Development of a live attenuated varicella vaccine. Biken J 18(1):25–33
234. Takahashi M, Otsuka T, Okuno Y, Asano Y, Yazaki T (1974) Live vaccine used to prevent the spread of varicella in children in hospital. Lancet 2(7892):1288–1290
235. Takao Y, Miyazaki Y, Onishi F, Kumihashi H, Gomi Y, Ishikawa T, Okuno Y, Mori Y, Asada H, Yamanishi K, Iso H (2012) The Shozu Herpes Zoster (SHEZ) study: rationale, design, and description of a prospective cohort study. J Epidemiol 22(2):167–174
236. Tang H, Moriishi E, Okamoto S, Okuno Y, Iso H, Asada H, Yamanishi K, Mori Y (2012) A community-based survey of varicella-zoster virus-specific immune responses in the elderly. J Clin Virol 55(1):46–50. doi:10.1016/j.jcv.2012.06.008
237. Thiele S, Borschewski A, Küchler J, Bieberbach M, Voigt S, Ehlers B (2011) Molecular analysis of varicella vaccines and varicella-zoster virus from vaccine-related skin lesions. Clin Vaccine Immunol 18(7):1058–1066. doi:10.1128/CVI.05021-11
238. Tillieux SL, Halsey WS, Thomas ES, Voycik JJ, Sathe GM, Vassilev V (2008) Complete DNA sequences of two oka strain varicella-zoster virus genomes. J Virol 82(22):11023–11044. doi:10.1128/JVI.00777-08
239. Tipples GA, Stephens GM, Sherlock C, Bowler M, Hoy B, Cook D, Grose C (2002) New variant of varicella-zoster virus. Emerg Infect Dis 8(12):1504–1505. doi:10.3201/eid0812.020118
240. Toyama N, Shiraki K, Dermatologists SotMP (2009) Epidemiology of herpes zoster and its relationship to varicella in Japan: a 10-year survey of 48,388 herpes zoster cases in Miyazaki prefecture. J Med Virol 81(12):2053–2058. doi:10.1002/jmv.21599
241. Trannoy E, Berger R, Hollander G, Bailleux F, Heimendinger P, Vuillier D, Creusvaux H (2000) Vaccination of immunocompetent elderly subjects with a live attenuated Oka strain of varicella zoster virus: a randomized, controlled, dose-response trial. Vaccine 18(16):1700–1706
242. Tsai SY, Lin CL, Wong YC, Yang TY, Kuo CF, Cheng JM, Wang JS, Kao CH (2015) Increased risk of herpes zoster following dermatomyositis and polymyositis: a nationwide population-based cohort study. Medicine (Baltimore) 94(28):e1138. doi:10.1097/MD.0000000000001138
243. Tseng HF, Chi M, Smith N, Marcy SM, Sy LS, Jacobsen SJ (2012a) Herpes zoster vaccine and the incidence of recurrent herpes zoster in an immunocompetent elderly population. J infect Dis 206(2):190–196. doi:10.1093/infdis/jis334
244. Tseng HF, Harpaz R, Luo Y, Hales CM, Sy LS, Tartof SY, Bialek S, Hechter RC, Jacobsen SJ (2016a) Declining effectiveness of herpes zoster vaccine in adults aged ≥60 years. J Infect Dis 213(12):1872–1875. doi:10.1093/infdis/jiw047
245. Tseng HF, Lewin B, Hales CM, Sy LS, Harpaz R, Bialek S, Luo Y, Jacobsen SJ, Reddy K, Huang PY, Zhang J, Anand S, Bauer EM, Chang J, Tartof SY (2015) Zoster vaccine and the risk of postherpetic neuralgia in patients who developed herpes zoster despite having received the zoster vaccine. J Infect Dis 212(8):1222–1231. doi:10.1093/infdis/jiv244
246. Tseng HF, Liu A, Sy L, Marcy SM, Fireman B, Weintraub E, Baggs J, Weinmann S, Baxter R, Nordin J, Daley MF, Jackson L, Jacobsen SJ, Vaccine Safety Datalink T (2012b) Safety of zoster vaccine in adults from a large managed-care cohort: a Vaccine Safety Datalink study. J Intern Med 271(5):510–520. doi:10.1111/j.1365-2796.2011.02474.x
247. Tseng HF, Luo Y, Shi J, Sy LS, Tartof SY, Sim JJ, Hechter RC, Jacobsen SJ (2016b) Effectiveness of herpes zoster vaccine in patients 60 years and older with end-stage renal disease. Clin Infect Dis 62(4):462–467. doi:10.1093/cid/civ930

248. Tseng HF, Schmid DS, Harpaz R, LaRussa P, Jensen NJ, Rivailler P, Radford K, Folster J, Jacobsen SJ (2014a) Herpes zoster caused by vaccine-strain varicella zoster virus in an immunocompetent recipient of zoster vaccine. Clin Infect Dis 58(8):1125–1128. doi:10.1093/cid/ciu058
249. Tseng HF, Smith N, Harpaz R, Bialek SR, Sy LS, Jacobsen SJ (2011a) Herpes zoster vaccine in older adults and the risk of subsequent herpes zoster disease. JAMA 305(2):160–166. doi:10.1001/jama.2010.1983
250. Tseng HF, Smith N, Sy LS, Jacobsen SJ (2011b) Evaluation of the incidence of herpes zoster after concomitant administration of zoster vaccine and polysaccharide pneumococcal vaccine. Vaccine 29(20):3628–3632. doi:10.1016/j.vaccine.2011.03.018
251. Tseng HF, Tartof S, Harpaz R, Luo Y, Sy LS, Hetcher RC, Jacobsen SJ (2014b) Vaccination against zoster remains effective in older adults who later undergo chemotherapy. Clin Infect Dis 59(7):913–919. doi:10.1093/cid/ciu498
252. Tsigrelis C, Ljungman P (2016) Vaccinations in patients with hematological malignancies. Blood Rev 30(2):139–147. doi:10.1016/j.blre.2015.10.001
253. Tyzzer EE (1906) The histology of the skin lesions in varicella. Philippine J Sci 1(4):349–375
254. Vandepapelière P, Horsmans Y, Moris P, Van Mechelen M, Janssens M, Koutsoukos M, Van Belle P, Clement F, Hanon E, Wettendorff M, Garçon N, Leroux-Roels G (2008) Vaccine adjuvant systems containing monophosphoryl lipid A and QS21 induce strong and persistent humoral and T cell responses against hepatitis B surface antigen in healthy adult volunteers. Vaccine 26(10):1375–1386. doi:10.1016/j.vaccine.2007.12.038
255. Veetil BM, Myasoedova E, Matteson EL, Gabriel SE, Green AB, Crowson CS (2013) Incidence and time trends of herpes zoster in rheumatoid arthritis: a population-based cohort study. Arthritis Care Res (Hoboken) 65(6):854–861. doi:10.1002/acr.21928
256. Vermeulen JN, Lange JM, Tyring SK, Peters PH, Nunez M, Poland G, Levin MJ, Freeman C, Chalikonda I, Li J, Smith JG, Caulfield MJ, Stek JE, Chan IS, Vessey R, Schodel FP, Annunziato PW, Schlienger K, Silber JL (2012) Safety, tolerability, and immunogenicity after 1 and 2 doses of zoster vaccine in healthy adults >/=60 years of age. Vaccine 30(5):904–910. doi:10.1016/j.vaccine.2011.11.096
257. Vesikari T, Hardt R, Rumke HC, Icardi G, Montero J, Thomas S, Sadorge C, Fiquet A (2013) Immunogenicity and safety of a live attenuated shingles (herpes zoster) vaccine (Zostavax(R)) in individuals aged >/= 70 years: a randomized study of a single dose vs. two different two-dose schedules. Hum Vaccine Immunotherap 9(4):858–864. doi:10.4161/hv.23412
258. von Bokay J (1909) Über den ätiologischen zusammenhang der varizellen mit gewissen fallen von herpes zoster. Wien Klin Wochenschr 22:1323–1326
259. Wang K, Lau TY, Morales M, Mont EK, Straus SE (2005) Laser-capture microdissection: refining estimates of the quantity and distribution of latent herpes simplex virus 1 and varicella-zoster virus DNA in human trigeminal Ganglia at the single-cell level. J Virol 79(22):14079–14087. doi:10.1128/JVI.79.22.14079-14087.2005
260. Ware J, Kosinski M, Keller SD (1996) A 12-Item Short-Form Health Survey: construction of scales and preliminary tests of reliability and validity. Med Care 34(3):220–233
261. Watson CP, Deck JH, Morshead C, Van der Kooy D, Evans RJ (1991) Post-herpetic neuralgia: further post-mortem studies of cases with and without pain. Pain 44(2):105–117 doi:0304-3959(91)90124-G [pii]
262. Watson CP, Morshead C, Van der Kooy D, Deck J, Evans RJ (1988) Post-herpetic neuralgia: post-mortem analysis of a case. Pain 34(2):129–138
263. Weibel RE, Neff BJ, Kuter BJ, Guess HA, Rothenberger CA, Fitzgerald AJ, Connor KA, McLean AA, Hilleman MR, Buynak EB (1984) Live attenuated varicella virus vaccine. Efficacy trial in healthy children. N Engl J Med 310(22):1409–1415
264. Weinberg A, Levin MJ, Macgregor RR (2010) Safety and immunogenicity of a live attenuated varicella vaccine in VZV-seropositive HIV-infected adults. Hum Vaccin 6(4):318–321
265. Weinberg A, Zhang JH, Oxman MN, Johnson GR, Hayward AR, Caulfield MJ, Irwin MR, Clair J, Smith JG, Stanley H, Marchese RD, Harbecke R, Williams HM, Chan IS, Arbeit RD, Gershon AA, Schodel F, Morrison VA, Kauffman CA, Straus SE, Schmader KE, Davis

LE, Levin MJ, Investigators USDoVACSPSPS (2009) Varicella-zoster virus-specific immune responses to herpes zoster in elderly participants in a trial of a clinically effective zoster vaccine. J Infect Dis 200(7):1068–1077. doi:10.1086/605611
266. Weinmann S, Chun C, Schmid DS, Roberts M, Vandermeer M, Riedlinger K, Bialek SR, Marin M (2013) Incidence and clinical characteristics of herpes zoster among children in the varicella vaccine era, 2005-2009. J Infect Dis 208(11):1859–1868. doi:10.1093/infdis/jit405
267. Weller TH (1983a) Varicella and herpes zoster. Changing concepts of the natural history, control, and importance of a not-so-benign virus. N Engl J Med 309(22):1362–1368. doi:10.1056/NEJM198312013092205
268. Weller TH (1983b) Varicella and herpes zoster. Changing concepts of the natural history, control, and importance of a not-so-benign virus. N Engl J Med 309(23):1434–1440. doi:10.1056/NEJM198312083092306
269. Weller TH, Witton HM (1958) The etiologic agents of varicella and herpes zoster; serologic studies with the viruses as propagated in vitro. J Exp Med 108(6):869–890
270. Weller TH, Witton HM, Bell EJ (1958) The etiologic agents of varicella and herpes zoster; isolation, propagation, and cultural characteristics in vitro. J Exp Med 108(6):843–868
271. Whitley RJ (2015) Editorial commentary: waning efficacy of the herpes zoster vaccine. Clin Infect Dis 60(6):910–911. doi:10.1093/cid/ciu922
272. Whitley RJ, Shukla S, Crooks RJ (1998) The identification of risk factors associated with persistent pain following herpes zoster. J Infect Dis 178(Suppl 1):S71–S75
273. Williams WW, Lu PJ, O'Halloran A, Kim DK, Grohskopf LA, Pilishvili T, Skoff TH, Nelson NP, Harpaz R, Markowitz LE, Rodriguez-Lainz A, Bridges CB (2016) Surveillance of vaccination coverage among adult populations – United States, 2014. MMWR Surveill Summ 65(1):1–36. doi:10.15585/mmwr.ss6501a1
274. Winthrop KL, Furst DE (2010) Rheumatoid arthritis and herpes zoster: risk and prevention in those treated with anti-tumour necrosis factor therapy. Ann Rheum Dis 69(10):1735–1737. doi:10.1136/ard.2010.133843
275. Wirgart BZ, Estrada V, Jackson W, Linde A, Grose C (2006) A novel varicella-zoster virus gE mutation discovered in two Swedish isolates. J Clin Virol 37(2):134–136. doi:10.1016/j.jcv.2006.06.007
276. World Health Organization (2014) Varicella and herpes zoster vaccines: WHO position paper, June 2014. Wkly Epidemiol Rec 89(25):265–288
277. Wu MY, Hsu YH, Su CL, Lin YF, Lin HW (2012) Risk of herpes zoster in CKD: a matched-cohort study based on administrative data. Am J Kidney Dis 60(4):548–552. doi:10.1053/j.ajkd.2012.03.018
278. Yamanishi K (2008) Molecular analysis of the Oka vaccine strain of varicella-zoster virus. J Infect Dis 197(Suppl 2):S45–S48. doi:10.1086/522122
279. Yang YW, Chen YH, Wang KH, Wang CY, Lin HW (2011) Risk of herpes zoster among patients with chronic obstructive pulmonary disease: a population-based study. CMAJ 183(5):E275–E280. doi:10.1503/cmaj.101137
280. Yao CA, Chen LK, Huang KC (2015) The immunogenicity and safety of zoster vaccine in Taiwanese adults. Vaccine 33(13):1515–1517. doi:10.1016/j.vaccine.2015.01.085
281. Yawn BP, Saddier P, Wollan PC, St Sauver JL, Kurland MJ, Sy LS (2007) A population-based study of the incidence and complication rates of herpes zoster before zoster vaccine introduction. Mayo Clin Proc 82(11):1341–1349
282. Yawn BP, Wollan PC, Kurland MJ, St Sauver JL, Saddier P (2011) Herpes zoster recurrences more frequent than previously reported. Mayo Clin Proc 86(2):88–93. doi:10.4065/mcp.2010.0618
283. Yun H, Yang S, Chen L, Xie F, Winthrop K, Baddley JW, Saag KG, Singh J, Curtis JR (2016) Risk of herpes zoster in auto-immune and inflammatory diseases: implications for vaccination. Arthritis Rheumatol 68(9):2328–2337. doi:10.1002/art.39670
284. Zerboni L, Sen N, Oliver SL, Arvin AM (2014) Molecular mechanisms of varicella zoster virus pathogenesis. Nat Rev Microbiol 12(3):197–210. doi:10.1038/nrmicro3215

285. Zhang J, Delzell E, Xie F, Baddley JW, Spettell C, McMahan RM, Fernandes J, Chen L, Winthrop K, Curtis JR (2011) The use, safety, and effectiveness of herpes zoster vaccination in individuals with inflammatory and autoimmune diseases: a longitudinal observational study. Arthritis Res Ther 13(5):R174. doi:10.1186/ar3497
286. Zhang J, Xie F, Delzell E, Chen L, Winthrop KL, Lewis JD, Saag KG, Baddley JW, Curtis JR (2012) Association between vaccination for herpes zoster and risk of herpes zoster infection among older patients with selected immune-mediated diseases. JAMA 308(1):43–49. doi:10.1001/jama.2012.7304
287. Zhao Y, Yang L, Wang X, Zhou Z (2014) The new insights from DPP-4 inhibitors: their potential immune modulatory function in autoimmune diabetes. Diabetes Metab Res Rev 30(8):646–653. doi:10.1002/dmrr.2530

Chapter 25
Conclusion: "All Roads Lead to Rome"

C. Peter N. Watson

Concluding this book is a fairly simple process. Herpes zoster will increase if unchecked as the population ages, and its complications are often severe and life changing. It permanently damages the peripheral and central nervous system. Postherpetic neuralgia, blindness, cosmetic facial scarring, meningitis, and encephalitis are some well-established catastrophes. There is, however, an increasingly protean role for varicella zoster virus in other diseases, such as the common conditions of stroke, myocardial infarction and granulomatous arteritis to rare conditions such as short-acting neuralgiform headache with conjunctival injection and tearing (SUNCT) one of the trigeminal autonomic cephalalgias (Nagel and Gilden in press 2016). The concept of a smouldering ganglionitis causing postherpetic neuralgia and zoster sine herpete in some patients is intriguing. Further research is necessary, but this might lead to successful treatment of postherpetic neuralgia and a variety of cryptic neuropathic conditions with antiviral drugs. Complications of herpes zoster to date are currently difficult to prevent by aggressively treating acute herpes zoster with antivirals, and severe postherpetic neuralgia can be impossible to treat or only partially relieved pharmacologically and surgically. After participating in editing this book, it is clear, especially from the difficulties in managing the varicella zoster virus when it recurs, that the future lies predominately with suppression by zoster prevention vaccines which appear to be increasingly effective and more broadly applicable. Another hope is that with improved clinical trial design of new drugs and perhaps with the promise of genetics for precision medicine, better drugs will be developed for neuropathic pain therapy in general, which will be applicable to treating postherpetic neuralgia and herpes zoster.

C.P.N. Watson
Department of Medicine, Division of Neurology, University of Toronto, Toronto, ON, Canada
e-mail: peter.watson@utoronto.ca

C.P.N. Watson et al. (eds.), *Herpes Zoster: Postherpetic Neuralgia and Other Complications,* DOI 10.1007/978-3-319-44348-5_25

Three Drawings by Susan Telling

Susan Telling was a patient of CPNW with intractable neuropathic facial pain for 34 years managed moderately well over these years with medications but also by expression of her suffering in her art and poetry. Many of her images and poems have been published in the journal of the Canadian Pain Society. She died suddenly in 2012. The three images here reflect her pain and the impact of this on her life and appropriately concludes this book in art as it began with a poem reflecting a patient's experience by Elizabeth MacCallum.

C.P.N. Watson et al. (eds.), *Herpes Zoster: Postherpetic Neuralgia and Other Complications,* DOI 10.1007/978-3-319-44348-5

STelling

STelling

Index

C.P.N. Watson et al. (eds.), *Herpes Zoster: Postherpetic Neuralgia and Other Complications,* DOI 10.1007/978-3-319-44348-5

MIX
Papier aus verantwortungsvollen Quellen
Paper from responsible sources
FSC® C105338

If you have any concerns about our products,
you can contact us on
ProductSafety@springernature.com

In case Publisher is established outside the EU,
the EU authorized representative is:
Springer Nature Customer Service Center GmbH
Europaplatz 3, 69115 Heidelberg, Germany

Printed by Libri Plureos GmbH
in Hamburg, Germany